ABNORMAL NEURONAL DISCHARGES

Abnormal Neuronal Discharges

Editors

Nicolas Chalazonitis, Ph.D., Vet. M.D.

Director of the Department of Neuromembrane Biophysics and Ultrastructure
Institute of Neurophysiology and Psychophysiology
CNRS, Marseilles, France

Michel Boisson, Ph.D.

Head of the Laboratory of Neurobiology
Scientific Center of Monaco
Principality of Monaco

Raven Press ■ New York

Raven Press, 1140 Avenue of the Americas, New York, New York 10036

Made in the United States of America

Library of Congress Cataloging in Publication Data
Main entry under title:

Abnormal neuronal discharges.

Papers presented at a symposium held in the principality of Monaco, July 12-13, 1977.
Includes bibliographical references and index.
1. Action potentials (Electrophysiology)–Congresses. 2. Neurons–Congresses. 3. Epilepsy–Congresses. I. Chalazonitis, Nicolas. I. Boisson, Michel. [DNLM: 1. Neuron–Physiology. 2. Epilepsy–Etiology. WL102.5 A153
QP363.A26 591.1 '88 76-58750
ISBN 0-89004-238-1

Preface

This volume summarizes the current state of our understanding of abnormal electrogenesis in terms of cellular and ionic mechanisms. Such mechanisms underlie the abnormal electrical activity that results in epilepsy, yet they are similar in many respects to normal rhythmic electrical patterns such as those seen during desynchronized sleep.

In addition to the neuronal changes that are involved in epileptogenesis, other types of abnormal discharges discussed include those involved in networks of general motor function, normal secretory neurons submitted to alterations in their normal chemical and electrical environment, and identifiable invertebrate neurons whose activity has been used as a model for the cellular events that occur during epileptogenesis.

New data are discussed that are leading to a greater understanding of the intrinsic oscillatory properties of abnormal discharging neurons after chemical (reversible) or pathological (irreversible) modifications. The "abnormal oscillator" is considered as the main generator of massive and repetitive discharge in altered networks from a physiological and neurochemical viewpoint. Finally, there is a discussion of the general oscillatory properties of model neuromembranes, and new mechanisms of abnormal ionic transmembrane movements are considered.

This volume, whose contributors range from clinical investigators to general physiologists, will be of interest to neurophysiologists, experimental neurologists, and biophysicists.

Nicolas Chalazonitis
Michel Boisson

Acknowledgments

On behalf of all participants of the symposium on Abnormal Neuronal Discharges held in the Principality of Monaco (July 12–13, 1977), we wish to express our deep gratitude to Dr. C. Solamito, President of the Scientific Center of Monaco, for his continuous encouragement and for the use of his facilities. We also thank the government of Monaco and Captain J. Y. Cousteau, Director of the Oceanographic Museum, and his staff for their kind social receptions.

The help and encouragement received from Professor J. Le Magnen, General Secretary of the Naturalia et Biologia Association, Professor J. Paillard, Director of the Institute of Neurophysiology and Psychophysiology, CNRS, and Professor J. Scherrer, organizer of the Twenty-Seventh International Congress of Physiological Sciences (Paris) are acknowledged.

We express our thanks to our secretary, Mrs. Michèle Andre, for her enthusiastic assistance during the colloquium and the preparation of this volume.

Finally, we are greatly indebted to our publishers, Raven Press.

The Editors

Contents

PATHOLOGIC NEURONAL ELECTROGENESIS INVOLVED IN MOTRICITY OF VERTEBRATES

OTHER ASPECTS OF BURST GENERATORS

CELLULAR MECHANISMS OF ABNORMAL ELECTROGENESIS IN NEURONAL MODELS OF INVERTEBRATES

IONIC MECHANISMS IN ABNORMAL NEUROMEMBRANES

Contributors

D. Albe-Fessard
Central Nervous System Physiology Laboratory
Pierre and Marie Curie University
75016 Paris, France

P. Andersen
Institute of Neurophysiology
University of Oslo
Oslo 1, Norway

E. Arnauld
Laboratory of Neurophysiology
University of Bordeaux II
33000 Bordeaux, France

M. Balerna
Biochemistry Center
Faculty of the Sciences
University of Nice
06034 Nice, France

J. L. Barker
Laboratory of Neurophysiology
National Institutes of Health
Bethesda, Maryland 20014

P. R. Benjamin
Ethology and Neurophysiology Group
School of Biological Sciences
University of Sussex
Brighton BN1 9 QG England

M. Boisson
Laboratory of Neurobiology
Scientific Center of Monaco
Principality of Monaco

W. H. Calvin
Department of Neurological Surgery
University of Washington
Seattle, Washington 98195

D. O. Carpenter
Neurobiology Department
Armed Forces Radiobiology Research Institute
Bethesda, Maryland 20014

H. Caspers
Physiology Institute
University of Münster
44 Münster, Germany

N. Chalazonitis
Department of Neuromembrane Biophysics and Ultrastructure
Institute of Neurophysiology and Psychophysiology
CNRS, Marseilles, France

D. Chapelle
Laboratory of Ontogenic Neurophysiology
University of Paris VI
UER 61, 75230 Paris Cedex 05, France

R. Chicheportiche
Biochemistry Center
Faculty of the Sciences
University of Nice
06034 Nice, France

G. Cordingley
Department of Physiology
Duke University Medical Center
Durham, North Carolina 27710

J. Del Castillo
Laboratory of Neurobiology
Bld Del Valle 201
San Juan, Puerto Rico 00901

A. M. Desroches
Laboratory of Ontogenic Neurophysiology
University of Paris VI
UER 61, 75230 Paris Cedex 05, France

J. J. Dreifuss
Department of Physiology
University of Geneva School of Medicine
1211 Geneva, Switzerland

C. Ducreux
Department of Neuromembrane Biophysics and Ultrastructure
Institute of Neurophysiology and Psychophysiology
CNRS, Marseilles, France

T. Dunaway
Department of Physiology
Duke University Medical Center
Durham, North Carolina 27710

P. Dunn
Department of Physiology
Duke University Medical Center
Durham, North Carolina 27710

D. S. Faber
Max Planck Institute for Brain Research
Department of Neurobiology
Frankfurt/Main, Federal Republic of Germany

M. Fosset
Biochemistry Center
Faculty of the Sciences
University of Nice
06034 Nice, France

B. H. Gähwiler
Biological and Medical Research Division
Sandoz Ltd.
4002 Basel, Switzerland

L. Gjerstad
Institute of Neurophysiology
University of Oslo
Oslo 1, Norway

M. Gola
Department of Neuromembrane Biophysics and Ultrastructure
Institute of Neurophysiology and Psychophysiology
CNRS, Marseilles, France

R. M. Gulrajani
Biomedical Engineering Group
Department of Physiology
University of Montreal
Montreal, Quebec, Canada

M. J. Gutnick
Unit of Comparative Medicine
Ben Gurion University of the Negev
BeerSheva, 84120 Israel

U. Heinemann
Department of Neurophysiology
Max Planck Institute for Psychiatry
D-8000 München 40 Federal Republic of Germany

C. B. Heyer
Department of Neurophysiology
Max Planck Institute for Psychiatry
D-8000 München 40 Federal Republic of Germany

J. Hoyer
Institute of General and Comparative Physiology
University of Wien
A-1090, Vienna, Austria

Y. Jacques
Biochemistry Center
Faculty of Sciences
University of Nice
06034 Nice, France

H. H. Jasper
Biomedical Engineering Group
Department of Physiology
University of Montreal
Montreal, Quebec, Canada

M. R. Klee
Max Planck Institute for Brain Research
Department of Neurobiology
Frankfurt/Main, Federal Republic of Germany

E. Labos
Semmelweis Medical University
First Department of Anatomy
1450 Budapest, IX, Hungary

P. Laget
Laboratory of Ontogenic Neurophysiology
University of Paris VI
UER 61, 75230 Paris Cedex 05, France

E. Lang
First Department of Anatomy
Semmelweis Medical University
1450 Budapest, IX, Hungary

M. Lazdunski
Biochemistry Center
Faculty of the Sciences
University of Nice
06034 Nice, France

E. Lothman
Department of Physiology and Pharmacology
Duke University Medical Center
Durham, North Carolina 27710

J. Louvel
Research Unit on Epilepsy
INSERM
75014 Paris, France

H. D. Lux
Department of Neurophysiology
Max Planck Institute for Psychiatry
D-8000 München 40, Federal Republic of Germany

P. A. Mathieu
Biomedical Engineering Group
Department of Physiology
University of Montreal
Montreal, Quebec, Canada

M. J. McCreery
Neurobiology Department
Armed Forces Radiobiology Research Institute
Bethesda, Maryland 20014

C. Ohye
Department of Neurosurgery
Gunma University
School of Medicine
Maebashi, 371 Gunma, Japan

M. Onozuka
Department of Physiology
Kanagawa Dental College
Yokosuka, Japan

M. R. Park
Department of Neurophysiology
Max Planck Institute for Psychiatry
D-8000 München 40, Federal Republic of Germany

O. Pompeiano
Institute of Human Physiology
University of Pisa
56100 Pisa, Italy

D. Poulain
Laboratory of Neurophysiology
University of Bordeaux II
33000 Bordeaux, France

D. A. Prince
Department of Neurology
Stanford University
School of Medicine
Stanford, California 94305

R. Pumain
Research Unit on Epilepsy
INSERM
75014 Paris, France

F. A. Roberge
Biomedical Engineering Group
Department of Physiology
University of Montreal
Montreal, Quebec, Canada

G. Romey
Biochemistry Center
Faculty of the Sciences
University of Nice
06034 Nice, France

P. Sandoz
Biological and Medical Research Division
Sandoz Ltd.
4002 Basel, Switzerland

P. A. Schwartzkroin
Department of Neurology
Stanford University School of Medicine
Stanford, California 94305

G. Somjen
Department of Physiology
Duke University Medical Center
Durham, North Carolina 27710

T. G. Smith
Laboratory of Neurophysiology
National Institutes of Health
Bethesda, Maryland 20014

E.-J. Speckmann
Institute of Physiology
University of Münster
Münster, Germany

A. Sugaya
Faculty of Pharmaceutical Sciences
Josai University
Sakado, Saitama, Japan

E. Sugaya
Department of Physiology
Kanagawa Dental College
Yokosuka, Japan

H. Takeuchi
Department of Neurochemistry
Institute for Neurobiology
Okayama University Medical School
Okayama, Japan

I. Tasaki
Laboratory of Neurobiology
National Institute of Mental Health
Bethesda, Maryland 20014

M. Usami
Research Laboratory
Chugai Pharmaceutical Co. Ltd.
Takatanobaba, Tokyo, Japan

J. D. Vincent
Laboratory of Neurophysiology and Physiopathology
University of Bordeaux II
Bordeaux, France

C. M. Woodbury
Neurobiology Department
Armed Forces Radiobiology Research Institute
Bethesda, Maryland 20014

P. J. Yarowsky
Neurobiology Department
Armed Forces Radiobiology Research Institute
Bethesda, Maryland 20014

C. Zuazaga de Ortiz
Laboratory of Neurobiology
and
Department of Pharmacology
Medical Sciences Campus
University of Puerto Rico
San Juan, Puerto Rico 00901

PATHOLOGIC NEURONAL ELECTROGENESIS INVOLVED IN MOTRICITY OF VERTEBRATES

Abnormal Neuronal Discharges, edited by N. Chalazonitis and M. Boisson. Raven Press, New York © 1978.

Nonsynaptic Mechanisms in Epileptogenesis

David A. Prince and Philip A. Schwartzkroin

Department of Neurology, Stanford University School of Medicine, Stanford, California 94305

Since the most important clinical issue related to abnormal cell discharges is epilepsy, it is appropriate to begin this volume with a consideration of the activities of neurons involved in epileptogenesis. It is our intention to discuss current hypotheses regarding cellular neurophysiological mechanisms of focal epileptogenesis in light of new data which require that we substantially modify previously held views. Although the emphasis is on electrophysiological studies, we recognize that "an adequate model cannot be provided by examining the phenomena of epilepsy from the parochial point of view of a single discipline in the neurosciences. Our goal is to view epilepsy from the broad base of neurobiology" (47).

The most extensive intracellular studies of experimental epilepsy have been done in foci produced by application of penicillin to cat neocortex and hippocampus (5,6,22,23,31–33; see refs. 1,34,36, and 37 for reviews). We deal here almost exclusively with the cellular phenomena underlying the resulting interictal epileptiform events which are analogous to the EEG "spikes" of the electroencephalographer.

Although in the past the synaptic alterations underlying focal epileptogenic discharges have received most attention, recent evidence suggests that nonsynaptic events play an important role. In this brief review, three examples of nonsynaptic mechanisms involved in epileptogenesis are discussed: these are (a) repetitive firing originating in intracortical axons, (b) effects of the ionic microenvironment on cortical neuronal activity, and (c) certain regenerative responses in hippocampal pyramidal cells that are involved in epileptiform discharge.

ANTIDROMIC SPIKING IN EPILEPTOGENESIS

The first direct evidence that nonsynaptically generated changes in excitability occurred in single neurons during epileptogenesis came from studies of thalamocortical relay (TCR) neurons during focal seizure discharges. It was hypothesized that repetitive firing might occur in cortical terminals of TCR cells, analogous to that produced in certain neuromuscular preparations by various drugs (34). "Backfiring" has now been demonstrated in intracortical axons of TCR cells in n. ventralis posterolateralis (VPL) (10,11) (Fig. 1A), n. ventralis lateralis (41), and lateral geniculate (45), as well as in axons of transcallosal neurons (40). These nonsynaptically generated spike bursts occur in both penicillin and strychnine foci (41). Examples of the antidromic burst firing phenomena are shown in Fig. 1. It

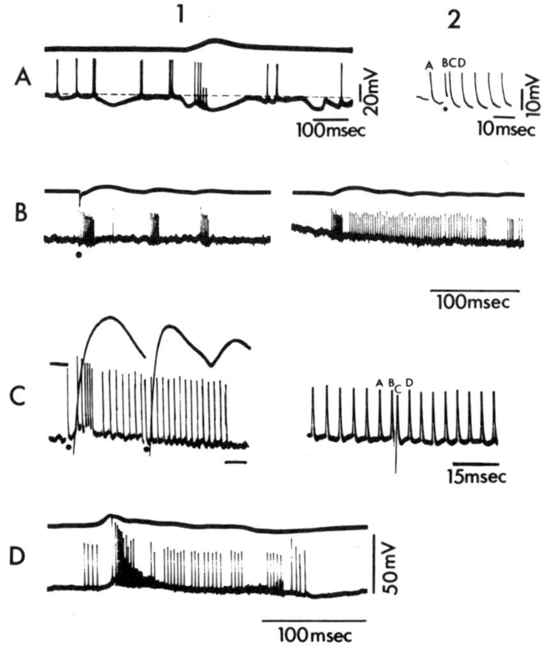

FIG. 1. Antidromic bursting originating in cortical axons. **A1:** Intracellular recording from a relay cell of n. VPL whose axon projects to penicillin focus in cortex; the spike burst arises below firing level (*dotted line*) during the IPSP, correlated with the cortical interictal EEG discharge. **A2:** In another relay cell, interposition of cortical stimulus during spontaneous burst similar to that in **A1** evokes an antidromic spike (*C*). Interval between spikes *B* and *C* is less than twice the latency for invasion of antidromic spikes triggered in the cortex, proving that burst spikes originate in cortex. **B:** Antidromic bursting in neuron during transition from interictal to ictal discharge. **B1:** During brief rhythmic EEG afterdischarge evoked by single cortical shock in penicillin focus, VPL relay neuron generates bursts of orthodromic spikes. **B2:** Shortly after onset of ictal episode provoked by several cortical stimuli, cell generates single orthodromic burst followed by train of regular interval spikes shown to be antidromic by collision technique. **C:** Antidromic burst follows orthodromic burst in TCR neuron projecting to penicillin focus **(C1).** Noncollision of known antidromic spike *C* with preceding one *B* shown in same neuron at faster sweep in **C2.** Stimuli to cortical surface. **D:** Spontaneous paroxysmal EEG and concurrent intracellular activity of VPL relay cell discharge during tonic phase of electrically evoked cortical seizure. Gaps in the antidromic spike train (midportion of segment) are multiples of the regular interspike interval in the antidromic burst. See text for further description. *Dots:* cortical stimuli. *Upper traces* in **A1, B, C1,** and **D:** EEG from cortical surface. Time calibration in **B2** for **B1** and **B2.** [**A** from ref. 10 (copyright 1972 by the American Association for the Advancement of Science); **B–D** from J. L. Noebels and D. A. Prince, *unpublished.*]

is known that penicillin per se will produce repetitive firing at axonal terminals in neuromuscular preparations (25); therefore, more recent findings (26) that intracortical axons generate prolonged bursts of antidromic spikes during electrically induced seizures (Fig. 1D) have particular significance. They suggest that the activities in any type of ictal episode may be sufficient to generate repetitive firing in axonal terminals. During the transition from interictal to ictal discharges, bursts of antidromic spikes may be recorded in VPL relay cells (12,26); these bursts increase in frequency and duration at the onset of ictal discharge (Fig. 1B, C), suggesting that antidromic bursting may contribute to the increasing excitability of cortical neurons. The underlying mechanism for nonsynaptic generation of such bursts is unknown. Significant changes in the ionic constituents of the extracellular space have been demonstrated during epileptogenesis (see below). This may be one factor leading to repetitive spike generation in axonal terminals. Because of the characteristics of the timing of antidromic bursts with respect to interictal and ictal EEG discharge, electrical interactions between cortical elements seem less likely to underlie antidromic burst generation. Whatever the mechanism, this phenomenon would provide a powerful excitation and synchronization of cortical neurons through the large arborizations of TCR axons. During antidromic bursting, the control for cell discharge is removed from the cell body where inhibitory synaptic influences might be expected and shifted to a site where no synaptic inhibition is present.

IONIC MICROENVIRONMENT

Significant changes in extracellular ionic concentrations of K^+, Ca^{2+}, and Cl^- might be expected during epileptogenesis. With repetitive activity, inhibitory postsynaptic potential (IPSP) generation might cause Cl^- to accumulate in cells, shifting the IPSP equilibrium potential and increasing excitability (20; see ref. 39 for review). Cell depolarization causes calcium to move intracellularly, either passively (2) or through regenerative spike mechanisms (17). This would tend to cause $[Ca^{2+}]_0$ to decrease during epileptogenesis. Calcium entry in turn produces an increased conductance for K^+ (3,14,19) which would contribute in part to accumulation of $[K^+]_0$, along with K^+ release during action potentials.

It was first proposed by Green (9) that alterations in the ionic microenvironment might account for the epileptogenic tendencies of hippocampus. The details of the potassium accumulation hypothesis and its various nuances have been recently reviewed (30,37,39) and are not discussed here. It is now known through both glial cell recordings and the use of potassium-sensitive microelectrodes that significant changes in $[K^+]_0$ from baseline levels of 3 to as high as 10 or 12 mM occur during epileptogenesis. Although most attention has been focused on K^+, recent measurements of $[Ca^{2+}]_0$ by Heinemann et al. (13) show that substantial changes in Ca^{2+} from resting levels of 1.3 mM to as low as 0.5 mM may occur during epileptiform discharges. It also seems likely that some decreases in $[Cl^-]_0$ can be expected during intense depolarizations, such as those occurring in epileptogenesis. These have not yet been measured in neocortex, but large changes in $[Cl^-]_0$ have been demonstrated in cerebellum during spreading depression (24).

One difficult question to resolve has been whether these changes which follow the onset of epileptiform discharges actually "feed back" onto neurons in the epileptogenic focus to affect their excitability. It has not been possible to evaluate effects of changes in $[K^+]_0$, $[Cl^-]_0$, or $[Ca^{2+}]_0$ on cells whose activity itself generates the changes in the first place. A first approximation of effects of such ionic alterations can be obtained *in vitro* with measurements from pyramidal cells in hippocampal slices. The advantages of this preparation for the study of epileptogenesis have been discussed elsewhere (28,42,43,49). In Fig. 2A, typical changes in $[K^+]_0$ seen *in vivo* in the hippocampus during interictal and ictal discharges are illustrated. In Fig. 2B, when hippocampal CA1 pyramidal cells are exposed to changes in $[K^+]_0$ similar to those occurring *in vivo*, significant increases in excitability occur. In Fig. 2B1–2, a CA1 neuron exposed to penicillin in a 3 mM $[K^+]_0$ medium generates only an exaggerated excitatory postsynaptic potential (EPSP) during orthodromic stimulation. When the $[K^+]$ in the medium is raised from 3 to 5 mM, the same stimulus regularly evokes depolarization shifts (DSs) in CA1 neurons and associated epileptiform extracellular field potentials (Fig. 2B3–4). A similar interaction between penicillin epileptogenesis and increases in $[K^+]_0$ has been reported in hippocampus *in vivo* (27). In fact, at concentrations of 10 mM $[K^+]_0$, it has been found that spontaneous epileptiform field potentials may occur even without penicillin in the hippocampal slice (29). The effects of increases of $[K^+]_0$ to 10 mM on field potentials and intracellular activities of hippocampal CA1 neurons *in vitro* are shown in Fig. 3. Spontaneous and evoked repetitive extracellular field potential discharges and neuronal DS-like potentials identical to those occurring during penicillin-induced epileptogenesis (cf. Figs. 3 and 5) replace "normal" activities under such conditions. These findings suggest that the changes in potassium ion occurring during epileptogenesis are sufficient to produce excitability changes both in the neurons that generate them and in other cells.

Alterations in excitability produced by variations in $[Ca^{2+}]_0$ would be difficult to predict since the ion has important divalent cation effects on cell membranes (8) and also effects on synaptic transmission (17) and dendritic spike generation in hippocampal neurons (44,48). In the hippocampal slice (Fig. 2C), the divalent cation effects on cell membranes appear to predominate since a decrease in $[Ca^{2+}]_0$ from the normal medium concentration of 2.0 mM to 1.0 to 0.75 mM regularly induces intense burst discharge in hippocampal neurons (Fig. 2C1–2), which is reversible when Ca^{2+} is increased again (Fig. 2C3–4). It should be noted, however, that a further lowering of

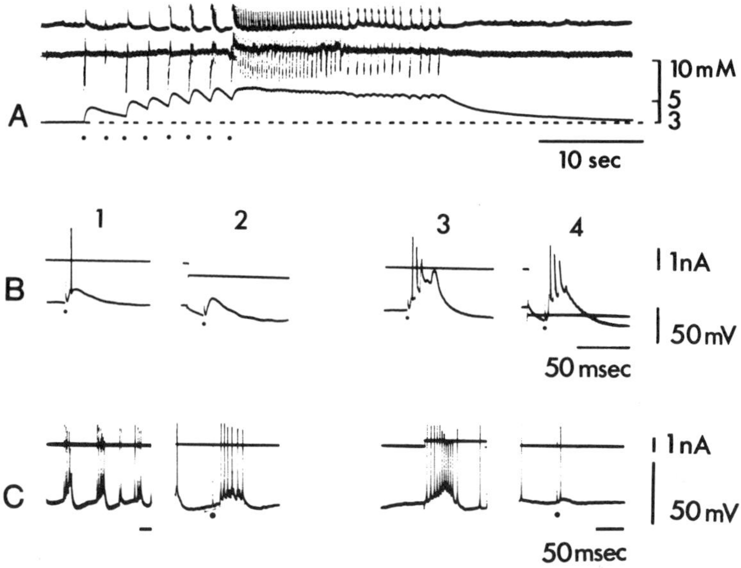

FIG. 2. A: Changes in $[K^+]_0$ during interictal and ictal penicillin discharge in cat hippocampus. *Upper trace,* EEG from hippocampal surface; *middle trace,* field potential from reference micropipette located 50 μm from K^+-sensitive electrode; *bottom trace,* K^+ activity signal derived from differential recording between reference electrode and ion-sensitive electrode; *dots,* stimuli to fornix; *dashed line in bottom trace,* 3 mM baseline level; *polarity,* negativity up in surface EEG and down in reference trace. **B:** Recordings from hippocampal CA1 neurons *in vitro.* **B1–2:** Typical responses to stratum radiatum stimulation (*dots*) in medium containing 3 mM $[K^+]$ and 2 mM penicillin. **B3–4:** Responses of another cell in same slice after perfusing with medium containing 5 mM $[K^+]$ and 2 mM penicillin. The same stimulus as in **B1–2** evokes typical depolarization shifts whose generation is dependent on the increased $[K^+]_0$. *Upper trace,* current monitor. **C:** Recordings from CA1 neuron in solution containing 0.75 mM Ca^{2+} **(C1–2)** and shortly after onset of exchange with solution containing 7.5 mM Ca^{2+} **(C3–4).** Bursting occurs spontaneously **(C1)** and is evoked by both orthodromic **(C2)** and intracellular **(C3)** stimuli. After 5 min of exchange with medium containing 7.5 mM Ca^{2+}, orthodromic stimulus (*dot*) evokes normal response—a single spike followed by a small depolarizing afterpotential. Spikes in **B–C** retouched for clarity. Time mark in **C4** for **C2–4;** 50 msec time mark under **C1.** [**A** from ref. 7a; **B** from P. A. Schwartzkroin and D. A. Prince, *unpublished;* **C** from J. R. Hotson and D. A. Prince, *unpublished.*]

$[Ca^{2+}]_0$ or introduction of a Ca^{2+} antagonist, such as manganese, results in a blockade of burst activity triggered in penicillin solutions by synaptic input (43) and also that due to intrinsic cell properties (48).

Some of the ionic alterations that may facilitate epileptogenic discharge have thus been demonstrated. Each change (e.g., decreased $[Ca^{2+}]_0$, increased $[K^+]_0$, probable decreased $[Cl^-]_0$) is in a direction that would increase excitability. The levels reached during ionic shifts *in vivo* are sufficient in the *in vitro* situation to produce significant effects on neural elements. Alterations in several ionic species simultaneously might act

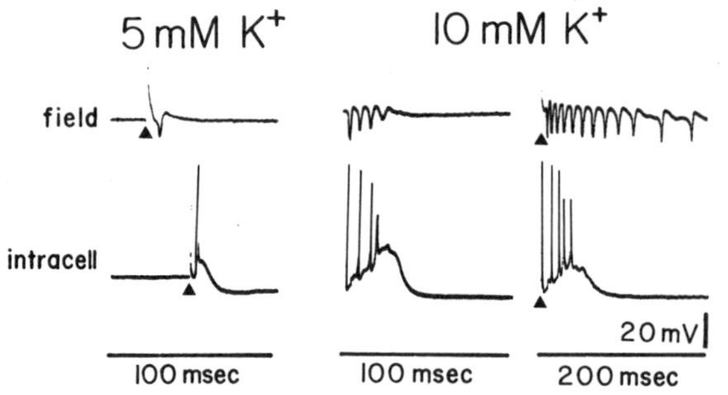

FIG. 3. Effects of increased $[K^+]_0$ on field potentials and intracellular activities in hippocampal CA1 region. *Upper sweeps:* field potentials recorded after stratum radiatum stimulation when medium contained 5 mM K^+ (first frame) and 10 mM K^+ (second and third frames). Sweep of second frame shows spontaneous burst internally triggered. *Lower sweeps:* responses of different CA1 pyramidal cells recorded in 5 mM K^+ (first frame) and 10 mM K^+ (second and third frames). Sweep of middle frame is triggered internally from spontaneous DS-like burst. Increased $[K^+]_0$ results in conversion of cell responses from those of simple EPSPs, which trigger spikes, to those of large amplitude prolonged depolarizations, which trigger spike bursts. Calibrations under bottom line for top and bottom frames of each vertical row. Voltage calibration is for all sweeps. Field potentials and cell discharges not recorded simultaneously. (From P. A. Schwartzkroin and D. A. Prince, *unpublished.*)

cooperatively. Thus, even though the measured $[K^+]_0$ changes alone might be insufficient to produce large changes in excitability, when taken in the context of significant decreases in $[Ca^{2+}]_0$ or $[Cl^-]_0$, repetitive firing of cells or axons could result (7). It is possible that such shifts are involved in the transition between interictal-ictal discharge.

REGENERATIVE RESPONSES IN HIPPOCAMPAL PYRAMIDAL CELLS *IN VITRO*

A third example of a nonsynaptic event contributing to epileptogenesis has recently been found in the burst responses of hippocampal pyramidal cells in the slice preparation. Matsumoto and Ajmone-Marsan (22) first described the large amplitude prolonged membrane depolarizations associated with high frequency bursts of spikes which occur in cortical neurons of penicillin foci coincident with interictal discharges. These DSs occur in neurons of all varieties of acute epileptiform foci thus far studied with intracellular recording techniques and are the hallmark of interictal epileptogenesis at the cellular level. There has been considerable discussion in the literature as to whether the DS represents a summated giant EPSP or is the result of some intrinsic abnormal response of involved neurons. This evidence has been discussed in detail elsewhere (37) and is only summarized here.

In favor of a synaptic origin for the DS have been the findings that these potentials are only evoked orthodromically in neocortical neurons and behave during current pulses as expected for EPSPs in the same cells (6,23,32). Furthermore, transitional forms exist between DSs and augmented EPSPs in penicillin foci of both immature (38) and mature (22,32) cortex. However, the identity of the DS with EPSPs has not been demonstrated, and attempts at showing inversion of DSs have been unconvincing (23,32). Unusual features of DSs suggesting differences from EPSPs include (a) demonstration of threshold behavior to gradually increasing stimuli, (b) stereotypy even with large latency shifts during increasingly intense stimuli, and (c) demonstrations that EPSPs and DSs have different excitability cycles in the same neuron (35). Some investigators (16) have suggested that DSs in motoneurons of spinal cord may represent regenerative responses in that they are produced by intense depolarizing pulses; this finding has not been confirmed by others (21). It has been known for some time that mammalian hippocampal neurons generate bursts of action potentials riding on a slow depolarization and that these bursts are intrinsic to the cell under study (15). There are of course many examples of DS-like bursts occurring in invertebrate neurons after exposure to convulsant agents, such as metrazol (e.g., refs. 4,18,46; see chapters by Speckman and Caspers, Chalazonitis, Takeuchi, and Sugaya, *this volume*). It has also been claimed that regenerative DS-like responses occur in hippocampal nerve cells in culture (50).

We have recently utilized the hippocampal slice preparation to further investigate DS generation (42,43). The typical effects of addition of penicillin to the bathing medium on field potentials and cell discharges are shown in Fig. 4. Penicillin (3.4 mM) changes the evoked field potential from a single spike to a repetitive discharge (cf. Fig. 4A1–2). These epileptiform events occur spontaneously, and simultaneous recordings in CA3 and CA1 regions show that they always lead in the CA3 region (Fig. 4A3) and can be blocked in CA1 if its connections with CA3 are severed (Fig. 4A4). Similar discharges do not occur in the granule cell area. Intracellular recordings show typical DS generation in hippocampal CA1 (Fig. 4B) and CA3 (43) neurons. Normally, CA1 neurons do not generate spontaneous bursts, but these may be evoked with intense depolarization. Bursts in normal solutions, however, are not accompanied by epileptiform field potentials. Bursts occur much more frequently in CA3 neurons spontaneously. To date, no significant differences between spontaneous CA3 bursts and those induced by penicillin exposure have been detected.

Although passive cell membrane properties are unchanged after penicillin application as measured by the intrasomatic electrode (see Fig. 5C–D), significant changes in response to orthodromic stimuli do occur (Fig. 5A–B). In this typical neuron, after 20 min of penicillin exposure, bursts of spikes appear to arise from the baseline without prepotential (Fig. 5B, right). This type of burst generation is further illustrated in Fig. 6. Here the effects of hyperpolarizing current are shown in two neurons during orthodromically

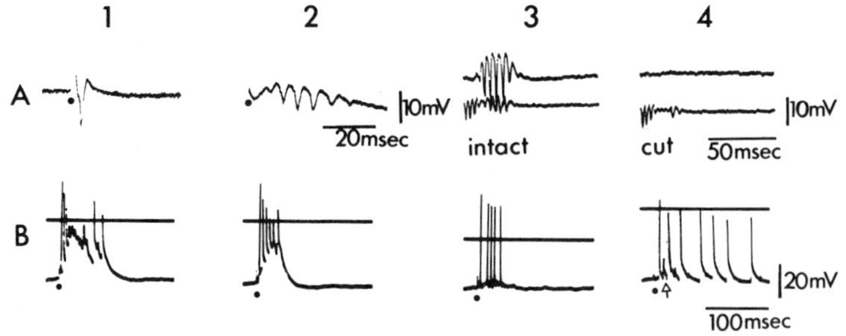

FIG. 4. Penicillin epileptogenesis in the hippocampal slice preparation. **A1:** Normal evoked field potential in CA1 region after stimulation of stratum radiatum (*dot*). **A2:** Same preparation exposed to medium containing 2 mM penicillin; less intense stimulation evokes a field potential with multiple peaks. **A3:** Spontaneous field potential bursts in CA1 (*upper trace*) and CA3 (*lower trace*) in preparation exposed to penicillin. **A4:** Persistence of spontaneous burst in CA3 but disappearance in CA1 after cutting connections between these areas. **B:** Evoked intracellular activities in CA1 neurons after exposure to penicillin. *Dots,* stratum radiatum stimulation. Cells of **B1–2** show DS-like potentials. Cells of **B3–4** show spike bursts with minimal underlying depolarization and small partial spikes (arrow in **B4**). (**A3–4** from ref. 43; **A1–2** and **B** from ref. 42.)

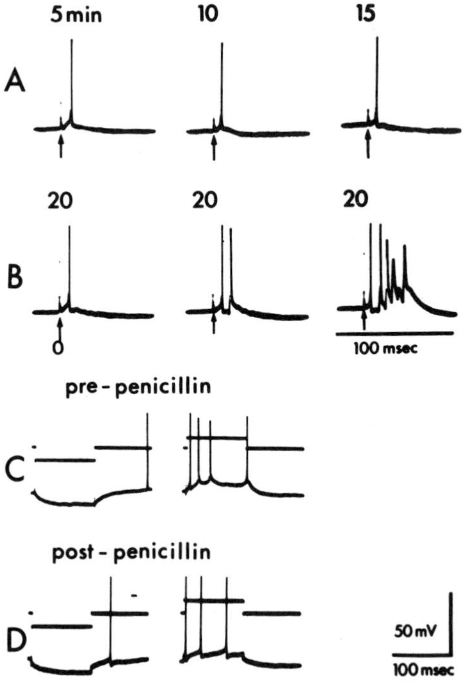

FIG. 5. Intracellular recordings from a CA1 hippocampal pyramidal neuron. **A–B:** Responses to orthodromic stimulation (*arrows*) at various times after penicillin exposure, showing development of spike burst arising without prepotentials (**B**, third frame). **C–D:** Comparison of responses to current pulses before and after penicillin. (Modified from ref. 43.)

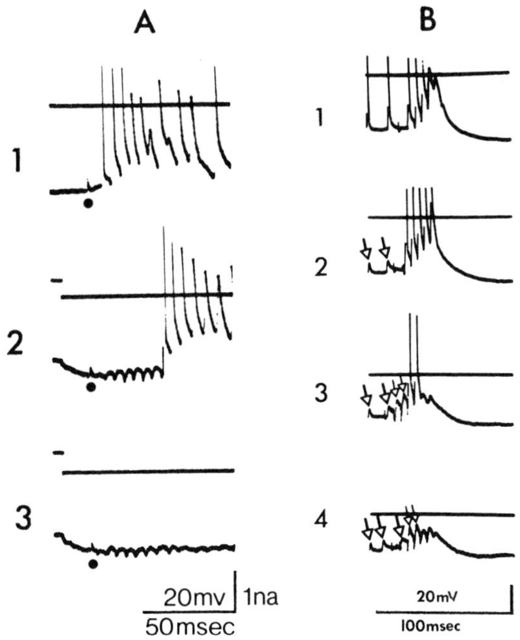

FIG. 6. Effects of hyperpolarizing current pulses on penicillin-induced DS generation in CA1 cells. **A1:** DS with multiple-sized burst spikes evoked by stratum radiatum stimulation (*dot*). Hyperpolarizing current pulses of increasing intensity in **A2** and **A3** delay the onset **(A2)** and completely block **(A3)** the DS, although epileptiform field potential seen intracellularly is present. **B:** Spontaneous DS generation at resting membrane potential **(B1)** and at increasing levels of DC intracellular hyperpolarization **(B2–4).** As hyperpolarization increases, spontaneous spikes are blocked, revealing underlying fast prepotentials (arrows in **B2**). Blockade of burst spikes **(B3–B4)** causes attenuation of DS slow envelope and reveals underlying rhythmic fast prepotentials (third and fourth arrows of **B3;** third, fourth, and fifth arrows of **B4**). Bursts of **B3–B4** contain some reflection of extracellular field potentials. Spikes in **B1** cut off at peaks. *Upper traces:* current monitor with zero level in **A1, B1.** Calibration under **B4,** 20 mV and 1 nA for sweeps of **B.** (**A** from ref. 42; **B** from ref. 43.)

triggered and spontaneous DS-like bursts. Notice that with increase in hyperpolarization, DSs may be delayed or completely blocked (Fig. 6A2–3), or gradually decreased in amplitude as spikes are progressively blocked (Fig. 6B) to reveal fast prepotentials. Using hyperpolarizing pulses, in no case were we able to uncover an EPSP of sufficient amplitude to account for the depolarizing envelope of the DS. These data have been interpreted as showing that hippocampal CA1 pyramidal cells are capable of intrinsic DS burst generation. The action of penicillin may be to increase EPSPs or depress IPSPs and thus give rise to more intense depolarizations of groups of cells that generate their own DSs and consequent field potentials. The slow depolarizations during DSs may be generated in part by small EPSPs, but large contributions must come from spike afterpotentials of bursts originating at a distance from the cell body. The difficulty described above in directly triggering DSs *in vivo* in the absence of orthodromic stimuli may be related to the site of DS generation rather than the requisite synaptic input. This is not to suggest that synaptic activities are unimportant for DS generation. When penicillin-treated slices are placed in high magnesium, low calcium solutions sufficient to block EPSPs, DSs are also blocked (43).

A final bit of evidence suggesting that intrinsic spike burst generating mechanisms are operant in hippocampal neurons comes from studies of CA3 pyramidal cells (48) in which bursts may occur spontaneously (Fig. 7A1–2) or be triggered with brief intracellular depolarizing pulses (Fig. 7A3–4). The later components of these bursts consist of several broad, low amplitude spikes, which occur in an all-or-none fashion and may be

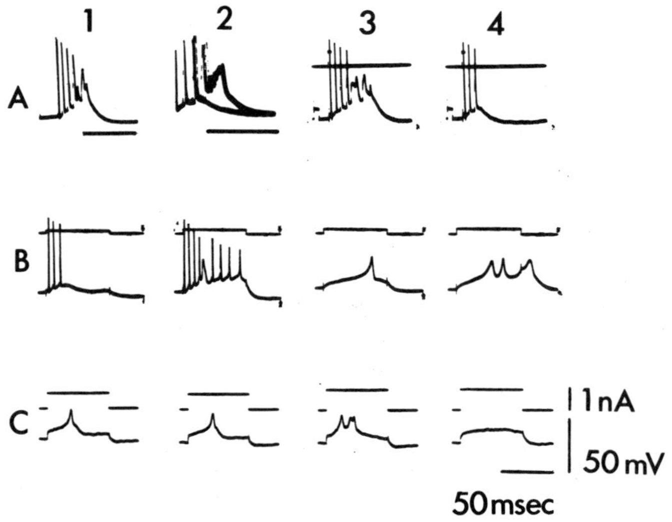

FIG. 7. Depolarizations and bursting behavior in three hippocampal CA3 pyramidal cells. In cell of **A,** spontaneous bursts **(A1–2)** and some triggered by 2 msec intracellular current pulses **(A3)** show initial phase with high amplitude, regular interval rapid spiking, followed by second phase with small amplitude slower spikes occurring at a higher threshold. **A2:** Superimposed sweeps of spontaneous burst activity triggered from first spike of the burst show that late phase is generated in all-or-none fashion. Similar behavior occurs in bursts triggered by intracellular depolarizing current pulses (cf. **A3** and **A4**). **B:** Another neuron showing brief burst during weak intracellular depolarizing current pulse **(B1)** and prolonged decrementing burst with an intermixed slow spike during stronger depolarization **(B2). B3** and **B4:** Single and repetitive high threshold, slow spikes evoked by intracellular depolarizations after bath perfusion with TTX (5×10^{-6} g/ml). Current pulses in **B2** and **B3** are identical; current increased in **B4. C:** Single **(C1–C2)** and repetitive **(C3)** TTX-resistant spikes in another neuron. Behavior prior to TTX was similar to that seen in **A1** and **A3. C4:** TTX-resistant spikes blocked after local application of 20 mM Mg^{2+} to CA3 region. Calibrations in **C4** for all sweeps, except for **A1–A2** where 50 msec time marks shown. Some spikes retouched for clarity; first and second spikes of bursts in **A2** cut off by scope face. (From R. K. S. Wong and D. A. Prince, *unpublished.*)

blocked by manganese perfusions. Similar-appearing spikes occur during intracellular depolarizations in the presence of tetrodotoxin (Fig. 7B3–4; C) and may be blocked by Mn^{2+}. These findings suggest that later portions of the spike burst are dependent on calcium spiking, which has also been demonstrated in CA1 pyramidal cells (44). It is of interest that hippocampal CA3 pyramidal cells have powerful intrinsic spike burst generating mechanisms and also serve as pacemakers for epileptogenic discharges after penicillin application. At present, it is unclear whether the drug has effects on intrinsic membrane events or merely on EPSPs and IPSPs which serve to synchronize populations and cause intrinsic DS generation.

CONCLUSIONS

These findings lead to the hypothesis that nonsynaptic mechanisms, such as repetitive firing of intracortical axons, changes in neuronal microenvironment, and generation of intrinsic neuronal DS-like responses, are important factors in epileptogenesis. These new data also seriously challenge the hypothesis that DSs are simply giant synaptic potentials. Finally, data from hippocampal pyramidal neurons suggest that bursting may be importantly related to calcium spike generation.

ACKNOWLEDGMENTS

This work was supported by grants NS06477 and NS12151 from the NINCDS, and by gifts from the Morris Research Fund and California Community Foundation. We thank Geraldine Chase for secretarial assistance. We are pleased to acknowledge the important contributions made to this work by Drs. Robert Fisher, Kin Futamachi, Michael Gutnick, John Hotson, Jeffrey Noebels, Timothy Pedley, and Robert Wong.

REFERENCES

1. Ajmone-Marsan, C. (1969): In: *Basic Mechanisms of the Epilepsies,* edited by H. H. Jasper, A. A. Ward Jr., and A. Pope, pp. 299–328. Little, Brown, Boston.
2. Baker, P. F., Hodgkin, A. L., and Ridgway, E. B. (1971): *J. Physiol.,* 218:709–755.
3. Barrett, E. F., and Barrett, J. N. (1976): *J. Physiol.,* 255:737–774.
4. Chalazonitis, N., and Takeuchi, H. (1968): *C. R. Soc. Biol. (Paris),* 162:1552–1556.
5. Dichter, M., and Spencer, W. A. (1969): *J. Neurophysiol.,* 32:649–662.
6. Dichter, M., and Spencer, W. A. (1969): *J. Neurophysiol.,* 32:663–687.
7. Falk, G., and Landa, J. F. (1960): *Am. J. Physiol.,* 198:1225–1231.

7a. Fisher, R. S., Pedley, T. A., Moody, W. J. Jr., and Prince, D. A. (1976): *Arch. Neurol.,* 33:76–83.

8. Frankenhaeuser, B., and Hodgkin, A. L. (1957): *J. Physiol.,* 137:218–244.
9. Green, J. D. (1964): *Physiol. Rev.,* 44:561–608.
10. Gutnick, M. J., and Prince, D. A. (1972): *Science,* 176:424–426.
11. Gutnick, M. J., and Prince, D. A. (1974): *J. Neurophysiol.,* 37:1310–1327.
12. Gutnick, M. J., and Prince, D. A. (1975): *Exp. Neurol.,* 46:418–431.
13. Heinemann, U., Lux, H. D., and Gutnick, M. J. (1977): *Exp. Brain Res.,* 27:237–243.

14. Hotson, J. R., Schwartzkroin, P. A., and Prince, D. A. (1977): *Neurosci. Abstr.*, 3:218 (Abstr. 680).
15. Kandel, E. R., and Spencer, W. A. (1961): *J. Neurophysiol.*, 24:243–259.
16. Kao, L. I., and Crill, W. E. (1972): *Arch. Neurol.*, 26:156–161.
17. Katz, B., and Miledi, R. (1969): *J. Physiol.*, 203:459–487.
18. Klee, M. R., Faber, D. S., and Heiss, W. D. (1973): *Science*, 179:1133–1136.
19. Krnjevic, K., and Lisiewicz, A. (1972): *J. Physiol.*, 225:363–390.
20. Llinas, R., Baker, R., and Precht, W. (1974): *J. Neurophysiol.*, 37:522–532.
21. Lothman, E. W., and Somjen, G. G. (1976): *Electroencephalogr. Clin. Neurophysiol.*, 41:337–347.
22. Matsumoto, H., and Ajmone-Marsan, C. (1964): *Exp. Neurol.*, 9:286–304.
23. Matsumoto, H., Ayala, G. F., and Gumnit, R. J. (1969): *J. Neurophysiol.*, 32:688–703.
24. Nicholson, C., and Kraig, R. P. (1975): *Brain Res.*, 96:384–389.
25. Noebels, J. L., and Prince, D. A. (1977): *Brain Res.*, 138:59–74.
26. Noebels, J. L., and Prince, D. A. (1977): *Neurosci. Abstr.*, 3:143 (Abstr. 443).
27. O'Connor, M. J., and Lewis, D. V. (1974): *Electroencephalogr. Clin. Neurophysiol.*, 36:337–345.
28. Ogata, N. (1975): *Exp. Neurol.*, 46:147–155.
29. Ogata, N., Hori, N., and Katsuda, N. (1976): *Brain Res.*, 110:371–375.
30. Pedley, T. A., Fisher, R. S., Futamachi, K. J., and Prince, D. A. (1976): *Fed. Proc.*, 35:1254–1259.
31. Prince, D. A. (1966): *Epilepsia*, 7:181–201.
32. Prince, D. A. (1968): *Exp. Neurol.*, 21:467–485.
33. Prince, D. A. (1968): *Exp. Neurol.*, 21:307–321.
34. Prince, D. A. (1969): In: *Basic Mechanisms of the Epilepsies,* edited by H. H. Jasper, A. A. Ward Jr., and A. Pope, pp. 320–328. Little, Brown, Boston.
35. Prince, D. A. (1971): *Electroencephalogr. Clin. Neurophysiol.*, 31:469–484.
36. Prince, D. A. (1976): In: *Brain Dysfunction in Infantile Febrile Convulsions,* edited by M. A. B. Brazier and F. Coceani, pp. 187–212. Raven Press, New York.
37. Prince, D. A. (1978): In: *Annual Review of Neuroscience,* edited by W. M. Cowan, Z. W. Hall, and E. R. Kandel, vol. 1, pp. 395–415. Annual Reviews, Palo Alto.
38. Prince, D. A., and Gutnick, M. (1972): *Brain Res.*, 45:455–468.
39. Prince, D. A., Pedley, T. A., and Ransom, B. (1977): In: *Dynamic Properties of Glia Cells,* edited by G. Franck, L. Hertz, E. Schoffeniels, and D. B. Tower. Pergamon Press, New York. (*In press.*)
40. Schwartzkroin, P. A., Futamachi, K. J., Noebels, J. L., and Prince, D. A. (1975): *Brain Res.*, 99:59–68.
41. Schwartzkroin, P. A., Mutani, R., and Prince, D. A. (1975): *J. Neurophysiol.*, 38:795–811.
42. Schwartzkroin, P. A., and Prince, D. A. (1977): *Ann. Neurol.*, 1:463–469.
43. Schwartzkroin, P. A., and Prince, D. A. (1977): *Brain Res.* (*In press.*)
44. Schwartzkroin, P. A., and Slawsky, M. (1977): *Brain Res.*, 135:157–161.
45. Scobey, R. P., and Gabor, A. J. (1975): *J. Neurophysiol.*, 38:383–394.
46. Speckman, E.-J., and Caspers, H. (1973): *Epilepsia,* 14:397–408.
47. Ward, A. A. Jr., Jasper, H. H., and Pope, A. (1969): In: *Basic Mechanisms of the Epilepsies,* edited by H. H. Jasper, A. A. Ward Jr., and A. Pope, pp. 1–12. Little, Brown, Boston.
48. Wong, R. K. S., and Prince, D. A. (1977): *Neurosci. Abstr.*, 3:148 (Abstr. 465).
49. Yamamoto, C. (1972): *Exp. Neurol.*, 35:154–164.
50. Zipser, B., Crain, S. M., and Bornstein, M. B. (1973): *Brain Res.*, 60:489–495.

Abnormal Neuronal Discharges, edited by
N. Chalazonitis and M. Boisson. Raven
Press, New York © 1978.

Microphysiology of Spinal Seizures

G. Somjen, E. Lothman, P. Dunn, T. Dunaway,
and G. Cordingley

*Department of Physiology, Duke University Medical Center,
Durham, North Carolina 27710*

CONVULSIONS OF SPINAL ORIGIN

The spinal cord has been a perenially useful testing ground of ideas concerning the functions and malfunctions of mammalian central gray matter. In the investigation of seizures, however, it has been rather neglected, perhaps because human epilepsy is customarily thought of as a disease of the brain. Yet severed from the brain, the spinal cord is capable of convulsive activity. Even in the absence of overt paroxysms, in the unstimulated spinal cord there are signs of irregular interneuron activity which is much more lively in high spinal than in decerebrate preparations (53). Intense electrical stimulation of the first cervical segment causes rhythmic contractions of the skeletal musculature during low frequency stimulation and tonic paroxysmal discharge thereafter (21). Transient hypoxia can cause persistent spontaneous discharges, as all practitioners of spinal cord electrophysiology sooner or later learn to their grief. Of the numerous stimulant drugs, strychnine has long been considered the spinal convulsant par excellence.

The minimal convulsant dose of strychnine is about the same for intact animals and decerebrate preparations as for decapitate spinal cords. Pentylenetetrazole and bemegride, on the other hand, have been considered to act primarily on the brainstem, because the minimal dose required to induce paroxysmal activity in the severed spinal cord is about 20 times higher than in preparations with intact brainstem (27,28). Yet both drugs can induce characteristic convulsive activity in spinal preparations, albeit at a high dose level (28) (Fig. 1). One may ask, therefore, whether these drugs have the same or a different mode of action in brain as in spinal cord. If the basic mechanism is identical, do the different dose levels reflect (a) a greater "affinity" for cellular elements or for subcellular organelles in the brain than in cord, (b) the same degree of affinity to the same cellular components, of which, however, there exist larger numbers in the brainstem than in the spinal gray matter, or, least probably, (c) that the drug penetrates more readily from blood into brainstem than into spinal cord?

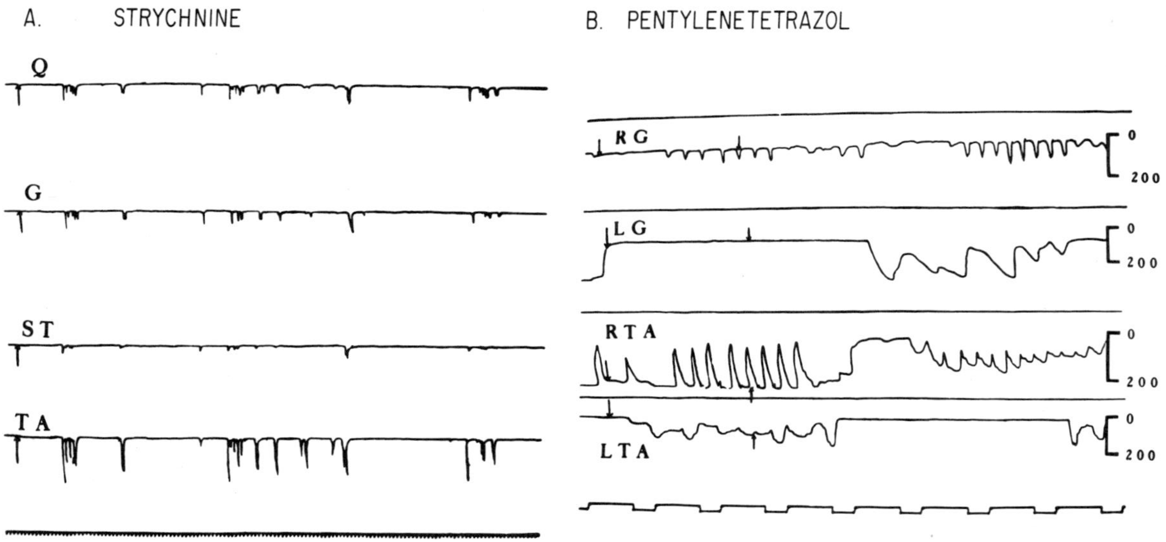

FIG. 1. Patterns of convulsion induced by strychnine **(A)** and pentylenetetrazole **(B)** in decapitate spinal cords. The contractions of individual muscles were recorded as downward deflections on a smoked paper kymograph. Q, quadriceps muscle; G, gastrocnemius; ST, semitendinosus; TA, tibialis anterior; RG, right gastrocnemius; LG, left gastrocnemius; RTA, right tibialis anterior. Time in sec. Arrows mark simultaneity in four traces. (Slightly modified from ref. 28.)

Besides relative dose level, strychnine also differs from pentylenetetrazole and bemegride in the pattern of the convulsions it induces. Cocontraction of antagonistic muscles is the rule with strychnine. The two other drugs cause convulsions that erupt now in flexors then in extensors and occasionally alternate in opposing muscle groups in rhythmic sequences (Fig. 1).

Cocontraction of antagonistic muscles under the influence of strychnine may be explained by blockade of reciprocal inhibition by the drug (7). But cocontraction is also characteristic of (a) the seizures induced by penicillin (37), a drug that leaves postsynaptic inhibition in the spinal cord unimpaired (14,17,19,39) and (b) electrically provoked seizures in intact cats (22). Yet penicillin resembles strychnine in one more respect; namely, the minimal convulsant dose is similar for intact cats (47,48,54), high spinal preparations (37), and cats rendered comatose by bilateral lesion in the midbrain reticular formation (T. Dunaway, *unpublished*). Convulsive activity of spinal origin has been reported not only in cats but also in human patients treated with massive doses of penicillin following injury causing brain damage and EEG silence (50). Large enough doses cause grand mal seizures in intact animals (48,54,58) and in decapitate spinal cords alike (37).

Since the same amount of penicillin induces seizures in intact cats as in animals with midbrain lesion and in high spinal preparations, one must wonder at which end of the neuraxis penicillin convulsions originate in an intact nervous system. Myoclonic twitching and "petit mal" type 4 to 5 per sec "runs" of waves or spike-wave complexes were characterized as centrencephalic by Prince and Farrell (47) but attributed to probably cortical origin by Gloor and Testa (24,55).

In the unanesthetized preparation with midbrain lesion, penicillin-induced tonic seizures usually appear in recordings from motor cortex before they erupt in the lumbosacral spinal cord (Fig. 2A); but in a sizeable minority of cases, the sequence is the reverse (not illustrated). During clonic convulsions, cortical waves sometimes lead, but frequently lag behind, discharges in spinal ventral root (Fig. 2B–D). The statistical incidence of cortical and spinal leading in two different experimental periods within the same preparation is illustrated in Fig. 3. During interictal discharges, motor cortex and lumbosacral spinal segments similarly alternate in leading and lagging.

Temporal sequencing need not reflect a causal relationship. Both cortical and spinal discharges may be driven by a subcortical pacemaker, e.g., in the diencephalon. This was almost certainly so whenever the cortical discharge preceded the spinal by less than 5 msec. Such short intervals, seen frequently during clonic seizures, are too brief for corticospinal conduction to take place (for cerebrospinal conduction times, see, e.g., ref. 44). Spinocortical lag times, however, usually amounted to 5 to 70 msec (modal values in different experiments between 10 and 50 msec) and are best explained by seizure activity initiated in spinal cord and conducted to cortex. For a subcortical intracranial focus to drive lumbosacral spinal segments 10 to

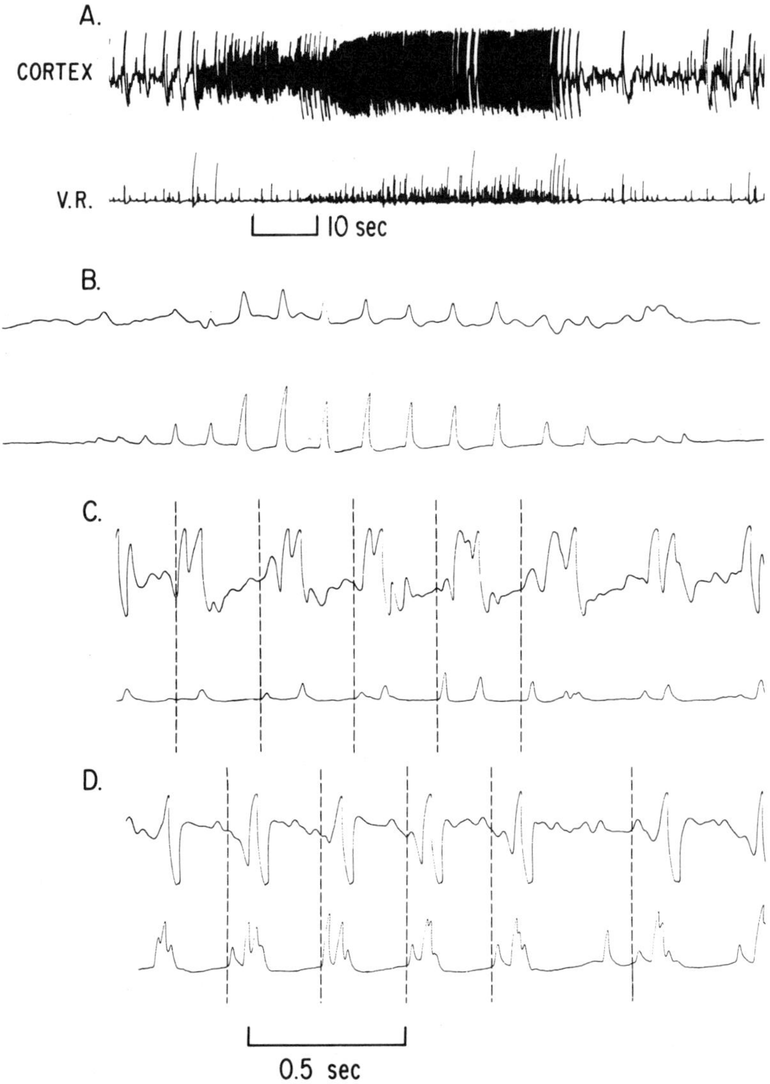

FIG. 2. Paroxysmal activity recorded simultaneously in the motor cortex representation of lumbar spinal cord and the (contralateral) L7 spinal ventral root. Recording on polygraph by EEG amplifiers. Cat rendered comatose by bilateral electrolytic lesion of midbrain reticular formation. **A:** Development of tonic seizure followed by brief clonic activity; seizure appearing first in cortex. **B:** Brief clonic episode, beginning in spinal cord. **C** and **D:** Clonic seizure activity. Vertical broken lines mark onset of seizure events. Note that sometimes the cortical discharge and at other times the spinal discharge occurs first. (T. Dunaway, unpublished experiment.)

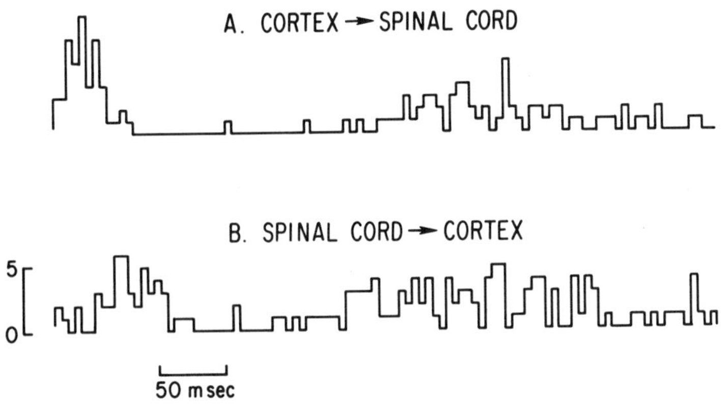

FIG. 3. Incidence distribution of clonic seizure events, similar to the ones illustrated in Fig. 2C and D. Diagrams generated by poststimulus histogram program of PDP 8E computer. **A:** Cortical clonic discharge used as "stimulus" trigger; histogram shows incidence of spinal discharges occurring after cortical discharges. **B:** Histogram of cortical discharges following spinal discharges. (T. Dunaway, unpublished experiment.)

50 msec before cortex, the velocity of conduction from the hypothetical pacemaker to spinal cord would have to exceed the conduction velocity to cortex by an improbably large margin.

EPILEPTIFORM PHENOMENA INDUCED BY PENICILLIN IN SPINAL CORD

Convulsive activity may erupt spontaneously in spinal cords treated with penicillin or they may be provoked by afferent stimulation. Gentle ruffling of hairs was remarkably effective in triggering seizures.

Spontaneous discharges conducted in spinal ventral roots may be used as an absolute diagnostic criterion of heightened excitation; in the acutely decapitate spinal cord, alpha motoneurons do not discharge action potentials when not stimulated, although they do show subliminal excitability fluctuations (see above).

Whether penicillin is administered topically or systemically, all or nearly all the motoneurons in a segment depolarize during a seizure (39). Interneurons may occasionally be subject to paroxysmal hyperpolarization and sometimes do not take any part in a seizure. Synchronization of paroxysmal activity occurs between segments and between the two sides of the cord; the shorter the distance, the more nearly perfect the synchrony (Fig. 4).

Convulsive paroxysmal activity in ventral roots is associated with electrotonically conducted negative shifts of potential in dorsal roots, which can be shown to be caused by episodic depolarization of primary afferent terminals (38). During such paroxysmal dorsal root potentials (DRPs), showers of antidromically conducted action potentials are also recorded,

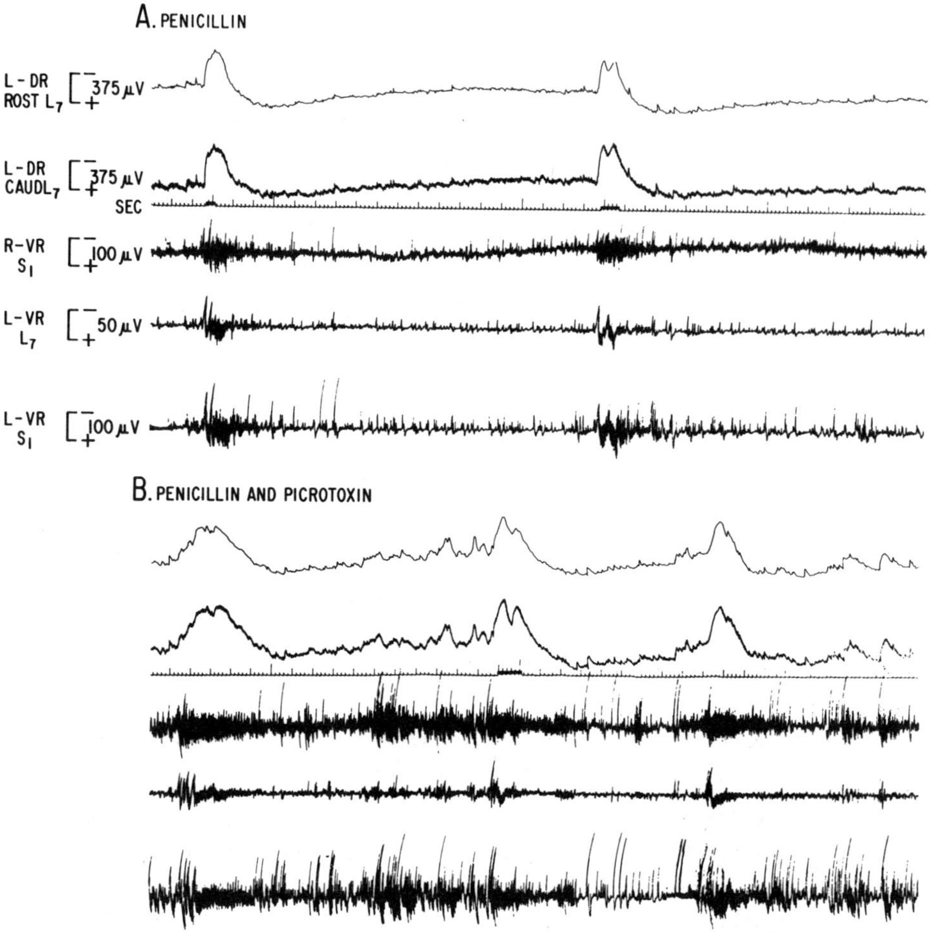

FIG. 4. Seizure activity recorded simultaneously in two dorsal root (DR) filaments, in L7 ventral root (VR), and in S1 VR of same and contralateral sides of high spinal cat treated with convulsant dose of penicillin. Ink-writer recordings by polygraph, direct coupled amplification for DR, and condenser-coupled amplification of VR discharges. (E. Lothman, unpublished experiment.)

which may be analogous to the spontaneous antidromic impulses reported to occur in thalamocortical fibers (26) and in motor axons in the phrenic nerve (42) when the terminal region is treated with penicillin. Paroxysmal DRPs are not suppressed by picrotoxin (Fig. 4).

During interictal convulsive discharges, one may record monophasic positive or diphasic positive-negative wavelets from the dorsal surface of the spinal cord. During prolonged ictal activity, the potential of the cord dorsum shifts in the negative direction (37). This is in contrast to the classic afferent-evoked negative DRP (6,34), which is associated with a positive cord dorsum potential (20); but it is similar to the negative shift of the potential recorded during prolonged afferent tetanic stimulation (36) from the dorsal surface of spinal cords not seized by convulsions.

In motoneurons of the spinal cord, interictal discharges are associated with paroxysmal depolarizing shifts, tonic seizures with protracted de-

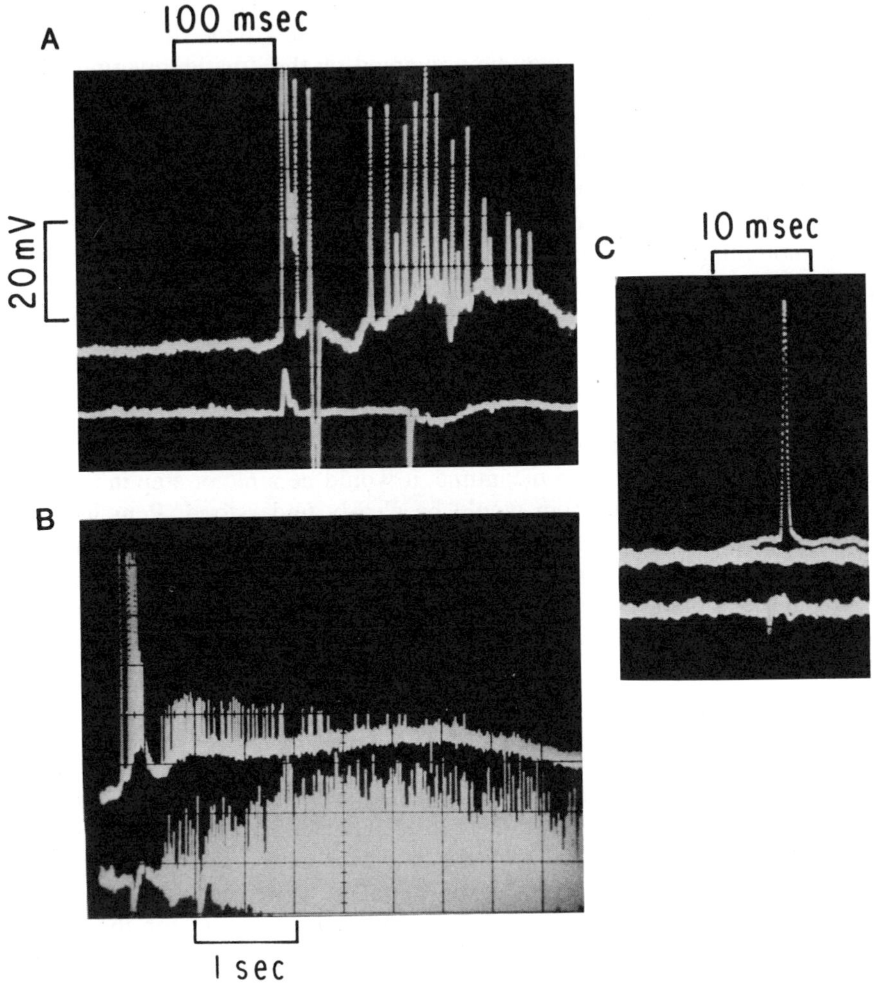

FIG. 5. Onset of tonic convulsive activity of motoneurons and of the ventral root of their segment. **A** and **B:** Same recording, played back from tape twice and recorded on a storage oscilloscope on different time bases (see time calibrations). **C:** From another cell. *Upper trace,* intracellular recording, amplification identical in **A, B,** and **C;** *lower trace,* electrical activity of ventral root. Note unusually low threshold for initial spike discharge, variable apparent threshold for subsequent action potentials, and absence of afterhyperpolarization. Note also that initially spikes are of normal amplitude but later vary irregularly.

polarization, and clonic activity with membrane oscillations synchronized with the population discharge (39). Paroxysmal depolarizations may remain subliminal, may trigger spikes sometimes in high frequency bursts, and sometimes may inactivate impulse activity. The spike potentials recorded from ventral horn cells during paroxysmal activity may have abnormally variable amplitudes (Fig. 5). Abnormal spikes may reveal anomalous membrane properties. The impulses may also be generated at a variable distance from the site of recording and may not be conducted to the site of recording. Invasion of the cell soma may be prevented by the drastic lowering of the input resistance of the cell (19). Action potential may be triggered at an unusually distant site by the intense current associated with the paroxysmal depolarization of the somadendritic membrane. The customary afterhyperpolarizations of soma spikes may also be shunted by the breakdown of membrane resistance. During seizure events, high frequency discharge of spike potentials of normal amplitude and configuration may be recorded from motor axons with intracellular electrodes (39).

PRESENT STATE OF THE SEARCH FOR A MECHANISM OF PENICILLIN-INDUCED SEIZURES

Eventually, an explanation will have to be sought for the action of all epileptogenic agents. In the meantime, it would be a major step in the right direction if any one of them would be clearly understood. Penicillin and pentylenetetrazole are probably the most investigated two convulsants, yet there is no consensus concerning the mechanism of action of either. The following is a brief synopsis of some of the theories concerning penicillin which have received attention in recent years with arguments for and against their validity.

The Potassium Hypothesis

There is no doubt that sufficient quantity of potassium ions introduced from an exogenous source into the extracellular fluid of central gray matter can cause convulsions (23). At issue, however, is the role of endogenous potassium in causing either toxic or traumatic epilepsies (40). As far as penicillin is concerned, there is no experimental support for the theory (38,43). Even though potassium does accumulate in extracellular fluid during seizures, its rise tends to follow rather than lead the discharge in ventral root. Furthermore, penicillin does not seem to impair the clearing of potassium from extracellular fluid (35,38). Cardiac glycosides, however, are known to impede the active reuptake of potassium into cells. If there is a drug that induces convulsions by causing the accumulation of K^+, it should be digitalis and its derivatives. We (12) have indeed found that after the administration of digitoxigenin, the clearing of elevated $[K^+]_0$ was retarded, and its "resting" level was increased in the spinal cord. Seizures erupted from $[K^+]_0$ levels, which varied widely, and not from a definable threshold level, and usually at K^+ concentrations which untreated spinal cords could readily reach without triggering seizures. As with penicillin, an intense rise of potassium occurred after, not before, the appearance of convulsive discharges.

While the above observations suggest that in the case of seizures induced by both penicillin and digitoxigenin, the extracellular accumulation of potassium was a consequence and not a cause of the seizure discharge, it is nevertheless possible that, once accumulated in excess, $[K^+]_0$ may influence the patterning of convulsive activity. Excess $[K^+]_0$ might have an influence on the release of transmitter from presynaptic terminals (11,41) and may also contribute to the depolarization of postsynaptic membranes. These effects may be small compared to other influences buffeting the cell (36).

Penicillin and Resting Membrane Resistance

Either an increase or a decrease of membrane resistance could lead to phenomena resembling seizures. Increased resistance of the resting membrane (reduced leak current) would enhance the electrotonic spread beyond synaptic regions of potential changes caused by transmitter agents. It may also cause repetitive firing in the wake of single impulses (4). Reduced membrane resistance could lead to depolarization and hence to enhanced excitability.

Increased resting membrane resistance was indeed reported after treatment with penicillin of a number of invertebrate excitable cells (2,3,29). In spinal motoneurons, however, resting membrane resistance appeared unchanged when it was compared to the untreated control state and after penicillin administration measured in the seizure-free intervals (19). The expected enhancement of monosynaptic excitatory postsynaptic potentials (EPSPs) was also not observed (14,19). Admittedly, a small change could have remained undetected in these experiments; it would not, however, explain the violent convulsions.

Membrane Instability or Negative Resistance

Pacemaker activity and spontaneous burst-generation of molluscan neurons was attributed to voltage- and time-dependent nonlinearities of the current-voltage function of the cell membrane by several teams of investigators (25,52,57). Treatment with pentylenetetrazole causes the appearance of irregularly recurrent episodic depolarizing shifts in nonpacemaker neurons and enhanced pacing and bursting in pacemaker cells (8,9). Although the pentylenetetrazole-induced voltage shifts differ in several important ways from the pacemaker activity of normal cells, they nevertheless appear to be generated by a negative slope-resistance region in the steady-state current-voltage curves induced by the convulsant drug in cells which normally do not have this characteristic (15,18).

Schwindt and Crill (51) recently reported that in the penicillin-treated spinal cord, some but not all motoneurons show negative resistance regions of the current-voltage function investigated by the voltage clamp technique. Such an observation could be reconciled with several unrelated observations. Pentylenetetrazole-induced paroxysmal depolarizations can also be provoked by brief depolarizing pulses, as well as by excitatory synaptic potentials (18). Thus the convulsive potentials in motoneurons could be triggered by reflex excitatory input. Furthermore, it may be inferred, although it is not experimentally confirmed, that during the plateau of the pentylenetetrazole-induced voltage shift, there should be a reduced apparent input impedance, as was indeed found in the penicillin-treated cord (19) and cortex (1). In this respect, membrane instability would be indistinguishable from an exaggerated EPSP.

There are, however, some experimental observations that are difficult to reconcile with the notion that inherent instability of the membrane is the basis of the paroxysmal depolarizations induced by penicillin. In Aplysia neurons, Pellmar and Wilson (45) found no effect on excitability and current-voltage functions, even when penicillin was applied in concentrations greatly in excess of that required to affect synaptic functions (see below). Even with pentylenetetrazole, synaptic effects were seen at a concentration five times lower than that needed to induce negative resistance region in the current-voltage function (46). The drugs may affect Aplysia cell differently from cats. In the case of neurons in mammalian nervous systems, observations conflict concerning the triggering of paroxysmal potential shifts in individual cells by injecting current through intracellular electrodes. If membranes were unstable because of a negative resistance characteristic, this should be easily achieved (18).

Kao and Crill (31,32) report that paroxysmal depolarizing shifts can indeed be triggered after penicillin treatment; but it appears that complex, double-step currents had to be used, and even these were not always successful. During paroxysmal activity, it may be difficult to distinguish a triggered depolarization from a spontaneous one, unless discharges in ventral root are concurrently recorded to monitor the behavior of the neuron population. In the experience of our laboratory, it is difficult or impossible to trigger a paroxysmal wave in a motoneuron when there is no simultaneous intercurrent discharge in the motoneuron pool (19,39). Other investigators (1) report similar failure in cortical penicillin foci. Furthermore, Schwindt and Crill (51) found nonlinearities in the current-voltage functions of some of the motoneurons in spinal cords not treated with penicillin. Therefore, to make a convincing case, even though these are technically very difficult experiments, it will be necessary to render statistically valid proof that penicillin treatment has altered the excitability characteristics of these cells.

Penicillin may cause instability of presynaptic terminals instead of postsynaptic membranes. The resulting burst-discharges would amplify transmitter output by a very large factor. Repetitive discharges at the nerve terminal should be conducted antidromically into dorsal roots. Unfortunately for this version of the theory, the antidromic bursts known as the dorsal root reflex are depressed rather than enhanced by penicillin (17,38).

Penicillin and Postsynaptic Inhibition Mediated by GABA

There are several concordant reports indicating that penicillin interferes with the inhibitory effects of gamma-aminobutyric acid (GABA) (10,13). This may or may not be the decisive factor in the generation of seizures in the brain. In the spinal cord, the role of GABA in postsynaptic inhibition is uncertain; and postsynaptic inhibition is unimpaired by penicillin (17,39). Although recurrent inhibition was somewhat depressed by penicillin, the effect was not impressive (39).

However, GABA is considered by several authors to be the mediator of presynaptic inhibition in the spinal cord. Possible interference with this function is the theory to be discussed next.

Penicillin, Negative DRP, and Presynaptic Inhibition

According to a report by Davidoff (16,17), penicillin depresses the negative DRPs evoked by afferent stimulation in the spinal cord of frogs (16) and cats (17). This observation would put penicillin into the same class of convulsant as picrotoxin and bicucullin, sharing with them the blocking effect of postsynaptic inhibition in brain and presynaptic inhibition in spinal cord.

Disconcertingly, in our hands (38) penicillin did not suppress afferent-evoked negative DRPs, nor the excitability change attributable to primary afferent depolarization, nor indeed the reflex inhibition commonly interpreted as presynaptic. While the amplitude of DRPs remained unchanged, their rate of rise was undoubtedly depressed, as was the antidromic discharge known as the dorsal root reflex. We found two possible reasons for our findings (38) to differ from those of Davidoff (17). During convulsions, spontaneous paroxysmal DRPs occluded the afferent-evoked DRP; and, especially with AC-coupled recording and electronically computed averages, such occlusion could be mistaken for depression. More important, in the presence of an anesthetic dose of pentobarbital, penicillin became a very effective depressant. How the two drugs interact is not clear, but it will be recalled that pentobarbital in moderate doses augments and in high doses depresses the negative DRP.

These observations seemed to rule out blockade of presynaptic inhibition as an explanation of the convulsant action of penicillin but maintained the impression of interference with some aspect of presynaptic function of the drug. This impression was reinforced by the somewhat paradoxical finding of depressed posttetanic potentiation by convulsant amounts of penicillin (33). Because of this persistent belief of presynaptic involvement, we decided to renew our investigation of the effect of penicillin on the negative DRP.

As a point of new departure, we took the observation that in the penicillin-treated spinal cord the best stimulus to provoke a seizure is the gentle but persistent ruffling of hairs. We argued that testing by single shocks may not reveal a defect which is manifest only during prolonged stimulation. Instead of the customary isolated volleys, we are therefore recording DRPs during and in the wake of prolonged repetitive stimulation. There is no posttetanic potentiation of negative DRPs which, instead, are subject to posttetanic fatigue (6,30). In some but not all experiments, posttetanic depression of afferent-evoked DRPs appeared greatly aggravated by penicillin treatment. We are in the process of investigating whether the exaggerated fatigue of DRPs is in any way correlated with the generation of seizures (C. Gray and G. G. Somjen, *work in progress*).

Blockade of Chloride Ionophores by Penicillin

According to Hochner et al. (29), the enhanced membrane resistance (i.e., reduced leak conductance) of crustacean muscle and the depression of GABA-mediated inhibition could both be explained by a single mechanism if penicillin blocked not the GABA-receptor but the carrier of chloride ions in the membrane. Pellmar and Wilson (45) report that in neurons of Aplysia, penicillin suppresses all chloride-dependent synaptic processes, regardless of transmitter. Penicillin seems to share this property with several other convulsant drugs, including pentylenetetrazole (46; see also chapter by Carpenter, *this volume*).

Impairment of the chloride conductance in the mammalian nervous system could cause suppression of inhibitory postsynaptic potentials (IPSPs) and repetitive discharge at presynaptic terminals (for this mechanism, see ref. 4). Iteration of spiking was reported at motor nerve terminals under the influence of penicillin (42,49); but at primary afferent terminals in the spinal cord, the reflexive discharge (DR reflex) is depressed rather than enhanced (38) (see also previous section). Whereas GABA-mediated inhibitory responses in brain are subject to depression by penicillin, the supposedly chloride-dependent but glycine-mediated spinal IPSPs are not. Yet the negative DRP, upon which penicillin has an effect under certain conditions, (see above) may in part be dependent on chloride ions (5).

PROSPECTUS

In sum, here are six characteristics in search of an explanation.

1. Administered either topically or systemically, penicillin induces convulsive activity in the spinal cord, which is synchronized over several segments. In a convulsing segment, all or nearly all alpha motoneurons but not nearly all interneurons depolarize. The convulsive dose is not higher for cord than for cortex. Paroxysmal discharges in spinal segments sometimes drive convulsive activity in the brain.

2. Paroxysmal depolarization of motoneurons is accompanied by a precipitous decline of membrane resistance. In seizure-free intervals, the resting membrane retains its normal impedance and voltage.

3. Monosynaptic EPSPs and disynaptic IPSPs evoked in spinal cord by synchronized afferent volleys are not affected by penicillin.

4. Presynaptic inhibition and the amplitude of negative DRPs evoked by single synchronous afferent volleys are unchanged after penicillin. The rate of rise of the negative DRP is usually reduced, and posttetanic fatigue of the DRP is aggravated. The dorsal root reflex is invariably depressed by penicillin, as is posttetanic potentiation of the monosynaptic reflex.

5. During convulsions, intraspinal primary afferents depolarize. This effect is more noticeable in afferent fibers of muscle nerves than in cutaneous fibers. Frequently but not invariably, paroxysmal DRPs begin earlier than seizure activity of motoneurones. Paroxysmal primary afferent depolarization gives rise to showers of antidromic impulses in sensory fibers.

6. Potassium activity rises in extracellular fluid during seizures, more so in ventral horn than in intermediated or dorsal gray matter. Paroxysmal potassium responses always trail behind, never lead, seizure activity. The release of potassium ions associated with evoked neural responses is also enhanced by penicillin, more noticeably in ventral horn than elsewhere.

Although the patterning of seizures induced by penicillin in high spinal preparations resembles epileptiform convulsions seen in intact animals, there is one important difference. Grand mal attacks can readily be provoked by electric stimulation of otherwise perfectly normal brains. The spinal cord requires a convulsant drug to be able to generate a full ictus. The seizure activity evoked by electric stimulation of cords not treated by drugs (21) is only rudimentary by comparison. It would seem almost that penicillin introduces into the spinal cord an element already present in the healthy forebrain, albeit in a latent state.

The conclusion is almost inescapable: seizure activity requires self-reexcitation. Tonic convulsions seem to be the result of unbridled positive feedback. In the oscillations that drive clonic beating, there must be an element of negative feedback with a significant delay in the feedback loop. A theory of cortical seizure activity (1) is based on recurrent excitation of pyramidal cells. In spinal cord, a recurrent excitatory pathway is not believed to exist, but only recurrent disinhibition (56). Accordingly, antidromic stimulation of the ventral root of penicillin-treated spinal cords does not initiate seizure activity (E. W. Lothman, *unpublished observation*). The local circuits providing positive feedback may of course exist in the concentration of short axon interneurons in the intermediate gray matter; these may provide the drive for motoneurons of the ventral horn. Feedback need not be neural, however, but could be biochemical. Chemical feedback was the basic idea in the potassium theory, which, as we saw earlier, is currently losing ground.

Almost all the facts relating to penicillin-induced spinal seizures could be tied together with the least strain of logic by the following assumptions. The primary convulsive event may be inward current welling up simultaneously in many afferent terminals. Paroxysmal DRP may be a manifestation of such current. If carried by both Na^+ and Ca^{2+} ions (see also chapter by Heinemann et al., *this volume*), it would bring about the release of a massive amount of transmitter, with the consequent widespread and synchronous depolarization of motoneurons, the drop in their membrane resistance, and the release of large amounts of potassium from both pre- and postsynaptic sources. A cause for the synchronous inward current of presynaptic terminals remains to be found. The theory would be helped greatly by the discovery of a metabolite, released into extracellular space in toxic amounts under the influence of penicillin, and in smaller amounts by nerve excitation. An amino acid with a nonselective effect on membrane permeability could fill the role.

We know that during convulsions, the demand for oxidative energy greatly exceeds that which normally is consumed at equivalent concentrations of potassium (35). The excess energy may be needed in part to pump back potassium through a membrane that has become too permeable for potassium. Additional extra energy may be spent in oxidizing the postulated toxic metabolite.

The author knows that the reader knows that the key elements postulated for this theory are not yet demonstrated experimentally. Still, the theory could organize observed facts, which otherwise seem disparate.

ACKNOWLEDGMENT

The authors' research was supported by USPHS grant NS 11933.

REFERENCES

1. Ayala, G. F., Dichter, M., Gumnit, J., Matsumoto, H., and Spencer, W. A. (1973): *Brain Res.*, 52:1–17.
2. Ayala, G. F., Lin, S., and Vasconetto, C. (1970): *Science*, 167:1257–1260.
3. Ayala, G. F., Spencer, W. A., and Gumnit, R. G. (1971): *Science*, 171:915–917.

4. Barchi, R. L. (1975): *Arch. Neurol.*, 32:175–180.
5. Barker, J. L., Nicoll, R. A., and Padjen, A. (1975): *Personal communication.*
6. Barron, D. H., and Matthews, G. H. C. (1938): *J. Physiol.*, 92:276–321.
7. Bradley, K., Easton, D. M., and Eccles, J. C. (1953): *J. Physiol.*, 122:474–488.
8. Chalazonitis, N., Ducreux, C., and Arvanitaki, A. (1972): *J. Physiol.*, 65:212A.
9. Chalazonitis, N., and Takeuchi, H. (1968): *C. R. Soc. Biol.*, 162:1552–1556.
10. Clarke, G., and Hill, R. G. (1972): *Br. J. Pharmacol.*, 44:435–441.
11. Cooke, J. D., and Quastel, D. M. J. (1973): *J. Physiol.*, 228:435–458.
12. Cordingley, G. E., and Somjen, G. G. (1977): *In preparation.*
13. Curtis, D. R., Game, C. J. A., Johnston, G. A. R., McCullock, R. M., and McLachlan, R. M. (1972): *Brain Res.*, 43:242–245.
14. Davenport, J., Schwindt, P. C., and Crill, W. E. (1976): *Soc. Neurosci.*, Abstr. 367.
15. David, R. J., Wilson, W. A., and Escueta, A. V. (1974): *Brain Res.*, 67:549–554.
16. Davidoff, R. A. (1972): *Brain Res.*, 36:218–222.
17. Davidoff, R. A. (1972): *Brain Res.*, 45:638–642.
18. Ducreux, C., and Gola, M. (1975): *Pfluegers Arch.*, 361:43–53.
19. Dunn, P., and Somjen, G. G. (1977): *Brain Res.*, 128:569–574.
20. Eccles, J. C., Magni, F., and Willis, W. D. (1962): *J. Physiol.*, 160:62–93.
21. Esplin, D. W., and Freston, J. W. (1960): *J. Pharmacol.*, 130:68–80.
22. Esplin, D. W., and Laffan, R. J. (1957): *Arch. Int. Pharmacodyn.*, 113:189–202.
23. Feldberg, W., and Sherwood, S. L. (1957): *J. Physiol.*, 139:408–416.
24. Gloor, P., and Testa, G. (1974): *Electroencephalogr. Clin. Neurophysiol.*, 36:499–515.
25. Gola, M., and Romey, G. (1973): *J. Physiol.*, 67:227.
26. Gutnick, M. J., and Prince, D. A. (1972): *Science*, 176:424–426.
27. Hahn, F. (1960): *Pharmacol. Rev.*, 12:482–530.
28. Heath, C. J. (1962): Thesis. University of Otago, Dunedin.
29. Hochner, B., Spira, M. E., and Werman, R. (1976): *Brain Res.*, 107:85–103.
30. Holobut, W., and Niechaj, A. (1973): *J. Physiol.*, 230:15–27.
31. Kao, L. I., and Crill, W. E. (1972): *Arch. Neurol.*, 26:156–161.
32. Kao, L. I., and Crill, W. E. (1972): *Arch. Neurol.*, 26:162–168.
33. LaManna, J., Lothman, E., Rosenthal, M., Somjen, G., and Younts, W. (1977): *Epilepsia*, 18: 317–329.
34. Lloyd, D. P. C., and McIntyre, A. K. (1949): *J. Gen. Physiol.*, 32:409–443.
35. Lothman, E. W., LaManna, J., Cordingley, G., Rosenthal, M., and Somjen, G. G. (1975): *Brain Res.*, 88:15–36.
36. Lothman, E. W., and Somjen, G. G. (1975): *J. Physiol.*, 252:115–136.
37. Lothman, E. W., and Somjen, G. G. (1976): *Electroencephalogr. Clin. Neurophysiol.*, 41:237–252.
38. Lothman, E. W., and Somjen, G. G. (1976): *Electroencephalogr. Clin. Neurophysiol.*, 41:253–267.
39. Lothman, E. W., and Somjen, G. G. (1976): *Electroencephalogr. Clin. Neurophysiol.*, 41:337–347.
40. Lux, H. D. (1973): *Mitt. Max Planck Gesellsch. Heft*, 1:34–52.
41. Morris, M. E., and Krnjević, K. (1976): *Advances in Pain Research and Therapy, Vol. 1*, edited by J. J. Bonica and D. Albe-Fessard, pp. 117–122. Raven Press, New York.
42. Noebels, J. A., and Prince, D. A. (1976): *Soc. Neurosci.*, Abstr. 377.
43. Pedley, T. A., Fisher, R. S., Futamachi, K. J., and Prince, D. A. (1977): *Fed. Proc.*, 35:1254–1259.
44. Pellmar, T. C., and Somjen, G. G. (1977): *Brain Res.*, 120:179–183.
45. Pellmar, T. C., and Wilson, W. A. (1977): *Brain Res. (In press.)*
46. Pellmar, T. C., and Wilson, W. A. (1977): *Science (In press.)*
47. Prince, W. A., and Farrell, D. (1969): *Neurology*, 19:309–310.
48. Raichle, M. E., Kutt, H., Louis, S., and McDowell, F. (1971): *Arch. Neurol.*, 25:232–239.
49. Raines, A., and Dretchen, R. K. L. (1975): *Epilepsia*, 16:469–476.
50. Sackellares, J. C., and Smith, D. B. (1977): *Am. EEG Soc.*, Abstr. 28, p. 33.
51. Schwindt, P. C., and Crill, W. E. (1976): *Soc. Neurosci.*, Abstr. 381.
52. Smith, T. C., Barker, J. L., and Gainer, H. (1975): *Nature*, 253:450–452.
53. Somjen, G. G., and Heath, C. J. (1966): *Exp. Neurol.*, 15:79–99.

54. Sutten, G. G., and Oldstone, M. B. A. (1969): *Neurology,* 19:859–864.
55. Testa, G., and Gloor, P. (1974): *Electroencephalogr. Clin. Neurophysiol.,* 36:517–524.
56. Wilson, V. J., and Burgess, P. R. (1962): *J. Neurophysiol.,* 25:392–404.
57. Wilson, W. A., and Wachtel, H. (1974): *Science,* 186:932–934.
58. Wyant, J. D. (1967): *J. Thorac. Cardiovasc. Surg.,* 54:579–581.

Abnormal Neuronal Discharges, edited by
N. Chalazonitis and M. Boisson. Raven
Press, New York © 1978.

A Cortical Epilepsy Model *In Vitro*

P. Andersen, L. Gjerstad, and I. A. Langmoen

The Institute of Neurophysiology, University of Oslo, Norway

Despite a large effort, many of the mechanisms underlying epileptic behavior of cortical neurons are not known. Clearly, the complicated structure and function of cortical tissue make the analysis of this pathological state difficult. Therefore, one would like to study epileptic behavior in a simple environment; this is the rationale for the use of models like the crayfish stretch receptor (3,9), crayfish excitatory junction potentials (6), giant synapse of the squid (4), abdominal ganglion of Aplysia (15), Mauthner fiber/giant fiber synapse in fish (14). Against the advantage of the isolated preparation, one can raise the objection that the behavior of these systems does not necessarily replicate the processes occurring in an epileptic focus in cortical gray matter.

With the advent of isolated central nervous tissue preparations (12,16), this situation is changed. In particular, when the transverse hippocampal slice was developed (13), it was possible to study a piece of cortical tissue *in vitro* in which all major afferent and efferent connections were intact. Admittedly, the section of the tissue naturally involves the cutting of a certain number of afferents. Because of the lamellar arrangement of the hippocampal cortex (2), however, the main elements of the normal tissue are maintained. In the *in vitro* situation, it is possible to control the environment in a way that is impossible in intact animals. The absence of blood brain barrier is one of the main advantages. In addition, there is an excellent mechanical stability, which allows the making of high quality, long-lasting intracellular recordings. Finally, all electrodes can be positioned under direct visual control.

With this new preparation, we wanted to study the behavior of hippocampal neurons to known epileptic agents and how that behavior is changed. The idea is to study in more detail than has been possible what particular conditions are necessary for developing epileptogenic activity and to compare this activity with data observed from intact preparations of the same structure (5).

METHODS

Adult guinea pigs (250 to 300 g) were anesthetized with ether. The brain was removed and the hippocampus quickly dissected out by freeing the septal and temporal ends and rolling out gently with a spatula. Sections (350 to 400 μm thick) were cut nearly transversely to the longitudinal axis of the hippocampus. The sections were transferred to a tissue bath where they were placed on a net with artificial cerebrospinal fluid from beneath and humidified gas (95% O_2, 5% CO_2) from above. The composition of the artificial cerebrospinal fluid was (in mM): NaCl 124, KCl 5, KH_2PO_4 1.25, $MgSO_4$ 1.2, $CaCl_2$ 2, $NaHCO_3$ 26, glucose 10. The pH of the solution was between 7.35 and 7.45.

The bath was kept at a temperature of 32 to 34°C. With higher temperatures, there was an unwanted tendency to increased excitability. To induce repetitive firing, 3.4 or 8.5 mM sodium benzyl penicillin (2,000 to 5,000 IU/ml was added to the artificial cerebrospinal fluid. Insulated and sharpened tungsten microelectrodes were used for stimulation cathodes. Recordings were made by fiber-containing micropipettes filled with 4M potassium citrate solution. The resistance was 120 to 200 megohms for intracellular work and 20 to 40 megohms for extracellular work. The input amplifier allowed current to be passed through the recording electrode. In this way, depolarizing or hyperpolarizing pulses could be injected across the soma membrane while recording the responses to the pulse itself or to orthodromic stimulation of several afferent pathways impinging on various parts of the dendritic tree.

RESULTS

The advantages of the transverse hippocampal slice are not only owing to the possibility of working *in vitro*. Because of the particular organization of the hippocampal cortex, a number of experiments can be made that are difficult to perform in other types of tissue. An essential point is the parallel arrangement of the afferent fibers. Figure 1A shows a diagram of a transverse hippocampal slice with the single layer of cell bodies (pyr) marked by a double line. The main axis of the dendrites is found normal to this layer, and the arrangement of the parallel apical dendrites in CA1 gives the structure a striped appearance, as the name, str. radiatum (rad), implies. On the other side of the cell bodies, the basal dendrites are found in a layer called str. oriens (or).

For our purpose, an important histological feature is the orientation of most afferent fibers parallel to the layer of the pyramidal cells. The boxed-in area in Fig. 1A is shown enlarged in Fig. 1B. Here, a single afferent fiber is symbolized by the line with open circles. The circles indicate *en passage* boutons of the afferent fibers. Because of this arrangement, it is possible to selectively stimulate afferent fibers to any part of the dendritic tree. Furthermore, because of the similarity of their diameters, a stimulus produces a synchronous afferent volley which can be recorded if the electrodes are placed not too far apart. In this way, the preparation allows the direct re-

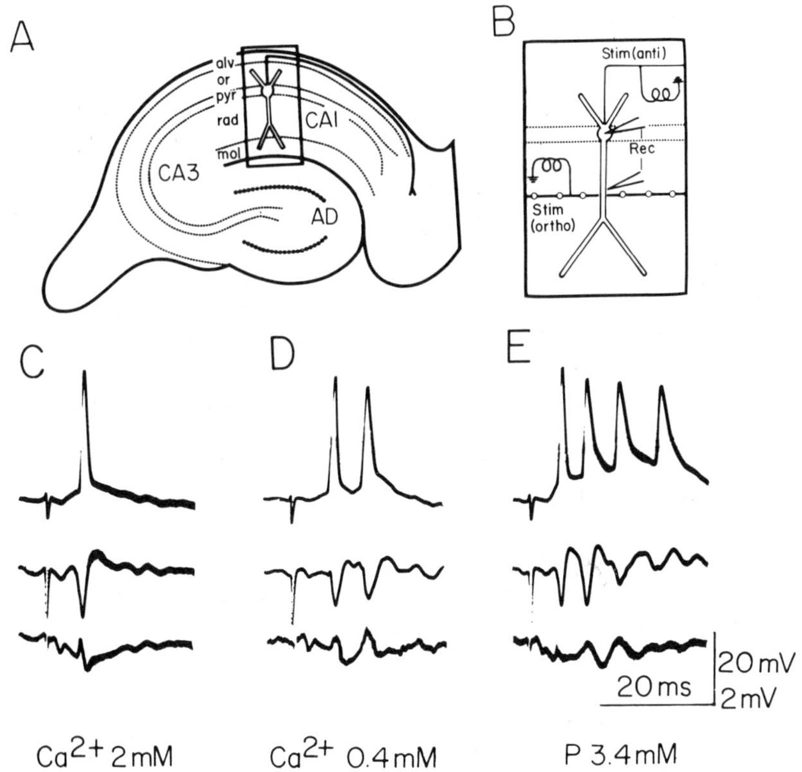

FIG. 1. A: Schematic diagram of the transverse hippocampal slice with the main partitions CA1, CA3, and dentate area (AD), and the layers of CA1 indicated. Alv, alveus; or, stratum oriens; pyr, stratum pyramidale; rad, stratum radiatum; mol, stratum moleculare. The boxed-in area is shown in greater detail in **B,** where the symbols indicate possible routes for stimulation (orthodromic and antidromic) and two recording sites, one intracellular in the soma and one extracellular at the point of afferent input. **C:** *upper trace,* intracellular record from a CA1 pyramidal cell; *middle trace,* extracellular trace taken by another adjacent electrode; *lower trace,* extracellular record from a third electrode located at the activated synapses in the dendritic field. **D:** As **C,** but the calcium concentration was reduced from 2 to 0.4 mM **E:** As **C,** but with 3.4 mM sodium benzyl penicillin added to the medium.

cording of the presynaptic afferent volley, local dendritic synaptic field potentials, the extracellular population spikes, and intracellular recording of pyramidal cell responses to orthodromic or direct stimulation.

In Fig. 1C–E, the upper row shows intracellular recordings from a CA1 pyramidal cell; the middle row shows the extracellular recording in the pyramidal cell layer, whereas the lower traces show the extracellular field potentials recorded from the synaptic layer in the dendritic field. In Fig. 1C, the responses are normal. The lower trace shows an initial diphasic deflection followed by a slower negative wave on which a diphasic spike is superimposed. The slower negative wave represents the extracellular field excitatory postsynaptic potential (EPSP), and the spike is a reflection of the population spike from the cell body layer. The middle row shows a slow wave on which a large negative population spike is seen. This is the extracellular sign of the nearly synchronous discharge of the large number of pyramidal cells (1). The upper row shows the intracellular record with an EPSP and the occurrence of a spike. Under normal conditions, an afferent impulse volley gives rise to only one or occasionally two spikes. With increased temperature, however, the response often shows more spikes. Similarly, following a reduction of the calcium concentration in the artificial cerebrospinal fluid (Fig. 1D), there was an increased number of population spikes. Intracellularly, (upper trace) two action potentials emerged, although the presynaptic volley had the same size as in Fig. 1C. Addition of 3.4 mM sodium benzyl penicillin gave much more developed repetitive behavior (Fig. 1E). Both the extracellular dendritic and pyramidal layer responses show a ripple of waves, signifying that a number of cells are discharged repetitively and synchronously. The upper trace shows how the penetrated cells now deliver four spikes in rapid succession. Interestingly, the spikes are riding on a large depolarizing wave. Furthermore, the action potentials show a successive decline in amplitude and a broadening of their duration, compared with the relatively unchanged initial action potential. This behavior was typical for virtually all cells recorded under penicillin conditions. Clearly, the response is reminiscent of that shown in cortical penicillin foci, and which has been called depolarizing shifts (DSs) (8,10).

The input resistance of the cell as measured by hyperpolarizing current pulses did not show any obvious change, nor did the response to a depolarizing current pulse. Therefore, the increased reactivity of the cell to synaptic activation cannot be explained by a general increase of the excitability of the membrane, nor by a change of the resistance of the membrane.

The increased responses could not be explained by an augmented presynaptic volley. Fig. 2A shows the presynaptic volley (arrow) and the first part of the field EPSP (asterisk) before penicillin, whereas Fig. 2B shows the same response during, and Fig. 2C after, penicillin administration. The graph in Fig. 2D indicates that the relationship between the stimulation current and the presynaptic volley was not changed by penicillin, nor was the relationship between the presynaptic volley and the field EPSP (Fig. 2E).

Following the development of the penicillin effect with intracellular recording, it was apparent that the underlying cause for the increased number of discharges was a slow wave of depolarization, clearly the DS of Prince (10). Since this could occur without the occurrence of superimposed spikes, it cannot be described as a spike afterpotential. Rather, the prolonged depolarization seems to be a reaction by itself and of primary importance for the understanding of the epileptic behavior of the neurons

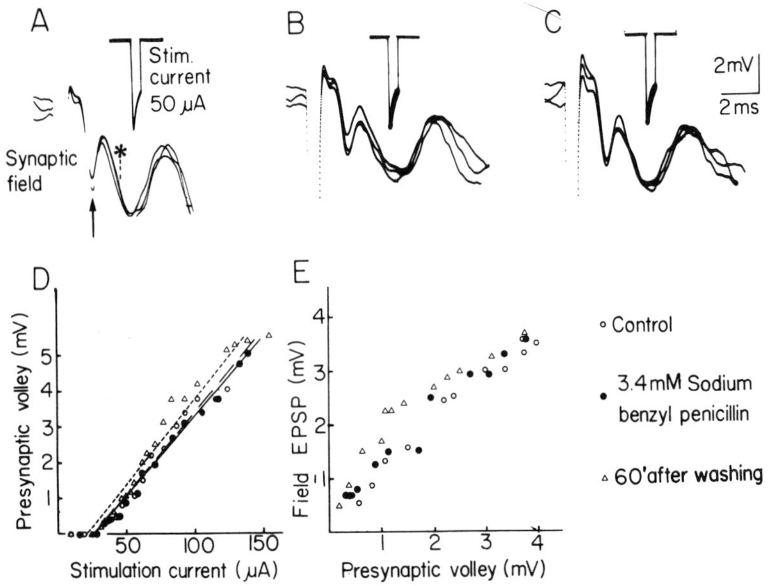

FIG. 2. A: Extracellular records from the layer of synaptically activated dendrites with a presynaptic volley (*arrow*) followed by a negative extracellular field EPSP (*asterisk*). The stimulus current is shown above on expanded sweep. **B:** Similar to **A** but taken during application of 3.4 mM sodium benzyl penicillin which gives rise to vigorous activity in the cell body layer. The records in **C** were taken 60 min after washing had started. **D:** Graph showing the relationship between the stimulation current (*abscissa*) and the size of the presynaptic volley (*ordinate*). *Open circles,* responses before penicillin treatment; *filled circles,* effect during penicillin treatment; *open triangles,* results 60 min after washing was started. **E:** Relationship between the presynaptic volley (*abscissa*) and the extracellular field EPSP (*ordinate*) during the same procedure as in **D.**

under study. The DS can be facilitated by slight depolarization of the cell. In Fig. 3A, the normal cell response is an ordinary spike taking off from a small EPSP followed by a depolarizing wave of moderate duration. On depolarizing the cell by injection of a 0.4 nA steady anodal current, however, the postspike depolarization increased in both amplitude and duration. As expected, the action potential size was reduced and its width increased during the depolarization. By further depolarization (0.6 nA) corresponding to a reduction of the membrane potential from −65 to −48, the action potential was nearly blocked, whereas a large depolarizing shift was produced. Conversely, the long depolarization can be counteracted by hyperpolarizing the membrane. The cell shown in Fig. 3B shows responses just around the threshold for spike initiation. The initial portion of the EPSP led twice to a spike with varying latency. In three trials, however, there was no action potential. Despite this, there was a long-lasting depolarization, DS. With hyperpolarization of the cell, there was a reduction of the DS so that at

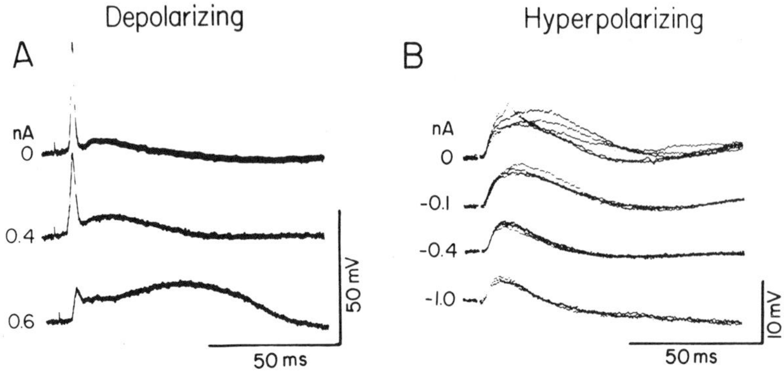

FIG. 3. A: Intracellular records from a CA1 pyramidal cell under the influence of 3.4 mM sodium benzyl penicillin in response to afferent stimulation of radiatum fibers. The second and third record show the result when a steady depolarizing current of 0.4 and 0.6 nA was passed. **B:** Intracellular records from another CA1 pyramidal cell showing the long-lasting depolarizing response following orthodromic stimulation. On passing hyperpolarizing currents of −0.1 to −0.4 nA, the slow depolarizing potential was reduced or even abolished and a simple monosynaptic EPSP remained. On further hyperpolarization by passing −1.0 nA, no additional change of the EPSP was observed.

−0.4 nA the slow depolarization had been removed and the ordinary EPSP was the only remaining response. Further hyperpolarization (−1.0 nA) did not change the EPSP further.

In other cases, when either the DS was more developed or stronger orthodromic activation was used, hyperpolarization could not entirely remove the long-lasting depolarization. Some of the later action potentials superimposed on the DS were blocked. Thus hyperpolarization of the cell is only partially effective in rectifying the hyperactivity of the cell.

DISCUSSION

The main finding in the present investigation was that the transverse hippocampal slice lends itself favorably to studies of the effect of penicillin on the activity of cortical neurons.

A major experimental finding has been the lack of the effect of penicillin on the presynaptic afferent volley. This is in contrast to the increased reactivity of myelinated afferent thalamocortical fibers which have been found in intact preparations (7). The fibers in the present case are unmyelinated, whereas the former have been myelinated. Whether there is a real difference between these two types of cortical fibers remains to be investigated.

The fact that completely normal EPSPs can be recorded in situations where the cell shows a definite increase of reactivity indicates that the initial orthodromic event is not changed, nor is it responsible for the hyperactivity.

The unchanged initial part of the extracellular EPSP in the present study indicates that the time course and size of the injected synaptic current do not seem to be responsible for the increased reaction. Furthermore, since hyperpolarization of the cell could remove the DS without changing the synaptic EPSP, this indicates that the DS is not just an increased and prolonged EPSP, a conclusion also reached in a similar preparation by Prince and Schwartzkroin (11).

Because the repetitive discharge cannot be produced by antidromic activation or by injection of a depolarizing current pulse across the soma membrane, it does not appear likely that the response takes place at or near the soma. All in all, the data suggest that the response consisting of a long-lasting depolarization with superimposed spikes may be generated somewhere in the dendritic tree by a membrane whose ionic channels may be changed because of the penicillin intoxication. Further information on the detailed mechanism underlying this reaction, which seems crucial for the understanding of the epileptic behavior of the neuron, will probably be gained from studies with manipulations of the extracellular and intracellular ionic environment of the cell, as well as interference with the appropriate ionophores.

SUMMARY

The processes underlying the repetitive activity of hippocampal neurons following penicillin application have been followed in the transverse hippocampal slice *in vitro.*

Application of 3.4 and 8.5 mM sodium benzyl penicillin did not change the presynaptic afferent volley of unmyelinated fibers, nor did penicillin change the initial part of the extracellular EPSP, indicating a normal time course and amplitude of synaptic current. However, both extracellular and intracellular recordings showed that the responses to orthodromic impulses changed from a single spike to a number of discharges.

Intracellularly, these repetitive discharges were seen to arise from a long-lasting depolarization similar to a DS seen in intact preparations. The DS could occur without a preceding spike, indicating that it differs from a depolarizing afterpotential. The DS could be facilitated by depolarization of the cell but could be reduced or even removed by hyperpolarization of the cell.

The hyperactivity could not be elicited by antidromic activation or by direct stimulation by passing a depolarizing current pulse across the soma membrane. The membrane potential and the membrane resistance only showed small and insignificant changes following application of penicillin.

In conclusion, the long-lasting depolarization is of great importance for the development of the pathological repetitive discharge. It seems to be generated in the dendrites somewhere between the soma and the synaptic parts of the cell, possibly by a changed reactivity of the dendritic membrane caused by the penicillin.

REFERENCES

1. Andersen, P., Bliss, T. V. P., and Skrede, K. K. (1971): *Exp. Brain Res.,* 13:208–221.
2. Andersen, P., Bliss, T. V. P., and Skrede, K. K. (1971): *Exp. Brain Res.,* 13:222–238.
3. Ayala, G. F., Lin, S., and Vasconetto, C. (1970): *Science,* 167:1257–1260.
4. Ayala, G. F., Spencer, W. A., and Gumnit, R. J. (1971): *Science,* 171:915–917.
5. Dichter, M., and Spencer, W. A. (1969): *J. Neurophysiol.,* 32:649–662.
6. Futamachi, K. J., and Prince, D. A. (1975): *Brain Res.,* 100:589–597.
7. Gutnick, M. J., and Prince, D. A. (1972): *Science,* 176:424–426.
8. Matsumoto, H., and Ajmone-Marsan, C. (1964): *Exp. Neurol.,* 9:286–304.
9. Meyer, H., and Prince, D. A. (1973): *Brain Res.,* 53:477–482.
10. Prince, D. A. (1968): *Exp. Neurol.,* 21:467–485.
11. Prince, D. A., and Schwartzkroin, P. A. (1978): *This volume.*
12. Richards, C. D., and Sercombe, R. (1968): *J. Physiol. (Lond.),* 197:667–683.
13. Skrede, K. K., and Westgaard, R. H. (1971): *Brain Res.,* 35:589–593.
14. Spira, M. E., and Bennet, M. V. L. (1972): *Brain Res.,* 43:235–241.
15. Wilson, W., and Escueta, A. V. (1974): *Brain Res.,* 72:168–171.
16. Yamamoto, C., and McIlwain, H. (1966): *J. Neurochem.,* 13:1333–1343.

Abnormal Neuronal Discharges, edited by
N. Chalazonitis and M. Boisson. Raven
Press, New York © 1978.

Rhythmic Discharges Related to Tremor in Humans and Monkeys

C. Ohye and D. Albe-Fessard

Department of Neurosurgery, Gunma University School of Medicine, Gunma, Japan; and Laboratoire de Physiologie des Centres Nerveux, Université Pierre et Marie Curie, Paris, France

For several years, we have been interested in the rythmic burst discharges observed in the central nervous system related or not related to tremor. In humans, such abnormal discharges have been studied in the ventrolateral part of the thalamus in the course of stereotactic surgery for the treatment of various types of tremor, such as static (mainly in Parkinson disease), postural (in posttraumatic disorder), and intentional tremor (in cerebellar disease or posttraumatic disorder). All these procedures were done with local anesthesia in the awake state; therefore, several observations in cases of cerebral palsy operated on in the narcotized state were also considered for comparison.

On the other hand, we have been producing parkinsonian-like spontaneous tremor in monkeys by a selective lesion in the mesencephalic ventromedial tegmental (VMT) area with radiological and physiological techniques, and have tried to clarify the nervous system mediating the spontaneous, sustained impulses of tremor thus produced by extracellular recording of the unitary activity mainly in the cerebellum.

In this chapter, a summary of results obtained in humans as well as in monkeys is presented and discussed in relation to the neural mechanism underlying the tremor.

RHYTHMIC AND NONRHYTHMIC BURST DISCHARGES IN THE HUMAN THALMUS

In the ventrolateral part of the thalamus, Albe-Fessard and her associates (1) found for the first time rhythmic burst discharges related to peripheral tremor in cases of Parkinson disease. Furthermore, systematic study showed that the thalamic ventralis intermedius (Vim) nucleus and its dorsal part and part of lateralis posterior nucleus was specifically concerned with the tremor mechanism in that rhythmic burst discharges apparently correlated with peripheral tremor were found preferentially in these areas;

their rhythmicity persisted even when the peripheral tremor was arrested by an active or a passive maneuver (2).

In a previous study (18), we confirmed these findings by an analysis of approximately 800 spontaneous discharges recorded extracellularly by a concentric bipolar electrode (outer diameter, 0.6 mm; tip, 10 μm; resistance, ~ 100 KΩ) from 80 cases of thalamic surgery. The spontaneous burst discharges were classified into two types: rhythmic (regular) and nonrhythmic (irregular) according to the regularity of interburst interval, the value of coefficient of variation (CV = SD/M, where SD = standard deviation, M = mean), 0.345 being chosen as the critical one. At that time, however, we did not consider those burst discharges that were evoked by the peripheral tremor and therefore time-locked to it. We examined another 47 recently operated cases (mainly Parkinson disease in the awake state) in which 46 burst type neurons, including 21 tremor time-locked burst discharges, were recorded on FM magnetic tape from 27 cases (the remaining 20 cases did not show burst discharges). These burst discharges were analyzed statistically by NOVA minicomputer of 12 KW core memory (19).

These observations showed that in the ventrolateral part of the thalamus, probably in the Vim nucleus, the ventrolateral nucleus, and their adjacent area, three types of burst discharges could be distinguished in the awake state. One is discharging in bursts, time-locked with tremor (Fig. 1A). This type of neuron responded to peripheral natural stimuli, such as passive or active movement of the corresponding contralateral part of the body (Fig. 1B). Human Vim neurons receiving kinesthetic afferent were studied carefully and the results already presented elsewhere (16); it is not surprising to find some Vim neurons that reflect the peripheral tremor, the neuron discharging in response to the mechanical movement of the trembling muscle or joint. The tremor synchronous burst discharge consists of four to five irregular repetitive spikes within about 100 msec. Interspike interval histograms revealed a gamma distribution during the initial 100 msec, within which some tendency of rhythmicity could be detected, and a second mode at about 200 to 250 msec, which corresponds to the interburst interval (Fig. 2). The fact that tremor-dependent rhythmic bursts existed was more clearly demonstrated by autocorrelograms of the spikes in a single unit. During the absence of tremor in peripheral muscles, the autocorrelogram showed only a flat distribution; during the tremor phase, however, periodic distribution was demonstrated.

The second type of burst discharge is rhythmic but its rhythm is not clearly related to that of the tremor. Although the frequency of the burst discharge is often very close to that of the tremor, it is not necessarily time-locked with peripheral tremor nor driven by the natural stimuli applied to the contralateral extremities. This type of rhythmic discharge persists independently of the peripheral tremor, whether or not it is there. The rhythmicity of these units is generally so constant, independent from the level of con-

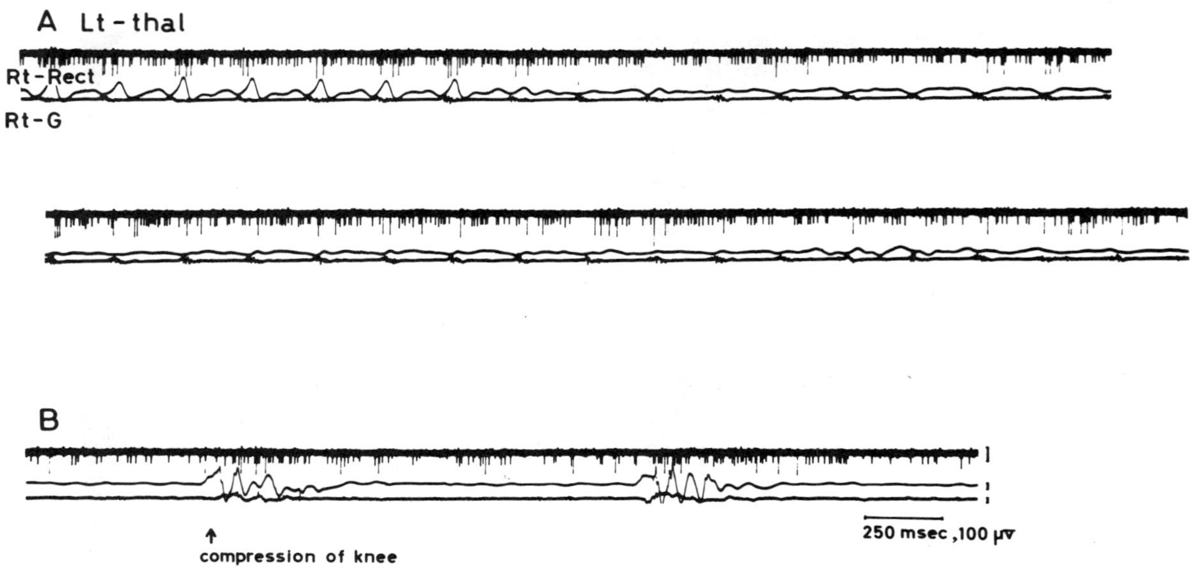

FIG. 1. A: Left thalamic (Lt-thal) discharges time-locked with monitor EMGs of the peripheral tremor in the contralateral rectus (Rt-Rect) and gastrocnemius (Rt-G) muscles. **B:** The same neurons responding to compression of the contralateral knee, indicated by an arrow and EMG artifact. Note that two (large and small) spikes are recorded simultaneously. The larger one is more sensitive to the peripheral stimuli as well as to the spontaneous tremor movement.

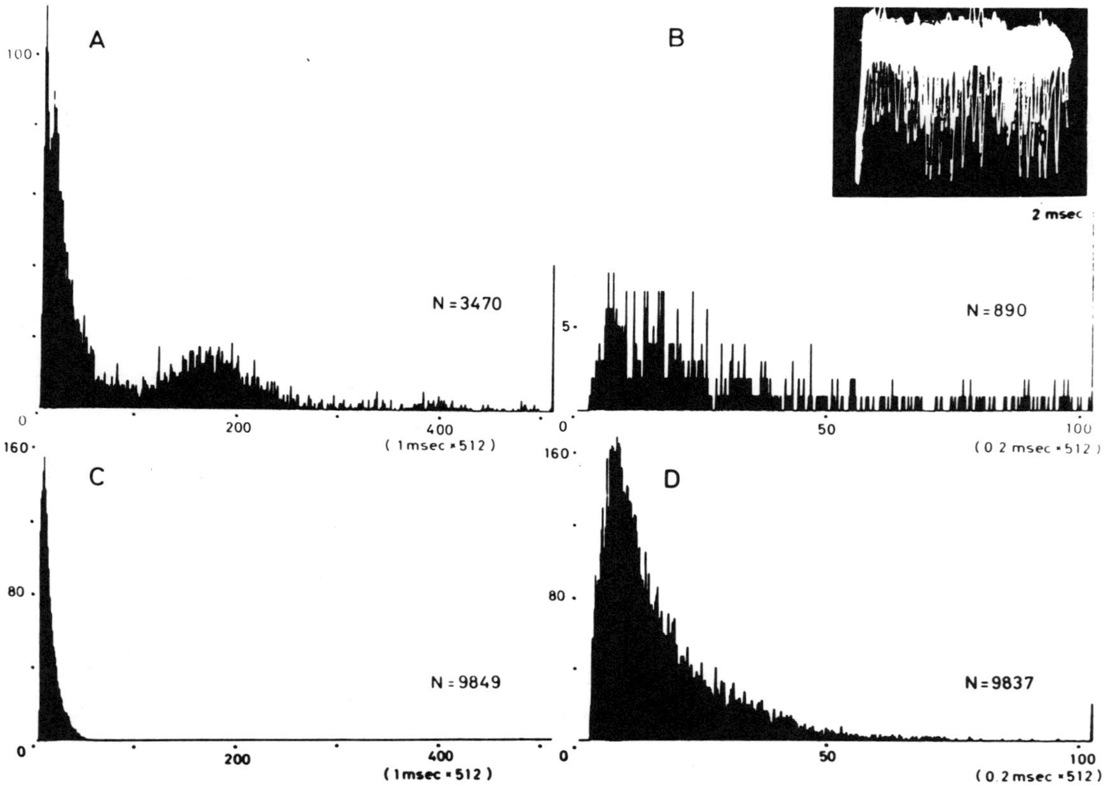

FIG. 2. Interspike interval histograms at different times of the large (**A** and **B**) and small (**C** and **D**) spikes shown in Fig. 1. *Inset:* Enlarged original spikes; tracings are triggered by the larger spike.

sciousness as far as tested, that it is reminiscent of the so-called "biological clock."

In contrast to the above-mentioned tremor-related burst, this type of burst discharge consists of four to five spikes within a short duration (about 10 msec) (Fig. 3). In general, the interval between the first and the second spike is quite fixed, being about 2 to 3 msec; the intervals of the following spikes fluctuate – the later the spike, the more dispersion. It is also a general rule that the second and later spikes are smaller in amplitude than the first. As seen in Fig. 3, the interspike interval histogram demonstrates two modes; the first is a type of gamma distribution corresponding to intraburst spike interval, and the second corresponds to interburst interval. The distribution around the second mode is almost symmetrical, indicating that the burst is rhythmic.

The third type of burst discharge is irregular in its burst-to-burst interval and quite independent of the tremor. One burst consists of four to five spikes within a short duration (about 10 msec). The characteristics of the spikes in one burst are essentially the same as those mentioned above. It is not influenced by the peripheral stimuli or by mental or physical tasks. In several cases, however, strong stimuli applied to the contralateral peripheral part suppressed or facilitated the burst discharges. For example, in one case, strong pressure on the contralateral first finger specifically suppressed the irregular burst discharge during the stimuli. In another case, tapping on the popliteal fossa induced burst discharges; the burst discharge recruited by the second stimuli outlasted the cessation of the stimuli (Fig. 4). Relative independence of such burst discharge is presumed to be due to indirect connection with the afferent input, if any, suggesting that they might be interneurons.

In the ventral thalamic area, these three different types of burst discharge could be recorded at random, the three being intermingled during lowering of the electrode. In one case, the tremor-independent rhythmic burst was found; then, 1.5 mm away, the tremor time-locked rhythmic burst responsible for passive dorsiflexion of the ankle joint was found. In fact, from 27 recent cases (21 cases of Parkinson disease, five of essential or postural tremor, and one of torticollis), 46 burst discharges, including tremor time-locked (21 neurons) and independent (25 neurons), were found intermingled in and around the Vim nucleus if superimposed on the corresponding atlas of the human thalamus (20), as shown in Fig. 5.

In contrast to what is seen in the awake state, burst discharges were more frequently found in the narcotized state [57.8% of the spontaneous discharges versus 27.8% in the previous study (18)]. The burst discharges were certainly independent of tremor movements and were observed in patients having no tremor. Some of them, however, were quite regular in their interburst interval, and even the frequency could be very close to that of Parkinson tremor. Moreover, the distribution of the burst discharges

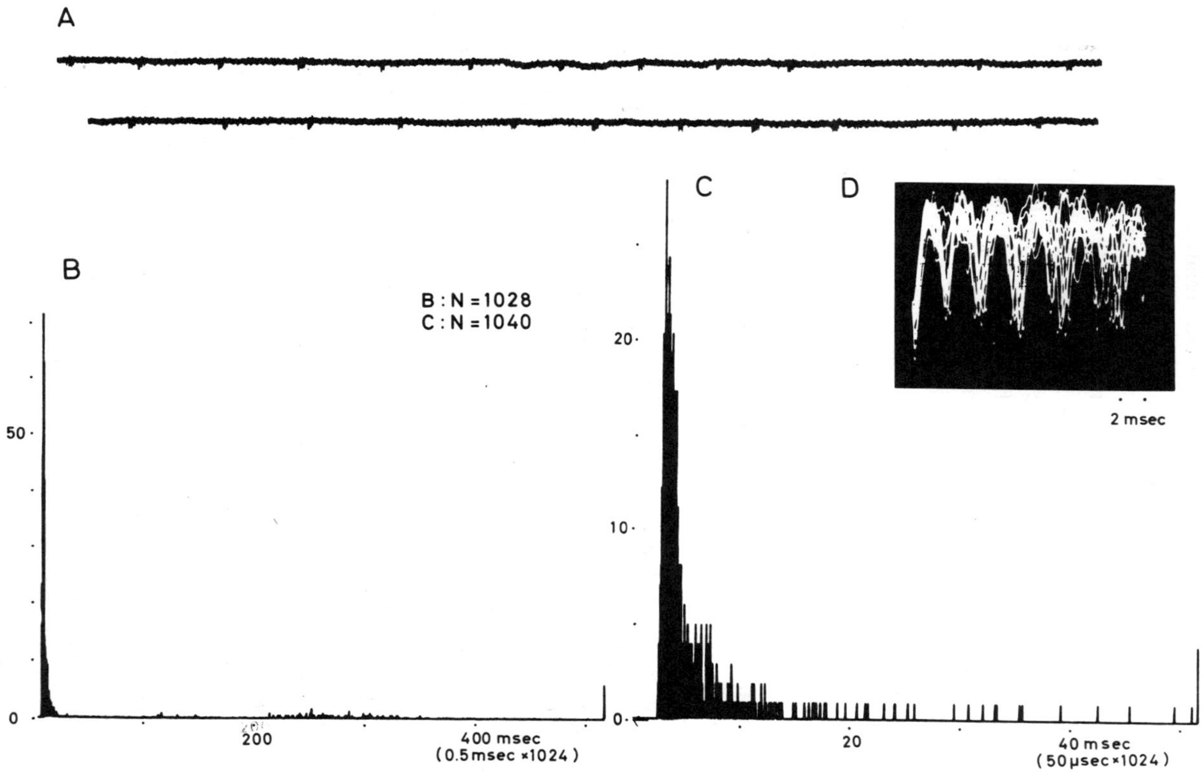

FIG. 3. Interspike interval histograms **(B** and **C)** at two different times of a rhythmic but tremor-independent burst discharge **(A).** About 10 bursts are superimposed, triggered by the first spike **(D).**

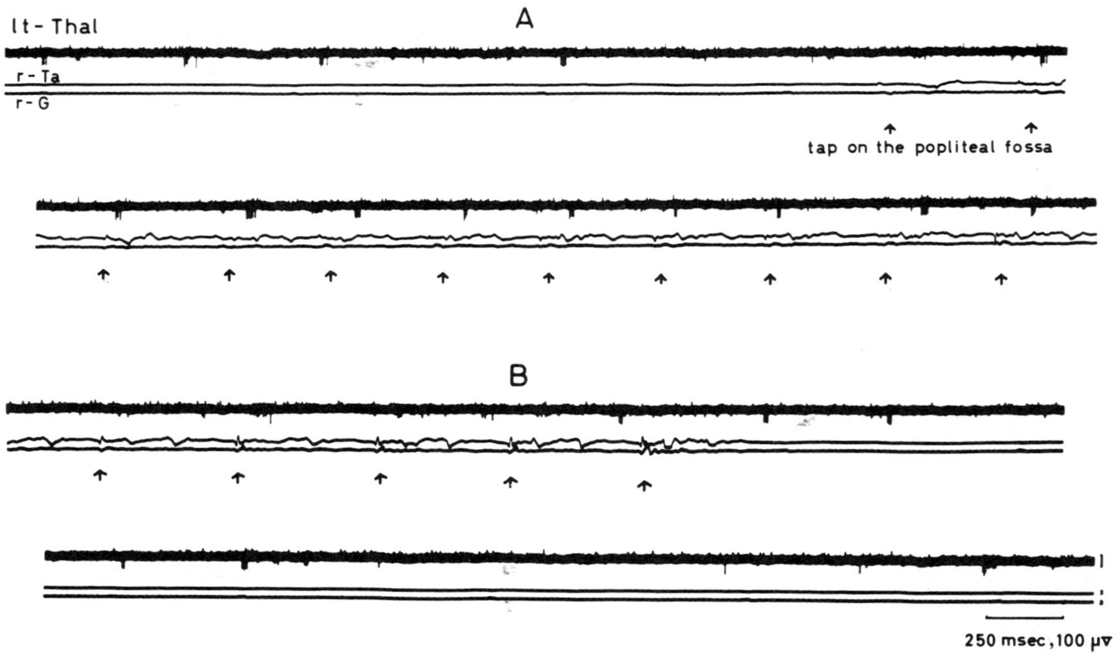

FIG. 4. Irregular burst discharge (*top trace*) responding to a tap on the contralateral popliteal fossa. Approximate time of stimulation is shown by arrows and artifact in EMGs (*lower two traces*). lt-Thal, left thalamic discharge; r-Ta, EMG of right tibialis anterior; r-G, EMG of right gastrocnemius muscle.

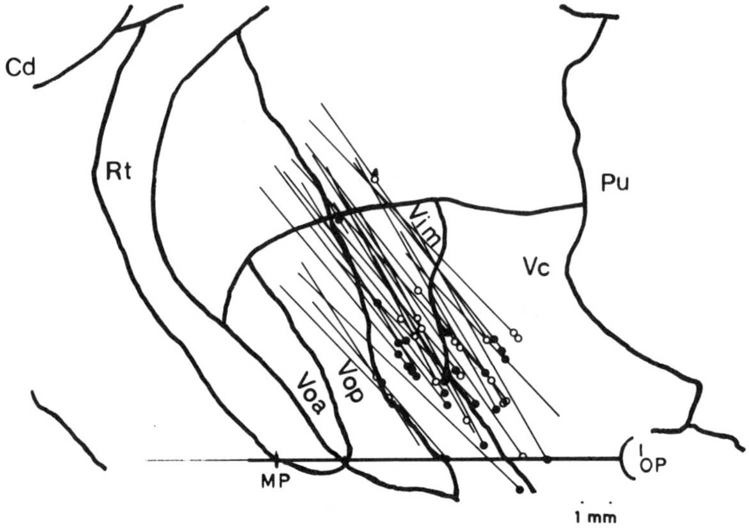

FIG. 5. Burst discharges recorded from 27 recent cases are plotted with respective trackings reconstructed from X-ray film (lateral view) and superimposed on a parasagittal section of the human thalamus (20). *Open circles,* tremor time-locked burst discharge; *filled circles,* tremor-independent burst discharge. Cd, caudate nucleus; CP, posterior commissure; Rt, reticular nucleus; Voa, ventralis oraris anterior nucleus; Vop, ventralis oraris posterior nucleus; Vim, ventralis intermedius nucleus; Vc, ventralis caudalis nucleus; Pu, putamen. *Thick horizontal line,* intercommissural line with its midpoint (MP).

within the thalamus was also quite similar to that found in awake parkinsonian cases, mostly occupying the Vim nucleus.

Summary and Comment

We found three types of burst discharges in the human thalamus in the awake state. They were (a) rhythmic burst, time-locked with peripheral tremor and responding to natural stimulation (probably kinesthetic in origin) of the contralateral side, (b) rhythmic burst, independent of tremor but with frequency close to it, as well as independent of other types of natural stimuli, and (c) nonrhythmic burst, not related to tremor but rather to the sleeping state. These three types of burst discharges are located preferentially in and around the thalamic Vim nucleus and its dorsal part. As the latter two types of tremor-independent burst discharges are also found in the same thalamic area in cerebral palsy patients without clinical rhythmic movements, they pertain to nonpathological systems. Although their functional significance in relation to the tremor mechanism is still difficult to interpret, one possibility is that they play an important role in modifying voluntary smooth movement into uneven tremor movements. In fact, in several cases with tremor of unknown etiology, we found an extraordinarily high incidence of burst discharges (*unpublished observation*). A functional or organic deafferentation might be responsible, as suggested by other investigators (3,5).

Finally, we should add that the important role of this region is demonstrated by the fact that a small stereotactic lesion in this same Vim area abolishes various kinds of tremor without causing neurological deficit (13,15).

RHYTHMIC DISCHARGES IN THE CEREBELLUM OF A MONKEY WITH EXPERIMENTAL TREMOR

In monkeys (*Macaca irus*), we have succeeded in constantly producing the spontaneous parkinsonian-like tremor by a small lesion in the mesencephalic VMT area with the aid of radiological and physiological localizing methods (14,17). In view of further research on the tremor mechanism, we have developed a special recording technique with which the deep-seated structures could be explored stereotactically in the awake, trembling monkey (4).

Lamarre et al. (7,9) have studied extensively the rhythmic neurons of the thalamus in monkey manifesting the spontaneous tremor after the same VMT lesion. They found the rhythmic neurons related to the peripheral tremor in almost the equivalent thalamic ventrolateral area that we found in human cases and in motor cortex.

Considering the above-mentioned observations that thalamic Vim neurons are specifically involved in tremor in humans as well as in monkeys, we have explored the cerebellum, this being one step in the pathway before the thalamus (22). By stereotactic horizontal or oblique approach, posterior lobes of four monkeys with spontaneous sustained tremor were explored. Using tungsten or glass micropipettes, a total of 73 tracks were made; we found groups of rhythmic burst discharges in 24 cases.

At the cortical surface area, spontaneously firing large spike discharges in continuous fashion were quite abundant, but their activity was mostly independent of the tremor. Among them, presumed single and complex spikes suggesting Purkinje cell discharges (21) could be distinguished with no relation to the tremor. Only several unidentified neurons were found firing rhythmically, in the tremor phase. Advancing the electrode further, at about 15 mm from the surface a cluster of neurons was found which were discharging time-locked with tremor (Fig. 6A). In general, the negative spikes were of short duration and small amplitude. The rhythmic neurons responded also to passive movement of the corresponding peripheral joint or when taps were applied to the muscle body; they did not respond to light touch of skin or hair (Fig. 6B). From a provisional histological examination, they might represent neurons of cerebellar nuclei, presumably of interpositus nucleus. At the cerebellar level, these cells were restricted to the ipsilateral

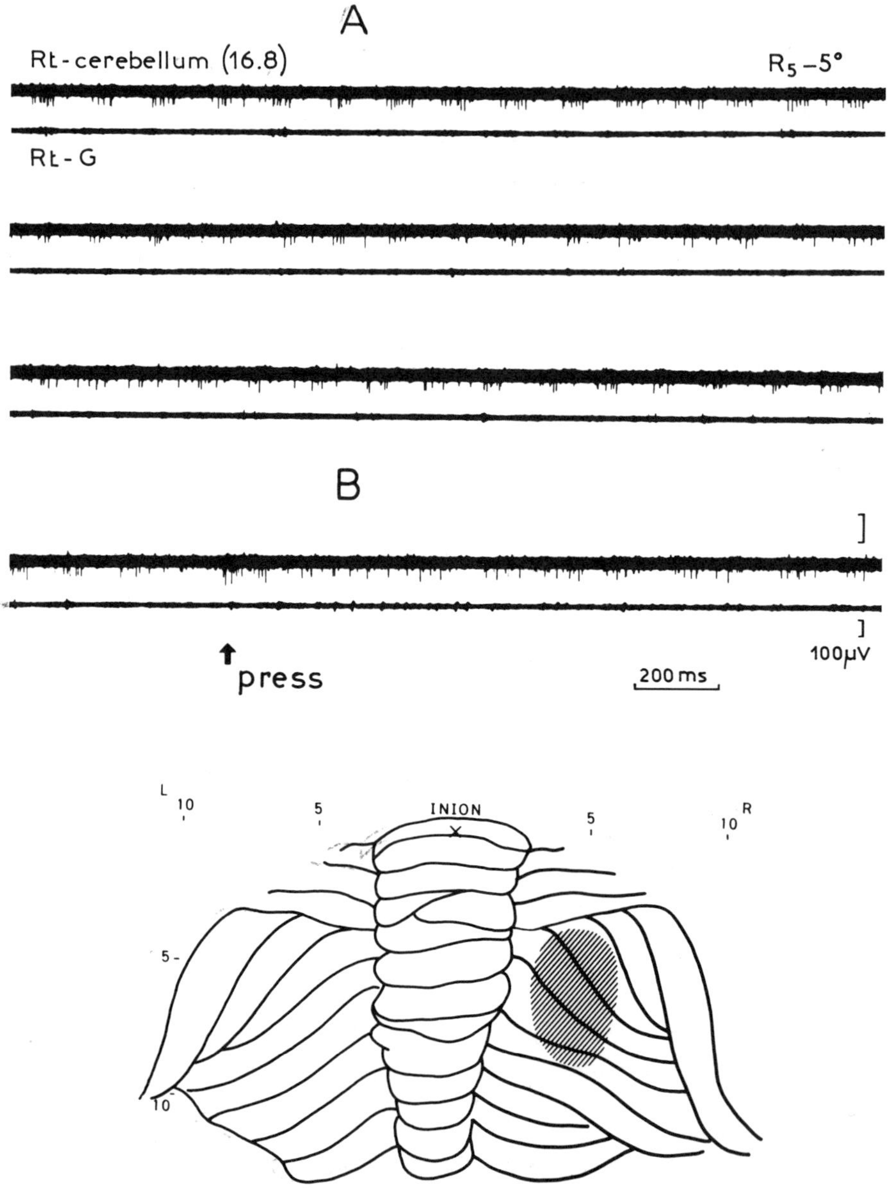

FIG. 6. A: Rhythmic discharges connected with tremor, recorded from a deep cerebellar nucleus (Rt-Cerebellum), 16.8 mm from the surface by horizontal approach, in a monkey with experimental tremor. EMG of the ipsilateral gastrocnemius (Rt-G) muscle is shown for monitoring tremor. **B:** The same neuron responded to pressure (press) on the right knee. **C:** Posterior view of the posterior lobe of cerebellum showing the extent of the entrance of the electrodes which have recorded rhythmic discharges (*shaded area*) at the cerebellar level, time-locked with tremor of the ipsilateral extremities. Numbers (5 and 10) indicate distances in millimeters referred to an inion.

side of the tremor and at about 5 mm from the midline, corresponding to the depth of the lobulus paramedianus where a part of the spinocerebellar tract was thought to project (Fig. 6C). One burst consists of five to ten spikes within several milliseconds; they are similar to the tremor time-locked burst of the human thalamus. We have not yet seen short repetitive discharges in the cerebellum of the tremor monkey similar to the second type of activity we described in humans.

Summary and Comment

In the cerebellum of monkeys with spontaneous tremor after VMT lesion, several neurons of the cerebellar nuclei, probably of interpositus nucleus and some surface neurons, were firing synchronously with the tremor. They were restricted to the side ipsilateral to the tremor, at a laterality corresponding to the position of the lobulus paramedianus. As they responded also to peripheral natural stimuli, they were thought to be produced by peripheral tremor. In this sense, neurons of the cerebellar posterior lobe of a monkey with experimental tremor seemed to fire only passively in response to the mechanical movement of tremor.

Cerebellar contribution to tremor generation has been suggested for harmaline-induced tremor (8,10,12), and participation of the midline structures, such as vermis and fastigeal nucleus, was then emphasized. Our study of the posterior lobe of the cerebellum demonstrated a different contribution to the tremor produced by destructive lesions, midline structures not being involved.

GENERAL COMMENT

At present, interrelationships of three types of burst discharges in the human thalamus and the cerebellar rhythmic discharges in experimental tremor are not fully understood as a whole in the tremorogenic mechanism. Rather, a simple afferent pathway is indicated, from peripheral mechanoreceptor to the thalamic Vim nucleus via cerebellum, the latter being involved only as a relay station. Larochelle et al. (11) have shown that the cerebellar nuclei, at least dentate and interpositus nuclei, are the structures whose destruction exaggerates the tremor production. The apparently passive role of the cerebellum presented above might be on the same line as the latter assumption.

The thalamic Vim nucleus or its adjacent structure is presumed to play a more active role in the production of tremor. We found many burst-type neurons, tremor time-locked or tremor independent in this particular area. Moreover, selective stereotactic lesion of this area abolishes various kinds of tremor movement.

We agree with the hypothesis presented by Lamarre (6), who suggested the role of reciprocal thalamocortical system in the production of coarse tremor and of the brainstem-cerebellar system for the production of fine physiological tremor.

ACKNOWLEDGMENTS

Most of the data presented here were obtained at the Neurological Clinic for Nervous Disease and Stereotaxy in Tokyo. The authors are grateful for the courtesy of Dr. Narabayashi, Director of the Clinic and Dr. Saito, who undertook some of the statistical analysis.

REFERENCES

1. Albe-Fessard, D., Arfel, G., and Guiot, G. (1963): *Ann. Chir.,* 17:1185–1214.
2. Albe-Fessard, D., Guiot, G., Lamarre, Y., and Arfel, G. (1966): In: *The Thalamus,* edited by D. P. Purpura and M. D. Yahr, pp. 237–253. Columbia University Press, New York.
3. Anderson, P., Olsen, L., Skrede, K., and Sveen, O. (1969): In: *Third Symposium on Parkinson's Disease,* edited by F. J. Gillingham and I. M. L. Donaldson, pp. 112–118. Livingstone, Edinburgh.
4. Feger, J., Ohye, C., Gallouin, F., and Albe-Fessard, D. (1975): In: *Advances in Neurology, Vol. 10: Primate Models of Neurological Disorders,* edited by B. S. Meldrum and C. D. Marsden, pp. 35–45. Raven Press, New York.
5. Joffroy, A. J., and Lamarre, Y. (1971): *Brain Res.,* 27:386–389.
6. Lamarre, Y. (1975): In: *Advances in Neurology, Vol. 10: Primate Models of Neurological Disorders,* edited by B. S. Meldrum and C. D. Marsden, pp. 23–34. Raven Press, New York.
7. Lamarre, Y., and Cordeau, J. P. (1967): *Actual. Neurophysiol.,* 7:141–166.
8. Lamarre, Y., and Dumont, M. (1972): In: *Medical Primatology,* edited by E. I. Goldsmith and J. Moor-Jankowski, pp. 274–281. Karger, Basel.
9. Lamarre, Y., and Joffroy, A. J. (1970): In: *L-Dopa and Parkinsonism,* edited by A. Barbeau and F. M. McDowell, pp. 163–170. Davis, Philadelphia.
10. Lamarre, Y., de Montigny, C., Dumont, M., and Weiss, M. (1971): *Brain Res.,* 32:246–250.
11. Larochelle, L., Bédard, P., Boucher, R., and Poirier, L. J. (1970): *J. Neurol. Sci.,* 11:53–64.
12. Llinás, R., and Volkind, R. A. (1973): *Exp. Brain Res.,* 18:69–87.
13. Meyermann, R., Müller, D., Orthner, H., and Roeder, F. (1974): In: *Central-Rhythmic and Regulation,* edited by W. Umbach and H. P. Koepchen, pp. 391–396. Hippotrates-Verlag, Stuttgart.
14. Ohye, C., Fukamachi, A., Imai, S., Miyazaki, M., Isobe, I., Nakajima, H., and Shibazaki, T. (1977): *Submitted for publication.*
15. Ohye, C., Maeda, T., and Narabayashi, H. (1977): *Appl. Neurophysiol.* (*In press.*)
16. Ohye, C., and Narabayashi, H. (1977): *Submitted for publication.*
17. Ohye, C., and Narishige, E. (1977): *Clin. Physiol.,* 7:170–176.
18. Ohye, C., Saito, Y., Fukamachi, A., and Narabayashi, H. (1974): *J. Neurol. Sci.,* 22:245–259.
19. Saito, Y., and Ohye, C. (1974): *Confin. Neurol.,* 36:318–325.
20. Schaltenbrand, G., and Bailey, P. (Eds.) (1958): *Introduction to Stereotaxis, with Atlas of the Human Brain.* Georg Thieme, Stuttgart.
21. Thach, W. T. (1968): *J. Neurophysiol.,* 31:785–797.
22. van Vuren, J. M. (1976): *J. Neurosurg.,* 45:37–48.

Abnormal Neuronal Discharges, edited by N. Chalazonitis and M. Boisson. Raven Press, New York © 1978.

Reexcitation in Normal and Abnormal Repetitive Firing of CNS Neurons

William H. Calvin

Department of Neurological Surgery, University of Washington, Seattle, Washington 98195

Synapses exhibit various alterations in their sensitivity to spikes; they may facilitate, depress, posttetanically potentiate, and so on. It is reasonable to suspect more enduring changes in synaptic function as a physical substrate for memory and learning. It is less commonly recognized, however, that the sensitivity of spike production mechanisms can also change (3,4) and that intermittent conduction between the spike trigger zones and an output synapse on a branch of a long axon may change with the history of the cell (28).

I have recently reviewed alterations in the sensitivity of pacemaker-like rhythmic firing processes at normal sites of spike initiation (3); elsewhere, I have reviewed pathological repetitive firing from ectopic sites (6). The purpose of this chapter is to concentrate those sensitivity alterations involving extra spikes at normal sites of spike initiation, the reexcitation mechanism that underlies extra spikes, the historic factors associated with extra spike production, and the pathophysiological situations in which extra spikes can be identified.

SENSITIVITY ALTERATIONS IN CNS REPETITIVE FIRING

While spikes may begin at an initial segment trigger zone, the retrograde invasion of the somadendritic region creates a large, long afterhyperpolarization that exerts a strong control over the time to the next spike (5,10) and thus the basic sensitivity of the pacemaker-like rhythmic firing process.

Cat spinal motoneurons exhibit sudden changes in sensitivity to depolarizing current injected through the recording microelectrode. Granit and coworkers (19) found that the slope of the low end of the current-to-rate curve is two to six times less sensitive to current than the high end. The Heyer and Llinás (20) experiments can be interpreted to suggest that the participation of the dendrites in the retrograde invasion changes at the point where the sensitivity alters.

Besides this alteration in the pacemaker-like rhythmic firing process (22), the sensitivity of current-to-rate curves may also be augmented by extra spikes (Fig. 1). During sustained rhythmic firing to a steady depolarizing

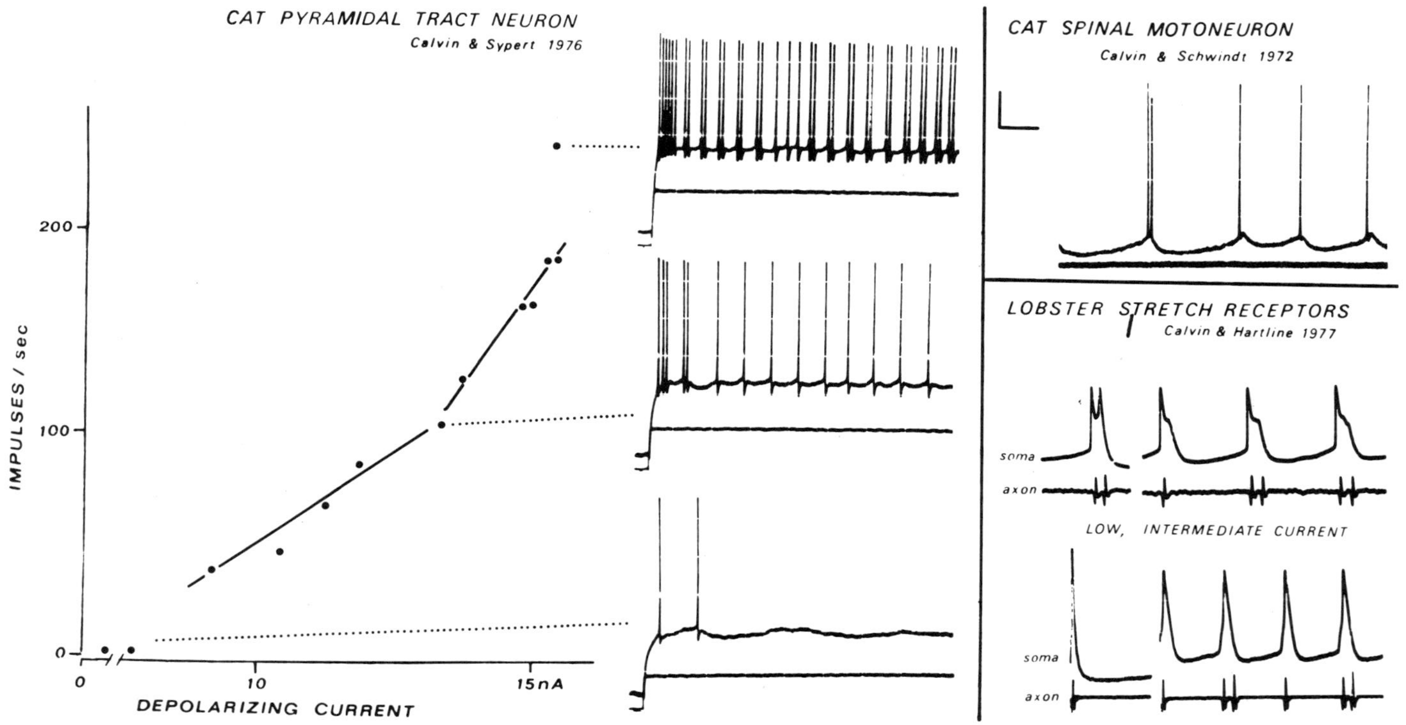

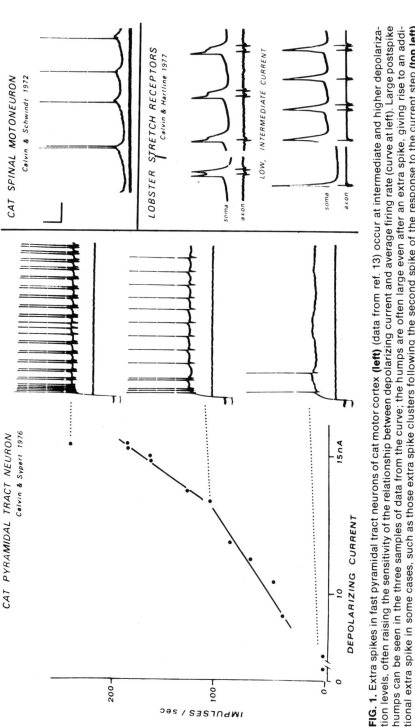

FIG. 1. Extra spikes in fast pyramidal tract neurons of cat motor cortex **(left)** (data from ref. 13) occur at intermediate and higher depolarization levels, often raising the sensitivity of the relationship between depolarizing current and average firing rate (curve at left). Large postspike humps can be seen in the three samples of data from the curve; the humps are often large even after an extra spike, giving rise to an additional extra spike in some cases, such as those extra spike clusters following the second spike of the response to the current step **(top left)** (see also Fig. 3). Similar postspike humps are often seen in cat spinal motoneurons **(upper right)** (data from ref. 10) but cause extra spikes only at low current levels (current increases slightly during trace). Note fluctuation in size of postspike hump. A similar disappearance of extra spikes with rising current levels is seen in many lobster stretch receptor neurons **(middle right)** (data from ref. 5). Samples are shown from low **(left)** and intermediate **(right)** levels of depolarizing current. Extra spikes arise from shoulder-like postspike humps. *Lower traces,* extracellular recordings made simultaneously from the axon, demonstrating that extra spikes begin downstream from the soma and do not always retrogradely invade the soma. In other lobster stretch receptor neurons, fatigue-like situations may delay the retrograde invasion of the soma **(lower right);** the initial segment trigger zone recovers its excitability while the spike is still in progress next door in the soma, and the reexcitation results in an extra axon spike. Calibration bar: 20 mV, 10 nA, 20 msec for motor cortex; 10 mV, 50 msec for the motoneuron; 25 mV intrasomatic, 0.25 mV extraaxonal, 2.5 msec for lobster neurons.

current where, for example, one expects a spike 100 ± 5 msec following the previous spike, one may be surprised to see extra spikes intermittently appearing 2.0 msec after an expected spike. Depolarizing currents can often be found where extra spikes occur after every expected spike, and other currents can be found where extra spikes totally disappear. Extra spikes tend to raise the average firing rate, although there is sometimes a compensatory lengthening of the following interspike interval (3).

In cat pyramidal tract neurons with fast conduction speeds, the upper portion of the current-to-rate curve is steeper; but this is often due to extra spikes rather than to changes in the pacemaker-like rhythmic firing process, as in motoneurons. With more current, extra spikes are more likely to be seen after expected spikes and, in some cells, multiple extra spikes are seen (3,11–13). Cortical neurons are unusual in this regard. In cat spinal motoneurons (2,21), dorsal column nucleus neurons (7), and primary vestibular neurons (32), extra spikes are more likely to be found at low rhythmic firing rates and to disappear as the rhythmic firing rates are driven higher with more input. In lobster stretch receptors (3,5), examples of both low- and high-end augmentation of sensitivity can be seen (Fig. 1). When extra spikes drop out with increasing current, one may see a negative sensitivity region in the current-to-rate curve at this transition (3,5).

Extra spikes typically arise from postspike humps, as in most examples shown in Fig. 1. Many CNS neurons possess postspike humps, although they are usually not large enough to intersect the falling threshold at the end of the relative refractory period and reexcite the cell (1). In some cases (Fig. 1, lower right), extra spikes are seen in the axon when the retrograde invasion of the somadendritic region is delayed or prolonged (5). Basic to a discussion of the mechanisms underlying reexcitation is an understanding of what happens when a spike, beginning at an initial segment trigger zone, tries to retrogradely invade the soma and the dendrites.

MECHANISMS FOR REEXCITATION

Although the properties of a unit patch of membrane probably differ between the initial segment and the soma (2,10), the flare in diameter alone is a major concern. The reason may be seen in several computer simulations of flaring axons with uniform membrane patch properties (17,30) when a spike is propagating from small to large segments. The spike becomes markedly wider at the flare, sometimes two to three times longer in duration than it is in either small or large segments distant to the flare zone. Furthermore, characteristic notching of the spike is seen at the flare, not unlike the spike components seen in CNS neurons. Notches appear on the falling as well as the rising phase of the spike; sometimes a double-peaked spike is seen, analogous to the recordings that early workers believed were from the initial segment (15).

One of the most elementary conditions for reexcitation is when a region recovers its excitability while there is still a spike in progress immediately next door, e.g., the two- to threefold wider spike seen at the flare zone in the Goldstein and Rall (17) simulation. In the flare simulations, the small diameter segment is reexcited and a new spike propagates backward at the same time as the original spike is propagating forward from the flare. This reexcitation is sometimes termed "reflection" for obvious reasons (8). Figure 1 (lower right) shows simultaneous recordings from lobster stretch receptor soma and axon; when retrograde invasion of the soma is slowed at intermediate depolarizations (notching on the rising phase of the soma spike), extra spikes intermittently occur in the axon, just as in the computer simulations of the axon flare.

The extra spikes seen in crustacean stretch receptor neurons seem to originate from the same initial segment trigger zone as do the expected spikes (16,18). These trigger zone localization experiments are most convincingly done with bipolar extracellular recording electrodes which can be moved along the region to find where the triphasic spike waveform reverses polarity. With this method, Ringham (31) has even seen small proximal movements of the trigger site at high firing rates. In CNS neurons where such mapping is most difficult, the trigger zone can only be approximately located; it seems to be downstream from the soma.

It is possible, however, to infer whether extra spikes start from the same general region as expected spikes. The notch on the rising phase of the spike seen with an intrasomatic recording electrode is not the only notch, as electrically differentiated records show (Fig. 2). In intracellular recordings from cat pyramidal tract neurons (PTNs), there is an equally characteristic notch on the falling phase (24) which can also be identified in many cat spinal motoneuron recordings (25). Whether this notch is due to a three-compartment (axon/soma/dendrites) version of the flaring axon geometric effect or to inhomogenous membrane properties, the characteristic notching of antidromically elicited spikes may be compared to those seen during repetitive firing, both expected and extra spikes. If, for example, spikes were to originate in the dendritic tree as occurs in some pathological cases (25), a different notching sequence might be expected for extra spikes and expected spikes.

Figure 2 shows recordings from three cat PTNs, which Sypert and I (11–13) obtained during our study of fast and slow PTN repetitive firing. Susan Johnston and I have recently analyzed tape recordings of these data using electrical differentiation of the spikes. At left in Fig. 2 are antidromic spikes, elicited by stimulating the medullary pyramids. The characteristic notches on the rising and falling phases can always be seen most easily in antidromic activations, where the soma and dendrites are at their resting potential and thus require longer to reach threshold than when using orthodromic activation, which "predepolarizes" the somadendritic region. The

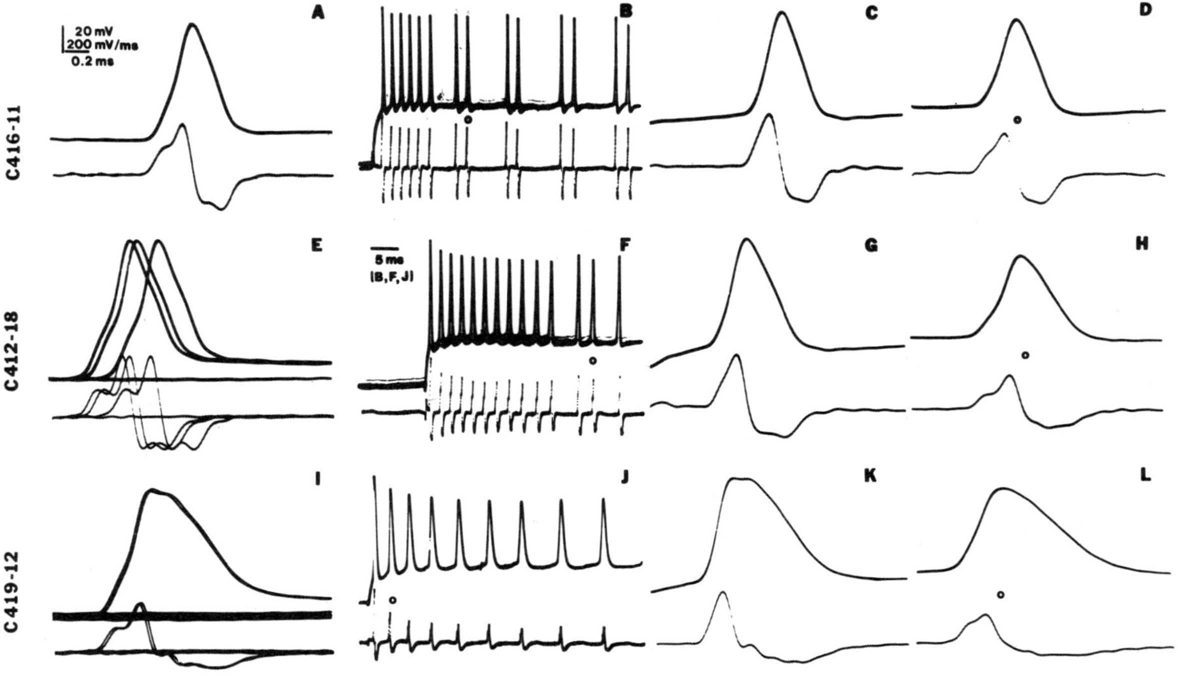

FIG. 2. Spike components studied using electrical differentiation in three cat PTNs (W. H. Calvin, G. W. Sypert, and S. Johnston, *unpublished data*). Antidromic spikes (**A, E,** and **I**) elicited when the cells were at their resting potential show the characteristic "IS-SD" notch on their rising phases, especially visible in the dV/dt trace below the spikes. In PTNs, there is an equally characteristic notch on the falling phase of the spike. For each PTN, the repetitive firing response to an intermediate-sized current step is shown in **B, F,** and **J;** the dV/dt trace in these slow sweeps demonstrates the changes in maximal dV/dt and notching which occur during repetitive firing. The first spike of the repetitive response (expanded by fast sweeps in **C, G,** and **K**) usually has minimal notching on the rising phase because the depolarizing current "predepolarizes" the soma and facilitates retrograde invasion. In the slow PTN shown in the bottom row, the second rhythmic spike is expanded in **L,** demonstrating the marked changes in spike parameters which occur during sustained rhythmic firing in slow PTNs. The two fast PTNs (middle and top rows) were selected because they exhibited extra spikes during sustained repetitive firing. The extra spike marked with an open circle in **B** has been expanded in **D** to show the dV/dt waveform of an extra spike. The fast PTN in the middle row has a burst of 10 extra spikes following the first spike and then settles into intermittent extra spiking; one such extra spike is marked with an open circle and expanded in **H.** Extra spikes **(D, H)** do not differ from antidromic spikes **(A, E)** in the basic notching sequence and have a dV/dt waveform much like that of other spikes preceded by a short interspike interval.

second column of Fig. 2 utilizes a slow sweep so that a number of successive spikes may be compared with each other, following the application of a medium-sized current step through the recording microelectrode and bridge circuit.

The bottom row of pictures in Fig. 2 is from a slow PTN, characteristically free of extra spikes. The notching of the first spike of the rhythmic train is seldom as pronounced as that seen in antidromic spikes. Notching of the rising phase is seen in subsequent spikes, especially at shorter interspike intervals, indicating slower retrograde propagation of subsequent spikes. The two right columns are expanded sweeps of two of the spikes from the repetitive firing seen in the second column, the first spike and one of the subsequent spikes.

The two top rows in Fig. 2 are from two different fast PTNs, each with prominent extra spikes during sustained repetitive firing. While extra spikes are easily identified during sustained firing, e.g., 200 msec after the step begins, they are also often part of the initial response to the current step; indeed, many of the spikes in the slow sweeps (Fig. 2B,F) are probably extra spikes. The circles identify an extra spike which is clearly part of a doublet; pictures in the far right column show such an extra spike expanded. In each case, the extra spikes have the same notching sequence as the antidromic spikes. This evidence makes it highly unlikely that extra spikes propagate from the dendritic tree toward the axon. Thus extra spikes in fast PTNs probably originate from approximately the same downstream trigger zone as the expected spikes, just as in the crustacean stretch receptor neurons.

HISTORIC FACTORS IN REEXCITATION

Postspike humps are probably, in many respects, simply a version of notching on the falling phase of the spike; all evidence thus far indicates that they too are part of the retrograde invasion of the dendritic tree and not recurrent synaptic potentials (2).

Extra spikes may be identified during sustained repetitive firing in 25% of our fast PTNs, usually at an interval of about 2 msec following an expected "rhythmic" spike. Spikes also occur with the same characteristic interspike intervals during the initial burst of spikes following the current steps; on this basis, we (13) estimated that 50% of our fast PTNs had extra spikes. Thus extra spikes augment the initial response of many PTNs. The extra spikes do not, however, necessarily begin following the first spike of the train in response to the current step. As may also be seen in cat spinal motoneurons and lobster stretch receptors (1,2,5), there is a distinct tendency for extra spikes to begin following the second spike of a rhythmic train in PTNs, corresponding to a similar tendency for postspike humps to be small following the first spike in some cells (Fig. 1, left; Fig. 3).

There is thus the possibility that an antecedent spike "primes" the extra

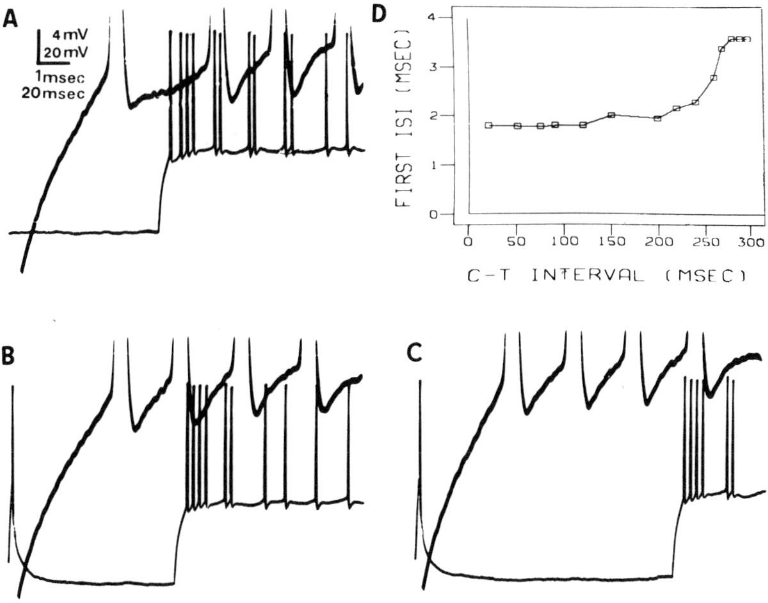

FIG. 3. Tendency of extra spikes to occur following the second rather than the first spike of a rhythmic response to a current step suggested that the first rhythmic spike served to "prime" the extra spike process. To examine this hypothesis, we (data from ref. 12) conditioned the response to the current step by eliciting a single spike some time in advance of the application of the current step. This single spike was produced by a very brief pulse of current through the same intracellular recording microelectrode used to produce current steps, thus avoiding the problems of interpretation associated with antidromic activations. A "test" current step level was selected which, in the absence of the "conditioning" spike, produced a 3.7 msec interval between the first and second spikes of the rhythmic response; the extra spikes following the second rhythmic spike occurred at a characteristic 1.8 msec interspike interval. If the conditioning spike occurred within several hundred milliseconds of the beginning of the test current step, the interval between the first and second spikes of the response shortened to the 1.8 msec characteristic extra spike spacing. *Upper right:* Graph shows that this first interspike interval could be graded by C-T intervals between 200 and 300 msec. *Upper left:* Unconditioned response, with a magnification of the subthreshold region of the first 10 msec of the response superimposed on the low gain, slow sweep picture. *Lower:* Data for two different C-T intervals. Same neuron as in Fig. 1 (left) and Fig. 2 (top row).

spike process. George Sypert and I (12) set out to investigate this in some of our experiments by using a brief current pulse to elicit a single "conditioning" spike, followed some time later by the current step, which elicited repetitive firing. In a neuron where extra spikes were present after the second (but not the first) rhythmic spike of the step-evoked train, the time course of the priming effect of our conditioning spike could be clearly demonstrated. In Fig. 3, one sees that the first interspike interval shortens down to the 1.8 msec characteristic extra spike spacing for this neuron if the priming spike is present anywhere within several hundred milliseconds before the current step. This is much longer than the afterhyperpolarization following a PTN spike; the use of a brief intracellular current pulse to elicit the conditioning spike tends to minimize the possibility of a recurrent synaptic explanation. Thus the extra spike process would seem to exhibit some "memory."

While a single spike may have substantial effects on the extra spike process in fast PTNs, one can also see cases in peripheral nervous systems where extra spikes become more likely as the result of many antecedent spikes. This fatigue-like effect can be seen for the reexcitations at dorsal root ganglia and focally demyelinated regions (22) and can be studied intracellularly in the lobster stretch receptor neuron (3,5). As noted earlier, anything that broadens the duration of a spike without similarly lengthening the spike (and refractory period) in an adjacent region provides the basic setup for reexcitation. In lobster stretch receptor neurons, one can see a slight broadening of the intrasomatically recorded spike with use; at some firing rates, extra spikes will tend to occur (Fig. 1, lower right).

One can even produce (Fig. 3B in ref. 5) a metastable form of a latchup in a few cells due to a tendency to produce extra spikes at low rates but only following a history of higher rates. Before and during a current increment, purely rhythmic firing can be seen; following the return of the current to its original value, extra spikes may begin after each rhythmic spike. This condition may last only a few seconds and only certain currents usually produce it; thus this is hardly a robust example of a neuronal flip-flop.

While "use" can obviously augment extra spikes, one can also see examples where a long silence augments the probability of seeing extra spikes upon resumption of rhythmic firing (5). Indeed, it can be hypothesized (1) that some neurons use their extra spike processes as a type of automatic sensitivity control; e.g., following a long silence, one might get two spikes for the price of one.

EXTRA SPIKES IN PATHOPHYSIOLOGY

In various pathological states, neurons can be seen "spontaneously" firing in high frequency bursts of spikes. The extremely stereotyped intervals between spikes seen in some cases suggest that extra spikes are involved. There have been several types of pathophysiology where we have been able to explicitly identify the reexcitation process at work. In focally demyelinated mammalian axons, we were able to establish that the "extra" spike was indeed propagating backward from the pathological region (22), and we have hypothesized that such reflections could help explain the trigger phenomena in tic douloureux (6,8). In several other pathophysiological cases, we have been able to study normal CNS neurons and compare the characteristic features of their extra spike processes with the bursts from the pathological version of approximately the same neuron species.

Deafferentation in External Cuneate

The external cuneate nucleus (ECN) of the cat has several aspects that make it most useful for our purpose: (a) Neurons there normally exhibit extra spikes during rhythmic firing; indeed, we estimated that perhaps half of the neurons from which we recorded extracellularly in ECN had extra spikes at a characteristic interspike interval ranging in different neurons from 0.8 to about 2.0 msec (7). (b) The ECN can be partially deafferented by multiple level dorsal rhizotomies; indeed, all of the round vesicle-type synaptic terminals can be eliminated, leaving only the other 20%, which contains flattened vesicles (23).

On severing the input axons, it is not surprising to see that the profound "spontaneous activity" of the ECN disappears (23) for several days. What is surprising is that it returns, although such firing is no longer a function of forelimb position. Indeed, it returns with a vengeance: in addition to the normal doublet and triplet spike clusters, neurons may be seen with 10 to 12 spikes in a tight, stereotyped cluster, all at the characteristic extra spike spacing for ECN. Such firing patterns are seen even 90 days later.

One might presume that the ECN neurons have augmented their extra spike production so extra spikes produce still more extra spikes in a vicious cycle. We have been unable to determine whether that is directly related to the loss of synaptic terminals from the cell or secondary to the disuse which must result from the loss of tonic proprioceptive spike trains from the forelimb. Normal ECN neurons, however, do exhibit extra spikes only at the lower input levels; slowly moving the forelimb to drive the rhythmic firing rate up will result in the dropping out of the extra spikes (7).

Epileptogenic Foci

Neurons in chronic epileptogenic foci of cerebral cortex have long been thought to be partially deafferented, partly on the basis of the depopulation of such regions and partly from the loss of spines from the dendritic trees of the remaining cells (33,34).

Neurons in the chronic foci, whether human foci (9,27) or alumina cream-induced foci in monkey cortex (14,35), exhibit bursting firing patterns spontaneously, typically with high firing rates within the burst (200 to 500/sec). Unlike the ECN case, however, this comes as no surprise since the EEG spikes have long been considered to indicate massive synaptic bombardment, recurring at intervals and synchronously affecting many adjacent neurons at once. Normal current-to-rate conversion by a cortical neuron (as in Fig. 1) should thus result in high rates. There might be nothing abnormal about the neuron under study, only an abnormality in its synaptic input. As in the acute penicillin focus, such a situation results in some difficulties in sorting out how much of a burst is due to abnormalities in the neuron itself and how much is just a big input which needs tracing back to antecedent neurons.

There are several situations in chronic epileptogenic foci where the bursting firing patterns cannot be ascribed so readily to massive inputs in this manner. Adjacent neurons, simultaneously recorded extracellularly by a single microelectrode, do not have overlapping bursts in ECN (7); this is also sometimes the case in chronic cortical foci (9,27). Although this could be ascribed to multiple sources of massive inputs, such independent bursting makes it seem more likely that the burst process is endogenous to the neurons.

A second reason for suspecting extra spike bursts is that some firing patterns from chronic epileptogenic foci have such stereotyped interspike intervals that successive bursts can be virtually superimposed on one another. This could be accomplished by a synaptic bombardment of extraordinary repeatability or by a depolarization so large that, although variable, it always forces the repetitive firing process to its high rate limit. Or, as in the bursts from deafferented ECN, the stereotyped bursts might result from an extra spike vicious cycle.

A third line of evidence from extracellular recordings has been the so-called "long first interval" burst pattern seen in cerebral cortex (but never in ECN) where a very stereotyped burst occurs just as in the case above but only following an initial spike 5 to 15 msec earlier (14). The interval between the initial spike and the stereotyped "afterburst" typically varies with each successive repetition of the burst, even though the afterbursts remain quite superimposable. Since our discovery of this structured burst pattern in alumina monkey foci (14), this burst type has been characteristically seen and studied in a wide variety of chronic foci (reviewed in ref. 35); only the fast conducting pyramidal tract neurons exhibit long first interval stereotyped "epileptic" bursts. An early theory, regarding a role of backfiring and axon conduction time in the interval separating the initial spike and the afterburst, has been displaced by more recent evidence linking the somadendritic region with both events (35).

As noted earlier, we now know that normal cat PTNs (only fast ones, again) readily exhibit extra spikes. Many exhibit a propensity for extra spikes to begin only following the second rhythmic spike of the train evoked by a depolarizing current step (Fig. 3), as if the first spike of the train "primed" the extra spike process so that it could be evoked by the second rhythmic spike.

This suggests that only a moderate level of synaptic depolarizing current is needed to evoke a long first interval burst, e.g., that depolarization corresponding to the 5 to 15 msec initial interspike interval (200–67/sec rates) on the current-to-rate curve. This now implies that the "epileptic neuron" may have augmented its sensitivity via extra spikes so that low or moderate input levels elicit firing rates which would otherwise be associated only with massive input levels.

DO CNS NEURONS HAVE AUTOMATIC SENSITIVITY CONTROLS?

There are a number of chronic pain disorders associated with sensory loss, which has long suggested that CNS neurons might exhibit a denervation supersensitivity mechanism similar to that seen in muscle, where the extrajunctional membrane develops acetylcholine receptors analogous in most respects to those at the neuromuscular junction itself. These extrajunctional receptors are partially under the control of trophic factors from the presynaptic axon but are also regulated by the electrical activity of the muscle itself (26); stimulating the muscle electrically over a period of days can inhibit the development and maintenance of extrajunctional receptors secondary to denervation or anesthetic block of the axon. The electrical activity similarly plays a role in preventing the development of fibrillation-type pacemaker activity in the muscle (29). If mere muscle has these two mechanisms for augmenting sensitivity when normal input is reduced or lost, why not CNS neurons? There is, unfortunately, little direct evidence as yet that CNS neurons develop extrasynaptic receptors for transmitters and much less evidence that postsynaptic electrical activity serves to control the process; a similar situation exists for pacemaker activity induced into CNS neurons by deafferentation/disuse (but see ref. 7).

DISCUSSION

Extra spike production is a different type of sensitivity control available to CNS neurons (1). As we have seen, there is some evidence for the augmentation of extra spiking in deafferentation/disuse and suspected deafferentation disorders, such as epilepsy. If the extra spike mechanism serves as an automatic sensitivity control in CNS neurons, what is the time course over which the mechanism integrates past electrical activity? It is probably not seconds or acute recording experiments would probably have noted an increased propensity for extra spikes to occur after a brief rest (as occasionally noted in lobster stretch receptors). It is easy to imagine that hours or days of inactivity might augment extra spiking or produce pacemaker activity, but this is quite difficult to investigate experimentally.

The mechanisms underlying extra spiking seem to be those of reexcitation: the usual trigger zone is reexcited by the depolarization remaining in the somadendritic region from the antecedent spike. Various factors might improve the chances for reexcitation. (a) The somadendritic spike might repolarize more slowly. (b) The active extent of retrograde invasion might increase to include more of the somadendritic region. (c) The retrograde conduction velocity might slow down. Endogenous mechanisms, such as the extent of sodium inactivation, could affect such changes. All these factors are also potentially under synaptic control; i.e., an input synapse might not only contribute current to the trigger zone but might also affect the retrogradely invading spike and thus reexcitation possibilities. For example, active dendritic synapses might so load the spike mechanism that active invasion would cease at the soma instead of continuing out the dendrites. Synaptic pathways with preferential effects on extra spiking have not yet been observed, although there is some discussion of synaptic effects on motoneuron "secondary ranges" by Granit et al. (19).

Sensitivity alterations in repetitive firing, whether "automatic" or under synaptic control, hold promise for understanding functional plasticity which is traditionally ascribed to the synapse, as noted at the beginning of this chapter. Neurons were once viewed as rather fixed computing units (4), so that functional plasticity had to reside at the sites where they were tied together. Sensitivity alterations in the repetitive spike production process, together with alterations in intermittent conduction in the axons which transmit trigger zone spikes to the output synapses (28), now give the rest of the neuron some potential for intrinsic plasticity.

ACKNOWLEDGMENTS

This research was supported by NIH research grants NS 09677 and NS 04053, awarded by USPHS. George W. Sypert collaborated on the experiments reported in Figs. 2 and 3, and Susan Johnston assisted in the data analysis.

REFERENCES

1. Calvin, W. H. (1974): *Brain Res.,* 69:341–346.
2. Calvin, W. H. (1975): *Brain Res.,* 84:1–22.
3. Calvin, W. H. (1977): *Fed. Proc.* (*In press.*)
4. Calvin, W. H., and Graubard, K. (1978): In: *The Neurosciences, Fourth Study Program,* edited by F. O. Schmitt and F. G. Worden. MIT Press, Cambridge (*In press.*)
5. Calvin, W. H., and Hartline, D. K. (1977): *J. Neurophysiol.,* 40:106–118.
6. Calvin, W. H., Howe, J. F., and Loeser, J. D. (1977): In: *Pain in the Trigeminal Region* edited by D. J. Anderson and B. Matthews. Elsevier/North-Holland, Amsterdam (*In press.*)
7. Calvin, W. H., and Loeser, J. D. (1975): *Exp. Neurol.,* 48:406–426.
8. Calvin, W. H., Loeser, J. D., and Howe, J. F. (1977): *Pain,* 3:147–154.
9. Calvin, W. H., Ojemann, G. A., and Ward, A. A., Jr. (1973): *Electroencephalogr. Clin. Neurophysiol.,* 34:337–351.
10. Calvin, W. H., and Schwindt, P. C. (1972): *J. Neurophysiol.,* 35:297–310.
11. Calvin, W. H., and Sypert, G. W. (1975): *Brain Res.,* 83:498–503.
12. Calvin, W. H., and Sypert, G. W. (1975): *Neurosci. Abstr.,* 1:715.
13. Calvin, W. H., and Sypert, G. W. (1976): *J. Neurophysiol.,* 39:420–434.
14. Calvin, W. H., Sypert, G. W., and Ward, A. A., Jr. (1968): *Exp. Neurol.,* 21:535–549.
15. Coombs, J. S., Curtis, D. R., and Eccles, J. C. (1957): *J. Physiol.* (*Lond.*), 139:198–231.
16. Edwards, C., and Ottoson, D. (1958): *J. Physiol.* (*Lond.*), 143:138–148.

17. Goldstein, S. S., and Rall, W. (1974): *Biophys. J.*, 14:731–757.
18. Grampp, W. (1966): *Acta Physiol. Scand.* [*Suppl.*], 262:3–36.
19. Granit, R., Kernell, D., and Lamarre, Y. (1966): *J. Physiol.* (*Lond.*), 187:401–415.
20. Heyer, C. B., and Llinás, R. (1977): *J. Neurophysiol.*, 40:480–488.
21. Hoff, H. E., and Grant, R. S. (1944): *J. Neurophysiol.*, 7:305–322.
22. Howe, J. F., Calvin, W. H., and Loeser, J. D. (1976): *Brain Res.*, 116:139–144.
23. Kjerulf, T. D., O'Neal, J. T., Calvin, W. H., Loeser, J. D., and Westrum, L. E. (1973): *Exp. Neurol.*, 39:86–102.
24. Koike, H., Mano, N., Okada, Y., and Oshima, T. (1972): *Exp. Brain Res.*, 14:449–462.
25. Kuno, M., and Llinás, R. (1970): *J. Physiol.* (*Lond.*), 210:807–821.
26. Lømo, T., and Rosenthal, J. (1972): *J. Physiol.* (*Lond.*), 221:493–513.
27. Ojemann, G. A., Calvin, W. H., and Ward, A. A., Jr. (1975): *Electroencephalogr. Clin. Neurophysiol.*, 38:555.
28. Parnas, I. (1978): In: *The Neurosciences, Fourth Study Program*, edited by F. O. Schmitt and F. G. Worden. MIT Press, Cambridge (*In press.*)
29. Purves, D., and Sakmann, B. (1974): *J. Physiol.* (*Lond.*), 237:157–182.
30. Ramon, F., Joyner, R. W., and Moore, J. W. (1975): *Fed. Proc.*, 34:1357–1363.
31. Ringham, G. L. (1971): *J. Neurophysiol.*, 34:773–784.
32. Rupert, A., Moushegian, G., and Galambos, R. (1962): *Exp. Neurol.*, 5:100–109.
33. Scheibel, M. E., Crandall, P. H., and Scheibel, A. B. (1974): *Epilepsia*, 15:55–80.
34. Westrum, L. E., White, L. E., Jr., and Ward, A. A., Jr. (1964): *J. Neurosurg.*, 21:1033–1046.
35. Wyler, A. R., Fetz, E. E., and Ward, A. A., Jr. (1975): *Brain Res.*, 98:1–20.

Abnormal Neuronal Discharges, edited by N. Chalazonitis and M. Boisson. Raven Press, New York © 1978.

Ectopic Action Potential Generation: Penicillin Action on Pyramidal Tract Fibers

J. Louvel and R. Pumain

Unité de Recherches sur l'Epilepsie INSERM, Paris, France

Many chemical agents induce epileptiform waves when directly applied to the cerebral cortical surface. Among these agents, penicillin has stood out in investigations of recent years. Within a few minutes after topical application to a restricted area of the cerebral cortex, focal epileptiform waves occur in a regular fashion and in some cases give way to ictal discharges. Intracellular recordings performed in these foci showed that the most prominent feature is the development of paroxysmal depolarization shifts (PDS), which coincide with every interictal epileptiform wave. These shifts consist of a long-lasting depolarization usually giving rise to action potentials (APs) that are rapidly inactivated by cathodal blocking.

The exact nature of the PDS still remains to be determined. According to some observations, they may be the result of an alteration of neuronal membrane properties at the somatic (4,5,6,8,11,17,25) or at the dendritic level (2,27). Nevertheless, most available data suggest that in the mammalian cortex, PDS are giant excitatory postsynaptic potentials (EPSP) or summated EPSP (1,3,7,21,22). The mechanisms underlying the generation of PDS are not entirely clear, although several explanations are proposed by data collected in various experimental situations:

(a) A steady-state depolarization in the epileptic neuronal aggregate could enhance the excitability of individual neurons (22). In these conditions, a normal direct or recurrent excitatory drive might generate the PDS. This hypothesis is upheld by experiments showing that convulsant drugs induce steady depolarizations, either by modifying ionic conductances or by interfering with the Na-K pump (10,16). However, it is difficult to detect such depolarizations in the mammalian cerebral cortex.

(b) An increase in subsynaptic membrane responsiveness might be possible, but is unlikely because of the generation of normal EPSP and PDS in the same neuron. To our knowledge, no experimental evidence is yet at hand to support this hypothesis.

(c) An increase in presynaptic mechanism efficiency is still another possibility. It has been recently demonstrated that epileptiform discharges are concomitant with repetitive ectopic AP generation (EAPG) in presynaptic thalamocortical fibers or endings. Orthodromic propagation of these bursts would produce an intense synaptic bombardment of the focal neurons. On the other hand, these EAP are conducted antidromically to the thalamic cell bodies where they can be characterized as antidromic by the collision test (17,26,28). Since EAPG was shown with strychnine as well as with other drugs in several simpler systems (22), this phenomenon may well be one of the factors explaining convulsive activity. Thus far, EAPG in the central nervous system (CNS) has been demonstrated only on thalamocortical fibers; the question may be raised whether or not it could occur on other intracortical fibers or endings. Furthermore, it would be of interest to show that such a phenomenon may indeed exist on recurrent collaterals of pyramidal tract (PT) axons since these collaterals make direct excitatory synaptic contacts with many PT neurons and thus provide a possible functional basis for the synchronization of neuronal activity during an EEG spike (30).

For this purpose, we tested the effect of penicillin on PT cell activity when applied to the bulbar level of PT in rats. We assumed that if EAPG was shown at this level, i.e., either on the axons or on the endings of PT fiber collaterals, then such a phenomenon would be likely to occur on intracortical collaterals of PT cell axons within a cortical penicillin focus. This experiment could also provide some clues to another question: Is EAPG induced by an alteration of membrane properties under the direct action of penicillin or by any nonspecific intense excitatory drive?

METHODS

Animals were anesthetized using a mixture of methoxyflurane, nitrous oxide, and oxygen. Surface recordings (band pass 0.2 c/sec to 5 kc/sec) and extracellular unit recordings were performed using standard techniques. A bipolar stainless steel electrode was positioned stereotactically into the PT (P = 0.8; L = 0.8; H = −7) according to the atlas of Pellegrino and Cushman (20) at the bulbar level.

Using electrical PT stimulations (0.06 to 0.3 mA, 0.05 to 0.1 msec), PT cells could be characterized by antidromic activation according to the following criteria (11): (a) short and constant response latency following a single PT stimulation, (b) ability of the response to follow high frequency stimulation (250/sec), and (c) occurrence of collision with spontaneous spikes.

Criteria for Antidromic Generation

Under normal conditions, any spontaneously occurring AP recorded close to a PT cell body is orthodromic; i.e., it is generated in the cortex and will follow the corresponding PT axon away from the cell body. If this AP

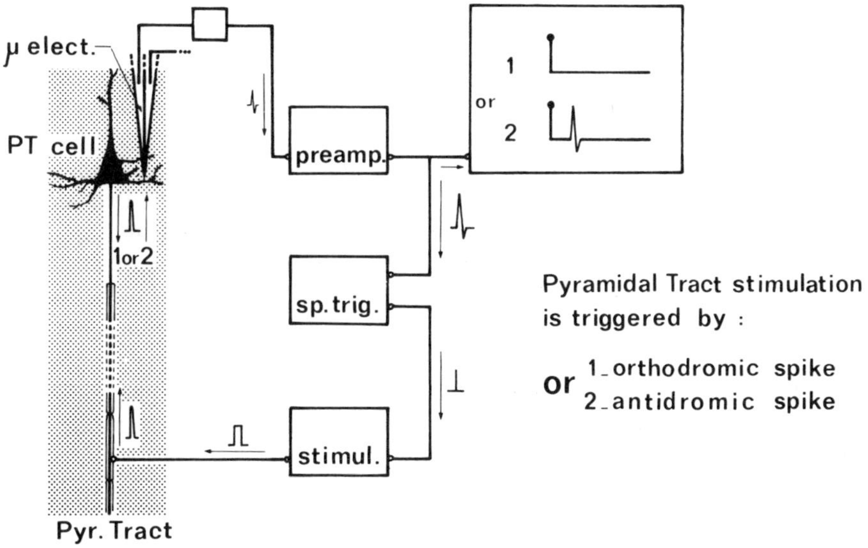

FIG. 1. Extracellular AP are recorded near a PT neuron. **1:** One of these AP may trigger, without delay, an electrical PT stimulation. The test antidromic AP evoked by the stimulus will collide with any orthodromic triggering AP. **2:** If the triggering AP was antidromically conducted, the test antidromic AP will not collide with the triggering AP and will therefore be recorded.

triggers a PT stimulation, no antidromic AP in response to the stimulation will be recorded in the cortex within a period of time equal to twice the conduction time between the stimulated site and the cortex (plus the axon refractory period) because of collision (Fig. 1.1). If such a response can be detected during this time interval, the triggering AP must then itself have been conducted antidromically (Fig. 1.2). Paired pulses were delivered to the PT as a control of the stimulation effectiveness in case of collision. Orthodromic AP could be elicited at any time by ionotophoretic Na DL-homocysteate or Na L-glutamate applications through the other barrels of the microelectrode, thus providing a control for the collision test.

RESULTS

Cellular Activity

Weak electrical stimulation evoked a small positive-negative field potential on the ipsilateral cortical surface at a mean latency of 1 msec. In microelectrode recordings, the positive waves began to reverse in polarity at about 0.6 mm below the cortical surface and became negative at the depth where antidromic unit responses could be recorded (29). The mean latency of antidromically evoked AP was 1.4 to 4.3 msec. Only units activated by a weak stimulus were selected, and the present results are therefore more representative of fast rather than slow PT neurons (31). Figure 2 illustrates the responses of a PT cell recorded prior to and following a penicillin application. The spontaneous AP, which triggered the PT stimulation in Fig. 2, must have been antidromically conducted since the evoked antidromic AP was recorded with latency far less than twice the conduction time plus the axon refractory period. In some cases, we failed to record any spontaneously occurring antidromic AP, even after repeated applications of penicillin. The reason for this could be an inadequate localization of the recording electrode at the cortical surface or a failure of penicillin to be effective because of the concentration and/or the injection site. The delay of occurrence of the first spontaneous antidromic AP following penicillin applications was quite variable (up to 40 min). The unit discharge was then usually modified; the mean frequency discharge rate was

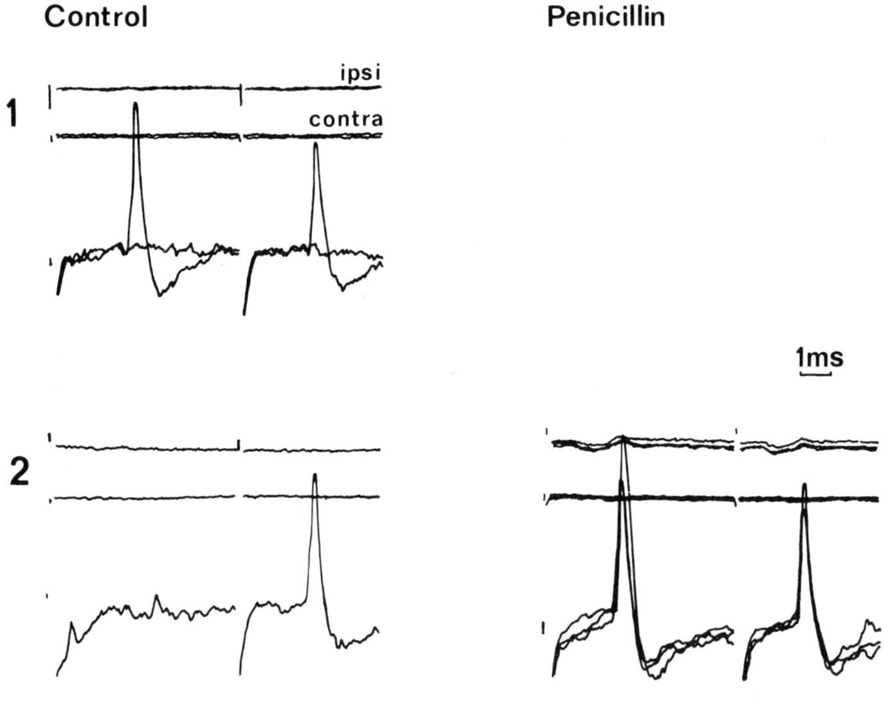

FIG. 2. *Upper and middle traces:* Surface recordings near the site of microelectrode penetration and on contralateral homotopic cortex, respectively. *Lower traces:* Extracellular PT cell recording. **1:** Antidromic activation following a double-shock PT stimulation. For the second sweep, the stimulation strength (80 μA) was slightly decreased and did not give rise to AP. **2:** PT stimulation was triggered by a spontaneously occurring AP (not on record). *Left,* before PT penicillin injection; *right,* 30 min after PT penicillin injection (80 IU), three successive sweeps. The stimulus strength had to be slightly increased because of shunting of the stimulating electrode by penicillin solution. AP were retouched for clarity.

often increased, and the timing pattern was biased toward a bursting behavior.

The number of spikes within these bursts was not constant for any one unit (usually three to four) and the mean interspike interval was 3 to 4 msec. In most cells, spontaneously occurring orthodromic and antidromic AP were intermingled, but in a few cases only antidromic AP could be detected after a while. The tested antidromic AP were often the first of an AP burst. We did not perform the collision test for the subsequent AP of these bursts.

Gross potential recordings were sometimes performed through the stimulating electrode; in some cases, no sign of paroxysmal activity could be detected when spontaneous antidromic AP were generated.

Cortical Surface Recordings

Spontaneous interictal EEG spikes could often be recorded with a variable delay (from a few minutes to 1 hr) on the ipsilateral cortical surface following PT penicillin application (which sometimes had to be repeated).

These EEG spikes usually exhibited a diphasic negative-positive shape with a varying amplitude and had a duration of 30 to 40 msec (see Figs. 3 and 5). Their frequency, although irregular, was in a range of 3 to 10/sec. Electrical PT stimulations failed to trigger such EEG spikes, but their frequency was altered during repetitive stimulation. Once established, this activity could last 2 or 3 hr. The cortical area where these EEG waves could be recorded is indicated in Fig. 4. This area could be even more restricted for each single experiment since Fig. 4 has been drawn from the results of several experiments. It is more or less possible to superimpose this area with the area of projection of the antidromic field potential.

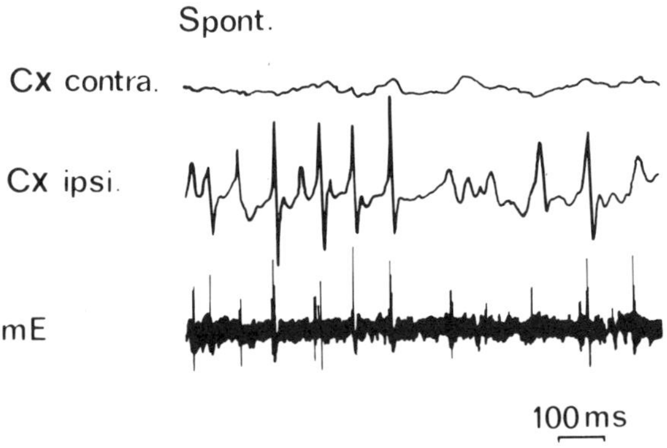

FIG. 3. *Lower trace:* Extracellular recording of a PT neuron exhibiting nearly exclusively antidromically conducted AP (see text). Each negative limb of surface paroxysms is associated with a cellular discharge.

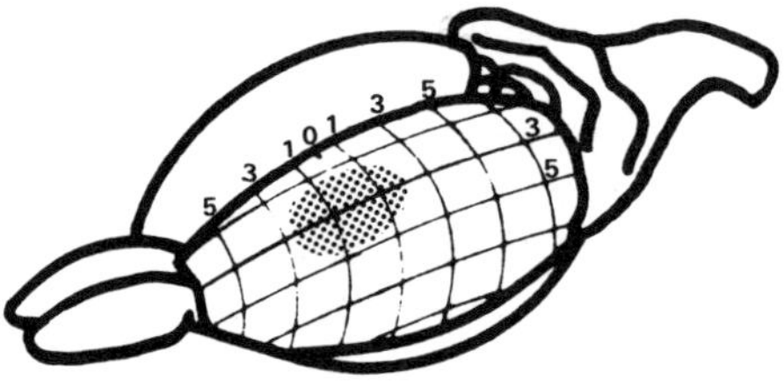

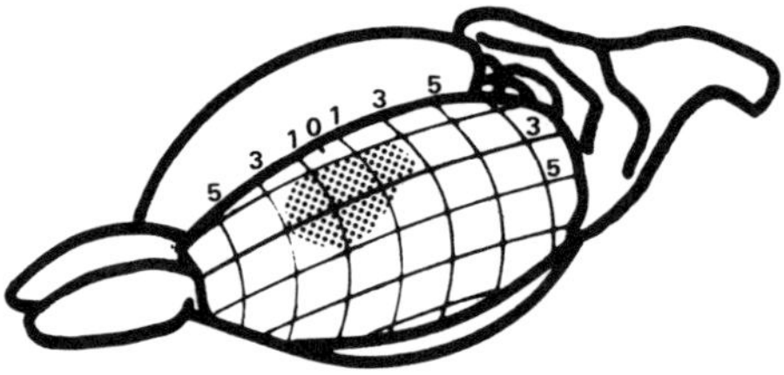

FIG. 4. Top: Antidromic response area. Cortical area where antidromic field potentials could be recorded following moderate (no more than 0.3 mA) PT stimulation. **Bottom:** Spiking area. Cortical area where surface paroxysms following PT penicillin application could be recorded. 0, bregma position.

Relationship Between EAPG and Cortical Paroxysms

In some cases, EAPG could be demonstrated in the absence of any surface epileptiform event, especially when small doses of penicillin had been applied to the PT; the reverse, however, was never observed.

The time relationship between EAP and cortical paroxysms could be studied in only a limited number of neurons because the collision test could not be used during this analysis; the frequency of both cellular and surface activities was altered during repetitive antidromic activation. Therefore, only cells exhibiting nearly exclusively spontaneous antidromic activation following PT penicillin application before and after the period of analysis were selected. For these cells, AP generation was roughly coincident in time with cortical paroxysms (Fig. 3).

Some AP or AP bursts occurred independently of the recorded surface transients; in most cases, however, the cellular activity was time-locked with the EEG spikes (Fig. 4). The closest correlations were found when surface waves had a sharply negative rising limb. In many instances, AP bursts appeared to lead the surface events or to be coincident with the negative phase (Fig. 5); but other bursts of the same neurons may lag behind the surface negativity.

DISCUSSION

EAPG

EAPG in thalamic nuclei sending axons to a cortical penicillin focus has been quite convincingly demonstrated by Gutnick and Prince (15) for the ventralis posterolateralis nucleus, by Rosen et al. (24) and by Scobey and

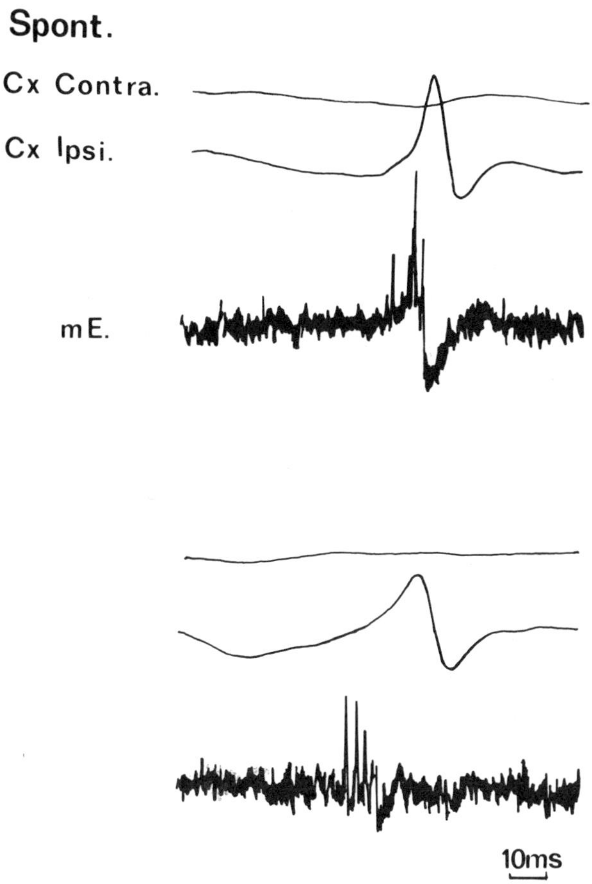

FIG. 5. Note the variations in time relationships between surface and cellular activities. The band pass of cellular recordings has been modified to show slower phenomena.

Gabor (28) for the lateral geniculate nucleus, and by Schwartzkroin et al. (26) for the ventrolateral nucleus. The present study shows that penicillin induces the same effect on PT fibers or on their bulbar collaterals. Therefore, the same phenomenon is likely to occur on PT collateral fibers or terminals within a cortical penicillin focus.

It is improbable that antidromic PT cell activation was the result of an injury discharge of PT fibers due to the positioning of the stimulating electrode and of the cannula because (a) EAPG could be recorded only some time after penicillin injection, and once demonstrated could be recorded for hours in the same cell; and (b) similar injections of control solutions resulted in no effect. Since no axoaxonal contacts have been described on PT fibers, synaptic mechanisms are unlikely to be involved in EAPG.

Specificity of Penicillin Action

This effect is in all probability related to penicillin properties, since similar injection of the same penicillin solution preincubated with penicillinase (Institut Pasteur Production) lent negative results (14). Moreover, under our conditions, EAPG does not seem to be the result of an intense, nonspecific excitatory drive altering axonal properties, since spontaneous antidromic AP were generated after penicillin application when no paroxysmal event could be recorded at the injection site. Nevertheless, such EAPG induced by paroxysmal or even by more physiological events may occur in other experimental situations (12,23).

Location of the Ectopic Generator

It is not possible to determine the exact location of the ectopic generator; if we assume, however, that the site of EAPG is situated between the cortex and the injection site, and that orthodromic and antidromic conduction velocities are identical, then calculations analogous to those of Scobey and Gabor (28) allow us to predict that it should be situated on the first third of any PT fiber next to the injection site. Furthermore, similar penicillin injections administered 1.5 mm dorsally to the usual injection site were ineffective in generating EAP. It can therefore be concluded that the ectopic generator is located near the injection site.

Site of Penicillin Action

Since we usually selected the more excitable and faster conducting fibers, it can be assumed that penicillin induces EAPG on myelinated fibers; but we do not know whether the site of action is situated on the PT fibers (at the level of nodes of Ranvier, for instance) or on the endings of their bulbar collaterals. In a recent study, we unsuccessfully investigated the effect of exposing an isolated node of Ranvier of frog sciatic nerve fiber (in collaboration with J. M. Dubois) to solutions of penicillin of increasing concentrations in voltage clamp experiments. Noebels and Prince (19) tested the action of penicillin on an isolated rat phrenic nerve-hemidiaphragm preparation by recording phrenic nerve AP across a sucrose gap. While prolonged exposure of the nerve trunk did not produce any repetitive activity, AP bursts generated near or at the motor nerve terminals following an orthodromic stimulation could be detected. Moreover, Lothman and Somjen (18) demonstrated that during penicillin-induced epileptiform activity in the spinal cord of curarized decapitate cats, the paroxysmal discharges recorded from peripheral nerves persisted when ventral roots had been cut but ceased with both ventral and dorsal roots cut. They concluded that this activity "did not originate in the peripheral nerve but rather within the cord and was conducted via dorsal and ventral roots in the periphery."

These experiments suggest that the preferential site of penicillin action may be the nerve terminals, but further studies are required to elucidate this problem, at least from the standpoint of the CNS.

Surface Epileptiform Waves

Gloor et al. (13) demonstrated that injections of penicillin into various subcortical structures in cats and monkeys resulted in cortical epileptiform activity only when penicillin leaked into CSF spaces or diffused to the overlying cortex along the needle tract. Thus the occurrence of epileptiform waves, which we did not expect, on the ipsilateral cortical surface following PT penicillin application on one side was interpreted as being attributable to penicillin seepage into the cortex. However, several arguments pleaded against this interpretation: (a) The small amount of penicillin injected and the distance between the injection site and the cortical surface made it unlikely that a critical concentration for eliciting paroxysmal waves was reached at the cortical level. (b) The cortex was not in direct contact with the CSF (it was covered by an agar-agar gel), and the cisterna magna was opened. (c) Such epileptiform events could be recorded only within a restricted area of the cortical surface.

The following control experiments were then performed: (a) Penicillin was injected into the PT. The resulting surface transients were recorded on one side. The cortex was then carefully undercut with a silver spatula, and the epileptiform waves disappeared. However, it was still possible to create an acute epileptogenic focus by direct penicillin application onto the undercut cortex. (b) When an epileptogenic focus is produced by cortical penicillin application, subsequent penicillinase application onto the focus suppresses the epileptiform activities within 10 to 20 min (14).

If penicillinase is applied to a cortical focus resulting from PT penicillin application, then the time course of the discharge is not reduced when compared to control. It is thus unlikely that such foci are due to diffusion of penicillin from the injection site.

Neuronal Activity and Cortical Paroxysms

The present data show that the surface epileptiform transients vary with time in size, shape, and frequency. The phase relationship existing between cellular discharges and surface potentials is not always very precise. These findings contrast with the much more regular occurrence and stereotyped appearance of surface waves and the higher degree of synchronization between cell firing and EEG spikes observed in an acute focus created by direct cortical penicillin application.

The mechanisms underlying the generation of cortical paroxysms following PT penicillin injection is not to be easily explained; any given interpretation can only be speculative.

The occurrence of cortical paroxysms seems to be related, in a more or less loose manner, with EAPG. Such AP generation, occurring in a sufficient number of PT fibers in a relatively asynchronous fashion, could produce at times "effective antidromic patterns" able to give rise to surface waves, with probable reinforcement of excitatory or inhibitory intracortical mechanisms.

There could be disturbances of such a mechanism by descending impulses following cortical paroxysms, providing a possible explanation for their observed fluctuations. On the other hand, a possible diffusion of penicillin to lemniscal fibers overlying the PT cannot be ruled out. EAP generated on these fibers would produce ascending impulses. The convergence of both antidromic and orthodromic volleys onto the same cortical areas (since there is some overlapping of motor and sensory cortices in the rat) could generate such surface paroxysms.

Nevertheless, further studies, including intracellular recordings, are required for a better understanding of the mechanisms involved.

SUMMARY

Extracellular recordings were made from PT neurons in rat cerebral cortex. An antidromic test AP evoked by electrical PT stimulation regularly collided with orthodromic AP recorded in the cortex. After a penicillin injection in the PT, many antidromic test AP did not collide with AP recorded in the cortex. It is concluded that these spontaneously occurring AP have been conducted antidromically to the cortex and that they must have been generated at an ectopic site. Therefore, such a phenomenon is likely to occur on intracortical collaterals of PT cell axons within a cortical penicillin focus. Moreover, under the same experimental conditions, interictal surface paroxysms appeared on a restricted cortical area and were most often coincident with penicillin-induced antidromic AP. Possible underlying mechanisms are discussed.

REFERENCES

1. Ajmone Marsan, C. (1969): In: *Basic Mechanisms of the Epilepsies,* edited by H. H. Jasper, A. A. Ward, and A. Pope, pp. 299–319. Little Brown, Boston.
2. Andersen, P. (1978): *This volume.*
3. Ayala, G. F., Ditcher, M., Gumnit, R. J., Matsumoto, H. and Spencer, W. A. (1973): *Brain Res.,* 52:1–17.
4. Ayala, G. F., Liw, S., Vasconetto, C. (1970): *Science,* 167:1257–1259.
5. Chalazonitis, N., and Takeuchi, H. (1968): *C.R. Soc. Biol.* (*Paris*), 162:1552–1556.
6. David, R. J., Wilson, W. A., and Escueta, A. V. (1974): *Brain Res.,* 67:549–554.
7. Dichter, M., and Spencer, W. A. (1969): *J. Neurophysiol.,* 32:649–662.
8. Ducreux, C. H., Takeuchi, H., and Chalazonitis, N. (1972): *C.R. Acad. Sci.* (*Paris*), 275:1907–1909.
9. Faugier, S., and Willows, A. O. D. (1973): *Brain Res.,* 52:243–260.
10. Freeman, A. R. (1973): *J. Neurobiol.,* 4:567–582.

11. Fuller, J. M., and Schlag, J. D. (1976): *Brain Res.,* 112:283–298.
12. Gabor, A. J., and Scobey, R. P. (1975): *J. Neurophysiol.,* 38:395–404.
13. Gloor, P., Hall, G., and Coceani, F. (1966): *Exp. Neurol.,* 16:333–348.
14. Gutnick, M. J., and Prince, D. A. (1971): *Neurology,* 21:759–764.
15. Gutnick, M. J., and Prince, D. A. (1972): *Science,* 176:424–425.
16. Hochner, B., Spira, M., and Werman, R. (1976): *Brain Res.,* 107:85–103.
17. Klee, M. R., Faber, D. S., and Heiss, W. D. (1973): *Science,* 179:1133–1135.
18. Lothman, E. W., and Somjen, G. G. (1975): *Electroencephalogr. Clin. Neurophysiol.,* 41:237–252.
19. Noebels, J. L., and Prince, D. A. (1976): *Neurosci. Abstr.,* 2:264.
20. Pellegrino, L. J., and Cushman, A. J. (1967): In: *A Stereotaxic Atlas of the Rat Brain,* edited by R. M. Elliott, G. Lindzey, and K. Mac Corquodale. Appleton-Century Crofts, New York.
21. Prince, D. A. (1968): *Exp. Neurol.,* 21:467–485.
22. Prince, D. A. (1969): In: *Basic Mechanisms of the Epilepsies,* edited by H. H. Jasper, A. A. Ward, and A. Pope, pp. 575–603. Little Brown, Boston.
23. Rosen, A. D., and Vastola, E. F. (1977): *J. Neurophysiol.,* 40:9–15.
24. Rosen, A. D., Vastola, E. F., and Hildebrand, Z. I. M. (1973): *Exp. Neurol.,* 40:1–11.
25. Speckmann, E. D., and Caspers, H. (1973): *Epilepsia,* 14:397–408.
26. Schwartzkroin, P. A., Mutani, R., and Prince, D. A. (1975): *J. Neurophysiol.,* 38:795–811.
27. Schwartzkroin, P. A., and Prince, D. A. (1976): *Neurosci. Abstr.,* 2:266.
28. Scobey, R. P., and Gabor, A. J. (1975): *J. Neurophysiol.,* 38:383–394.
29. Stone, T. W. (1972): *Exp. Neurol.,* 35:492–502.
30. Szentagothai, J. (1975): *Brain Res.,* 95:475–496.
31. Towe, A. L., and Harding, G. W. (1970): *Exp. Neurol.,* 29:366–381.

Abnormal Neuronal Discharges, edited by N. Chalazonitis and M. Boisson. Raven Press, New York © 1978.

The Generation of Rhythmic Discharges During Bursts of REM

Ottavio Pompeiano

Istituto di Fisiologia Umana, Cattedra II, Università di Pisa, Pisa, Italy

PHASIC MOTOR EVENTS DURING DESYNCHRONIZED SLEEP

During desynchronized sleep, in addition to the main tonic events characterized by low voltage, fast cortical activity (5) and atonia in the posterior cervical muscles (20,38), there is the sudden eruption of phasic motor events, which can be classified under two headings: (a) rapid eye movements (REM) (5,6), and (b) rapid contractions of the facial and limb musculature, which occur synchronously with bursts of REM and resemble the myoclonic twitches induced by epileptic discharges (12).

The pattern and organization of these twitches have been studied in unrestrained cats by recording simultaneously the EMG activity from flexor and extensor muscles of proximal and distal parts of both hindlimbs (12). The twitches are more likely to occur in association with the most intense and prolonged bursts of REM. On the other hand, no phasic change in muscular activity appears during isolated ocular movements. Myoclonic jerks are generally more frequent in distal than in proximal muscles, and they are more prominent in flexor than in extensor muscles. Although gamma motoneurons may fire during the bursts of REM (23), the twitches are basically due to phasic excitatory influences acting on alpha motoneurons. In fact, section of the dorsal roots does not influence the frequency nor the pattern of these muscular contractions (12). Experiments of spinal cord section indicate that the effective pathways course along the dorsolateral funiculi.

There are several descending tracts located in this region, which contribute to these muscle contractions; one of these is the corticospinal tract, which, in the cat, mainly influences the flexor motoneurons via the interneurons of the flexion reflex pathway (26). Destruction of the corticospinal tract does not prevent the occurrence of myoclonic activity during desynchronized sleep (28), indicating that other supraspinal descending pathways contribute to this activity; however, experiments of unit recording have shown that single pyramidal tract neurons located in the precentral gyrus of unanesthetized monkeys undergo profound changes in their temporal pattern of discharge during sleep. In particular, while the discharge is quite regular during relaxed wakefulness, bursts of activity separated by long intervals of complete inactivity occur during desynchronized sleep (7–10). We have seen that in cats the increase in unitary activity as well as in the integrated discharge of the pyramidal tract is strikingly related in time with the bursts of REM (28,33). Parallel to this phasic increase in the pyramidal discharge, there is also an increase in unit activity as well as in the integrated discharge from the red nucleus synchronously with REM (13) (Fig. 1A). Moreover, experiments of extracellular recording of single neurons in the spinal cord of unrestrained, unanesthetized cats have clearly shown that the responses of single units to stimulation of the flexion reflex afferents greatly increase during desynchronized sleep, particularly at the time of REM bursts (25; see ref. 39). These units, which have the characteristic behavior of spinal cord interneurons, are located in laminae V to VIII of Rexed (48), i.e., in the same region of the spinal cord that receives

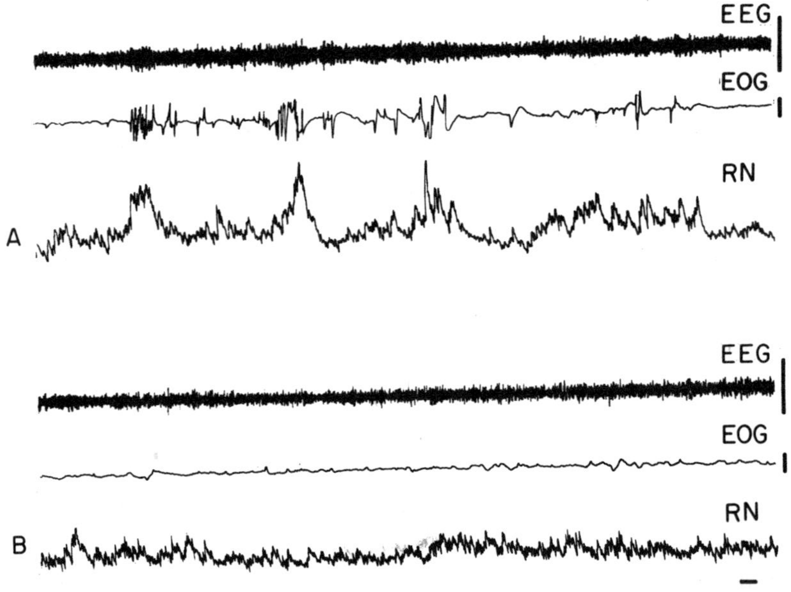

FIG. 1. Phasic increase in the rubral activity during the REM of desynchronized sleep and its abolition following vestibular lesion. Unrestrained, unanesthetized cat; bipolar records. EEG, electroencephalogram (left parietooccipital); EOG, electrooculogram; RN, integrated red nucleus activity. Time calibration, 1 sec; voltage calibration, 0.5 mV. **A:** Experiment made 3 days after chronic implantation of the electrodes. During desynchronized sleep, there are large phasic increases in the rubral discharge which are strikingly related in time with the bursts of REM. **B:** Same animal as in **A.** Experiment made 7 days after chronic implantation of the electrodes and 2 days after bilateral lesion of the medial and descending vestibular nuclei. Absence during desynchronized sleep of the large bursts of REM with related phasic enhancements of rubral discharge. (From ref. 44.)

the terminals of several supraspinal descending pathways, including the corticospinal and the rubrospinal tracts (37).

It is of interest that the discharge pattern of the pyramidal tract neurons during REM bursts differs from that which occurs during wakefulness in relation to a brief extension of the wrist (9,10). Movement is generally associated with relatively smooth modulation of the discharge frequency of the pyramidal tract neurons. Moreover, whereas some units show an increase in their activity, other units have decreased activity, and adjacent units show reciprocal relationships. During REM, however, the discharge patterns of adjacent neurons tend to be positively correlated; neurons tend to fire synchronously, and the highly differentiated temporospatial pattern associated with waking movements disappears. With the extremely intense pyramidal tract bursts characteristic of desynchronized sleep, the output which descends via the pyramidal tract closely resembles that of the convulsive discharge set up by electrical stimulation or strychninization of the cortex.

The great similarity between the jerks of convulsive activity and twitches occurring during REM suggests that some common mechanisms are involved. The close relationships between REM sleep, REM deprivation, and seizure susceptibility strongly support this conclusion (see ref. 39).

We will not discuss here the intracortical neuronal mechanisms that may lead to an increased excitability of corticospinal neurons during desynchronized sleep; we do discuss the central mechanisms that generate the synchronous ascending volleys responsible for REM and the related bursts of high frequency discharge involving pyramidal and extrapyramidal tract neurons.

RHYTHMIC DISCHARGES OF PONTINE NEURONS RELATED TO PHASIC MOTOR EVENTS DURING DESYNCHRONIZED SLEEP

Monophasic potentials have been recorded from different brain structures during the REM of desynchronized sleep, attributable to synchronous discharges originating from or triggered by neurons located in the dorsolateral part of the pontine reticular formation (RF) (see refs. 20,21, and 24). Both isolated and clustered pontine waves, related, respectively, to the *irregular* appearance of isolated ocular jerks and bursts of REM, can be differentiated (22,35; see also refs. 4 and 11).

REM and related pontine waves appear not only in the intact preparation during desynchronized sleep but also during the cataplectic episodes that occur spontaneously following total decortication or chronic decerebration (11,19,36; see ref. 20). In these instances, however, the oculomotor pattern is characterized by the *regular* occurrence of bursts of horizontal REM.

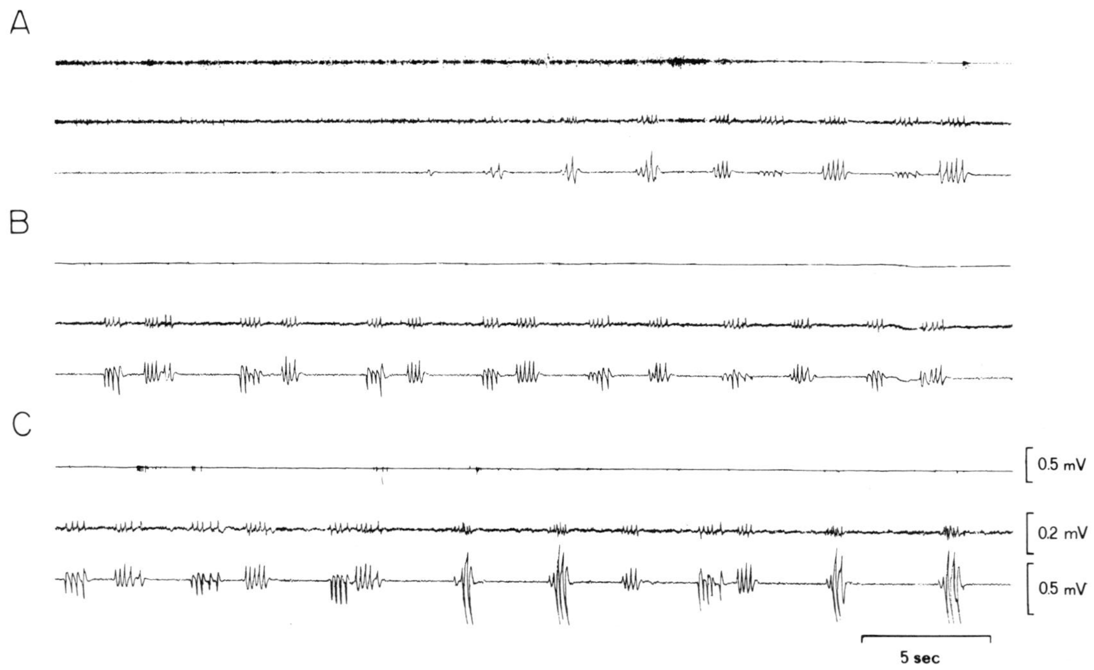

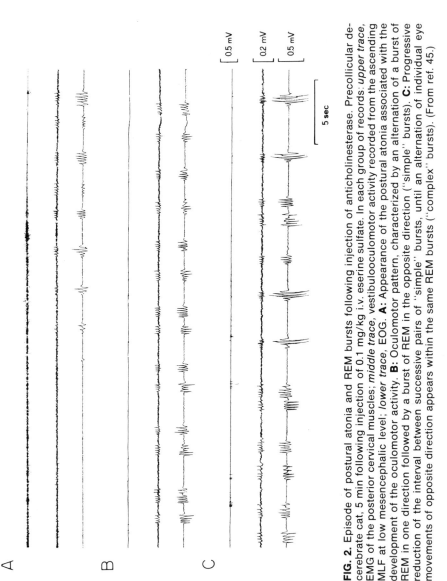

FIG. 2. Episode of postural atonia and REM bursts following injection of anticholinesterase. Precollicular decerebrate cat, 5 min following injection of 0.1 mg/kg i.v. eserine sulfate. In each group of records: *upper trace,* EMG of the posterior cervical muscles; *middle trace,* vestibulooculomotor activity recorded from the ascending MLF at low mesencephalic level; *lower trace,* EOG. **A:** Appearance of the postural atonia associated with the development of the oculomotor activity. **B:** Oculomotor pattern, characterized by an alternation of a burst of REM in one direction followed by a burst of REM in the opposite direction (''simple'' bursts). **C:** Progressive reduction of the interval between successive pairs of ''simple'' bursts, until an alternation of individual eye movements of opposite direction appears within the same REM bursts (''complex'' bursts). (From ref. 45.)

which are associated with clustered pontine waves, while the isolated ocular jerks and the related pontine waves are absent.

There is evidence that similar cataplectic episodes can also be induced in acute decerebrate animals by systemic administration of centrally active drugs, producing either an increase in level of acetylcholine (27,29,31) or a decrease in monoamines (see ref. 21). In both instances, REM are oriented along the horizontal plane. However, while the REM produced by small doses of an anticholinesterase (eserine sulfate) are always grouped in bursts, those elicited by drugs producing a depletion of monoamines at monoaminergic synapses in the brainstem (reserpine) may also appear as isolated ocular jerks.

Figure 2 illustrates the typical oculomotor pattern that occurs in acute

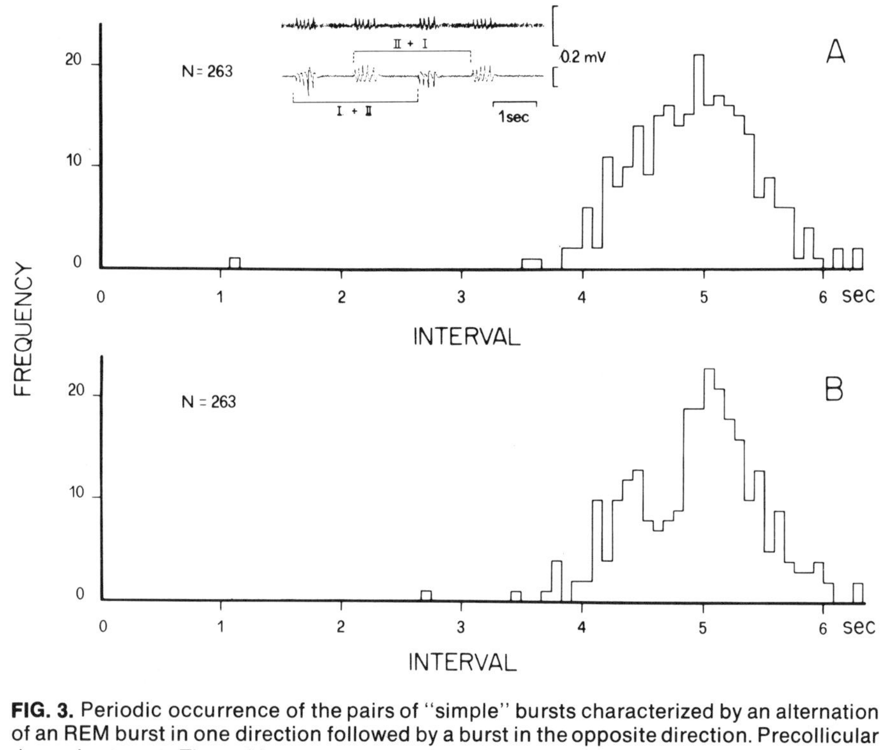

FIG. 3. Periodic occurrence of the pairs of "simple" bursts characterized by an alternation of an REM burst in one direction followed by a burst in the opposite direction. Precollicular decerebrate cat. These histograms summarize the results obtained during 20 periods of 100 sec duration characterized by the occurrence of REM episodes. These episodes were elicited by injection of 0.1 mg/kg i.v. eserine sulfate, which was repeated several times at the regular intervals of about 30 min. Each REM episode usually lasted several minutes and was then followed by prolonged periods of ocular quiescence. **A:** Frequency distribution of the intervals evaluated between the beginning of an REM burst to the left side and the beginning of a successive burst oriented to the same direction (I + II). **B:** Frequency distribution of the intervals between successive bursts of REM oriented to the right side (II + I). (From ref. 31.)

decerebrate cats following intravenous injection of an anticholinesterase. In particular, a burst of REM in one direction is followed after a regular interval of ocular quiescence by a burst of REM in the opposite direction (Fig. 2A and B). Moreover, a progressive reduction of the interval between successive bursts of REM may also appear, until an alternation of the individual eye movements occurs within the single bursts (Fig. 2C).

Whether the animal has been chronically or acutely decerebrated, or whether cholinomimetic or monoaminolytic drugs are injected, both the isolated ocular jerks and the bursts of REM oriented in a given direction occur at regular intervals, as shown by the sequential analysis of these REM (30,31) (Fig. 3). This finding suggests that a common mechanism which is triggered by cholinergic facilitatory structures but inhibited by monoaminergic structures is responsible for the periodic occurrence of REM in decerebrate cats.

GENERATION OF RHYTHMIC DISCHARGES DURING BURSTS OF REM

We have developed a deterministic linear mathematical model which may account for the two fundamental features of the REM episodes induced by intravenous injection of an anticholinesterase in acute decerebrate cats: (a) the abolition of the postural activity which lasts throughout the cataplectic episodes, and (b) the regular occurrence of the bursts of REM and the periodic configuration within each train of REM (45).

To describe mathematically the essential characteristics of these episodes, we postulate the existence in the pontine RF of a self-excitatory system, which may activate the bulbospinal inhibitory system responsible for the postural atonia (38). We also postulate the existence of an oscillatory system in the dorsolateral part of the pontine RF, which may transform the steady input originating from the self-excitatory network into a sinusoidally modulated output. The excitatory impulses originating from the oscillatory system of one side during the positive component of a sinusoidal cycle would then impinge on pontine reticular structures, which may transform the periodically modulated input into a rhythmic phasic output, due to the presence of an inhibitory feedback mechanism related to these structures. This system would finally act on the oculomotor neurons either directly or indirectly via reticular projections to vestibulooculomotor neurons (42).

The Self-Excitatory System

Experiments of microelectrode recording have shown that pontine reticular units undergo a striking increase in the discharge rate which occurs not only in intact, unanesthetized animals throughout desynchronized episodes (14–16) but also in decerebrate animals during cholinergically induced cataplectic episodes (17). In particular, among 251 units recorded from different pontine regions before, during, and after episodes of postural atonia induced in decerebrate animals after injection of 0.03 to 0.1 mg/kg i.v. of an anticholinesterase (eserine sulfate), 81 units (32.3%) were facilitated throughout the cataplectic episodes. Most of these neurons (68 units) were located in the pontine RF, namely in the gigantocellular tegmental field (GTF), the paralemniscal tegmental field (PTF), and the central tegmental field (CTF) (1), while 13 units were found in structures other than the pontine RF. The proportion of units whose discharge increased throughout the cataplectic episodes was significantly greater in the pontine RF (68/140, i.e., 49% of the total population) than in the remaining structures (13/70, i.e., 19%) ($p < 0.01$; chi-square test, 1 d.f.).

For each unit of the two main populations of neurons which showed a significant increase in frequency of discharge during the cataplectic episode with respect to the baseline, the tendency to fire selectively during the induced episodes of postural atonia (index of selectivity) was quantified by the ratio Fa-Fc/Fc, where Fa and Fc, respectively, represent the mean rate obtained during the cataplectic episode and during the control period of postural rigidity. The general mean of the index obtained from the two populations of units, which were significantly facilitated throughout the cataplectic episode, corresponded to 9.65 ± 25.44, SD, with an average value being greater for the units located in the pontine RF (10.81 ± 27.73, SD) than for the units located in the remaining structures (3.56 ± 3.37, SD). Moreover, the index of selectivity was on the average greater for the GTF units (18.69 ± 35.92, SD) than for the CTF (10.32 ± 36.47, SD) and PTF units (6.14 ± 9.12, SD). These findings indicate that the pontine RF, particularly the GTF, CTF, and PTF, represent the most likely candidates for the "executive mechanism" responsible for the suppression of postural activity.

Experimental data indicate that these neurons are cholinergic in nature (40) and are under the inhibitory control of monoaminergic structures (41).

The Oscillatory System

The oscillatory system transforms the tonic input originating from the self-excitatory system into a sinusoidal final output; we have used the model of the neuronal network analyzed by Rashevsky (47), which applies to our problem, with some slight changes (45).

It is known that in a neuronal system a cyclic output can be obtained by reciprocal interaction of two specialized cell groups, where the cells of the first group are excitatory to cells of the second group, and the cells of this last group are in turn inhibitory to those of the former group. As a result of this interaction, one may observe a sinusoidal modulation of the discharge rate of the excitatory neurons, which would then act on the structures controlling the oculomotor activity. The RF with its intrinsic anatomical organization (50) represents the most suitable structure to develop such a periodic activity.

Experiments of unit recording, performed in precollicular decerebrate cats during the cataplectic episode produced by injection of 0.03 to 0.1 mg/kg i.v. of eserine sulfate, have recently demonstrated the existence of neurons located in the pontine RF, which showed periodic changes in their discharge rate during the regular bursts of REM induced by the anticholinesterase (18). In particular, among the 251 pontine neurons recorded before, during, and after these cataplectic episodes, 32 (12.7%) of the tested units showed a sinusoidal increase in their discharge rate during the bursts of REM in one direction, followed by sinusoidal decreases in their discharge rate during the bursts of REM in the opposite direction.

Fig. 4A and B illustrates the polygraph record of one unit showing a deceleration of its discharge during the REM burst in one direction, followed by an acceleration during the REM burst in the other direction. Oscillographic records of this pattern of discharge are illustrated in Fig. 4D, while the correlation between changes in unit activity and development of the oculomotor pattern is shown as an average sequential pulse density histogram in Fig. 4C. It is clear from this figure that the average changes in unit activity were almost sinusoidal in shape. Control experiments indicated that central rather than peripheral mechanisms were responsible for these changes. This hypothesis was supported by the observation that both the half periods of the variations in unit discharge started before the occurrence of the corresponding burst of REM and began to decline before the end of the burst. Moreover, it was observed that within each side of the pontine RF there were two populations of neurons which showed reciprocal changes in their rate profiles during the bursts of REM oriented in a given direction (Fig. 5), suggesting that the cyclic alternation of their activity depended on their reciprocal interaction. It is assumed that these two populations of neurons represent the individual components of the oscillatory system postulated by the theory (45).

The hypothesis that the self-excitatory system may trigger the oscillatory system, thus playing an important role in the appearance of REM, is supported by the fact that while the index of selectivity of the self-excitatory pontine neurons corresponded on the average to 9.65 ± 25.44, SD (17), the average value obtained for the units which showed a periodic modulation of their discharge during the bursts of REM corresponded only to 4.83 ± 8.36, SD (18).

Generation of Rhythmic Activity

It has been postulated that the discharge occurring during the positive component of the oscillatory cycle does not impinge directly on the ocu-

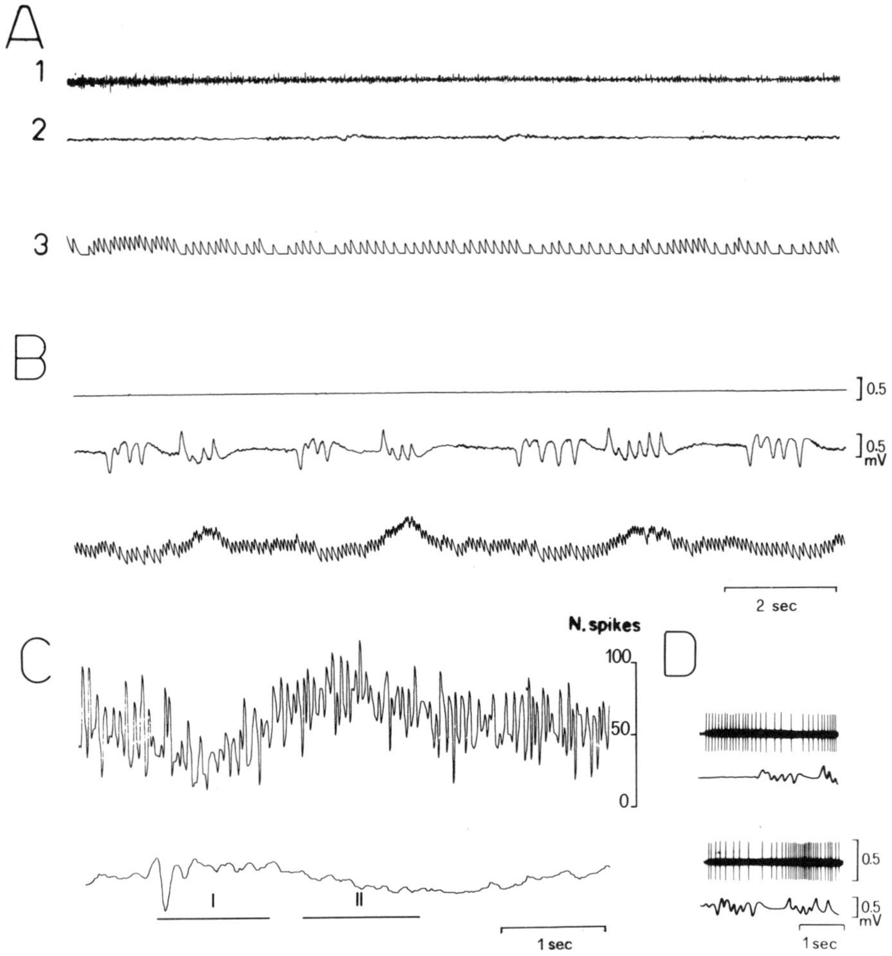

FIG. 4. Periodic modulation of the discharge rate of pontine reticular neurons during the bursts of REM induced by an anticholinesterase. Precollicular decerebrate cat. **A** and **B:** EMG of the ipsilateral triceps muscle of the brachium (*1*), the EOG (*2*), and the root mean square activity of a single neuron located in dorsal part of right CTF (*3*) were recorded with a polygraph. The pontine reticular neuron showed a low discharge frequency during the state of postural rigidity **(A)** and a steady increase in unit discharge during the episode of postural atonia following injection of 0.1 mg/kg i.v. eserine sulfate **(B).** Note the periodic decrease in unit discharge during the burst of REM oriented to the contralateral side (downward deflections in the EOG) followed by a periodic increase in unit discharge during the successive REM burst oriented to the ipsilateral side (upward deflections in the EOG). **C:** Computer record of another reticular unit located in the pontine tegmentum of the right side during the REM bursts induced by the same dose of anticholinesterase. *Upper trace,* sequential pulse density histogram of 50 averaged sweeps (256 bins with the dwell time of 20 msec per bin); *lower trace,* averaged record of the EOG. The scale next to the computer record represents the number of spikes per bin. The horizontal lines indicate the average duration of the bursts of REM oriented contralateral (I) and ipsilateral (II) to the recording side. Due to the variable intervals between successive eye movements, only the first eye movement within burst I is detectable in the averaged record. **D:** Oscillographic records of the same unit as in **C.** In each group of records: *upper trace,* corresponds to the unit activity; *lower trace,* to the eye movements. Note the periodic decrease in unit discharge during bursts of REM oriented toward the contralateral side (downward deflections of the EOG) and the increase in unit discharge during bursts of REM oriented toward the ipsilateral side (upward deflections of the EOG). (From ref. 18.)

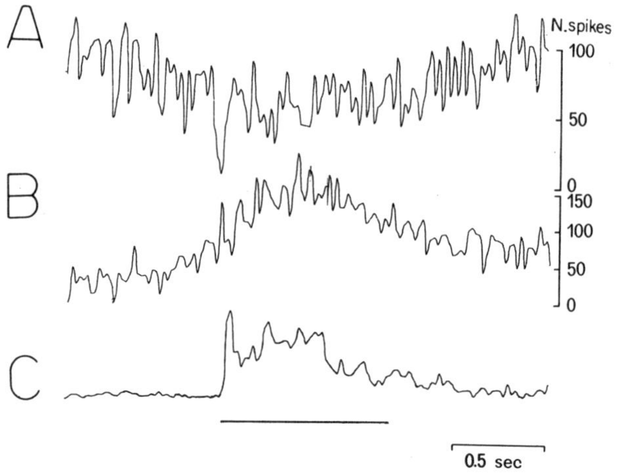

FIG. 5. Reciprocal behavior of two pontine reticular neurons showing periodic changes in the unit discharge during the bursts of REM. Precollicular decerebrate cat. The electrical activity of two pontine reticular neurons was recorded from the ventromedial part of the left GTF during the cataplectic episodes induced by injection of 0.1 mg/kg i.v. eserine sulfate. The neurons were located 100 μm away from each other, unit A being located dorsal to unit B. **A** and **B:** Computer records of the two individual units (100 averaged sweeps, 128 bins with the dwell time of 20 msec per bin). The scale next to the computer records represents the number of spikes per bin. The sweeps were triggered by large amplitude monophasic potentials recorded from the ascending MLF, which were related to eye movements oriented toward the left side. The monophasic potentials related to eye movements oriented to the right side were of small amplitude and were not used for triggering. **C:** Average of the large monophasic potentials recorded from the ascending MLF, which appeared synchronously with the eye movements oriented toward the recording side. The average pulse density histograms show the periodic decrease in unit discharge of the neuron in A and the reciprocal increase in unit discharge of the neuron in B, both preceding the trains of monophasic potentials related to the ipsilateral bursts of REM. (From ref. 18.)

lomotor neurons but rather indirectly via structures that sample and integrate the output of the oscillatory system by activating a fast phase generator (45). This generator would then produce rhythmic discharges due to the presence of an inhibitory feedback mechanism (2,49,51).

Among the 251 pontine neurons recorded before, during, and after the cataplectic episodes induced by the anticholinesterase, 60 units (23.9%) showed phasic rhythmic changes in their activity that coincided with or were closely related to the individual monophasic pontine potentials that preceded the corresponding REM (18). No attempt was made in this study to determine whether the periodically modulated neurons were connected with these distinct yet rhythmic pontine neurons. However, the fact that the index of selectivity was on the average lower for the rhythmic (2.48 ± 4.20, SD) than for the periodically modulated (4.83 ± 8.36, SD) neurons supports the contention that the latter population of neurons may drive the rhythmic units.

Premotor Structures Contributing to REM

The rhythmically discharging pontine reticular neurons may act on the oculomotor neurons not only directly but also by activating vestibulooculomotor neurons. Activation of the pontine reticular structures of one side causes contraction of the ipsilateral lateral rectus muscle, while just the opposite result is obtained by stimulating the vestibular nuclei of that side (46). We have postulated, therefore, that the pontine system produces horizontal eye movements to the ipsilateral side by exciting not only the ipsilateral reticular structures but also the contralateral vestibular nuclei (45).

The possibility that the vestibular nuclei are involved in the oculomotor activity during desynchronized sleep is supported by the results of lesion experiments. These show that complete bilateral destruction of the vestibular nuclei prevents the occurrence of the bursts of REM (43) and the related potentials in the brain typical of this phase of sleep (35), while the isolated ocular jerks and the corresponding potentials still persist after this lesion (35,43). The same lesion also suppresses the phasic increases in pyramidal and rubrospinal discharge (Fig. 1B), as well as the muscular twitches, which are synchronous with bursts of REM (34,44). These findings indicate that the pontine generator gives rise to rhythmic discharges in the vestibular nuclei, which contribute not only to REM but also to muscular twitches caused by activation of cortical and subcortical structures. The same pontine generator, however, is still able to trigger the oculomotor neurons in the absence of the vestibular nuclei.

The role played by the vestibulooculomotor neurons during REM is further illustrated by experiments of unit recording. These show that phasic changes in firing rate of vestibular nuclear neurons precede the REM not only during desynchronized sleep in intact animals (3) but also during the REM episodes produced in decerebrate preparations by an anticholinesterase (52) or following depletion of the monoaminergic nerve terminals in the brainstem produced by reserpine (32).

It is of interest that most of the neurons located in the medial vestibular nucleus, mono- or polysynaptically activated by stimulation of the ipsilateral labyrinth and antidromically excited by stimulation of the ascending MLF, undergo reciprocal changes in firing rate during the eye jerks oriented in both directions of the horizontal plane; the on-firing occurs during REM oriented either ipsilaterally (group a units) or contralaterally to the side of the recording (group b units). These phasic changes in discharge rate, which were of the burst-tonic (Fig. 6A) or the phasic type (Fig. 6B), occurred 11 to 15 msec prior to the activation of the corresponding lateral rectus muscle; they involved not only units located within the medial vestibular nucleus but also axons of vestibulooculomotor neurons located close to the abducens motoneurons. In addition to these units, a third population of units was recorded from the rostral part of the medial vestibular nucleus.

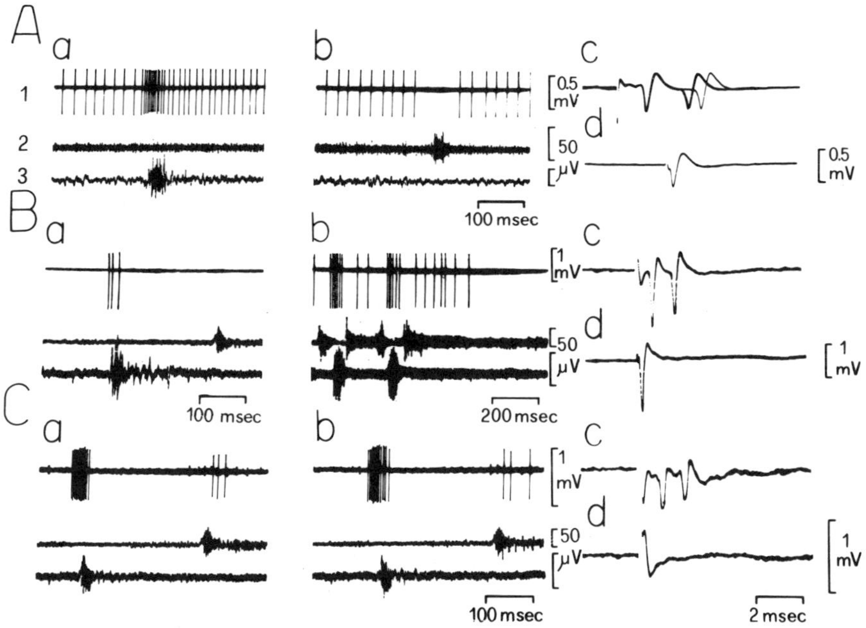

FIG. 6. Main types of discharge of group b units recorded from the left medial vestibular nucleus during the reserpine-induced REM in decerebrate cats. Precollicular decerebrate cat, pretreated with reserpine (0.5 mg/kg i.p. 1 day before decerebration). *Record 1,* unit discharge; *record 2,* EMG of the left lateral rectus muscle; *record 3,* EMG of the right lateral rectus muscle. **A:** Unit of the burst-tonic type showing a burst of discharge followed by an enhanced firing level during contraction of the right lateral rectus (*a*) and a pause during contraction of the left lateral rectus (*b*). **B:** Unit of the phasic type showing a burst of three spikes during contraction of the right lateral rectus (*a*). When a group of large REM appears after injection of 0.03 mg/kg i.v. eserine sulfate, the burst of unit discharge occurring during contraction of the right lateral rectus contains more spikes and is followed by an after discharge, which is interrupted by a pause during the eye jerk in the opposite direction (*b*). **C:** Unit of the burst type showing a burst of high frequency discharge 20 to 25 msec prior to the contraction of the right lateral rectus and a few spikes following the activation of the left lateral rectus (*a,b*). The three units showed orthodromic responses to electric stimulation of the ipsilateral labyrinth (*c*); however, antidromic responses to MLF stimulation affected only units A and B (*d*).

This population showed a higher firing rate and a longer lead (17 to 22 msec) with respect to the EMG activation of the corresponding lateral rectus muscle than the units of the burst-tonic and the phasic type (Fig. 6C). These burst neurons probably represent a step in a hierarchic organization of the structures involved in the REM prior to the vestibulooculomotor neurons.

These particular findings were obtained during the isolated ocular jerks occurring in chronically reserpinized cats, as well as during the bursts of REM elicited in the same preparations by small doses of eserine (32).

Alternated Activity of the Oscillatory Systems of Both Sides

It has already been reported (31) that during the cholinergic episodes a burst of REM in one direction is followed by a burst of REM in the opposite direction. To explain this finding, it has been postulated that the oscillatory systems of both sides show an independent although mutually related cyclic activity, which impinges on the preoculomotor reticular and vestibular structures of the corresponding sides. It is in the reciprocal interaction between the oscillatory systems of the two sides that one might find the explanation of the regular occurrence of successive bursts of REM oriented in opposite directions. Moreover, a progressive reduction of the phase shift between cycles of activity of the oscillatory systems of the two sides may account for the occurrence of bursts of REM characterized by an alternation of the individual ocular movements within the burst (45).

SUMMARY

During desynchronized sleep, there is the sudden occurrence of REM due to rhythmic discharges of neurons located in the pontine RF and the vestibular nuclei. REM similar to those that occur in intact preparations during desynchronized sleep have been observed during the episodes of postural atonia, which appear spontaneously in chronic decerebrate preparations. The same episodes can also be elicited in acute decerebrate animals by systemic administration of centrally active drugs, producing either an increase in level of acetylcholine or a decrease in monoamines. During desynchronized sleep, the oculomotor pattern is characterized by the irregular occurrence of isolated ocular movements and bursts of REM, whereas those occurring in decerebrate animals appear at very regular intervals.

A deterministic linear methematical model has been developed to account for the regular occurrence of rhythmic discharges, which lead to bursts of REM during the episodes of postural atonia occurring in decerebrate animals. The model postulates the existence of self-excitatory neurons particularly located in the GTF, which tend to fire at a high rate throughout these episodes. The steady discharge originating from these pontine neurons would then impinge on an oscillatory neuronal system located in the dorsolateral part of the pontine RF, which transforms the tonic input into a sinusoidal output. It has also been assumed that the excitatory impulses originating during the positive component of the oscillatory cycle impinge on a fast phase generator, which transforms the sinusoidally modulated input into a rhythmic output, due to the presence of an inhibitory feedback mechanism. This system would finally act on the oculomotor neurons not only directly but also by activating the vestibular nuclei. The experimental data so far accumulated are in agreement with these hypotheses.

ACKNOWLEDGMENTS

This investigation was supported by USPHS research grant NS 07685–09 from the National Institute of Neurological and Communicative Disorders and Stroke, and by a research grant from the Consiglio Nazionale dellé Ricerche, Italy.

REFERENCES

1. Berman, A. L. (1968): *The Brain Stem of the Cat.* University of Wisconsin Press, Madison.
2. Berthoz, A., Baker, R., and Goldberg, A. (1974): *Brain Res.,* 71:233–238.
3. Bizzi, E., Pompeiano, O., and Somogyi, I. (1964): *Arch. Ital. Biol.,* 102:308–330.
4. Chouvet, G., and Gadea-Ciria, M. (1974): *EEG Clin. Neurophysiol.,* 36:597–607.
5. Dement, W. (1958): *EEG Clin. Neurophysiol.,* 10:291–296.
6. Dement, W. C. (1964): In: *The Oculomotor System,* edited by M. B. Bender, pp. 366–416. Harper & Row, New York.
7. Evarts, E. V. (1964): *J. Neurophysiol.,* 27:152–171.
8. Evarts, E. V. (1965): In: *Aspects Anatomo-Fonctionnels de la Physiologie du Sommeil,* edited by M. Jouvet, pp. 189–212. CNRS, Paris.
9. Evarts, E. V. (1965): In: *Progress in Brain Research, Vol. 18: Sleep Mechanisms,* edited by C. Bally and J. P. Schadé, pp. 81–91. Elsevier, Amsterdam.
10. Evarts, E. V. (1967): *Res. Publ. Assoc. Nerv. Ment. Dis.,* 45:319–337.
11. Gadea-Ciria, M. (1972): *Étude Séquentielle des Pointes Ponto-géniculo-occipitales (PGO) an Cours du Sommeil Paradoxal chez le Chat Normal, et après Lésions Corticales et Sous-corticales. Ses Modifications Pharmacologiques.* Thèse de Doctorat de Spécialité, Lyon.
12. Gassel, M. M., Marchiafava, P. L., and Pompeiano, O. (1964): *Arch. Ital. Biol.,* 102:449–470.
13. Gassel, M. M., Marchiafava, P. L., and Pompeiano, O. (1965): *Arch. Ital. Biol.,* 103:369–396.
14. Hobson, J. A. (1974): In: *Advances in Sleep Research, Vol. 1,* edited by E. D. Weitzman, pp. 217–250. Spectrum, New York.
15. Hobson, J. A., McCarley, R. W., Freedman, R., and Pivik, R. T. (1974): *J. Neurophysiol.,* 37:1297–1309.
16. Hobson, J. A., McCarley, R. W., Pivik, R. T., and Freedman, R. (1974): *J. Neurophysiol.,* 37:497–511.
17. Hoshino, K., and Pompeiano, O. (1976): *Arch. Ital. Biol.,* 114:244–277.
18. Hoshino, K., Pompeiano, O., Magherini, P. C., and Mergner, T. (1976): *Arch. Ital. Biol.,* 114:278–309.
19. Jeannerod, M., Mouret, S., and Jouvet, M. (1965): *EEG Clin. Neurophysiol.,* 18:554–566.
20. Jouvet, M. (1967): *Physiol. Rev.,* 47:117–177.
21. Jouvet, M. (1972): *Ergeb. Physiol.,* 64:166–307.
22. Kawamura, H., and Pompeiano, O. (1970): *Pfluegers Arch.,* 317:10–19.
23. Kubota, K., Tanaka, R., and Tsuzuki, N. (1967): *Jap. J. Physiol.,* 17:613–626.
24. Laurent, J. P., Cespuglio, R., and Jouvet, M. (1974): *Brain Res.,* 65:29–52.
25. Lenzi, G. L., Pompeiano, O., and Rabin, B. (1968): *Boll. Soc. Ital. Biol. Sper.,* 44:44.
26. Lundberg, A., and Voorhoeve, P. (1962): *Acta Physiol. Scand.,* 56:201–219.
27. Magherini, P. C., Pompeiano, O., and Thoden, U. (1972): *Arch. Ital. Biol.,* 110:234–259.
28. Marchiafava, P. L., and Pompeiano, O. (1964): *Arch. Ital. Biol.,* 102:500–529.
29. Matsuzaki, M. (1969): *Brain Res.,* 13:247–265.
30. Mergner, T., Coulter, J. D., Magherini, P. C., and Pompeiano, O. (1977): *Proc. XXVII Int. Congr. Physiol. Sci., Paris,* XIII:501.
31. Mergner, T., Magherini, P. C., and Pompeiano, O. (1976): *Arch. Ital. Biol.,* 114:75–99.
32. Mergner, T., and Pompeiano, O. (1977): In: *Control of Gaze by Brain Stem Neurons,* edited by R. Baker and A. Berthoz, pp. 243–251. Elsevier, Amsterdam.
33. Morrison, A. R., and Pompeiano, O. (1965): *Arch. Ital. Biol.,* 103:538–568.

34. Morrison, A. R., and Pompeiano, O. (1966): *Arch. Ital. Biol.,* 104:214–230.
35. Morrison, A. R., and Pompeiano, O. (1966): *Arch. Ital. Biol.,* 104:425–458.
36. Mouret, J. (1964): *Les Mouvements Oculaires au Cours du Sommeil Paradoxal.* Thèse de Medicine, Lyon.
37. Nyberg-Hansen, R. (1966): *Ergeb. Anat. Entwgesch.,* 39:1–48.
38. Pompeiano, O. (1967): *Res. Publ. Assoc. Nerv. Ment. Dis.,* 45:351–423.
39. Pompeiano, O. (1969): In: *Basic Aspects of the Epilepsies,* edited by H. H. Jasper, A. A. Ward, Jr., and A. Pope, pp. 453–473. Little Brown, Boston.
40. Pompeiano, O. (1975): In: *Advances in Sleep Research, Vol. 3, Narcolepsy,* edited by C. Guilleminault, W. C. Dement, and P. Passouant, pp. 411–449. Spectrum, New York.
41. Pompeiano, O., and Hoshino, K. (1976): *Arch. Ital. Biol.,* 114:310–340.
42. Pompeiano, O., Mergner, T., and Corvaja, N. (1977): In: *The Structure and Function of the Cerebral Commissures,* edited by I. Steele Russell, M. W. van Hof, and G. Berlucchi. McMillan, New York (*In press.*)
43. Pompeiano, O., and Morrison, A. R. (1965): *Arch. Ital. Biol.,* 103:569–595.
44. Pompeiano, O., and Satoh, T. (1967): *Pfluegers Arch.,* 298:159–162.
45. Pompeiano, O., and Valentinuzzi, M. (1976): *Arch. Ital. Biol.,* 114:103–154.
46. Precht, W. (1975): In: *MTP International Review of Science. Neurophysiology,* edited by A. C. Guyton and C. C. Hunt, vol. 3, pp. 81–149. Butterworths, London.
47. Rashevsky, N. (1971): *Bull. Math. Biophys.,* 33:539–553.
48. Rexed, B. (1954): *J. Comp. Neurol.,* 100:297–380.
49. Robinson, D. A. (1971): *The Control of Eye Movements,* edited by P. Bach-y-Rita, C. C. Collins, and J. E. Hyde, pp. 519–538. Academic Press, New York.
50. Scheibel, M. E., and Scheibel, A. B. (1958): In: *Reticular Formation of the Brain,* edited by H. H. Jasper, L. D. Proctor, R. S. Knighton, W. C. Noshay, and R. T. Costello, pp. 31–68. Little Brown, Boston.
51. Schmid, R. M. (1973): In: *Fifth Symposium on the Role of the Vestibular Organs in Space Exploration, NASA-SP 314,* pp. 237–249. Office of Technology Utilization, NASA, Washington, D.C.
52. Thoden, U., Magherini, P. C., and Pompeiano, O. (1972): *Arch. Ital. Biol.,* 110:260–283.

Abnormal Neuronal Discharges, edited by N. Chalazonitis and M. Boisson. Raven Press, New York © 1978.

Ontogenesis of Cortical Unit Activity Located in a Penicillin-Induced Focus in the Rabbit

A. M. Desroches, D. Chapelle, and P. Laget

Laboratoire de Neurophysiologie Ontogénétique, Université Pierre et Marie Curie, Paris, France

Previous works (2,9,12) have been entirely devoted to the study of correlations between paroxysmal activities recorded on the cortex surface and the spontaneous unitary cell activities.

We are less informed about the respective role of cells in the different cortical layers in regard to triggering, maintenance, and arrest of critical activities, and the functional relationships that may occur during these activities. However, the ontogenetic study of various bioelectric cortical activities, such as evoked activities by peripheral stimulation (4,7,11), showed that maturation provokes important changes in the synaptic organization of the cortex. Consequently, we thought that an ontogenic approach to convulsive activities could provide interesting information on their intracortical organization.

We therefore studied the modifications, according to the ages, of neuronal unit activities in the different cortical layers within the epileptogenic foci, which were produced by a weak dose of topically applied penicillin.

MATERIAL AND METHODS

Experimental Animals and Age Groups

All experiments were performed on 57 young rabbits. This species was chosen because of the low degree of maturation of its central nervous system (CNS) at birth and the conveniently abundant litters.

These animals ranged in age from less than 1 day to 30 days and were divided into four age classes, according to the following criteria of development:

0 to 3 days: From the third day, peripheral stimulation provokes the appearance of the "arousal reaction" on the electrocorticogram (ECG)

4 to 8 days: The eighth day corresponds to the appearance of the early positive surface deflection of the somesthetic evoked potential (SEP) obtained by stimulation of a peripheral sensory nerve.

9 to 15 days: The fifteenth day corresponds to the achievement of maturation of the somesthetic and visual evoked potentials.

16 to 30 days: At the thirtieth day, the ECG has reached its adult aspect in regard to the maturation and frequency of arousal θ activities.

Surgical Procedure

All surgical procedures were carried out under light ether anesthesia with a tracheal cannula inserted. The animals were then immobilized with gallamine triethiodide (2 mg/100 g) and artificially ventilated. Afterward, they were partly immersed in water at 38°C. The head was fixed in a stereotaxic holder (Horsley-Clarke) of suitable dimension. All pressure contacts and incised tissues were infiltrated with a local anesthetic (xylocaine 1%). The skull was exposed in order to insert, in contact with the duramater, nine silver electrodes (ball tip diameter: 0.8 mm) for recording the global cortical activities. Four of these electrodes were placed on each hemisphere and another on the frontal bone used as reference. Electrocardiogram (EKG) was also monitored throughout the experiment.

Elicitation of a Cortical Epileptogenic Focus

The epileptogenic focus was elicited by a local microapplication of a very small amount of penicillin solution (concentration of IU per microliter) on the somethetic area of the right hemisphere. For each animal, the site was determined by detection of the cortical activity evoked by a weak electric stimulation of the left forelimb (0.02 msec and 12 V). The powdered penicillin (Specillin G) was dissolved in a Ringer solution containing methyl blue for further histological localizations of the application site.

To be sure that the focus was not elicited by the volume of the penicillin injection but by the specific action of this drug, we injected at the same site a 10 times larger volume of a solution containing no penicillin. This latter control never evoked abnormal activity.

Recording of Responses

Unit activities were recorded through glass micropipettes filled with a solution of methyl blue in 1 M K acetate. Their resistance ranged from 20 to 30 megohms. The electrode localizations were later stained by iontophoresis of the methyl blue and verified histologically. Simultaneously, the unitary and electrocorticographic activities were registered on both photographic film and magnetic tape in view of statistical analysis.

RESULTS

Determination of the Penicillin Threshold Doses

We used only threshold doses of penicillin to avoid paroxysmal activities spreading from remote cortical and subcortical areas of the same or the contralateral hemisphere.

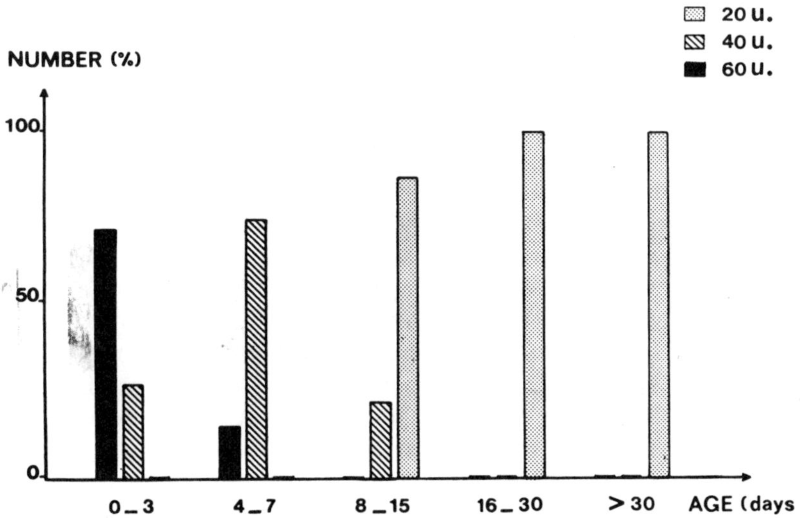

FIG. 1. Determination of penicillin threshold doses during maturation. The youngest animals need 3 times higher doses than adults. Whatever the age, doses remain very weak and do not exceed 20 to 60 IU.

First we determined the threshold dose for each animal, that is, a dose that produces on the ECG paroxystic activities only at the level of the focus: sharp waves, spikes, and waves. These appear only near the application point during the first minute after penicillin injection. In our previous work (1), the value of the threshold dose was determined for each developmental stage (Fig. 1).

0 to 3 days (11 animals): Eight animals (73%) required 60 IU, whereas three needed only 40 IU. The latter were 3-day-old rabbits.

4 to 7 days (14 animals): Two animals (15%), the youngest, required 60 IU and the 12 remaining (85%), only 40 IU.

8 to 15 days (20 animals): For 17 animals (85%) 20 IU were sufficient; three (15%) no older than 11 days needed 40 IU.

Beyond 15 days: For all these animals, 20 IU sufficed. Here we notice a well-known result—the higher resistance to convulsive drugs of the immature cortex (5,8,10). When the dose of penicillin was determined according to the above procedure, we proceeded to the unitary activities recorded at different cortical levels.

ECG: Morphology of Ictal and Interictal Changes—Evolution with Age

The penicillin threshold dose elicits cortical paroxystic anomalies, aspects which strongly differ according to the age of the animal. In all cases, however, anomalies appear 2 min after injection. They display alternate periods of intensive pathologic activities (ictal period duration about 30 sec to 1 min) and periods of weak activities (interictal period duration about 3 min). This cyclical pattern stops after 90 min on the average (minimum 60 min, maximum 120 min). The duration of this pattern varies from one animal to another but does not appear to be correlated to age.

Animals Younger than 8 Days

The first anomalies always appear very close to the point of application of the penicillin. They consist of rare slow spikes of variable amplitude but generally weak (40 μV); their duration fluctuates around 150 msec.

Both their frequency and spreading gradually increase; about 2 min after the appearance of the first paroxystic event, the ECG acquires the critical aspect, as it is usually observed in animals having a very immature cortex.

The slow spikes become more frequent (2 to 4/sec); their rhythm remains irregular and their amplitude, although larger than at the beginning of the seizure, does not exceed 60 μV (Fig. 2).

The slow waves spread first and only to the anterior cortical areas in the youngest animals, then to posterior areas for animals older than 4 days.

The propagation of anomalies to the contralateral hemisphere only occurs rarely and is limited to the largest slow spikes. In this case, they appear at the mirror image of the focus on the contralateral hemisphere.

When these paroxystic activities do not stay at the injection area of the penicillin but also spread to other homolateral cortical areas, the moment of appearance may considerably differ from one point to another. On the other hand, when they appear on the two hemispheres, they are synchronous on both.

The paroxystic activity after a maximum magnitude decreases, and after 30 sec to 1 min the interictal period begins when only some slow spikes remain (Fig. 2).

Animals Older than 8 Days

Just after the application of penicillin, the first signs of paroxystic activity appear in the shape of slow spikes with a duration and a frequency fairly comparable to those observed in younger animals. About 1 min after, however, the rhythm and the amplitude increase considerably (up to 120 μV), whereas their duration decreases. Furthermore, two new types of paroxystic patterns appear that are typical spike waves. From the tenth day, rapid, rhythmic postdischarges immediately follow a spike of large amplitude (Fig. 3) (14).

The acme crisis results in an increase of the number of spikes, spike waves, and postdischarges, as well as their tendency to cluster in large bursts only made of spikes.

On EEG recordings, the mono- and contralateral spread of paroxysmal activities in animals older than 1 week shows a nearly similar pattern to that

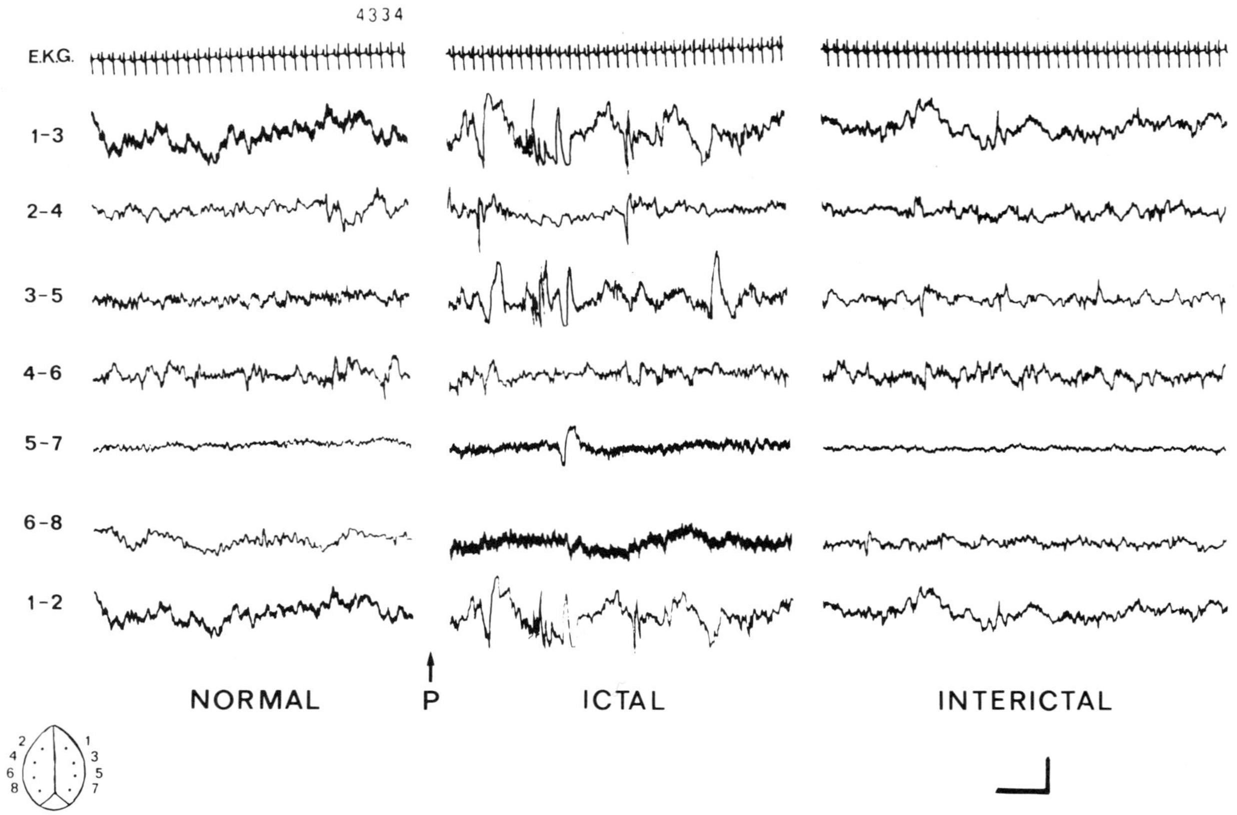

FIG. 2. Effects of a microdrop of penicillin solution (40 IU; threshold dose for a 4-day-old rabbit) deposited close to electrode 3. *Ictal period:* large and scarce but sharp waves, recorded between 1 to 3 and 3 to 5, spreading sometimes into occipital area or into mirror focus. *Interictal period:* slow waves of reduced frequency observed only close to the focus (1 to 3, 3 to 5). None of the paroxystic activities spread over other areas. Calibration, 1 sec, 12.5 μV.

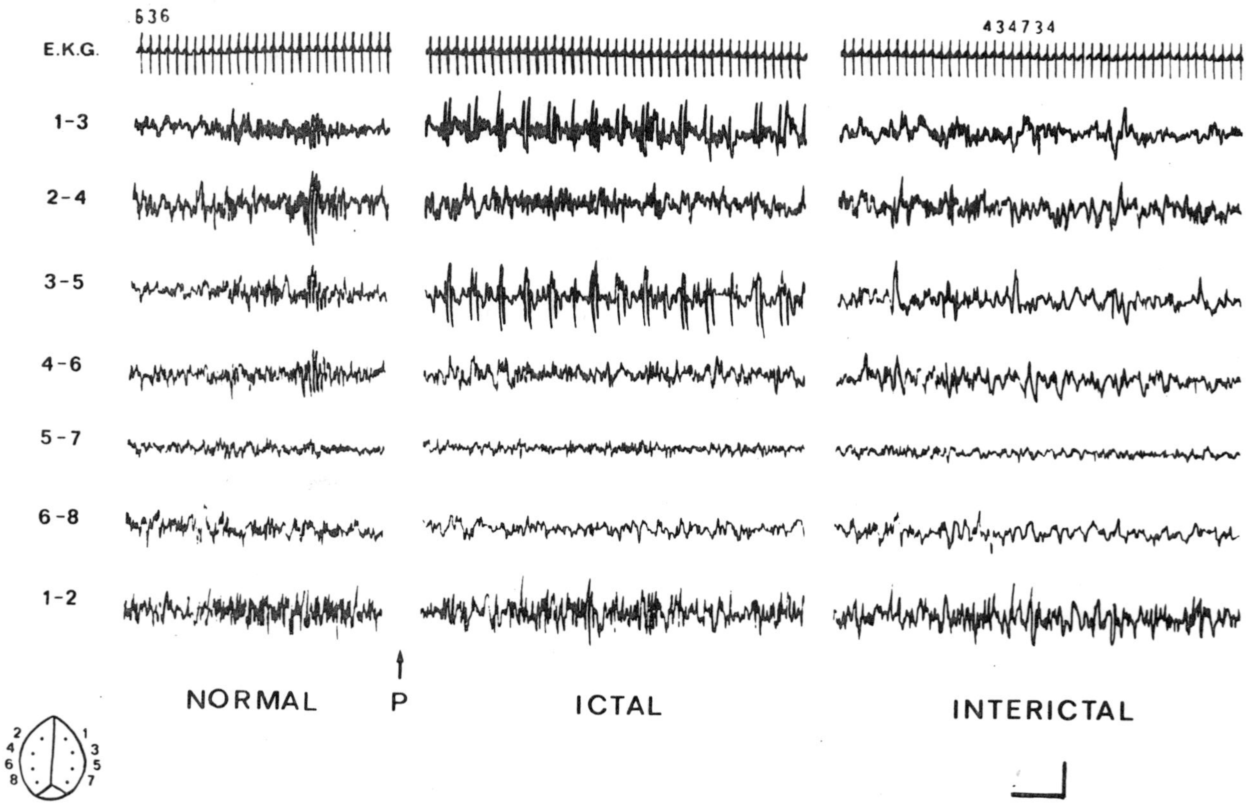

FIG. 3. Effects of a microdrop of penicillin solution (20 IU, threshold for a 16 day-old rabbit) deposited close to electrode 3. *Ictal period:* bursts of short-lasting spikes and rhythmical spike waves near the focus (1 to 3, 3 to 5). Sometimes spikes occurs at the mirror focus of the contralateral hemisphere and into occipital areas. *Interictal period:* isolated, asynchronous spikes in 1 to 3 and 3 to 5 (more rare in 2 to 4 and 4 to 6). No abnormal activity is recorded through the posterior derivations. Calibration, 1 sec, 50 μV.

in the youngest animals; we notice a trend of spikes and spikes and waves to synchronize largely in the homolateral hemisphere.

Cellular Activities

Animals Younger than 8 Days

In animals less than 8 days old, because of the small size of the neurons, only the firing of pyramidal cells (located in layers III and V) can be observed. It is also difficult to observe, in an extracellular record, the activity of a same neuron, especially in layer III. This explains the small number of neurons we could keep in the ictal interictal phases and before the application of penicillin.

Before Penicillin

In layer V, we studied 51 neurons with an average frequency of 2 to 3 spikes/sec. These showed a clear tendency to group in bursts of 2 to 3 spikes, separated by relatively long periods of silence.

In layer III, 43 neurons have been recorded. Average frequency of firing

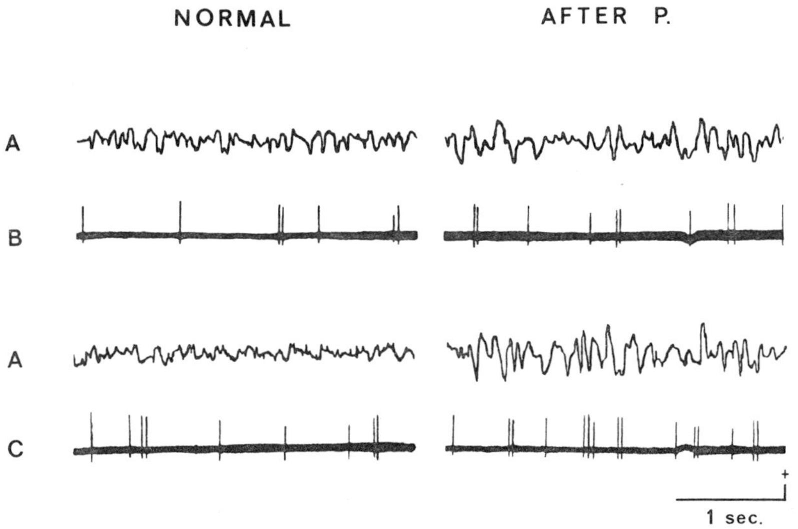

FIG. 4. Effects of 40 IU of penicillin on rhythm of unitary discharges in a 4-day-old rabbit. *A,* surface ECG recorded between 1 and 3; *B,* recording through microelectrode which is located close to the focus which appears near the surface electrode 3. The microelectrode is pushed into layer III of the small pyramidal cells; *C,* same as *B* but microelectrode is inserted into layer V of the large pyramidal cells. When spikes appear in the ECG, the frequency of unitary activity increases slightly in layers III and V, with a tendency to organization of spikes in bursts. Vertical calibration: *A,* 12.5 μV; *B* and *C,* 1 mV.

was slightly lower than for the neurons of layer V and did not exceed 2 spikes/sec. Their appearance was absolutely irregular and they did not show any trend to grouping (Fig. 4).

After Penicillin

We distinguish between ictal and interictal periods.

Ictal periods: In layer V, 35 neurons have been recorded. They all showed an increase in their firing rate to 3 to 5 spikes/sec. At the same time, their trend to bursting is more pronounced and their number increases. For 29 neurons (83%), bursts become synchronous with surface-spikes [the cells in this chapter are referred to as "positive cells" (+)]. We recorded 30 neurons in layer III. They showed an increase in their average firing rate (3 spikes/sec) but in a smaller proportion than those of layer V. Their behavior remains very irregular and shows no obvious relationship between the unitary firing and the surface paroxysmal events.

Interictal periods: In layer V, all of the 27 recorded cells indicated a slight decrease in the firing rate (3 to 5 spikes/sec to 3 to 4 spikes/sec). The bursts are less numerous and of a shorter duration; but for 22 of these cells (82%), the bursts remain synchronous with the surface paroxysms. In layer III, we noticed no significant change in the firing of the 25 recorded neurons (only a decrease in the average frequency of the firing rate).

Animals Older than 8 Days

Before Penicillin

In layer V, 60 neurons have been recorded. An increase of the firing rate (4 to 6 spikes/sec) expresses the maturation; at that stage we also notice a trend to grouping of spikes in numerous bursts; the older they are, the more the bursts contain spikes. In layer III, the 55 cells showed an increase of the firing rate with a higher frequency than in layer V (5 to 6 spikes/sec); therefore, this firing was irregular and without any tendency to grouping.

After Penicillin

Ictal period: In layer V, all the 45 neurons display an increase of the firing frequency (10 to 14 spikes/sec) as well as of the number and density of the bursts; 40 cells (89%) were, as denominated above, "positive" cells. In layer III, for the 37 neurons, the discharge frequency sharply decreased to 1 to 2 spikes/sec. Above all, 32 cells (86%) immediately stopped firing when a spike appeared on the ECG ["negative cells" (–)]. The remaining cells did not show any correlation with surface activities.

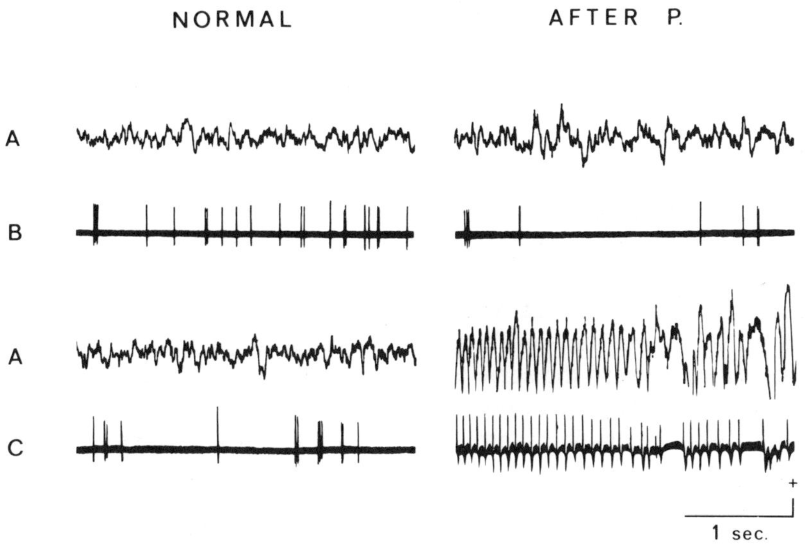

FIG. 5. Effects of 20 IU of penicillin on rhythm of unitary discharges in an 18-day-old rabbit. Same as for Fig. 4. During seizures, decreasing unitary activity of layer III cells with "negative" response pattern always occurs. On the contrary, from 8 to 10 days of age, there is increase of activity in layer V. Vertical calibration: *A*, 50 μV; *B* and *C*, 1 mV.

Interictal period: In layer V, the firing rate of the 28 recorded cells was inferior to that recorded during the ictal phase (5 to 6 spikes/sec) and similar to that observed before application of penicillin. On the other hand, bursts of spikes were still in accordance with surface spikes (25/28 positive cells, 89%). In layer III, the firing frequency of the 43 recorded cells has considerably increased (3 to 4 spikes/sec) although remaining lower than that observed before application of penicillin; but 86% of these neurons stopped firing, consequent to the appearance of the surface spikes ["negative cells" (–)].

DISCUSSION

Convulsive Effect of Penicillin: Advantage of a Weak Dose

We applied a microdrop of about 1 μl containing 60 IU to the cortex of the newborn rabbit and only 20 IU to that of the adult animal. In contrast, Wright and Bradley (14) used a small patch of 1 mm^2 impregnated with a solution containing 100,000 IU/ml; hence the number of units actually in contact with the cortex is difficult to ascertain.

The dose we used was definitely weaker. The ictal periods we have thus elicited are perhaps somewhat shorter than those observed by Wright and

Bradley (14), but their results and ours are quite comparable. However, we believe that weak doses are to be preferred because the effects elicited are more narrowly localized, which is an obvious advantage.

Homolateral Propagation of Paroxystic Activities

Propagation of paroxystic activities (slow spikes) displays the same stereotyped pattern on all young animals. They spread toward the anterior cortical areas on the homolateral cortex. We had already observed (*unpublished data*), in agreement with other authors, that superficial cortical response (SCR) and the spreading depression propagate more easily on the young animal toward the same anterior region.

We conclude from these observations that intracortical and perhaps subcortical association fibers grow more rapidly toward frontal areas than toward occipital areas.

Contralateral Propagation Activities

Contralateral propagation occurs only when a slow spike elicited by penicillin application on one hemisphere displays a sufficient voltage. The paroxystic event on the contralateral hemisphere generally appears on a point symmetrical to the focus elicited by drug on the other hemisphere. The scarcity of contralateral propagation of slow-spikes therefore appears to be due to the lower number of neurons bursting synchronously and to the lesser excitability of immature structures.

These data confirm those of others and lead to the conclusion that interhemispheric callosal connections are already functional at birth, despite their very low level of myelinization (13). Furthermore, it can be added that the well-known homotopy of callosal projection on both hemispheres in the adult is also present in the first postnatal days.

Contralateral propagation of slow spikes in the youngest animals, however, occurs only if those spikes are of sufficient magnitude, that is, if a certain number of neurons discharge synchronously.

Evolution with Age of the Activities Elicited in Layers III and V by Penicillin

Most authors who have studied the firing of cortical neurons during seizure have noticed that the pyramidal cells of the deep layers (V and VI) displayed volleys of spikes synchronous of the surface seizures. This phenomenon occurs whatever the age of the animal, even on the first postnatal day. This might account for the motor reactions that sometimes appear during seizure in newborn animals, even before birth.

When these neurons mature, the number and frequency of those spikes increase in the volleys. This could explain the increase in amplitude and the shortening of the duration of the seizures as observed on the surface cortex. On the other hand, there are few papers concerning the unitary discharge in more superficial layers, while pathological events occur on the cortical surface.

In agreement with Ishijima (3), we have found in older animals some "negative" cells in the third layer. Unlike the "positive" pyramidal cells of layer V, these cells can stop their firing when a slow spike occurs on the ECG. This interruption appears in ictal and in interictal periods, in the elicited focus and in the mirror focus.

The most reliable explanation of the functional difference between "positive" and "negative" neurons of layers III and V is probably the existence of intracortical inhibitory circuits connecting them.

This functional pattern closely depends on a relative degree of maturation and does not appear before a definite stage. In animals younger than 8 days, neurons of layer III never display a functional pattern correlated with the ECG paroxystic events (any "positive" or "negative" cells), therefore without any synchrony with discharge pattern of neurons in the layer V in the ictal as well as in the interictal periods.

A possible explanation would be that in the immature mammalian cortex, the intracortical inhibitory pathways are not yet functional. On the other hand, it appears that the beginning of the second postnatal week in the rabbit is of critical importance for other cortical phenomena, for instance, the appearance of the early positive deflection of the evoked somesthetic potentials (ESP). In younger rabbits, the ESP shows only a late negative component (4,6,11) generated by synaptic connections in cortical layers more superficial than those responsible for the early positive deflection in older animals (layers V and VI).

In addition, we suggest that the disappearance of the late negative deflection with age would be partly due to the onset of inhibition exerted by deep pyramidal cells on more superficial elements, as, for example, neurons of layer III through newly formed intracortical circuits.

ACKNOWLEDGMENTS

The authors wish to thank Drs. N. Delhaye-Bouchaud and M. A. Thomson for useful comments during the experiments and preparation of the manuscript. They also thank Ms. Henri for histological assistance and Ms. Cado and Ms. Millot for typing the manuscript.

REFERENCES

1. Desroches, A. M., Chapelle, D., Karoui, H., and Laget, P. (1976): *C. R. Soc. Biol. (Paris)*, 170:54–59.
2. Enomoto, T. F., and Ajmone Marsan, C. (1959): *EEG Clin. Neurophysiol.*, 11:199–218.

3. Ishijima, B. (1972): *Epilepsia,* 13:561–581.
4. Laget, P., and Delhaye, N. (1962): *Actualités Neurophysiologiques,* edited by A. M. Monnier, pp. 259–284. Masson, Paris.
5. Majoie, B. (1966): *Tératogènes et Ontogenèse.* (*Personal communication.*)
6. Mallecourt, D., Ekholm, J., Verley, R., and Scherrer, D. (1974): *Rev. EEG Neurophysiol.,* 4:37–52.
7. Persson, H. E. (1973): *Acta Physiol. Scand.* [*Suppl.*], 394.
8. Prince, D. A. (1968): *Exp. Neurol.,* 21:307–321.
9. Prince, D. A., and Futamachi, K. (1968): *Brain Res.,* 11:681–684.
10. Purpura, D. P. (1969): In: *Basic Mechanisms of the Epilepsies,* edited by J. J. Jasper, A. A. Ward, and A. Pope, pp. 481–505. Little Brown, Boston.
11. Scherrer, J., and Oeconomos, D. (1959): *Etudes neonatales* (*France*), 3:199–214.
12. Sypert, G. W., Oakley, J., and Ward, A. A. (1970): *Exp. Neurol.,* 28:305–325.
13. Ulett, G., Dow, R. S., and Larsell, O. (1944): *J. Comp. Neurol.,* 40:1–10.
14. Wright, F. S., and Bradley, W. E. (1968): *Electroencephalogr. Clin. Neurophysiol.,* 25: 259–265.

OTHER ASPECTS OF BURST GENERATORS

Abnormal Neuronal Discharges, edited by N. Chalazonitis and M. Boisson. Raven Press, New York © 1978.

Bursting Activity in Relation to Neurosecretory Processes in Hypothalamic Neurosecretory Cells

J. D. Vincent, D. A. Poulain, and E. Arnauld

Laboratoire de Neurophysiologie et Physiopathologie, Université de Bordeaux II, Bordeaux, France

Mammalian hypothalamic supraoptic and paraventricular nuclei contain neurosecretory cells which produce the hormones vasopressin and oxytocin that are released from axon terminals found in the neurohypophysis. A most interesting feature to emerge from the electrophysiological studies of these magnocellular systems is the bursting pattern of electrical activity displayed by a number of their cells. The bursting activity, called by most authors phasic or low frequency burster activity, was first described in the rat (7,11,33), then in the monkey (16,17,32), and more recently in the ewe (15).

PARAMETERS OF THE BURSTING PATTERN

The characteristics of the bursting pattern are similar in those mammalian neuroendocrine neurons in which it has been recorded; this pattern consists of alternating periods of spike activity and silence that take place in a more or less regular way (Fig. 1A). A small number of spikes also occur during the silent periods. Because of the isolated spikes and because it is often difficult to accurately define the beginning and end of the bursts, the analysis of firing periodicity is best done by determining an expectation density function for each neuron (1). This function typically shows a sinusoidal curve (Fig. 1B) that corresponds to an overall firing rate (total number of spikes recorded/total recording time in seconds) that is rarely high (range, 2.5 to 15 spikes/sec). Other parameters of the bursting pattern can be estimated accurately only in those cells that exhibit clear silent periods. These parameters include burst duration (range, 4 to 100 sec), silence duration (range, 4 to 240 sec), intraburst firing rate (range, 3.5 to 17 spikes/sec), and number of bursts per minute (range, 0.2 to 14 bursts/min). An interspike interval histogram of bursting firing shows a skewed frequency distribution with a shift toward short intervals (mode between 20 and 100 msec) (1,35) (Fig. 1C).

A common feature of bursting neurons is that the duration of the individual bursts as well as the period of silence between the bursts can vary

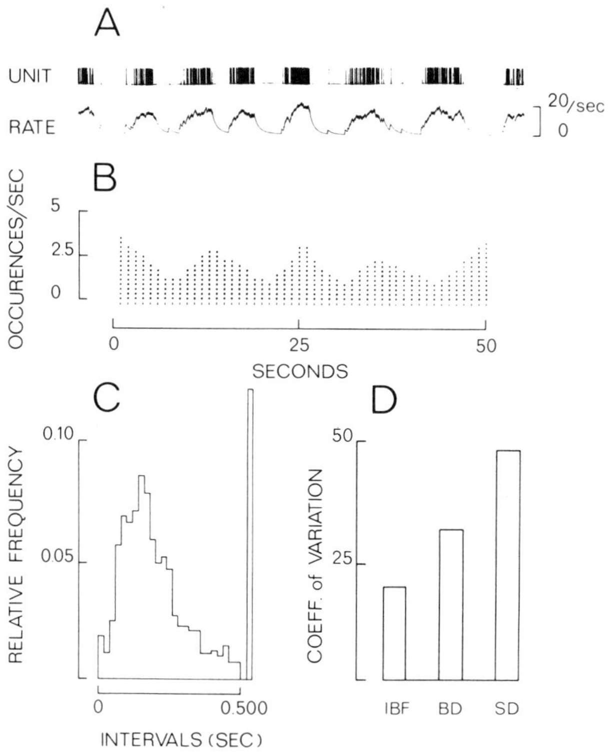

FIG. 1. Example of a bursting neurosecretory neuron recorded in the supraoptic nucleus of the conscious monkey (*Macaca mulatta*). **A:** Unit, polygraph recording of spike activity; Rate, analog output proportional to the firing rate (spikes/sec). **B:** Expectation density function correlogram showing a mean cycle period of 13 sec. **C:** Interspike interval histogram (bin width: 20 msec.) **D:** Coefficient of variation of the mean intraburst firing rate (IBF), burst duration (BD), and silent period duration (SD).

considerably within the same neuron. The coefficient of variation of these two parameters (SD/mean, expressed as percent) is high, ranging from 30 to 150%. The intraburst firing rate, on the other hand, remains much more constant, with a coefficient of variation of 10 to 20%. A high variability is also apparent between different bursting neurons under the same experimental conditions (1,35) (Fig. 2). This accounts for the absence of any obvious synchronization between bursting neurons.

RELATIONSHIP TO HORMONE RELEASE

The possible significance of bursting neuronal activity to particular neurohypophysial hormone release has until recently been poorly understood. Most studies have implicated this kind of firing with release of vasopressin; but since most of the stimuli utilized could induce the release of oxytocin

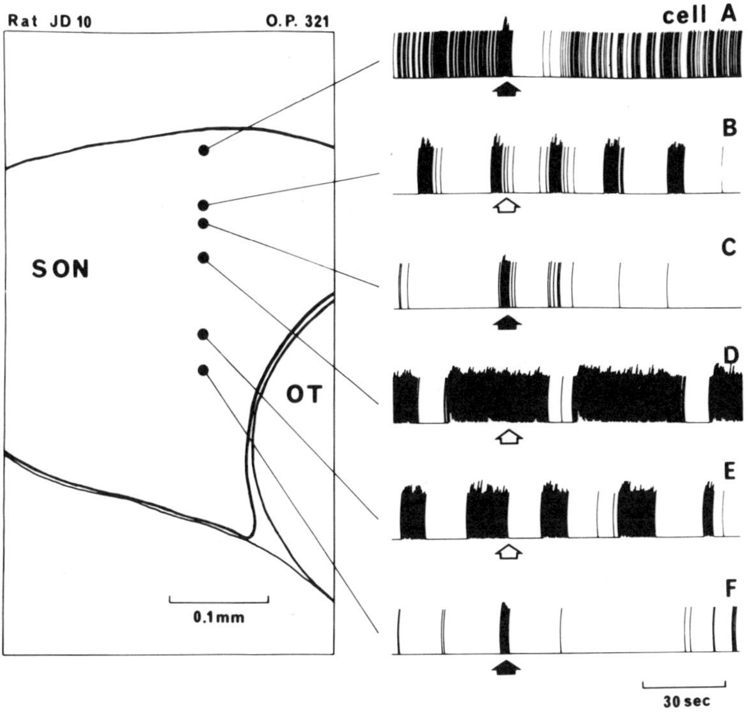

FIG. 2. Six neurosecretory neurons recorded in the supraoptic nucleus of the anesthetized lactating rat during suckling and after dehydration. Cells **A, C,** and **F** are oxytocin neurons identified by their brief high frequency discharge occurring (*black arrow*) 10 to 15 sec before reflex milk ejection. Cells **B, D,** and **E** do not show such high frequency discharge (*white arrow*), but present bursting patterns of electrical activity. All these cells have been recorded along the same electrode penetration. SON, supraoptic nucleus; OT, optic tract; OP, osmotic pressure in mOsm). (From ref. 35.)

(19), the results had to be interpreted with caution. In the conscious or anesthetized animal, only few neurosecretory cells exhibit the bursting pattern (10 to 15% in the monkey; 15 to 25% in the rat; 25% in the ewe).

It has been demonstrated that dehydration causes a progressive rise in plasma osmolality and vasopressin release concomitant with a progressive rise in the mean firing rate of neurons in the supraoptic nucleus. At the same time, the percentage of these cells firing in burst also increased (1) (Fig. 3). A similar phenomenon was shown to take place in the water-deprived rat (35,36). Hemorrhage, another potent stimulus of vasopressin release, was used to monitor changes in the firing pattern at the individual neuronal level. These experiments showed that the same neuron could switch from an irregular form of firing activity to the bursting pattern on application of the stimulus (28). The same results were obtained with another stimulus of vasopressin release, intraperitoneal injection of hypertonic saline (3).

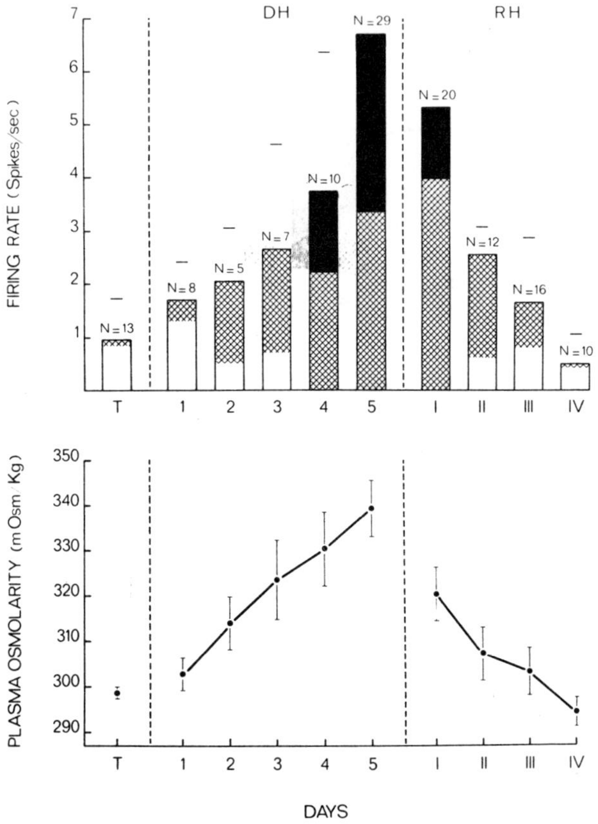

FIG. 3. Mean data observed during five successive days of water deprivation (*1, 2, 3, 4, 5*) and during recovery period thereof (*I, II, III, IV*) as compared with mean data in normally hydrated monkeys (*T*). **Upper:** Height of the blocks represents the mean firing rate of neurons recorded for each day of the experiments; horizontal bars represent the SD; SD for day 5 and 1 is 3.02 and 2.80/sec, respectively. The relative proportion of phasic cells is represented for each day by the cross-hatched part of the blocks. **Lower:** Mean plasma osmolarity ± SD. (From ref. 1.)

The use of the lactating rat as an experimental model provides a means to identify oxytocin and vasopressin neurons (23,24,34); this has made possible a clearer definition of the relationship of burst activity and vasopressin release. In the lactating rat at the time of milk ejection during suckling, oxytocin neurons show a brief high frequency discharge of action potentials; neither hemorrhage (28) nor intraperitoneal injections (3) of hypertonic saline were able to produce a bursting pattern of firing in these cells. Both stimuli, however, were effective in inducing such a pattern in the vasopressin neuron (Fig. 4). Furthermore, it has been demonstrated that after 6 hr of dehydration, 80% of the vasopressin neurons in the lactating rat display

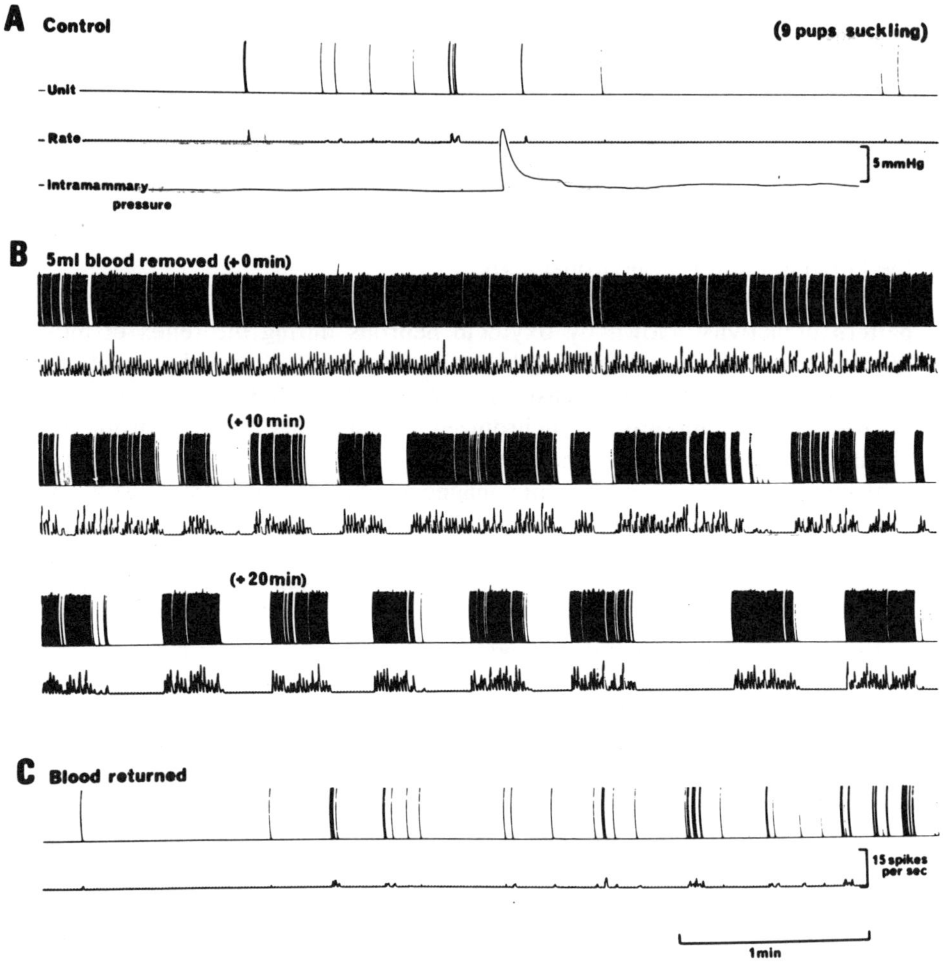

FIG. 4. Polygraph record of a vasopressin neuron recorded in the lactating rat. **A:** Control period during suckling. Note the slow, irregular pattern of firing and the absence of high frequency discharge before milk ejection (intramammary pressure recording). **B:** Three sections of the polygraph record starting at 0, 10, and 20 min after the removal of 5 ml of blood. Note the increase of firing rate with first, a continuous pattern, then a bursting pattern of activity. **C:** After returning the blood removed, the cell went back to its initial slow irregular pattern of activity. (From ref. 28.)

bursting firing; this percentage can increase up to 100% if dehydration is prolonged to 24 hr (35).

From these series of experiments, the following conclusions can be drawn: (a) The bursting pattern of neuronal firing encountered in mammalian neurosecretory cells is essentially a characteristic of vasopressin neurons. (b) The bursting pattern represents a functional state of activity of these neurons; this means that the vasopressin neurons are not always firing in bursts but only when the intensity of the stimulus is sufficient to lead to hormone release.

POSSIBLE MECHANISMS OF ACTION

At present, very little is known about the mechanisms responsible for the electrical activity displayed by neurosecretory cells and in particular the bursting pattern of neuronal activity.

That afferent inputs play an important role is evident from the temporal pattern of activity shown by oxytocin neurons during the reflex of milk ejection and from the contrasting behavior of the two neuronal populations under stimulation for the release of vasopressin (3,28,34). Also, bursting neurons can be temporarily synchronized by a stimulus, such as carotid occlusion (9). Nevertheless, the pathways in the central nervous system projecting onto the hypothalamic magnocellular neurons are mostly unknown, even though a large number of synapses, with either aminergic or cholinergic characteristics, have been seen to impinge on these cells (13,22, 25,39). Microiontophoretic application of monoamines (dopamine, serotonin, and especially norepinephrine) have a depressant effect on the activity of these neurons. On the other hand, acetylcholine can excite or inhibit the activity of the cell to which it is applied (2,7). These contrasting results may be reflecting inherent differences in the vasopressin and oxytocin cells themselves. Moreover, it cannot be excluded that the effects of a neurotransmitter depend on the ongoing activity of the cell on which it is acting. In bursting cells, for example, acetylcholine can trigger a burst if applied during a silent period or prolong the activity if applied during a burst. On the other hand, it inhibits neuronal activity in irregular firing neurons (B. Bioulac and J. D. Vincent, *unpublished observations*).

Antidromic stimulation is a commonly applied technique to identify the neurosecretory cells (1,2,6,8,11,15,16,23,26,30,32,37,38). One interesting observation arising from these studies is that after an antidromic shock, the cells display an inhibition of electrical activity lasting from 50 to 150 msec (6,16,20,21,26,38). Although the morphological evidence for collateral branching on the magnocellular neurons is scarce (4), the electrophysiological data have suggested that recurrent inhibition does take place. Moreover, vasopressin applied microiontophoretically has an inhibitory effect on neurosecretory cell activity (27; see also refs. 8 and 12). Recurrent inhibition could explain the long period of inhibition (lasting a few seconds) apparent in oxytocin neurons following high frequency discharge (23,34). It could also account for the similar inhibition displayed by neurosecretory cells in general after a volley of antidromic stimulation (5,31) or after activation by intracarotid injection of hypertonic saline (17,18,32). Its possible effects on bursting activity are still unclear. Antidromic stimulation can abolish firing when it acts during a burst, yet it can produce a burst when applied during the silent phase (10).

The bursting pattern and the conditions under which it evolves strongly suggest that the neurosecretory cells exhibiting this activity do so under the control of an endogenous oscillatory mechanism. In invertebrates, where bursting activities have been described, endogenous bursting pacemaker potentials have been shown to underly this kind of neuronal firing (14,29). Since all of our electrophysiological knowledge on bursting neurosecretory cells derives from extracellular recordings, further experiments utilizing intracellular techniques must be performed until we know whether such a mechanism takes place in vasopressin neurons.

ACKNOWLEDGMENTS

This work was supported by grants from INSERM (CR 176056.6) and CNRS (ERA 493). The authors are grateful to Dr. D. Theodosis for her assistance during the preparation of the manuscript.

REFERENCES

1. Arnauld, E., Dufy, B., and Vincent, J. D. (1975): *Brain Res.,* 100:315–325.
2. Barker, J. L., Crayton, J. W., and Nicoll, R. A. (1971): *J. Physiol. (Lond.),* 218:19–32.
3. Brimble, M. J., and Dyball, R. E. J. (1977): *J. Physiol. (Lond.),* 271:253–272.
4. Christ, J. F. (1966): *The Pituitary Gland, Vol. 3.* University of California Press, Berkeley.
5. Cross, B. A., Dyball, R. E. J., Dyer, R. G., Jones, C. W., Lincoln, D. W., Morris, J. F., and Pickering, B. T. (1975): In: *Recent Progress in Hormone Research,* pp. 243–293. Academic Press, New York.
6. Dreifuss, J. J., and Kelly, J. L. (1972): *J. Physiol. (Lond.),* 220:87–103.
7. Dreifuss, J. J., and Kelly, J. L. (1972): *J. Physiol. (Lond.),* 220:105–118.
8. Dreifuss, J. J., Nordmann, J. J., and Vincent, J. D. (1973): *J. Physiol. (Lond.),* 237:25–27.
9. Dreifuss, J. J., Harris, M. C., and Triboilet, E. (1976): *J. Physiol. (Lond.),* 257:337–354.
10. Dreifuss, J. J., Tribollet, E., Baertschi, A. J., and Lincoln, D. W. (1976): *Neurosci. Lett.,* 3:281–286.
11. Dyball, R. E. J. (1971): *J. Physiol. (Lond.),* 214:245–256.
12. Dyball, R. E. J. (1974): *J. Endocrinol.,* 60:135–143.
13. Fuxe, K., and Hökfelt, T. (1967): *Neurosecretion.* Springer-Verlag, Berlin.
14. Gainer, H. (1972): *Brain Res.,* 39:403–418.
15. Haskins, J. T., Jennings, J. P., and Rogers, J. M. (1975): *Physiologist,* 18:240.
16. Hayward, J. N., and Jennings, D. P. (1973): *J. Physiol. (Lond.),* 232:515–543.
17. Hayward, J. N., and Jennings, D. P. (1973): *J. Physiol. (Lond.),* 232:545–572.
18. Hayward, J. N., and Vincent, J. D. (1970): *J. Physiol. (Lond.),* 210:947–972.
19. Jones, C. W., and Pickering, B. T. (1969): *J. Physiol. (Lond.),* 203:449–458.
20. Kelly, J. S., and Dreifuss, J. J. (1970): *Brain Res.,* 22:406–409.
21. Koizumi, K., and Yamashita, H. (1972): *J. Physiol. (Lond.),* 221:683–705.
22. Leranth, C., Zaborsky, L., Marton, L., and Palkovits, J. (1975): *Exp. Brain Res.,* 22:509–523.
23. Lincoln, D. W., and Wakerley, J. B. (1974): *J. Physiol. (Lond.),* 242:533–554.
24. Lincoln, D. W., and Wakerley, J. B. (1975): *J. Physiol. (Lond.),* 250:443–461.
25. Morris, J. F. (1974): *J. Anat.,* 117:213.
26. Negoro, H., and Holland, H. C. (1972): *Brain Res.,* 42:385–402.
27. Nicoll, R. A., and Barker, J. L. (1971): *Brain Res.,* 35:501–511.

28. Poulain, D. A., Wakerley, J. B., and Dyball, R. E. J. (1977): *Proc. R. Soc. B.*, 196:367–384.
29. Strumwasser, F. (1973): *Physiologist,* 16:9–42.
30. Sundsten, J. W., Novin, D., and Cross, B. A. (1970): *Exp. Neurol.*, 26:316–329.
31. Vincent, J. D., and Arnauld, E. (1975): *Prog. Brain Res.*, 42:57–66.
32. Vincent, J. D., Arnauld, E., and Nicolescu-Catargi, A. (1972): *Brain Res.*, 45:278–281.
33. Wakerley, J. B., and Lincoln, D. W. (1971): *Brain Res.*, 25:192–194.
34. Wakerley, J. B., and Lincoln, D. W. (1973): *J. Endocrinol.*, 57:477–493.
35. Wakerley, J. B., Poulain, D. A., and Brown, D. (1978): *Brain Res.* (*In press.*)
36. Walters, J. K., and Hatton, G. I. (1974): *Physiol. Behav.*, 13:661–667.
37. Yagi, K., Azuma, T., and Matsuda, K. (1966): *Science,* 154:778–779.
38. Yamashita, H., Koizumi, K., and Brooks, C. M. (1970): *Brain Res.*, 20:462–466.
39. Zaborsky, L., Leranth, C., Makara, G. B., and Palkovits, M. (1975): *Exp. Brain Res.*, 22:525–540.

Abnormal Neuronal Discharges, edited by N. Chalazonitis and M. Boisson. Raven Press, New York © 1978.

Bursting Firing in a Mammalian Endocrine Neuron

J. J. Dreifuss, B. H. Gähwiler, and P. Sandoz

Department of Physiology, University of Geneva Medical School, Geneva; and Biological and Medical Research Division, Sandoz Ltd., Basel, Switzerland

Periodic bursting neurons are found in the nervous system of several invertebrates, e.g., in the sea hare (2), in land snails (3), and in the crayfish (13). In these species, periodic discharging of bursts of firing appears to be a property of cells which synthesize granule-enclosed neurohormones for secretion. Bursting activity in mammalian neuroendocrine cells has received less attention and is briefly reviewed (see also ref. 18).

IN VIVO STUDIES

In recent years, several workers have described the firing characteristics of hypothalamic supraotic and paraventricular neurons (for review, see refs. 11 and 21), the cells which synthesize the two neurohypophysial hormones oxytocin and vasopressin (17). These cells can be readily identified by antidromic invasion following stimulation of their axon at the level of the neurohypophysis. Periodic bursting activity went unrecognized in the early unit studies of these endocrine neurons. In 1971, however, Wakerley and Lincoln (19) showed that a proportion of paraventricular neurons in the rat were periodic bursters. Periodic bursting neurons or "phasic neurons" as they were to be routinely called, exhibit intermittent bursts of neuronal activity alternating with periods of a few seconds during which no action potentials occur. This discharge pattern may be specific of magnocellular endocrine neurons and may have functional significance. It has been observed in all species studied, in anesthetized as well as in conscious animals (12,14). The proportion of phasic neurons relative to the total population of magnocellular endocrine neurons studied varies between 10 and 50% and depends on experimental conditions. Thus an increase in the proportion of phasically firing neurons is seen during progressive dehydration, a condition characterized by increased neurophypophysial hormone release (1,20). Recent work (5,16,20) suggests that phasic activity may be confined to the vasopressin-producing neurons. Oxytocin-producing neurons do not show this tendency to fire in bursts (15).

Little is known of the origin of this peculiar firing pattern. Circumstantial evidence suggests that, as in invertebrate neurons (10), this activity is an intrinsic property of the cells, one not imposed on them by an external pacemaker. The evidence is as follows. First, when several phasic neurons are recorded in an animal along one recording electrode tract, they display different periodicities (20). Second, recordings have been obtained in a few instances from two adjacent phasic neurons through a single recording micropipette, and the cells were usually found not to fire synchronously (5,7; but see Fig. 5 in ref. 4). Third, typical bursts of firing can be triggered in a neuron if action potentials are generated during the quiescent phase by a train of electrical stimuli applied to its axon (6). Effective triggering of a burst is dependent on the stimulus train being suprathreshold for antidromic invasion of the cell. Subthreshold currents are ineffective, even when they activate many of the adjacent cells.

IN VITRO STUDIES

Since attempts to record intracellularly from supraoptic and paraventricular neurons *in vivo* appeared unrewarding, we have recently attempted

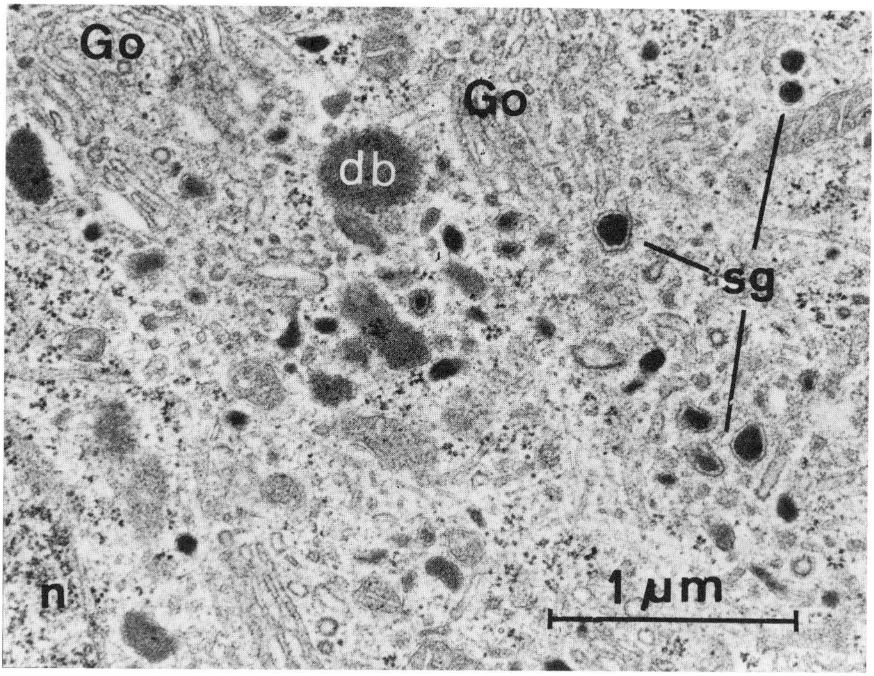

FIG. 1. Electron micrograph of part of the cell body of a neuron from the supraoptic area of a newborn rat; 31-day-old culture was fixed overnight at 4°C in 2.5% glutaraldehyde, postfixed for 2 hr in 2% OsO_4, embedded in Epon, and stained with uranyl acetate and lead citrate. Note the presence of membrane-bounded, dense-cored neurosecretory granules (sg) in association with Golgi cisternae (Go). Part of the nucleus of the cell is apparent (n); db, dense body.

to culture the hypothalamic supraoptic area from 1- to 7-day-old rats. The explants were grown on glass coverslips on a plasma clot, according to established techniques (8,9). The culture medium consisted of one part horse serum, one part Hanks' balanced salt solution, and two parts basal medium (Eagle) supplemented with glucose. After 2 to 11 weeks in culture, many large cell bodies, 20 to 30 μm in diameter, with a large nucleus and 1 to 2 prominent nucleoli were visible. Electron microscopic examination showed well-preserved neurons with secretory granules, 120 to 150 nm in diameter, both in perikarya, in association with Golgi cisternae (Fig. 1), and in axons. The cells were thus tentatively identified as magnocellular endocrine neurons. In electrophysiological experiments, many of the cells were found to be spontaneously active. Approximately 40% of the cells studied displayed a phasic pattern of firing (Fig. 2, upper). When the activity of two cells was recorded simultaneously with two extracellular microelectrodes

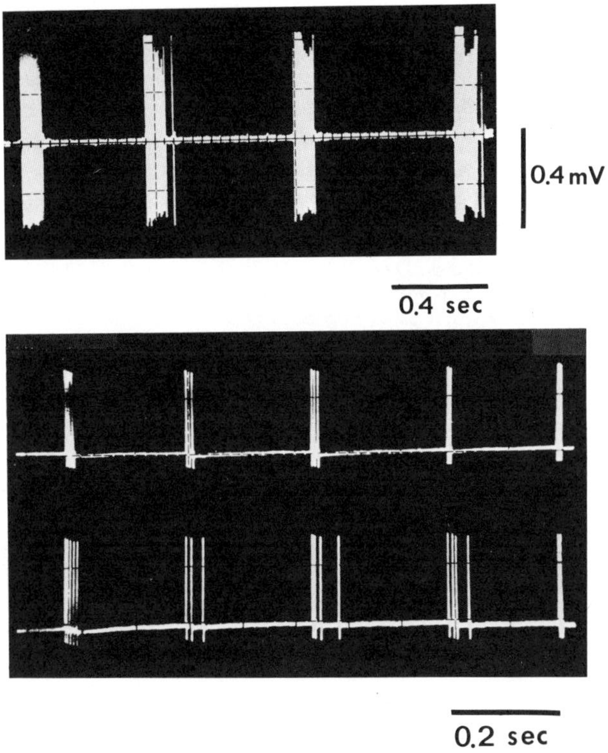

FIG. 2. Periodic bursting (phasic) pattern of neuronal firing in 27-day-old culture from the hypothalamic supraoptic area. **Upper:** Oscilloscope trace of extracellularly recorded action potentials from a neuron, with phases of neuronal activity alternating with periods of silence. *Lower:* Traces obtained with two independent single-barreled micropipettes; they show the activity of two neighboring neurons. (The deflections are the standard pulses delivered by the voltage discriminators.) Note that the onset of activity is synchronous in these two phasic cells.

located 50 to 500 μm apart, the neuron pair was often seen to discharge in phase (Fig. 2, lower). This suggests that entrainment of bursting activity occurs in cultured hypothalamic neurons. The origin of bursting activity of cultured hypothalamic neurons remains unknown. Further analysis will require intracellular recordings from phasic neurons. Such experiments are now in progress.

ACKNOWLEDGMENT

Part of this work was supported by grant 3.758.76 from the Swiss National Science Foundation.

REFERENCES

1. Arnauld, E., Dufy, B., and Vincent, J. D. (1975): *Brain Res.*, 100:315–325.
2. Arvanitaki, A., and Chalazonitis, N. (1964): *C. R. Soc. Biol.*, 158:1119–1122.
3. Barker, J. L., and Smith, T. G. (1978): *This volume.*
4. Cross, B. A., Dyball, R. E. J., Dyer, R. G., Jones, C. W., Lincoln, D. W., Morris, J. F., and Pickering, B. T. (1975): *Recent Prog. Horm. Res.*, 31:243–294.
5. Dreifuss, J. J., Harris, M. C., and Tribollet, E. (1976): *J. Physiol. (Lond.)*, 257:337–354.
6. Dreifuss, J. J. Tribollet, E., Baertschi, A. J., and Lincoln, D. W. (1976): *Neurosci. Lett.*, 3:281–286.
7. Dyball, R. E. J., and Pountney, P. S. (1973): *J. Endocrinol.*, 56:91–98.
8. Gähwiler, B. H. (1975): *Brain Res.*, 99:85–95.
9. Gähwiler, B. H., Mamoon, A. M., and Tobias, C. A. (1973): *Brain Res.*, 53:71–79.
10. Gainer, H. (1972): *Brain Res.*, 39:403–418.
11. Hayward, J. N. (1977): *Physiol. Rev.*, 57:574–658.
12. Hayward, J. N., and Jennings, D. P. (1973): *J. Physiol. (Lond.)*, 232:515–543.
13. Iwasaki, S., and Satow, Y. (1969): *J. Physiol. Soc. Jpn.*, 31:629–630.
14. Jennings, D. P., Haskins, J. T., and Rogers, J. M. (1978): *Brain Res. (In press.)*
15. Lincoln, D. W., and Wakerley, J. B. (1974): *J. Physiol. (Lond.)*, 242:533–554.
16. Poulain, D. A., Wakerley, J. B., and Dyball, R. E. J. (1977): *Proc. R. Soc. Lond. [Biol.]*, 196:367–384.
17. Vandesande, F., and Dierickx, K. (1975): *Cell Tissue Res.*, 164:153–162.
18. Vincent, J. D., Poulain, D. A., and Arnauld, E. (1978): *This volume.*
19. Wakerley, J. B., and Lincoln, D. W. (1971): *Brain Res.*, 25:192–194.
20. Wakerley, J. A., Poulain, D. A., and Brown, D. (1978): *Brain Res. (In press.)*
21. Yagi, K., and Iwasaki, S. (1977): *Int. Rev. Cytol.*, 48:141–186.

CELLULAR MECHANISMS OF ABNORMAL ELECTROGENESIS IN NEURONAL MODELS OF INVERTEBRATES

Abnormal Neuronal Discharges, edited by N. Chalazonitis and M. Boisson. Raven Press, New York © 1978.

Some Intrinsic and Synaptic Properties of Abnormal Oscillators

N. Chalazonitis

Institute of Neurophysiology and Psychophysiology, C.N.R.S., Marseilles, France

In the framework of this volume, neurocellular mechanisms dealing with the command of abnormal rhythmicities in motor networks are of primary importance. It is superfluous to consider some common examples of abnormal periodic motor activities (or seizures) in vertebrates, which are mainly illustrated in the reference section of the first chapter of this volume.

Neurocellular mechanisms are also elaborated on in many other chapters on neuronal pharmacology presently envisaged in the third chapter of this volume. Whereas a network of many neurons is necessary in order to elaborate on an abnormal rhythmic command of seizures, it has been considered worthwhile to explore single neuron abnormal activity.

In this chapter, some introductory remarks are focused on the intrinsic oscillatory properties of neuronal models elicited by oscillogenic molecules. Furthermore, the ability to oscillate (or oscillability) induced in neuronal membranes is examined as a very general phenomenon encompassing many different oscillogenic factors, particularly efficient on "predisposed neuromembranes."

Despite the importance of single cell exploration of altered (and/or pathological) motor command in vertebrates, many data of general significance have been gathered in experiments on elementary neuronal models. We briefly mention some experimental advantages of the molluscan ganglia (*Aplysia, Helix*) offering giant and identifiable neuronal somata. These perikarya are devoid of synaptic junctions on the soma. Consequently, they react with the oscillogenic factors with their pure somatic membrane primarily. However, secondary effects subsequent to transmitter release through oscillogenic molecules are extremely important, but they are not developed herein.

Moreover, under appropriate environmental conditions, it has been possible to analyze the properties of abnormally bursting cells during deafferented functioning, i.e., in the absence of spontaneous synaptic influences. It later became possible to explore the behavior of abnormally bursting cells when they are synaptically controlled by others; alternatively, one can explore the way in which the abnormally bursting cell can modulate the activity of other cells to which they are synaptically connected in elaborating seizures.

It is again emphasized that this chapter is mainly focused on mechanisms involved in the action of epileptogenic or more general convulsants in rhythmicities. Reciprocally, the investigation of mechanisms involved in the stabilization of rhythmic neurons already destabilized by abnormal factors is presented in the following chapter.

During the first era of "microelectrode exploration," the normal neuronal discharges could be classified in three types. This distinction of functioning has been possible for topographical recognition and subsequent identification of some neuronal models in Aplysia (3,13):

(a) *Stable neurons* are by definition resting normally, and once "activated" (i.e., depolarized by synaptic or other actions up to their firing level), they display only one adaptative discharge: a discharge with decremental spiking until extinction.
(b) *Regularly rhythmic neurons* or tonically rhythmic neurons normally display a spontaneous activity of constant frequency of spiking, even in the absence of any additional synaptic input.
(c) *Normal oscillators,* spontaneously active cells with burst activity, cells active on slow waves, or periodical discharge cells, and so on. (5,7,16,27).

Considering these three functional types of cells, it is easily predictable in which way pharmacological changes of their membrane potential (effected at any locus, i.e., somatic or axon hillock membrane or junctional membranes) will modify their normal activity. These "quantitative" influences in the neuronal output are generally considered either as concerning the direct excitability, when they are exerted on the somatic and axon hillock membranes, or as interfering with the mechanism of release of synaptic transmitters, if they interfere with the transmitter upon the appropriate postsynaptic receptors (i.e., antagonisms).

It is beyond the scope of this chapter to consider the neurochemical mechanisms of the epileptogenic molecules, which certainly contribute secondarily to the electrical properties of the membrane. Therefore, only direct actions on somatic nonjunctional membranes are considered here, particularly the ones eliciting "qualitative changes." By qualitative we mean any radical change, or "conversion" of the rhythmic functional type of the neuron, i.e., conversion of tonic activities into bursting ones, destabilization of stable neurons normally yielding "adaptative discharges" to bursting ones, and vice versa.

CONVERSION OF RHYTHMIC ACTIVITIES TO ABNORMAL BURST ACTIVITIES

By Epileptogenic Molecules

We do not repeat here in historic order all the references dealing with data obtained on single vertebrate neurons with different epileptogenic molecules. Many useful references of past studies are quoted throughout this volume. Some references to earlier publications are quoted from Chalazonitis and Arvanitaki (22) on convulsants and anticonvulsants on single neurons, as well as from pioneer studies on cortical neurons (36,38,40).

The conversion of tonically rhythmic neurons to abnormal bursting in the Aplysia ganglion allowed, for the first time, the possibility to be considered of a new molecular rearrangement in the somatic membrane by these molecules deeply modifying its bioelectrical properties. Such a modification, using either metrazol or bemegride, endows the membrane with intrinsic slow oscillatory properties and therefore "converts" the pattern of its activities from tonic to bursting endogenously.

Conversion of a Normal Regular Rhythmic Activity to a Paroxysmal Depolarization is Gradually Established

The application of epileptogenic molecules to soma of the rhythmic cells of *Helix* leads to the appearance of new patterns of activity, which are established during the following successive stages (26).

The first effect is a tendency toward grouping of the spikes according to the slow oscillations of the membrane potential; gradually, the amplitude of these waves increases the pattern of activity and only in a final stage does the "paroxysmal depolarization" or "the plateau of depolarization" appear on the "square-shaped slow waves." Illustrations of these conversions are given in Fig. 1. Details concerning latencies and concentrations of the drugs are given elsewhere.

Summarizing this conversion, we reiterate that tonic activity leads to slow waves and finally to square waves, or paroxysmal depolarization shifts (PDS). Elimination of the drug by washing reverses the establishment of the stages in the conversion.

In conclusion, the progressive adsorption of the pentylenetetrazol (PTZ) on the membrane macromolecules, which is very likely depicted kinetically by a saturation curve, establishes the different types of periodic discharges. The described successive stages, slow waves → "triangular waves" → PDS, progressively require the adsorption of increasing amounts of PTZ in the membrane. The paroxysmal depolarizations are but a final aspect resulting from a maximum saturation of the membrane for a given external concentration (Fig. 1).

Intrinsic Origin of the Periodic Activity on Slow Waves, Either Normal or Abnormal

The intrinsic or endogenous nature of such an oscillatory activity has been demonstrated by the following facts. First, it is possible to modify the frequencies of normal, as well as abnormal, slow waves by the application

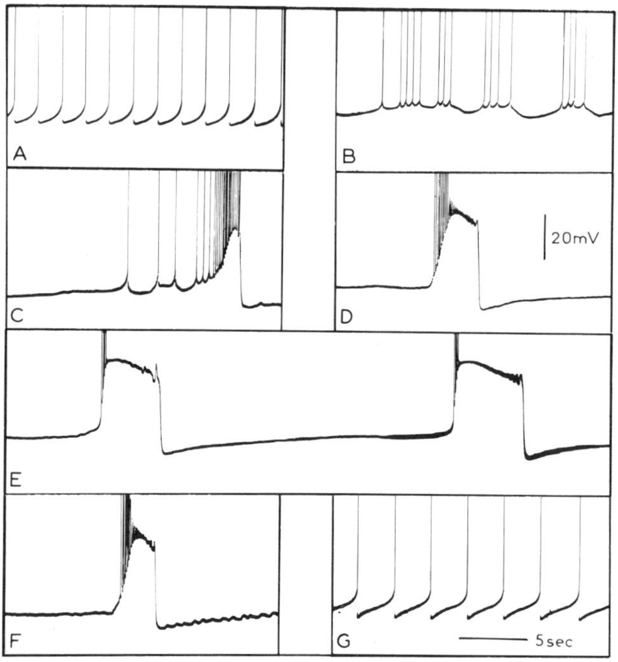

FIG. 1. PTZ-induced paroxysmal waves on Helix pacemakers. **A:** Pacemaker under normal conditions displays spontaneous activity of regular frequency. **B:** 5 min after PTZ application (0.65% in physiological solution), spike bursting on small slow waves. **C:** On the same neuron, 5 min action of a new PTZ solution (0.8%) induces "triangular" waves. **D:** Same neuron, under 0.8% PTZ during 13 min, appearance of the paroxysmal wave. **E:** Same neuron 27 min later, paroxysmal wave periodicity. **F:** Same neuron 7 min after repetitive washing with physiological saline. **G:** Complete recovery of the initial rhythm 60 min after repetitive washing.

of transmembrane direct current (DC) (15,26,41). Depolarizing DC accelerates the waving, whether the waving is normal or abnormal, and *vice versa* (Fig. 2).

Second, the abnormal waves or PDS may be recorded at high gain without significant superimposed postsynaptic potentials (except for some random synaptic noise) (25).

A third argument in favor of an endogenous origin of these repetitive PDS is that it is possible to record them in the soma after ligation of the neurite emerging from the neuronal soma. The isolated soma preparation was carried out on Aplysia cells after oscillogenic action effected with either cycloheximide (37) or PTZ (39).

Axonal Rhythmicities and Paroxysmal Depolarizations Obtained with Epileptogenic Molecules

PTZ application (0.5%) on crab unmyelinated nerves elicits in a few seconds synchronized rhythmical activities of the larger fibers ($\simeq$ 10 μm

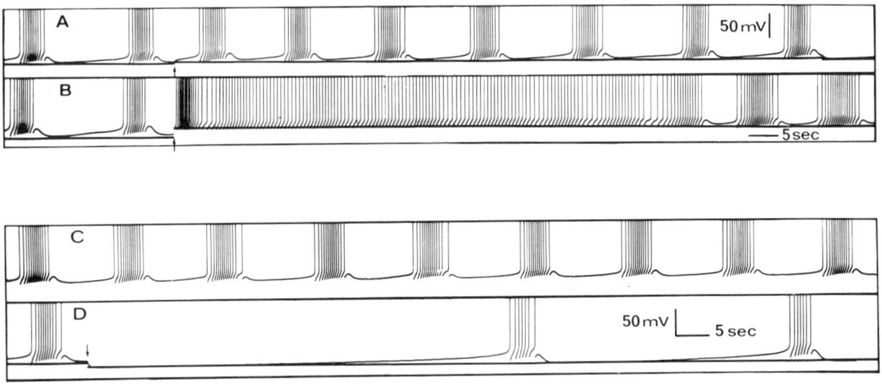

FIG. 2. DC modulation of a normal oscillator. **Upper:** Action of depolarizing DC. *A:* on an autoactive oscillator. *Arrow,* injection of 0.6 nA depolarizing DC. Acceleration of the waving, with subsequent accommodation. *B:* Injection of 3.9 nA, fusion of the waves, and finally accommodation. **Lower:** *C:* Spontaneous activity of the oscillator in normal conditions. *D: Arrow,* injection of hyperpolarizing DC (2.8 nA). Cessation of the activity and subsequent attenuation, i.e., decrease in waving frequency and in spiking on each wave.

diameter). After 60 sec treatment, a very long decaying plateau is conspicuous, prolonging the falling phase of the response. It may last approximately 200 to 250 msec (Fig. 3A) (31).

Cycloheximide, a powerful convulsant able to convert the stable giant neuron A (or R_2) (37) in an abnormal oscillator, also destabilizes the crab axonal membrane and finally elicits prolonged responses similar to those obtained by PTZ (31). The aforementioned conversions demonstrate the oscillogenic action of the epileptogenic molecules on membranes devoid of any synaptic junction.

Furthermore, the discrete destabilization of the axonal membrane may explain heterotopic axonal firing during a PDS displayed by the soma of a same neuron (see Chapter II, paragraph 2).

Functional Conversion of a Neuromembrane to an Abnormal Oscillator is a General Phenomenon

We define as an abnormal oscillator any rhythmic or even stable neuron that, once modified (pathologically or chemically, or even by physical means), may display periodic waves and discharges when brought on its firing level.

The conversion of tonically rhythmic activities to burst paroxysmal activities by epileptogenic molecules is a relatively easy conversion. Nevertheless, many other conversions to abnormal oscillators have been possible, not only on other functional types of neuromembranes but also with many different chemical or physical agents.

As early as 1956, Arvanitaki and Chalazonitis (2) demonstrated that iso-

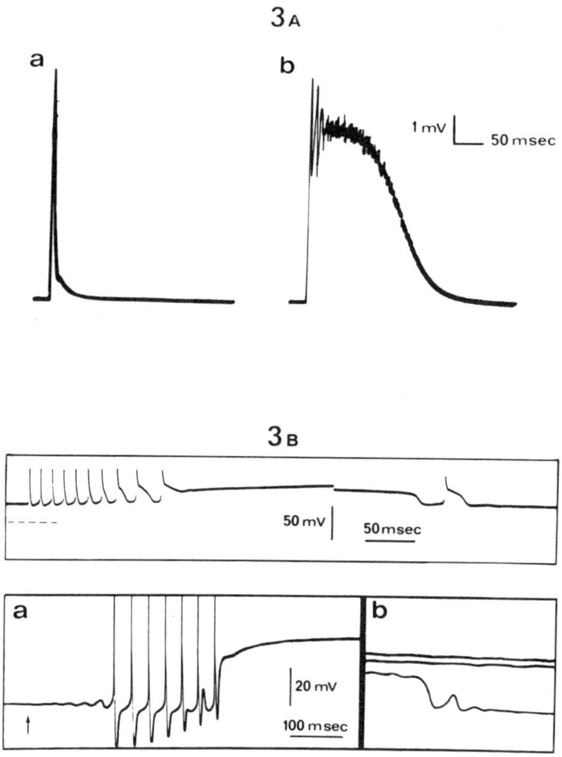

FIG. 3. Paroxysmal depolarization of axonal membranes. **3A:** Prolonged response of the unmyelinated nerve of Carcinus under PTZ (7 mg/ml). *a,* normal response; *b* under PTZ. **3B:** Paroxysmal waves on a neutral red-stained giant axon of Sepia. *Upper recording:* DC stimulation of rheobasic intensity and 50 msec duration (dashed line) on a neutral red-stained Sepia giant axon elicits a long-lasting paroxysmal wave of more than 3 sec duration. A spike discharge occurs during the depolarizing phase of the wave; the falling phase of the last spike gives rise to the wave "plateau." Continuation of the same recording (right side) 3 sec later: a "lengthened" spike on the spontaneous repolarizing phase of the wave. *Lower recordings: a:* Neutral red-stained axon. Next to the arrow, the "off" of a strong light stimulus of 1 sec duration and 0.5 10^{-2} cal g cm^{-2} sec^{-1} intensity elicits the paroxysmal wave. The firing goes on at "off" and the membrane potential abruptly reaches a depolarization plateau of more than 30 mV. *b:* Superimposed straight recordings show the continuous autorepolarization of the membrane at the sixth sec (upper trace) and the ninth sec (middle trace) following the maximum. The bottom trace shows the final abrupt repolarization at the fourteenth sec and a return toward the initial level. The total duration of the paroxysmal wave is about 15 sec.

lated Sepia giant axons incubated in seawater physiological saline containing neutral red (a phenazine vital dye) of about 10^{-5} M/liter displayed long-lasting paroxysmal waves, spontaneously in a few cases, but invariably after electrical or light stimulation (Fig. 3B).

On the other hand, Tasaki and Hagiwara (44), applying tetraethylammonium (TEA) salts to squid giant axon, demonstrated the same type of

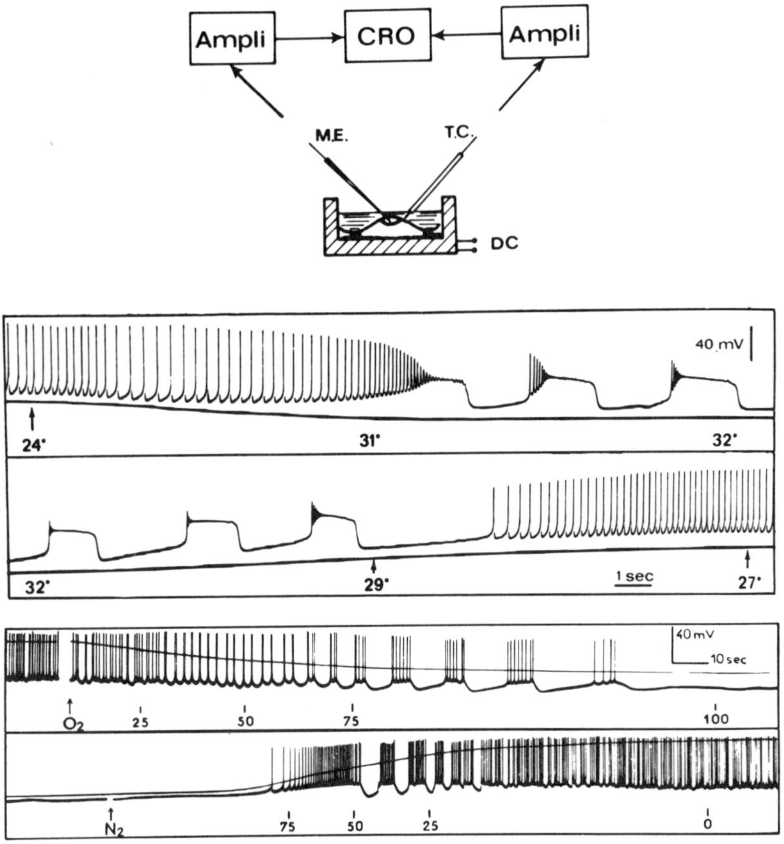

FIG. 4. Paroxysmal waves of a predisposed Aplysia pacemaker. **Upper schema:** Simultaneous recording of the electrical activity (by intracellular microelectrode, M.E.) and the temperature close to the cell, by miniaturized thermocouple (T.C.). Temperature transients effected by thermode (DC). **Top recordings:** *Upper recording,* the pacemaker neuron (Gen, or L_{11}, cell of Aplysia) displays paroxysmal waves in hyperthermic transient (temperature higher than 31°C). Initial activity is restored when the cell recovers on normothermia; *lower recordings,* the thin trace indicates temperature transients in centigrades. **Bottom recordings:** Gen or L_{11} cell of Aplysia, the upper thin trace is the cinespectrometric line of the intraneuronal hemoglobin instantaneous saturation, enabling the computation to intracellular pO_2 values. The pacemaker, initially in air, is submitted alternatively to oxygen or nitrogen atmosphere. Normal intracellular pO_2 is 6 mm Hg and corresponds to a 20% saturation of the hemoglobin. With hyperoxic intracellular conditions (pO_2 higher than 40 mm Hg, corresponding to an intracellular Hb saturation higher than 75%) paroxystic activity on slow waves is established. The lower part of the middle recording denotes the reversibility of the effect (disappearance of waving activity with return to normoxia). (Further details in ref. 17).

paroxysmal responsiveness. The fact that not only axons but also stable membrane potential neurons (which normally fire when synaptically activated, as some vertebrate cortical neurons) may be converted to oscillators by penicillin is well known in pioneer work (32,36).

With not only epileptogenic molecules but also many other agents on many different membranes, paroxysmal oscillatory activities have been demonstrated (11).

The first intracellular recording of a "paroxysmal wave in a single nerve cell of *Aplysia* treated with strychnin" was reported by Arvanitaki et al. (8; see also ref. 32).

Real paroxysmal waves with depolarization plateaus were demonstrated on the crayfish stretch receptor under strychnine and stretch action (46). In this neuron, it became conspicuous for the first time that the axon of this nerve cell is fully functional when the soma is inactivated during its paroxysmal depolarization plateau.

Not only epileptogenic (and generally convulsant) agents but even abnormal environmental conditions, such as hyperoxia, may convert a regular rhythmic to a paroxysmal activity (6,25). Besides hyperoxia, hyperthermia may also induce paroxysmal waves on the same functional type of cell (Fig. 4) (1,14). The interest of hyperthermic oscillogenic action at the neuronal level may be related to the well-known febrile convulsions of children (45).

Neurons Predisposed to Conversion in Paroxysmal Wave Activity

It has previously been reported that although the functional conversion of a given neuron to an abnormal oscillator is but a general phenomenon, the tonically firing neurons are the easiest to be converted in paroxysmal activity.

The predisposition of a given neuron to an abnormal oscillator is not only a peculiar sensitivity toward a specific oscillogenic agent; it is also the establishment of a new bioelectrical behavior toward many unspecific agents and particularly environmental factors (18).

An identifiable neuron is the Aplysia visceral ganglion called "Gen" (located near the emergence of the genital nerve) or L_{11}, displaying the conversion of its tonic rhythmic activity to an abnormal oscillating activity by many different factors: hyperthermia, hyperoxia, epileptogenic molecules, TEA, anesthetic vapors (ether, chloroform) (Fig. 4). Recently, Gola (34) found that heavy cations Ba^{2+}, Co^{2+}, and even tetrodotoxin may induce its paroxysmal activity. This activity corresponds to the appearance of rhythmic PDS or "square waves" similar to the ones recorded in cortical neurons under penicillin action. The predisposition of this particular neuron is demonstrated by the fact that other neurons (i.e., the giant neuron A, or R_2, of the same Aplysia ganglion), if treated under the same conditions as the Gen (or L_{11}) neuron, do not elicit any paroxysmal activity but only some slow changes of its membrane potential.

Biophysical Aspects of the Abnormal Oscillator Neuromembrane

It was demonstrated (5) that the electrical resistance (Rm) of a normal oscillator, the spontaneously bursting neuron Br (or R_{15}) of Aplysia, is not constant but periodically variable. The value of Rm reaches minima and maxima at well-defined points of the cyclic changes of the membrane potential (MP). Similarly, an abnormal oscillator displays even more conspicuous changes in Rm (6). Irrespective of the agent eliciting the paroxysmal depolarization, metrazol, bemegride, hyperthermia, and/or hyperoxia, the value of Rm is the lowest at two points: at the beginning and at the end of the plateau. Therefore, the Rm continuously increases during the plateau and during the interplateau interval. In summary, Rm, as a function of time and during repetitive PDS, displays two negative slopes corresponding to the beginning and end of the depolarization plateau.

These time-dependent negative slopes of the membrane Rm/f(t) have been also observed in experiments of clamped ramp potential. Under these conditions, the direct recording of the characteristic I/V, in clamped ramp (i.e., linearly rising potential as a function of time) also shows a negative change in the slope of the Rm. These ramp-clamped potential data gave (see Section 4, *this volume*) insight into ionic periodic permeability mechanisms of the abnormal oscillator (29,35).

In our opinion, the most important conclusion after ramp clamping data is the production of oscillatory transmembrane current corresponding to the overall changes in conductance during the ramp potential clamping.

Another interesting property of normal and/or abnormal oscillators is their adaptative behavior during long-lasting activation by transmembrane DC (7,15). The adaptation consists of a continuous decrease in the initial acceleration of the burst during depolarizing DC injection. Conversely, the abolition of spontaneous bursts, which occurs during hyperpolarizing DC injection, is temporary. After a given latency, which is a function of the DC intensity, the spontaneous bursts reappear, despite the continuous passage of transmembrane DC (Fig. 2).

Suggestions on Molecular Mechanisms Involved in Abnormal Bursting Activity

In previous studies, it was concluded that the membrane of a normal oscillator is endowed with a peculiar molecular framework, which is genetically determined. This conclusion arose because only a few neurons in the Aplysia ganglion display normal bursting activity; they are always the same neurons, easily identifiable for their location in the ganglion, identical size, pigmentation, and cytoplasmic ultrastructure (3,13,33). On the other hand, it was realized how easily a rhythmic neuron may be converted to an abnormal oscillator (26). Conversely, it has been possible to stabilize normal oscillators by catecholamines (10,28). Each of these conversions is reversible if the chemical agent is eliminated. This reversibility demonstrates that very small changes in the weak bond arrangement of the membrane macromolecules may be sufficient to confer oscillatory properties or vice versa, to stabilize a normal oscillator on a membrane.

It is also obvious from innumerable experiments that Ca^{2+} binding is always involved in many cases of oscillogenic conversion (9,12,22,42,47).

From a general point of view, whatever the period of a given oscillatory activity, fast or slow, the ability of a membrane to oscillate requires a peculiar molecular rearrangement.

We briefly report here that birefringence changes simultaneous to the duration of the nerve crab prolonged potentials have been recorded under either PTZ or TEA. They demonstrate membrane molecular rearrangements concomitant to the prolonged responses, which very likely imitate the neuronal membrane responses under PTZ or convulsant action. Therefore, the new membrane dynamics is displayed by rearrangements lasting as long as the slow oscillatory responses (21a).

The peculiar ionic slow conductances are but a new aspect of the oscillatory molecular framework. The already advanced mechanisms constitute the scope of the final section of this book. Such a new molecular dynamic structure concomitant to a PDS is necessary but not sufficient to elicit the repetition of the oscillatory activity.

Hypothesis of a Long-Lasting Endogenous Current Generator of the Repetitive Waving

The ability of a membrane to oscillate spontaneously for a long time with a constant waving frequency might require the continuous action of an endogenous transmembrane depolarizing current, bringing the oscillable membrane to its firing level. Such a current might be, as in many other circumstances (20), spontaneous and transcellular (see schema 19 in ref. 20). A flow of some ionic charge may enter an area of the somatic (inward direction) membrane and may be of outward direction through another area. The establishment of such a generator current may result from an imbalance of the two fundamental ionic processes participating in the equilibration of the membrane potential, the steady permeabilities and the speed of the ionic pumps (20,21). An oscillogenic chemical treatment could create and/or accentuate such an imbalance, the consequence of which could be the predominance of the pumps somewhere with a continuous exit of an ion, and the predominance of the passive entrance of the same ion through another patch of the membrane creating a continuous transcellular current. Another possibility would be a total transmembrane movement of a given ion in two periods: first, a slow outward movement depolarizing the membrane (for instance, during some minutes), followed by a reverse ionic movement of a comparable sluggishness, repolarizing the membrane. Under such conditions, the current would be only transmembrane and transitory.

Whatever the mechanism of the endogenous generator current formation (in unrestrained membranes, i.e., under normal conditions without clamping of the membrane potential), its simple occurrence cannot explain PDS rhythmicities of constant frequency (i.e., which do not display adaptation). Extracellular or intracellular application of depolarizing DC is always accompanied by PDS frequency adaptation (7). Therefore, a third requirement may be envisaged in order to maintain the constancy of repetitive PDS. It may be various environmental neurochemical influences, such as neuroamines continuously released during some abnormal "overflow" process, from neighboring neurons and continuously reaching the PDS discharging neuromembrane. Such a continuous overflow denotes a release condition, which is totally different from the well-known instantaneous and highly localized release during a synaptic transmission.

In conclusion, the functioning of an abnormal oscillator requires a peculiar rearrangement of its membrane molecules; the maintenance of oscillatory activity, spontaneously and for a long time, also requires the presence of a generator current, either transcellular or transmembrane, probably under some neurochemical control of exogenous or endogenous origin.

SOME SYNAPTIC PROPERTIES OF ABNORMAL OSCILLATORS

A number of synaptic properties already described in normal oscillators (16,19) are apparently valid in the case of abnormal oscillators. Nevertheless, it is obvious that many differences exist between normal and abnormal oscillators modified by an equivalent synaptic input.

Normal Synaptic Actions on Abnormal Oscillators

The conversion of a given rhythmic neuron to an abnormal oscillator, by epileptogenic or more generally oscillogenic molecules, permits in many cases the integrity of synaptic inputs on this neuron. The total disappearance of the postsynaptic potential has been observed (32) in only a few cases.

Therefore, despite the presence of an oscillogenic molecule, the intactness (or at least some preservation) of the synaptic inputs impinging on an abnormal oscillator will exert predictable modifications on its own oscillatory activity.

It will be recalled that an abnormal oscillator may be either spontaneously oscillating or even resting, according to the environmental conditions. Predictable action of synaptic control is considered from what is known about the normal oscillators (16,19).

Excitatory Synaptic Action

Experimental work on abnormal oscillators in elementary networks demonstrated that excitatory synaptic action may establish the repetitive oscillatory activity if the abnormal oscillator is initially at rest (Fig. 5).

It may also enhance the frequency of the oscillatory activity if the cell is initially spontaneously active. As a matter of fact, whatever the importance of a synaptic excitation input, only acceleration of the waving has been formed but never the so-called "fusion" of the waves to a continuous activity.

Inhibitory Synaptic Influences

On a given repetitively active abnormal oscillator, the summation of inhibitory postsynaptic potentials is equivalent to a hyperpolarizing DC and may therefore abolish the spontaneous activity of the oscillator. A further account of the inhibitory action is given in the following chapter.

Environmental Factors

Analogously, it is obvious that any other environmental factor, hyperpolarizing or depolarizing, will exert predictable changes in the frequency of

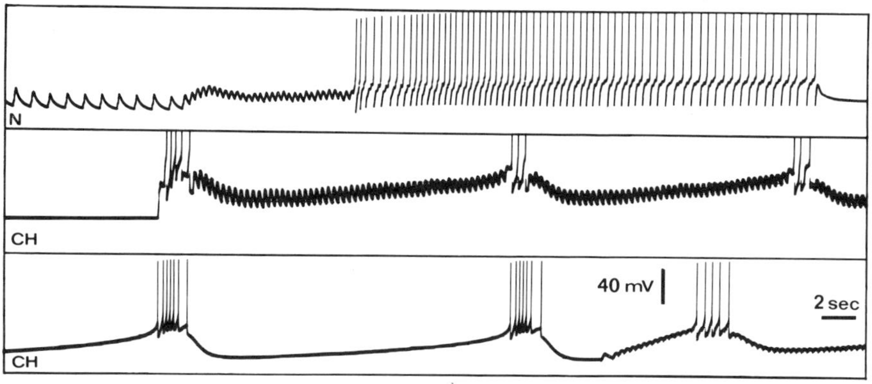

FIG. 5. Excitatory synaptic elicitation of abnormal slow waves. **N:** Intracellular recording of excitatory postsynaptic potentials (EPSPs) in the giant A (or R_2) silent neuron of Aplysia under normal conditions. The threshold summation of EPSPs finally elicits a continuous discharge of the adaptative type. **CH (middle):** The neuron treated with cycloheximide (3 μM/ml) remains silent. The summation of EPSPs then elicits slow waves. **CH (lower):** Same neuron treated with 5 μM/ml cycloheximide. It spontaneously displays slow waves. Finally, excitatory synaptic input accelerates the waving (decrease in the waving period).

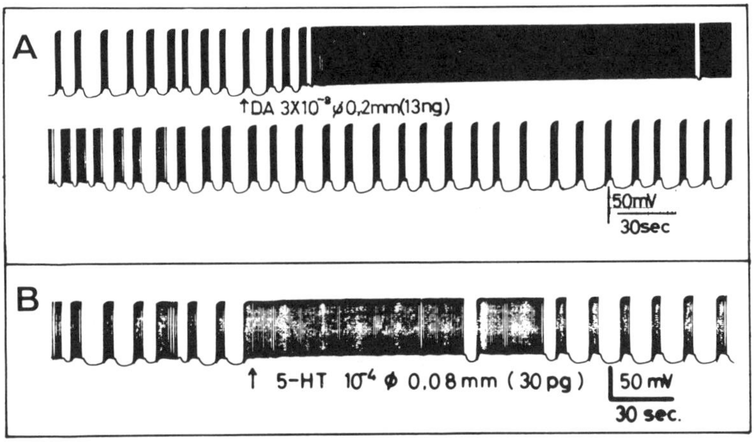

FIG. 6. Massive discharge of a normal oscillator by neuroamines (from ref. 43). **A:** Activity on slow waves under normal conditions of an oscillator of the snail *Achatina fulica. Arrow,* direct application on the oscillator of a microdroplet (200 μm diameter) of dopamine (3×10^{-3} M) corresponding to a quantity of 13 ng elicits depolarization and massive discharge (i.e., fusion of the waves for approximately 3 min). **B:** *Arrow,* direct application on a normal oscillator of a microdroplet (80 μm diameter) of 5-hydroxytryptamine (10^{-4} M) corresponding to a quantity of 30 pg elicits depolarization and fusion of the waves for 2.5 min. (Courtesy of Dr. H. Takeuchi).

the oscillator, as well as an equivalent DC of appropriate direction (16,19). If the environmental factors correspond to changes in pO_2 and pCO_2 in the physiological range, only modulations of waving frequency are possible. However, in some cases of molluscan normal oscillators, treatment with dopamine or 5-hydroxytryptamine elicited fusion of the waving (Fig. 6).

Efferent Discharges of the Abnormal Oscillators and the Heterotopic Axonal Output

When an abnormal oscillator (e.g., PTZ-treated soma) is spontaneously autoactive on slow waves, the sustained somatic discharge is totally propagated through the axon, without any decrement. In other words, each one of the somatic spikes may propagate along the axon (Fig. 7).

In all other cases, when the abnormal activity of the soma corresponds to a paroxysmal depolarization plateau, it is obvious that during that plateau the somatic membrane is inactivated by "excessive and long-lasting depolarization."

What, then, is the behavior of the axon emergency from the inactivated soma? To elucidate this point, in collaboration with Takeuchi, bioelectrical events were recorded simultaneously by intracellular microelectrodes inside the neuron and by extracellular electrodes on its axon, emerging in the efferent nerve conveying the neuronal information to the effectors.

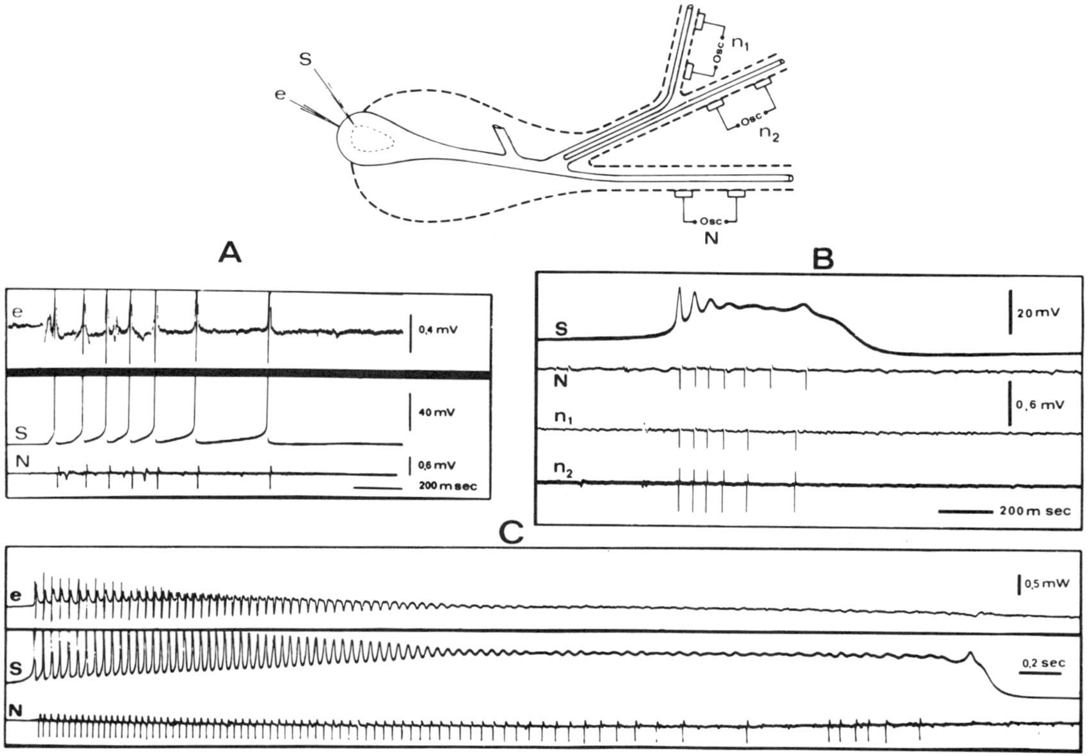

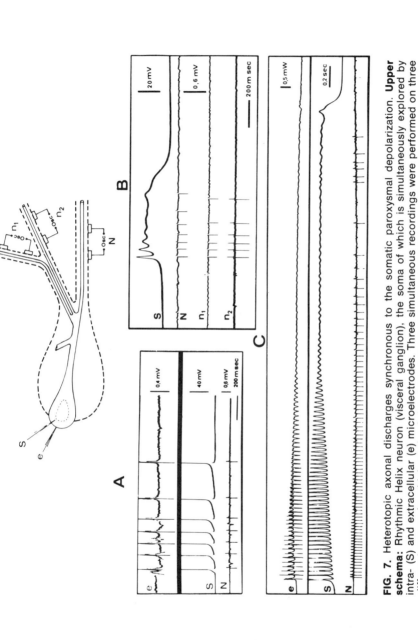

FIG. 7. Heterotopic axonal discharges synchronous to the somatic paroxysmal depolarization. **Upper schema:** Rhythmic Helix neuron (visceral ganglion), the soma of which is simultaneously explored by intra- (S) and extracellular (e) microelectrodes. Three simultaneous recordings were performed on three different nerves in which each of the three axonal branches of the neuron are engaged (n_1, n_2, and N). **A:** e, extracellular somatic recording; S, intrasomatic recording; N, axonal recording of slow waves elicited under weak PTZ actions. Each somatic spike propagates through the axon normally. **B:** Under further PTZ action, the soma (S) displays paroxysmal depolarization. During (PD) inactivation of the soma, each of the axonal branches displays a discharge. The axonal discharges are not always identical. **C:** Elongation of PDS after further action of PTZ; e and S as in **A.** During the somatic plateau, its main axon continues firing. N, extracellular recording on the main efferent nerve of the soma.

As shown in Fig. 7A, under normal conditions (before any epileptogenic treatment), the neuron is tonically autoactive at a constant frequency and each one of the somatic spikes emerges after a constant latency along the efferent nerve and is recorded there as a rapid axonal spike.

During the plateau depolarization of the soma, a burst of spikes is recorded propagating along its emerging axon (Fig. 7B, C). Therefore, during the somatic inactivation (plateau), a burst generator mechanism exists in the axon hillock synchronous to the plateau. The synchronous coexistence of the somatic PDS and the axonal discharge suggests that both activities might have the same cause. It is likely that the axonal discharge is elicited by some generator outward current through the axon hillock. Such a current would activate the axonal membrane, even if it is not yet altered by the oscillogenic molecules eliciting a discharge. On the other hand, the same outward current crossing the somatic membrane would establish its plateau depolarization (22–24).

It will be restated that the differential behavior between the somatic and axon hillock membranes of a same neuron may be imitated by injection of excessive depolarizing current through the neuronal membrane, even under normal conditions (i.e., without any oscillogenic treatment) (30).

It is obvious that during each "paroxysmal depolarization plateau," despite the somatic membrane inactivation, the generation of a burst of spikes by the axon hillock, which propagates through the emerging axon to the effector net up to the muscles, should be the cause of massive periodic discharge development in the networks producing seizures.

Persistence and/or Amplification of the Reciprocal (Inhibitory-Excitatory) Command

Activities of one pair of neurons simultaneously displaying contrasting signs of synaptic activation have been previously defined as "reciprocal" or "antergic" activities. Many examples of reciprocal command (excitatory and inhibitory) exerted synchronously on two postsynaptic neurons have been published (4,15). Reciprocal activation exerted by a given synaptic input on rhythmic neurons has been extensively described, particularly reciprocal behavior between a normal oscillator and a tonically rhythmic cell (22).

In the case of given synaptic input, dependent on two rhythmic cells already converted to abnormal oscillators, their reciprocal behavior will persist. Such a behavior is but a consequence of the above-signaled persistence of the synaptic inputs after appropriate treatment with convulsants. The reciprocally acting synaptic input will continue to exert excitatory effects on the first neuron and inhibitory ones on the second. The only conspicuous difference in their reciprocal output will be predictable from the fact that they are converted into abnormal oscillators (Fig. 8).

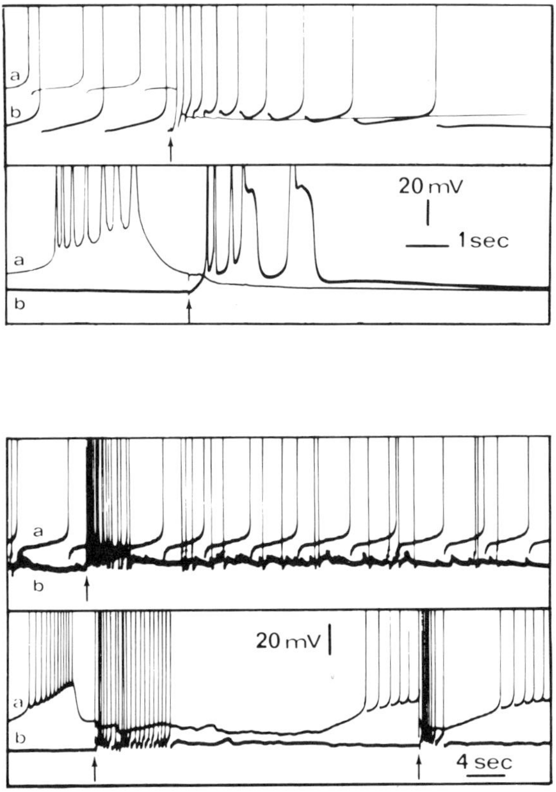

FIG. 8. Reciprocal postsynaptic command on a couple of abnormal oscillators. **Upper frame, upper recording:** *a* and *b:* Helix rhythmic neurons under normal conditions. *Arrow,* a synchronous synaptic afference to both neurons by nerve stimulation, which elicits, on the *a* neuron, a long-lasting inhibitory potential and on the *b* neuron, EPSPs giving rise to a spike discharge. **Lower recording,** *a, b:* Same neurons after strychnine action. Both pacemakers are converted to abnormal oscillators. *Arrow,* the synaptic afference (nerve stimulation) always elicits a long-lasting inhibitory potential on the upper cell and paroxysmal waves following an EPSP on the lower cell. **Lower frame, upper recording:** Reciprocal command on two autorhythmic neurons of Helix. *Arrow:* under normal conditions, a synaptic synchronous command on both, slightly inhibits the *a* neuron, but in contrast elicits a discharge on the *b* neuron. **Lower recording,** *a, b:* Same neurons under PTZ action. *a* is converted to an abnormal oscillator and the inhibitory action (*arrow*) is prolonged (amplified). *b:* This neuron under PTZ became silent, but the reciprocal excitatory command (*arrows*) persists and always elicits a discharge of variable duration.

Besides the persistence of the synaptic command exerted reciprocally on two neurons (already abnormal oscillators), the neuron commander itself may be also converted to an abnormal oscillator. In that case, the reciprocal command seems to be amplified, i.e., long-lasting. Such an amplification is subsequent to a long-lasting discharge burst of the commander on both subordinated neurons (Fig. 8).

CONCLUDING REMARKS

Once a normal cell has been converted to an abnormal oscillator the following arises: The more general function of an abnormal oscillator is the organization of a new rhythmogenesis of the bursting type, irrespective of the functional type of other monitored neurons, e.g., other oscillators, tonically rhythmic neurons, and/or stable membrane neurons.

Nevertheless, if an elementary network of neurons might be defined as a three-neuron system including an abnormal oscillator, on which another neuron impinges synaptic input, and from which a third neuron receives an abnormal output, some predictabilities could be possible concerning the overall output of this system. Any such predictability would require the exact type of conversion occurring in any one of these three neurons after a given abnormal change.

To determine definitely such overall conversions of an identifiable elementary network, further study is necessary.

Some particular aspects of abnormal bursting activity may be abridged as follows: (a) the role of a frequency multiplier, e.g., in eliciting a burst in response to a single excitatory afference (Fig. 8); (b) elicitation of spatio-temporal patterns, wherein reciprocal and simultaneous excitation and inhibition monitoring coupled identifiable neurons; (c) elicitation not only of alternate patterns but also of a massive discharge output, as in the case of oscillators sensitive to extrasynaptic neuroamines; and (d) a possible mechanism of massive discharge output has already been proposed elsewhere in the case of an elementary network but composed of five neurons (22).

ACKNOWLEDGMENTS

This work is dedicated to my wife and main collaborator, Dr. Arvanitaki. I am very pleased to acknowledge common experimentations with my co-workers, Drs. M. Boisson, R. Chagneux, C. Ducreux, M. Gola, E. Labos, T. Morales, H. Takeuchi, and Y. Watanabe.

REFERENCES

1. Arvanitaki, A. (1962): *C. R. Acad. Sci. (Paris)*, 255:1523–1525.
2. Arvanitaki, A., and Chalazonitis, N. (1956): In: *Microphysiologie Comparée des Éléments Excitables*. Colloque Int. C.N.R.S., pp. 154–158. Paris.
3. Arvanitaki, A., and Chalazonitis, N. (1958): *J. Physiol. (Paris)*, 50:122–125.
4. Arvanitaki, A., and Chalazonitis, N. (1961): *C. R. Acad. Sci. (Paris)*, 252:193–195.
5. Arvanitaki, A., and Chalazonitis. N. (1964): *C. R. Soc. Biol. (Paris)*, 158:1119–1123.
6. Arvanitaki, A., and Chalazonitis, N. (1965): *C. R. Soc. Biol. (Pàris)*, 159:1377–1393.
7. Arvanitaki, A., and Chalazonitis, N. (1967): In: *Neurobiology of Invertebrates,* edited by J. Salanki, pp. 169–199. Plenum Press, New York.
8. Arvanitaki, A., Dhalazonitis, N., and Otsuka, M. (1956): *C. R. Acad. Sci.*, 243:307–309.

9. Barker, J. L., and Gainer, H. (1974): *Brain Res.*, 65:516–520.
10. Boisson, M., and Chalazonitis, N. (1972): *Comp. Biochem. Physiol.*, 41:883–885.
11. Boisson, M., and Chalazonitis, N. (1973): *C. R. Acad. Sci. (Paris)*, 276:1025–1028.
12. Carpenter, D., and Gunn, R. (1970): *J. Cell Physiol.*, 75:121–128.
13. Chalazonitis, N. (1959): *Arch. Sci. Physiol.*, 13:41–78.
14. Chalazonitis, N. (1961): *J. Physiol.* (Paris), 53:289–290.
15. Chalazonitis, N. (1963): *Ann. NY Acad. Sci.*, 109:451–479.
16. Chalazonitis, N. (1967): In: *Neurobiology of Invertebrates,* edited by J. Salanki, pp. 201–226. Plenum Press, New York.
17. Chalazonitis, N. (1968): *Ann. NY Acad. Sci.*, 147:419–459.
18. Chalazonitis, N. (1976): *J. Physiol.* (Paris), 72:75A.
19. Chalazonitis, N. (1977): *J. Physiol.* (Paris), 73:441–452.
20. Chalazonitis, N. (1977): In: *Advances in Experimental Medicine and Biology, Vol. 78: Tissue Hypoxia and Ischemia,* edited by M. Reivich, R. Coburn, S. Lahiri, and B. Chance, pp. 85–100. Plenum Press, New York.
21. Chalazonitis, N., and Arvanitaki, A. (1970): In: *Advances in Biochemistry and Pharmacology, Vol. 2: Biochemistry of Simple Neuronal Models,* edited by E. Costa and E. Giacobini, pp. 245–284. Raven Press, New York.
21a. Chalazonitis, N., and Chagneux, R. (1978): *J. Physiol.* (Paris) (*In press.*)
22. Chalazonitis, N., and Arvanitaki, A. (1973): In: *International Encyclopedia of Pharmacology and Therapeutics, Section 19: Anticonvulsant Drugs, Vol. 2,* edited by J. Mercier, pp. 401–424. Pergamon Press, New York.
23. Chalazonitis, N., Ducreux, C., and Arvanitaki, A. (1972): *J. Physiol.* (Paris), 65:212A.
24. Chalazonitis, N., Ducreux, C., and Takeuchi, H. (1973): *J. Physiol.* (Paris), 67:257A.
25. Chalazonitis, N., Gola, M., and Arvanitaki, A. (1965): *C. R. Soc. Biol.*, 159:2451–2455.
26. Chalazonitis, N., and Takeuchi, H. (1968): *C. R. Soc. Biol.*, 162:1552–1554.
27. Chalazonitis, N., Watanabe, Y., and Arvanitaki, A. (1972): *J. Physiol.* (Paris), 65:375A.
28. Ducreux, C., and Chalazonitis, N. (1971): *C. R. Soc. Biol.*, 169:1350–1356.
29. Ducreux, C., and Gola, M. (1975): *Pfluegers Arch.*, 361:43–53.
30. Ducreux, C., Takeuchi, H., and Chalazonitis, N. (1972): *C. R. Acad. Sci. (Paris)*, 275: 1907–1909.
31. Ehilé, E., and Chalazonitis, N. *In preparation.*
32. Faber, D. S., and Klee, M. R. (1974): *Brain Res.*, 65:109–126.
33. Frazier, W., Kandel, E., Kupfermann, I., Wasiri, R., and Coggeshall, R. (1967): *J. Neurophysiol.*, 30:1288–1351.
34. Gola, M. (1976): *Experientia,* 32:585–587.
35. Gola, M., and Romey, G. (1973): *J. Physiol.* (Paris), 67:277A.
36. Matsumoto, H., and Ajmone-Marsan, C. (1964): *Exp. Neurol.*, 9:305.
37. Morales, T., and Chalazonitis, N. (1970): *C. R. Soc. Biol.*, 164:1792–1797.
38. Prince, D. A. (1969): In: *Basic Mechanisms of the Epilepsies,* edited by H. Jasper, A. Ward, and A. Pope, pp. 320–328. Little Brown, Boston.
39. Speckmann, E. J., and Caspers, H. (1973): *Epilepsia,* 14:397–408.
40. Sugaya, E., Goldring, G., and O'Leary, J. L. (1964): *Electroencephalogr. Clin. Neurophysiol.*, 17:661–669.
41. Sugaya, A., Sugaya, E., and Tsujitani, M. (1973): *Jap. J. Physiol.*, 23:261–274.
42. Takeuchi, H. (1978): *This volume.*
43. Takeuchi, H. (1978): Thesis. Université de Provence III.
44. Tasaki, I., and Hagiwara, S. (1957): *J. Gen. Physiol.*, 40:859–885.
45. Taylor, A., and Dunsted, C. (1971): *Epilepsia,* 12:33–45.
46. Washizu, Y., Bonewell, G. W., and Terzuolo, C. A. (1961): *Science,* 133:333–334.
47. Watanabe, Y., and Arvanitaki, A. (1970): *J. Physiol.* (Paris), 62:227–228.

Abnormal Neuronal Discharges, edited by N. Chalazonitis and M. Boisson. Raven Press, New York © 1978.

Attenuation and Stabilization of Oscillatory Activities

N. Chalazonitis

Institute of Neurophysiology and Psychophysiology, C.N.R.S., Marseilles, France

In the preceding chapter, we reported some intrinsic and synaptic properties of abnormal oscillators. Mainly, we described the conversion of some rhythmic neurons into abnormal oscillators and their role in abnormal rhythmogenesis: the mechanism of initiation of their efferent burst, their signification, when they command synaptically other neurons and *vice versa,* how the abnormal oscillators are monitored synaptically by others, and so on.

In this chapter, the reciprocal aspect of the abnormal ability to oscillate (or abnormal oscillability) is considered. It is important to investigate how an abnormal oscillator may be stabilized, i.e., definitively converted to a stable neuron.

The term stabilization is used here in the sense of abolition of any type of rhythmogenesis; a stabilized neuron, if depolarized by any means [direct current (DC) or excitatory synaptic input] should never display bursts and slow waves or repetitive paroxysmal depolarizations. The only type of discharge of a stabilized or naturally stable neuron is the adaptative type of discharge. Therefore, stabilization does not mean a simple hyperpolarization with temporary abolition of the burst activity.

Whereas the stabilization of abnormal oscillators as defined above constitutes the main theme of our investigation, all other possibilities that may establish long-lasting hyperpolarization and inhibition of the abnormal oscillator, obtained mainly by chemical factors, are considered extensively. By long-lasting inhibition we mean one whose duration is many times longer than the period of the burst activity (10,12,13). Under such conditions, it has been established (for example, if the long hyperpolarization is obtained by injection of DC in the oscillator; see Fig. 2 of the preceding chapter), that not only the frequency of the slow waves decreases but also the mean frequency of the spikes on each burst. Therefore, the total number of emitted spikes during the long-lasting inhibition (the duration of which is of the order of 1 or more min) decreases; conversely, the mean frequency of the spiking in the scale of minutes is substantially decreased. Such a modification in the burst activity is defined as an attenuation. In this case, there is no definitive abolition of the slow oscillatory activity of the neuron; hence there is no stabilization.

Some data are given as to phenobarbital action on abnormal oscillator. This molecule is representative among the current antiepileptogenic agents, and we examine some data concerning its action on the abnormal oscillator activity only. Many other aspects of the barbiturate action are found in the chapter by Takeuchi (38). On the same line of research, the possible action of endogenous neuroamines (as catecholamines) on abnormal oscillators is presented.

Finally, to complete the data concerning catecholamine effects on abnormal oscillatory activities, we concentrate on catecholamine action on normal oscillations. In this case, some new mechanisms, which may explain attenuation and/or real stabilization of normal oscillators, are suggested.

ATTENUATION AND/OR STABILIZATION OF ABNORMAL OSCILLATOR ACTIVITY

Attenuation Via Antagonistic Action of Phenobarbital

General antagonistic actions between barbiturates and pentylenetetrazol (PTZ) have been already reviewed (29,30,34,43). Toman (42) found depressive actions of barbiturates on nerve conduction. Some important antagonistic effects of barbiturates on oscillatory activity of nerve fibers initiated by decalcifant anions (phosphate) have been reported. Takeuchi (36), by applying phenobarbital to the stable giant neuron of Aplysia (A or R_2), found hyperpolarizing effects on the somatic membrane leading to a decreased synaptic activability. Barker and Gainer (6) confirmed Takeuchi's results and further reported a decrease in the amplitude of the excitatory postsynaptic potential (EPSP) using pentobarbital (see also ref. 35).

Abnormal oscillators, available after PTZ treatment and exposed to phenobarbital in the presence of PTZ, display a continuous hyperpolarization and lose their spontaneous activity. In such a case, according to Takeuchi, the injection of depolarizing DC restores the initial paroxysmal activity; therefore, the phenobarbital action corresponds to an attenuation and not to a real stabilization (38) (see also Fig. 1). In this recording, the attenuation by phenobarbital is obvious. Instead of an afferent discharge elicited by a single EPSP and composed of 30 spikes under metrazol, the phenobarbital action reduced the discharge to six spikes only.

The molecular mechanism of such an antagonism is unknown. As the somatic (and/or axon hillock) membranes are involved in oscillogenesis, the antagonism may be primarily localized in this membrane, irrespective of the kind of competition. Despite the fact that attenuation cannot be conspicuous before an exposure to phenobarbital for a few minutes, indirect action may also occur. In that case, some delayed release and action of

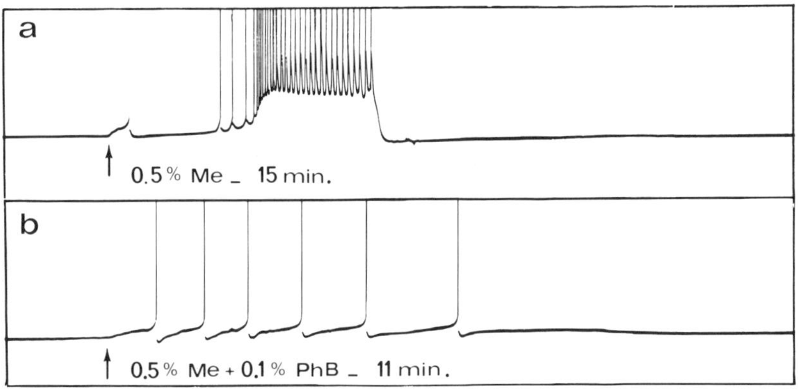

FIG. 1. Attenuation by phenobarbital of paroxysmal bursts elicited under PTZ. **a:** *Arrow,* elicitation of a paroxysmal burst by a single EPSP (*arrow*) of a neuron treated by PTZ (Me) for 15 min. **b:** Attenuated burst elicited by a same synaptic afference after application for 11 min of the same PTZ-containing saline supplemented with 0.1% phenobarbital.

neuroamines of endogenous origin could hyperpolarize the rhythmogenic membrane.

Attenuation by Phenobarbital Via Indirect Inhibitory Synaptic Action

In addition to the direct hyperpolarizing action of phenobarbital, indirect inhibitory action may occur via inhibitory interneurons by the following mechanism. Phenobarbital may convert some regularly tonic neurons to oscillators, following an evolution similar to the one described under PTZ (17,39). When the phenobarbital-converted neuron spontaneously displays paroxysmal depolarizations, these plateaus can still be synchronous to efferent axonal discharges, as in the case of PTZ plateaus (see preceding chapter). The inhibitory interneurons that command an abnormal oscillator might display these long-lasting discharges elicited by phenobarbital. Consequently, they will periodically exert efficient inhibitory action on the main abnormal oscillator, in addition to any other hyperpolarizing action exerted on the main oscillator directly by phenobarbital. An example of such a mechanism is given in Fig. 2.

Persistence of the Synaptic Inputs Under Phenobarbital

Besides the "direct-indirect" inhibition exerted by phenobarbital, other effects may be finally exerted in nets containing reciprocally innervated neurons. In the example illustrated in Fig. 3, neurons are reciprocally monitored by a same afferent pathway. The upper neuron, which is normally of the tonic type, is monitored excitatorily; the lower one, which normally is of the bursting type, is monitored inhibitorily. The phenobarbital action never

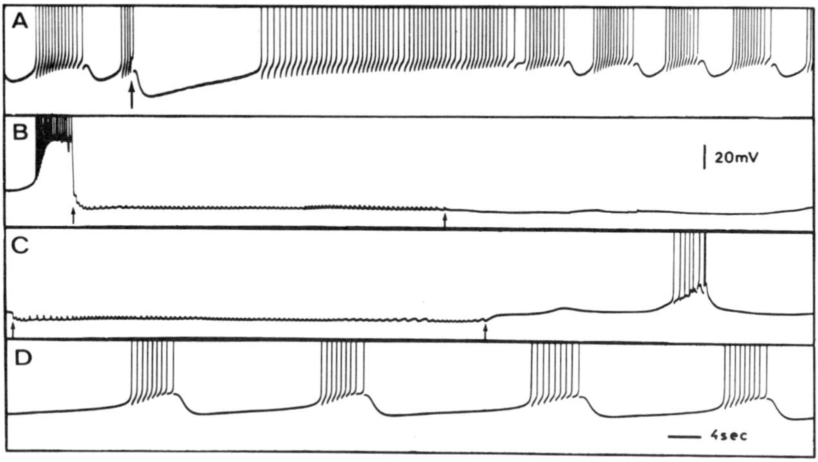

FIG. 2. Periodic indirect inhibition of a *Helix* oscillator under phenobarbital. **A:** Normal oscillator of *Helix. Arrow,* responds inhibitorily (long IPSP) to an appropriate afference. **B:** *Between arrows,* long inhibition established spontaneously and periodically by the paroxysmal discharges of the inhibitory interneuron converted to an abnormal oscillator under phenobarbital. **C:** Same description as in **B.** The reappearance of the waves after the spontaneous inhibitory discharge means that the hyperpolarizing direct action of the phenobarbital exerted on the oscillator is insufficient and of short duration. **D:** After repetitive washing with physiological saline, reestablishment of the burst activity of the oscillator.

abolishes the reciprocal command on each one of these neurons. Under phenobarbital, the first neuron displays paroxysmal depolarization, whereas the second neuron further hyperpolarizes via inhibitory postsynaptic potentials (IPSPs).

Attenuation and/or Stabilization of Abnormal Oscillators Under Exogenous Dopamine

The conversion of the stable giant neuron of *Aplysia* (A or R_2) by cycloheximide may or may not lead to a spontaneous burst activity, depending on the concentration (32). When the spontaneous burst activity is established (Fig. 4), the application of an approximately equivalent quantity of dopamine for several minutes does not give any stabilization. Despite a repetitive washing with seawater of the *Aplysia* visceral ganglion, cycloheximide cannot be eliminated, and the burst activity persists (31). A further application of dopamine alone (without any addition of cycloheximide) can lead to a hyperpolarization and arrest the burst activity. In such a case, the excitatory synaptic input is still valid, and liminal summation of EPSPS reestablishes the burst activity. Further washing of the cycloheximide by dopamine leads to a complete attenuation of the bursts. Dopamine does not seem to antagonize the cycloheximide effect, but does help in detaching

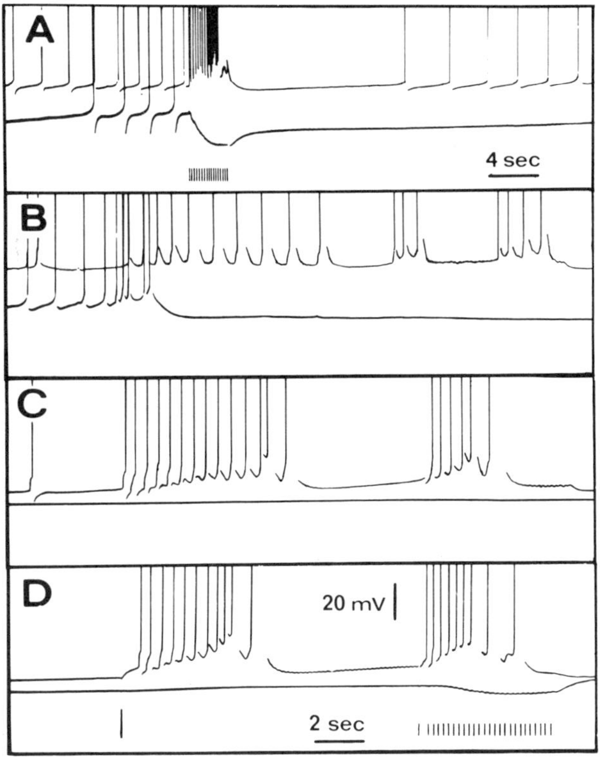

FIG. 3. Dual effects of phenobarbital on two different normal pacemakers. In all recordings: *Upper traces,* a tonically rhythmic neuron; *lower traces,* bursting pacemaker. **A:** In normal conditions, both neurons respond reciprocally to an afferent synaptic activation (vertical bars), the tonic neuron excitatorily, the bursting neuron (lower recording) inhibitorily. **B:** After 4 min phenobarbital (0.3%), the tonic neuron starts displaying bursts; the lower neuron ceases its burst activity by direct hyperpolarization. **C:** Eight minutes later, total conversion of the upper neuron to an abnormal oscillator. **D:** Ten minutes after phenobarbital application, a synaptic afference (vertical bars), single or repetitive, elicits extra bursts on the abnormal oscillator but inhibitory hyperpolarization on the lower neuron. In conclusion, the reciprocal synaptic activation still works despite the conversion.

cycloheximide under conditions in which the normal saline cannot. Therefore, dopamine may be considered under these conditions not as an antagonist stabilizer but rather as a weak eluter of cycloheximide.

In contrast to such a discrete action of dopamine on a cycloheximide abnormal oscillator, recent results obtained by Ducreux (21) show that under the same conditions dopamine can stabilize abnormal oscillators (Fig. 4B). Spontaneous paroxysmal depolarization elicited under PTZ on initially rhythmic neurons of *Helix* may be attenuated by dopamine. A further action of dopamine, always in the presence of PTZ, leads to a total cessation of the abnormal activity. When depolarizing current then activates the

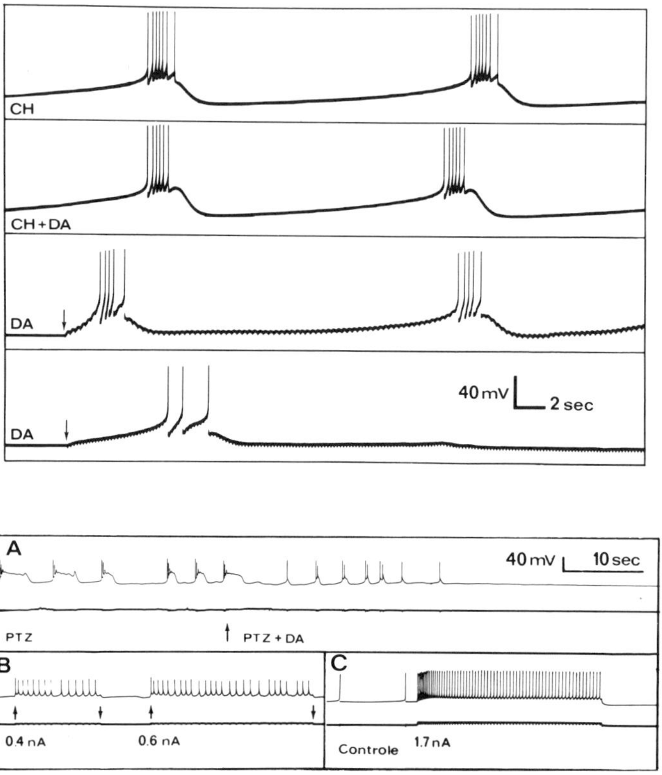

FIG. 4. Upper: "Antagonism," dopamine, and cycloheximide. *CH:* Spontaneous burst activity of the A (or R_2) cell of *Aplysia* converted to an abnormal oscillator by cycloheximide (2 mg/ml) since 1 hr. *CH + DA:* Application of a saline containing both cycloheximide (2 mg/ml) and dopamine (2 mg/ml) does not stabilize the oscillator. *DA:* Application of dopamine alone hyperpolarizes the oscillator and stops the bursts. At the arrow, an excitatory synaptic afference shows the persistence of the burst activity. *DA, last recording:* An attenuation of the burst activity is obtained after dopamine application during 30 min, probably by substitution of dopamine to cycloheximide, on the same receptors. An excitatory synaptic bombardment (*arrow*) elicits only one slow wave. *Lower:* "Antagonism" dopamine and PTZ (from ref. 21). **A:** PTZ elicited repetitive paroxysmal depolarization (PD) on a *Helix* neuron. *Arrow,* application of saline containing PTZ plus dopamine at same concentration. Attenuation of PDS and cessation of the activity. **B:** Injection of threshold depolarizing DC; indicated intensities elicit only a continuous activity. **C:** Washing out the neuron with physiological saline establishes the initial tonic activity.

neuron up to the firing level, the activity is no longer bursting but rather adaptative (Fig. 4B).

Stabilization of Abnormal Oscillator in Hyposodic Media

Morales-Cardona (31) demonstrated that when the *Aplysia* (A or R_2) giant stable neuron is converted to an oscillator by cycloheximide, the sub-

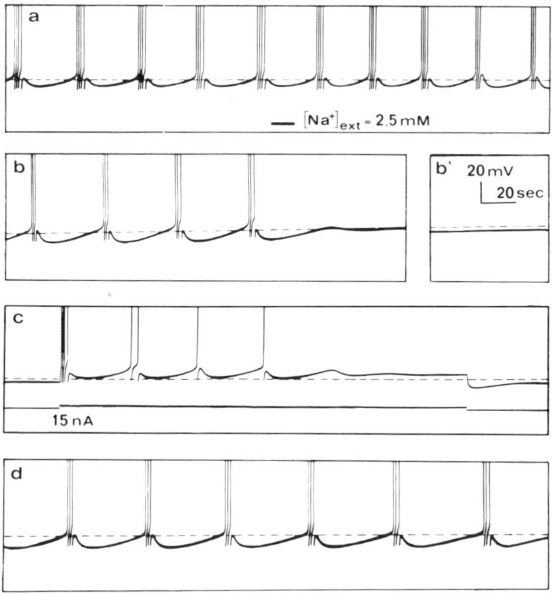

FIG. 5. Stabilization in low sodium media of an abnormal oscillator. **a:** Stable A neuron (or R_2) of *Aplysia* with spontaneous burst activity after cycloheximide treatment. At the signal (bar), application of a hyposodic saline (2.5 mM, Na^+ + 472.5 mM Tris-HCl, pH 7). **b:** Continuation of the recording. After 3 min the burst activity stops. **b′:** Fifteen minutes later, no recovery. **c:** Injection of depolarizing DC in the soma elicits a very attenuated burst activity. **d:** Recovery of the initial activity after 7 min application of physiological saline. (From ref. 31.)

stitution of a hyposodic saline for a normal one (artificial seawater), in which after substitution with tris [tris-hydroxymethylaminomethane] only 2.5 mM Na^+ is present, the burst activity is first attenuated and later abolished (hyperpolarization). If the cell is then activated by depolarizing DC, it does not display any repetitive burst activity. In conclusion, the substitution of tris for sodium in the external medium of the cell (and therefore in the molecular framework of its membrane) stabilizes the rhythmogenic area of the membrane (Fig. 5).

Conclusion

As a general conclusion, any attenuation and/or stabilization of an abnormal oscillator comes from a direct antagonism between an oscillogenic and a hyperpolarizing and/or stabilizing molecule. The localization of the competition between both molecules for sensitive membrane areas and *a fortiori* for membrane molecular receptors remains undetermined.

The ideal conditions for stabilization are not yet adopted. For instance, only abnormal oscillators irreversibly altered could constitute the correct preparation in studies of stabilizing action. Only such a type of neuronal models would imitate some pathological neurons establishing motor dysfunctioning.

ATTENUATION AND/OR STABILIZATION OF THE NORMAL OSCILLATOR

By Exogenous Dopamine and Norepinephrine

It is well known that the hyperpolarizing effects of catecholamines, as well as opposite effects (37), have long been an object of experimentation (4,8,22,24,28,41). Because exogenous dopamine, as well as the appropriate nerve stimulation, hyperpolarizes, both actions leading to inhibition of the *Aplysia* oscillator enabled Ascher et al. (4) to suggest that dopamine would be the normal transmitter on the Br (or R_{15}) oscillator.

Besides its transmitter role, dopamine is also a stabilizer; this property is particularly examined in this chapter. Ducreux and Chalazonitis (22)

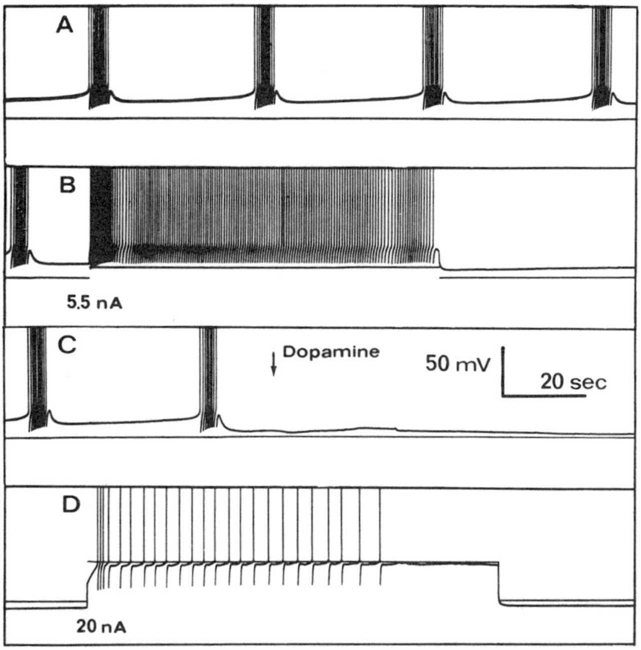

FIG. 6. Stabilization of the *Aplysia* normal oscillator (Br or R_{15}) under dopamine. **A:** Normal burst activity. **B:** Under normal conditions, depolarizing DC (5.5 nA) elicits the "fusion" of the waves, which is equivalent to a "massive discharge." **C:** *Arrow,* dopamine application (0.1 μm/ml) elicits, in some seconds, hyperpolarization and cessation of the activity. **D:** Under dopamine, a depolarizing DC (3.5 times higher than in **B**) is "threshold" and elicits only an adaptative discharge (stabilization).

described a possible stabilization of Helix normal oscillators by dopamine. Treated with dopamine, the oscillator is not merely hyperpolarized and stopped but, if reactivated by depolarizing DC, it cannot display burst activity (Fig. 6).

Similar results have been obtained by Boisson and Chalazonitis (7–9) on the *Aplysia* oscillator (Br or R_{15}) during treatment with norepinephrine (10^{-4} to 5.10^{-4} M). (Other experimental details are given in refs. 7–9).

The careful examination of the data illustrated in Fig. 6 permits a better definition of stabilization. Stabilization does not mean only a conversion of the bursting activity into a single adaptative discharge but also a substantial decrease in the mean value of spike frequency considered during the discharge. In other words, stabilization also means a decrease in the ability of spike elicitation after treatment of the oscillator by dopamine.

Furthermore, Sakharov and his group injected dopamine into the *Aplysia* oscillator (Br or R_{15}) and also found analogous conversion of the pattern of activity [quoted after Boisson and Chalazonitis (9)].

Prolonged Inhibitions and Stabilization of Normal Oscillators Elicited by Repetitive Synaptic Afference

The relative limitations in understanding membrane molecular mechanisms leading to attenuation and/or stabilization of abnormal oscillations led us to seek similar effects and mechanisms (attenuation and stabilization) observed in normal oscillators through a particular inhibitory afference.

Very early, the identification of a normal oscillator in *Aplysia* (Br or R_{15}) (1,2,11) made the study of environmental (14,19,25) and synaptic (13,16,33) influences possible.

The more important peculiarity of the normal oscillator of *Aplysia* is the presence of an inhibitory system that may inhibit its burst activity for minutes. The first long-lasting inhibitions on this neuron were observed on warming (or electrical stimulation of afferent inhibitory pathways) (12,13). Inhibitions of long duration (ILD) have been confirmed by Frazier et al. (23), Jahan-Parvar and Von Baumgarten (26), Parnas et al. (33), Boisson and Gola (10), and others (28). Not only oscillators but other neurons as well can display ILD, already described by Tauc (40).

Arguments for Parasynaptic Dopamine Release ("Overflow") During the Long-Lasting Inhibition of the (Br-R_{15}) Oscillator of *Aplysia*

Overflow usually means some further release of transmitter from a presynaptic terminal into the intercellular space (5). The following arguments are in favor of some dopamine overflow establishing a long-lasting inhibition.

The long inhibition of *Aplysia* oscillator has already been extensively described for either "biphasic" (40) or amphoteric (3a) postsynaptic potentials. The hyperpolarization created by those particular postsynaptic potentials is long lasting (Fig. 7). In Fig. 7, direct hyperpolarizing pulses through injection of brief DC cannot overlast the hyperpolarization, and the waves reappear at "off" (16,20). The careful observation of these recordings demonstrates that some persistence of the inhibitory transmitter release occurs after the end of the hyperpolarizing phase of the biphasic inhibitory potential (BIP). The repetition of BIPs accentuates the duration of the hyperpolarization, and the metasynaptic action ("meta" has here a temporal significance) of the release is likely to follow a few minutes later. The question of some parasynaptic ("para" has here a topographic significance, meaning around the synaptic junctions) could be suggested, explaining the delayed persistence of the hyperpolarization.

Among the electrophysiological arguments of dopamine release by overflow are the indirect, hyperthermically triggered, long-lasting inhibitions of the oscillator (12). In that case, many recent recordings confirm the previ-

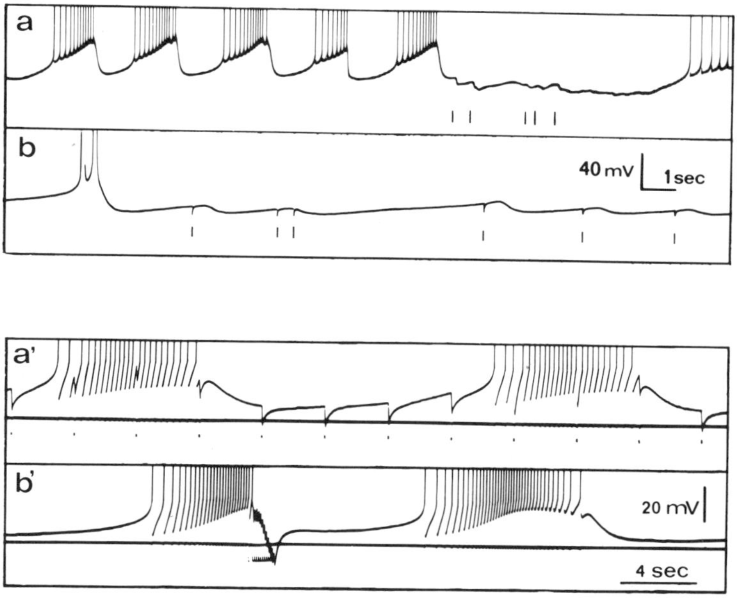

FIG. 7. Inhibition and attenuation of the normal burst activity (Br or R_{15} cell of Aplysia). **Upper:** *a:* Normal burst activity. Some sparse repetitive inhibitory afferences. Vertical bars establish an overlasting inhibition. *b:* When the cell spontaneously reaches its waving level, each burst is "attenuated." Vertical bars indicate single inhibitory afference leading to BIPs. **Lower:** *a':* On a normal burst activity, repetitive hyperpolarizing pulses (10 msec and 0.3 frequency) of direct inward current. *b':* Same short pulses (10 msec) of hyperpolarizing DC at higher frequency (10/sec). At "off," the waving reappears and there is no overlasting inhibition and/or attenuation.

ous results and demonstrate that during hyperthermic transients, an impinging of inhibitory bursts occurs, which is followed by long persistence of the hyperpolarization. On the other hand, the demonstration that some tonic pacemakers (such as the Gen or L_{11} cell) are converted to burst pacemakers in hyperthermia favors the concept that the hyperthermic bursting activity of the inhibitory interneuron acting on the Br-R_{15} oscillator might be a pacemaker similar to the Gen-L_{11} cell.

In addition, it is known that the oscillator (Br-R_{15}) along with its probable inhibitor cell (Gen-L_{11}) both contain dense core vesicles (DCV) (16a,29); otherwise they are both aminergic neurons. This peculiarity of both neurons may validate electron microscope pictures (15,27), suggesting the possibility of inhibitory transmitter synaptic release and overflow.

Morphologic Arguments

The first argument comes from the very frequent presence in the *Aplysia* neuropile, as well as in the *Helix* neuropile, of presynaptic terminals, containing both (classic) vesicles (clear or gray) near the junction, and the presence of very numerous DCV in which cathecholamines or indolamines would exist. Therefore, whatever the kind of transmitter released by the classic vesicles, the DCV would participate, releasing their own neuroamines, possibly with delayed kinetics (see Fig. 8).

The second argument concerns direct and/or densitometric evaluations of the membrane layers of axons containing only DCV. In that case, and particularly under hyperthermic conditions, the unit membrane of the "varicose axons" seems altered. It has been designed as "collapsed" due to the relative disappearance of its interlayer clear space (18) (Fig. 9). As the membrane of these varicose axons and their branches, which are filled with DCV, are particular, one may consider not only the possibility of "uptakes" but also of parasynaptic releases, i.e., overflows. Frazier et al. (23) have already described small axonal "digitation" filled with DCV in contact of the Br-R_{15} cell neurite.

Considering these arguments, and above all the long-lasting inhibition found by exogenous dopamine application, the following hypothesis can be formulated: The inhibitory neurons commanding the normal oscillators release some transmitter responsible for the appearance of a biphasic potential. As soon as this release is "off," an overflow of hyperpolarizing transmitter occurs mainly through nonjunctional and/or parajunctional areas of the presynaptic terminals. The overflow occurs during a hyperthermic transient or after exhaustive repetitive stimulation of the interneuron. During this overflow, the maximum concentration locally reached might be of the same order as that of the exogenous catecholamine (10^{-4} M) used in stabilization. Consequently, the oscillator will be hyperpolarized and stabilized (for a few minutes).

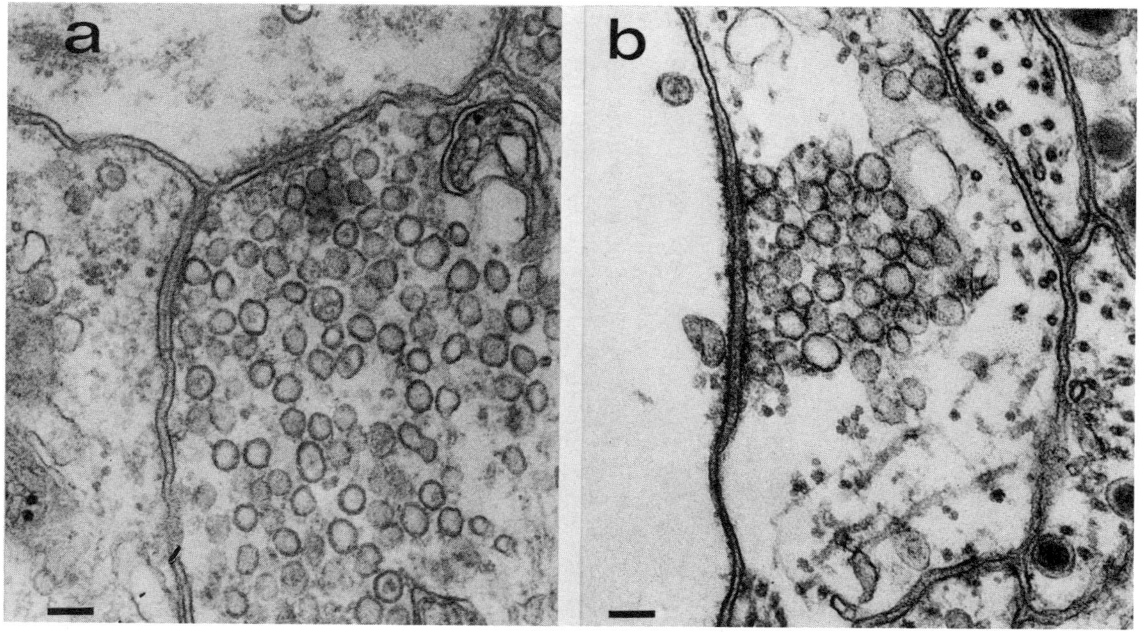

FIG. 8. Two types of prejunctional terminals in Aplysia ganglion. **a** and **b:** Gray vesicles (600 Å diameter) of the "classic" type in presynaptic endings of the *Aplysia* ganglion neuropile.

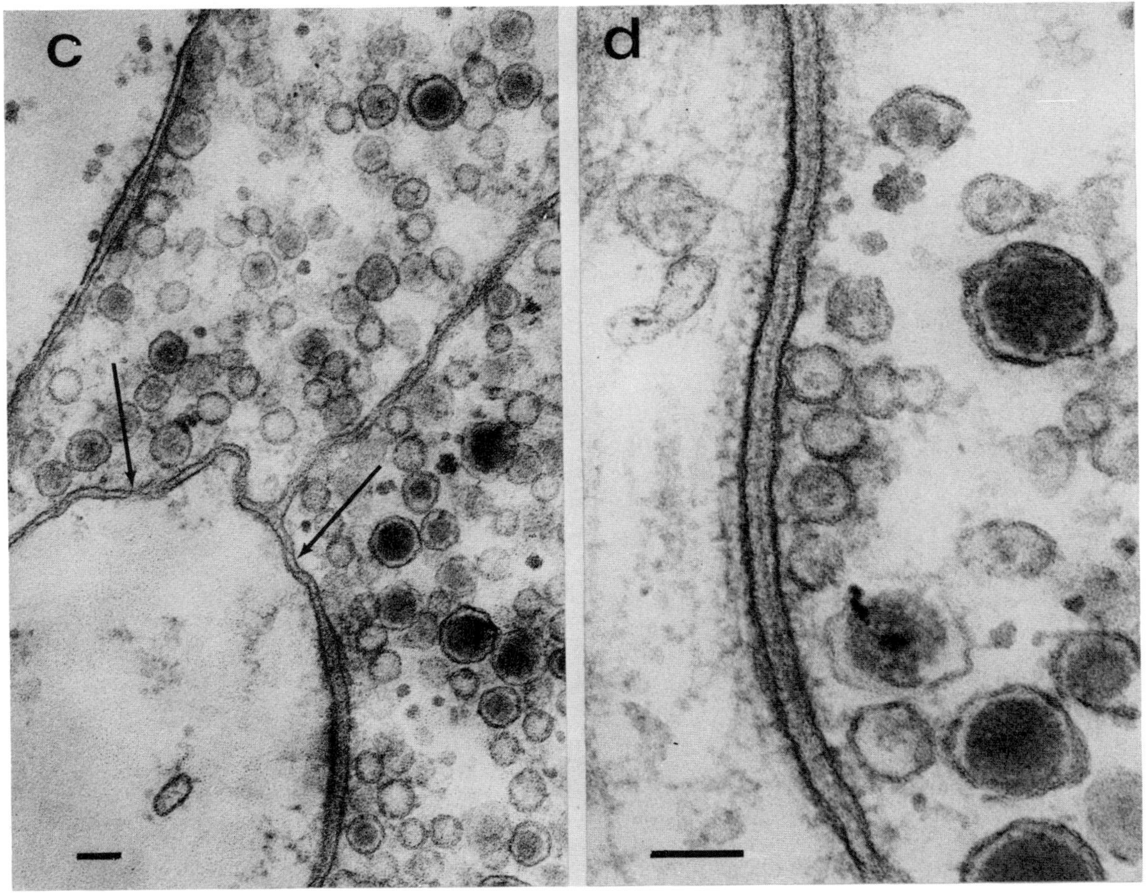

FIG. 8. c and **d**: Presence of both gray vesicles and DCV in presynaptic endings of the Aplysia ganglion neuropile. In **c,** arrows indicate parasynaptic membrane areas in which overflow might occur.

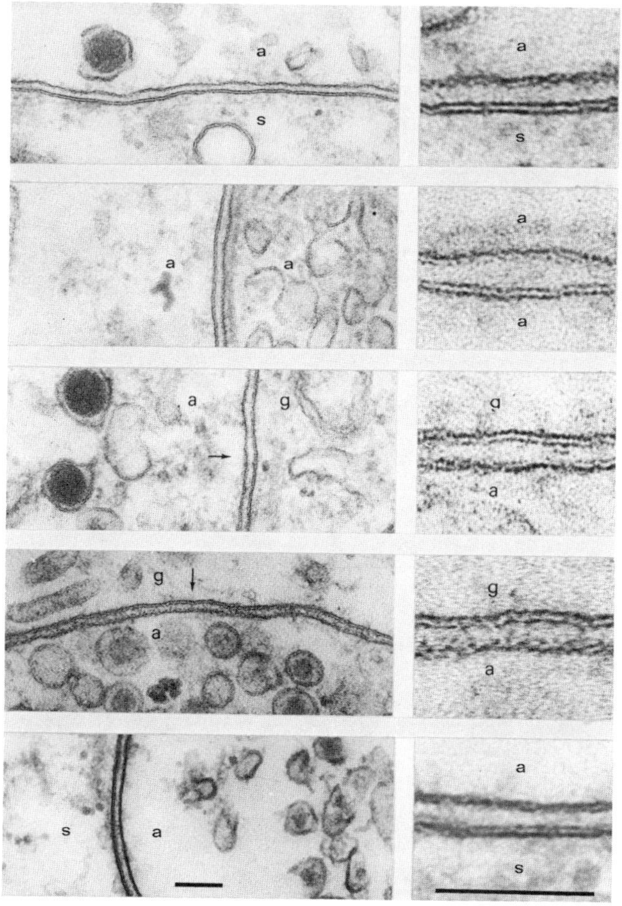

FIG. 9. Differentiated membrane of axonal varicosities filled with DCV in *Helix* ganglion neuropile. a, Axonal space (cytoplasm) filled with DCV; s and g, respectively, somatic and glial cytoplasm of somata and glia neighboring the axonal "varicosities." The figures of the right column are enlargements of the corresponding figures of the left column. Only the unit membranes of the DCV-containing axons are of the "collapsed" type.

Stabilization of the Normal Oscillator in Hyposodic Media

First, Watanabe and Arvanitaki-Chalazonitis (44) demonstrated that when sodium is replaced by cholinium in the physiological saline, the slow waves but not the spiking disappear. Whereas the oscillator becomes silent, even after partial substitution of the sodium, the injection of depolarizing DC elicits but a continuous adaptative discharge of spikes (Fig. 10).

The survival of the spike discharge after the disappearance of the wave might last more than 1 hr in some cases. Therefore, the hyposodic saline action on the normal oscillator affects the normal oscillator under com-

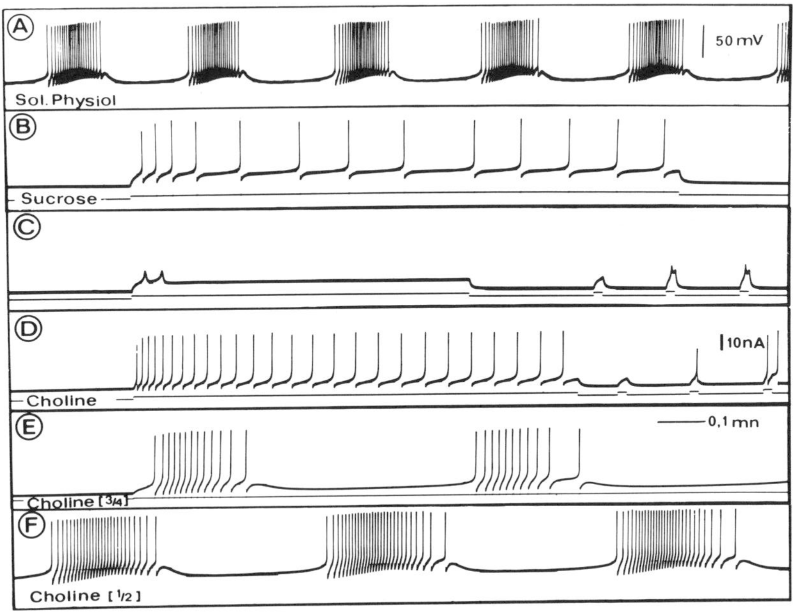

FIG. 10. Earlier disappearance of the slow waves after substitution of sodium by sucrose or choline. **A:** Oscillator of Aplysia (Br or R_{15}) cell in normal conditions. **B:** Substitution of seawater NaCl by isotonic sucrose leads to the cessation of the bursting. Injection of depolarizing DC elicits an adaptative discharge only. **C:** Under sucrose 20 min later, even the spiking disappears. **D:** A total substitution of sodium by cholinium abolishes the bursting. Under depolarizing threshold, DC, the soma elicits an adaptative discharge only. **E** and **F:** Indicated partial substitution of sodium leads merely to an attenuation of the slow waves.

parable conditions to those of the abnormal oscillator (cycloheximide-treated cell) (Fig. 5).

CONCLUSION

The stabilization of oscillators, whether normal or abnormal, is a difficult task. Attenuation of periodic discharge is less difficult. It has been pointed out that attenuation may be due to the initiation of long-lasting hyperpolarizations. The chemical agent performing long-lasting hyperpolarization, attenuation, and even in some cases real stabilization, may exert direct changes on the oscillating membrane when exogenously applied. Moreover, the same agents may provoke delayed indirect or secondary actions if they are able to release, particularly through overflow endogenous molecules, which exert long-lasting inhibitory effects on neighboring neuromembrane.

The cellular scale analysis of attenuation and long-lasting inhibition cannot be expected to contribute to any improvement in the choice of efficient attenuators. The only interest of such an analysis of oscillogenic and antioscillogenic action seems to lie in the presentation of new potentialities of neuromembranes performing normal or abnormal rhythmogeneses in given neuronal networks.

REFERENCES

1. Arvanitaki, A., and Chalazonitis, N. (1958): *J. Physiol. (Paris)*, 50:122–125.
2. Arvanitaki, A., and Chalazonitis, N. (1964): *C. R. Soc. Biol.*, 158:1119–1123.
3. Arvanitaki, A., and Chalazonitis, N. (1965): *C. R. Soc. Biol.*, 159:1783–1793.

3a. Arvanitaki, A., and Chalazonitis, N. (1964): *C. R. Soc. Biol.*, 158:1674–1677.

4. Ascher, P., Kehoe, J. S., and Tauc, L. (1967): *J. Physiol. (Paris)*, 59:331–332.
5. Bacq, Z. M. (1976): *J. Physiol. (Paris)*, 72:371–542.
6. Barker, J. L., and Gainer, H. (1973): *Science*, 182:720.
7. Boisson, M. (1973): Thesis. University of Paris.
8. Boisson, M., and Chalazonitis, N. (1972): *Comp. Biochem. Physiol.*, 41A:883–885.
9. Boisson, M., and Chalazonitis, N. (1976): In: *Neurobiology of Invertebrates. Gastropoda Brain*, edited by J. Salanki, pp. 425–435. Akademiai Kiado, Budapest.
10. Boisson, M., and Gola, M. (1976): *Comp. Biochem. Physiol.*, 54C:109–113.
11. Chalazonitis, N. (1959): *Arch. Sci. Physiol.*, 13:41–78.
12. Chalazonitis, N. (1962): *C. R. Acad. Sci.*, 255:1652–1653.
13. Chalazonitis, N. (1967): In: *Neurobiology of Invertebrates*, edited by J. Salanki, pp. 201–226. Plenum Press, New York.
14. Chalazonitis, N. (1968): *Ann. NY Acad. Sci.*, 147:419–459.
15. Chalazonitis, N. (1969): In: *Cellular Dynamics of the Neuron*, edited by S. H. Barondes, pp. 229–243. Academic Press, New York.
16. Chalazonitis, N. (1977): *J. Physiol. (Paris)*, 73:441–452.

16a. Chalazonitis, N. *Unpublished.*

17. Chalazonitis, N., and Arvanitaki, A. (1973): In: *International Encyclopedia of Pharmacology and Therapeutics, Section 19: Anticonvulsant Drugs, Vol. II.*, edited by J. Mercier, pp. 401–424. Pergamon Press, New York.
18. Chalazonitis, N., and Costa-Chagneux, H. (1970): *C. R. Soc. Biol.*, 164:2067–2070.
19. Chalazonitis, N., Gola, M., and Arvanitaki, A. (1965): *C. R. Soc. Biol.*, 159:2451–2455.
20. Chalazonitis, N., Watanabe, Y., and Arvanitaki, A. (1972): *J. Physiol. (Paris)*, 65:375A.
21. Ducreux, C. *Unpublished.*
22. Ducreux, C., and Chalazonitis, N. (1971): *C. R. Soc. Biol.*, 169:1350–1353.
23. Frazier, W. T., Kandel, E. R., Kupfermann, I., Waziri, R., and Coggeshall, R. E. (1967): *J. Neurophysiol.*, 30:1288–1351.
24. Gerschenfeld, H. M. (1973): *Physiol. Rev.*, 53:1–119.
25. Gola, M. (1969): Thesis. University of Marseilles.
26. Jahan-Parvar, M., and Von Baumgarten, R. (1969): *Am. J. Physiol.*, 215:1246–1257.
27. Jourdan, F. (1972): Thesis. University of Lyon.
28. Kerkut, G. A., Horn, N., and Walker, R. J. (1969): *Comp. Biochem. Physiol.*, 30:1061–1074.
29. Kirsten, E. B., and Schoener, E. P. (1972): *Neuropharmacology*, 11:591–599.
30. Krupp, P., and Monnier, M. (1973): In: *International Encyclopedia of Pharmacology and Therapeutics, Section 19: Anticonvulsant Drugs, Vol. II.*, edited by J. Mercier, pp. 371–400. Pergamon Press, New York.
31. Morales-Cardona, T. (1971): Thesis. University of Provence, Marseilles.
32. Morales, T., and Chalazonitis, N. (1970): *C. R. Soc. Biol.*, 165:1792–1797.
33. Parnas, I., Armstrong, D., and Strumwasser, F. (1974): *J. Neurophysiol.*, 37:594–608.
34. Singh, P., and Huot, J. (1973): In: *International Encyclopedia of Pharmacology and Therapeutics, Section 19: Anticonvulsant Drugs, Vol. II*, edited by J. Mercier, pp. 427–507. Pergamon Press, New York.
35. Somjen, G. G., and Gill, M. (1963): *Pharmacol. Exp. Ther.*, 140:19–30.
36. Takeuchi, H. (1968): *C. R. Soc. Biol.*, 162:488–490.
37. Takeuchi, H. (1978): Thesis. University of Marseilles III.

38. Takeuchi, H. (1978): *This volume.*
39. Takeuchi, H., and Chalazonitis, N. (1968): *C. R. Soc. Biol.*, 162:491–493.
40. Tauc, L. (1960): In: *Inhibition of the Nervous System and Gamma-Aminobutyric Acid*, edited by E. Roberts, pp. 85–89. Pergamon Press, Oxford.
41. Tauc, L. (1967): *Physiol. Rev.*, 47:521–593.
42. Toman, J. E. P. (1952): *Pharmacol. Rev.*, 4:168–218.
43. Toman, J. E. P. (1969): *Epilepsia*, 10:179–192.
44. Watanabe, Y., and Arvanitaki-Chalazonitis, A. (1970): *J. Physiol. (Paris)*, 62:227–228.

Abnormal Neuronal Discharges, edited by N. Chalazonitis and M. Boisson. Raven Press, New York © 1978.

Modifications of the Convulsant-Induced Abnormal Biopotentials of a Molluscan Giant Neuron by Drugs, Divalent Ions, and Temperature Change

Hiroshi Takeuchi

Department of Neurochemistry, Institute for Neurobiology, Okayama University Medical School, Okayama, Japan

Abnormal plateau formation of the neuronal biopotential accompanied by grouped spike discharges has been produced by the application of convulsants, including metrazol (pentylenetetrazole) and strychnine, to identifiable neurons of lower animals, such as mollusca (6,8,10,11,12,14,16,20, 24,25) and crustacea (23). The abnormal biopotentials of these neurons closely resemble the biopotentials from epileptogenic neurons in the mammalian brain, which correspond to the electroencephalographic spikes induced acutely by metrazol (7) and penicillin (1,9,15,17) and chronically by aluminum hydroxide gel (18).

In the present study, a regularly firing giant neuron (~200 μm in diameter) was identified in the suboesophageal ganglia (on the dorsal surface of the right parietal ganglion) of a European edible snail (*Helix pomatia*); and reoccurring abnormal plateau formation of the biopotential was induced by bath application of either metrazol or strychnine. The present experiment was an attempt to define the effects of application of barbiturates and theophylline and of changes in the concentration of divalent ions (Ca^{2+} and Mg^{2+}) in the medium and temperature on such abnormal plateau formation.

EFFECTS OF METRAZOL AND STRYCHNINE ON THE SOMATIC BIOPOTENTIAL AND AXONIC IMPULSE OF AN IDENTIFIED NEURON

The neuron used sends axonic branches in at least four peripheral nerves issuing from the suboesophageal ganglia: the right pallial nerve, the anal nerve, and two branches of the left pallial nerve. The neuron, therefore, could be easily identified. One or two microelectrodes (filled with 2.5 M KCl) were implanted in the neuronal soma, one to record the intracellular somatic biopotential, the other to alter the membrane polarization as necessary. Axonic impulses of the neuron were recorded from the right pallial nerve with bipolar silver wire electrodes (the interval between the two wires was about 3 to 5 mm). The amplitude of the axonic impulses, recorded synchronously with the somatic spikes, was so large that the impulses of the neuron examined were easily distinguishable from those of other neurons. The physiological solution of the snail used as the medium in the experiment has been described by Cardot (5). The temperature of the experimental room was maintained at 20 to 23°C.

Convulsants examined were classified into two categories, as follows: drugs producing prolonged abnormal plateau formation of the biopotential (lasting for several seconds), e.g., metrazol, bemegride, coramine, and ethamivan; and drugs resulting in a shortened duration of the abnormal plateau formation (several hundred milliseconds), e.g., strychnine and harmine.

Figure 1 shows the effect of metrazol in producing plateau formation of the biopotential of the identified neuron. In the physiological state, a somatic action potential corresponded in time to an axonic impulse recorded from the right pallial nerve. The application of metrazol in a concentration such as 1.2×10^{-2} kg/liter produced a slow depolarizing oscillation of the somatic biopotential which was accompanied by grouped spike discharges. In this state of neuronal excitability, axonic impulses still corresponded to the somatic spikes. When the plateau formation of the somatic biopotential was well developed in the presence of metrazol, however, this concordance of the axonic impulses with the local oscillations of biopotential of somatic neuromembrane was lost.

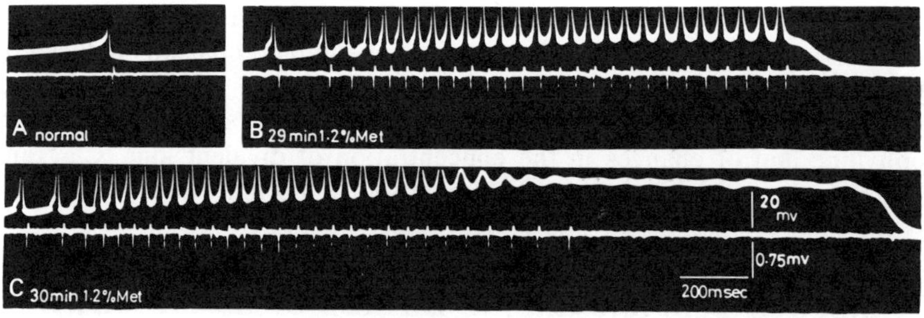

FIG. 1. Effects of metrazol (pentylenetetrazole) on the somatic biopotential and the axonic impulse of the neuron examined. **A–C:** *Upper traces,* intracellularly recorded somatic biopotential; *lower traces,* axonic impulses recorded in the right pallial nerve. **A:** In the normal state, the axonic impulse was concordant with the somatic spike. **B:** 29 min after the application of 1.2×10^{-2} kg/liter metrazol, abnormal slow depolarizing oscillation of biopotential with grouped spike discharges was produced, axonic impulses still corresponding to the somatic spikes. **C:** 30 min after 1.2×10^{-2} kg/liter metrazol, plateau formation of the somatic biopotential was well developed; the axonic impulses were not concordant with the local oscillations of the somatic biopotential during the plateau period. *Upper vertical bar,* calibration of somatic biopotential (20 mV); *lower vertical bar,* calibration of axonic impulse (0.75 mV); *horizontal bar,* time course (200 msec).

Figure 2 shows abnormal plateau formation of short duration caused by strychnine. In a concentration of 10^{-3} kg/liter, for instance, strychnine produced a double spike discharge of the somatic biopotential. In this situation, there was no second axonic impulse, suggesting that the initiation site of the axonic impulse may have been in a refractory state. The application of strychnine in a concentration such as 3×10^{-3} kg/liter resulted in the appearance of short-lasting plateau formation of the somatic biopotential with two or three axonic impulses.

Washizu et al. (22,23), using the crayfish stretch receptor neuron, also demonstrated that multiple axonic impulses of the neuron examined were discharged during the somatic abnormal plateau formation, and that these axonic impulses sometimes were not concordant with the somatic local responses which occurred during plateau formation of the somatic biopotential. The plateau formation in their experiments had been induced by either strychnine or the intracellular injection of cations.

The plateau formation of the somatic biopotential caused by the above-mentioned convulsants would appear to be due not to synaptic activation but to some change in the nature of the somatic neuromembrane since the occurrence of the plateau formation could be controlled by the membrane potential change caused by the intracellular current injection (6,11).

In the presence of the convulsants, bombardment of excitatory postsynaptic potentials (EPSPs) from another neuron sometimes resulted in grouped spike discharges of the neuron examined (Fig. 3A). Furthermore,

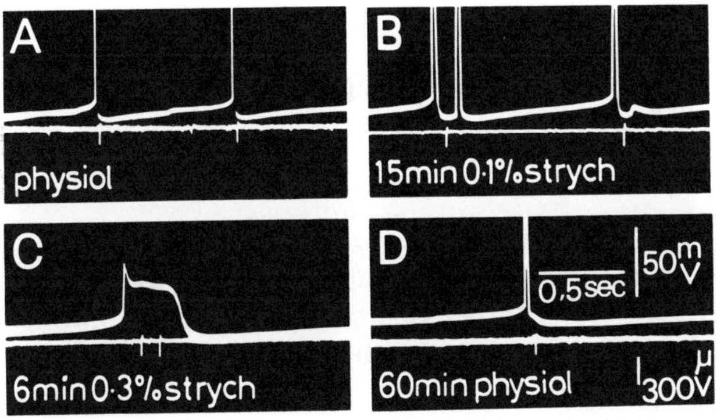

FIG. 2. Effects of strychnine on the somatic biopotential and the axonic impulse of the neuron examined. **A–D:** *Upper traces,* intracellular somatic biopotential; *lower traces,* axonic impulses recorded in the right pallial nerve. **A:** Normal state. **B:** 15 min after the application of 10^{-3} kg/liter strychnine, somatic double spikes appeared, the second somatic spike lacking the corresponding axonic impulse. **C:** 6 min after 3×10^{-3} kg/liter strychnine, the somatic biopotential showed a plateau of short duration, axonic impulses not corresponding to local responses of the somatic membrane in the plateau period. **D:** 60 min after washing with the physiological solution, the effects of strychnine completely disappeared. *Upper vertical bar,* calibration of somatic biopotential (50 mV); *lower vertical bar,* calibration of axonic impulse (300 μV); *horizontal bar,* time course (0.5 sec).

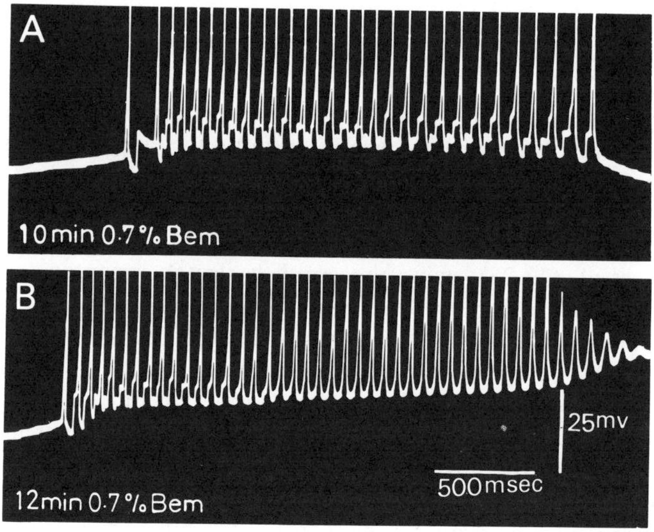

FIG. 3. Effects of EPSP bombardment from another neuron on plateau formation of the somatic biopotential of the neuron examined in the presence of bemegride. **A:** 10 min after the application of 7×10^{-3} kg/liter bemegride, a bombardment of EPSPs resulted in grouped spike discharges. **B:** 12 min after 7×10^{-3} kg/liter bemegride, a bombardment of EPSPs shifting the somatic neuromembrane toward depolarization triggered plateau formation of the biopotential. *Vertical bar,* calibration of somatic biopotential (25 mV); *horizontal bar,* time course (500 msec).

the depolarizing shift of the neuromembrane caused by the EPSP bombardment in the presence of the convulsants could trigger plateau formation of the somatic biopotential (Fig. 38).

EFFECTS OF BARBITURATES ON THE OCCURRENCE OF THE METRAZOLIZED PLATEAU FORMATION

Blocking of the depolarizing response to acetylcholine and of the EPSP have both been reported in molluscan neurons using certain barbiturates (2,3,21). Furthermore (19), pentobarbital in high concentrations has been demonstrated to cause hyperpolarization of Aplysia neurons.

Barbiturates, such as amobarbital, secobarbital, and phenobarbital, in high concentrations were also shown to cause hyperpolarization of the somatic membrane of the neuron examined in the present study and to result in a markedly decreased frequency of its spike discharges. The critical concentrations of these drugs at which the spike discharges of the neuron disappeared were, in order, 3 to 5×10^{-4} kg/liter for amobarbital, 5×10^{-4} to 10^{-3} kg/liter for secobarbital, and 1 to 2×10^{-3} kg/liter for phenobarbital.

Figure 4 shows the effects of phenobarbital on the repetitive metrazolized plateau formation. The application of 10^{-2} kg/liter metrazol produced the

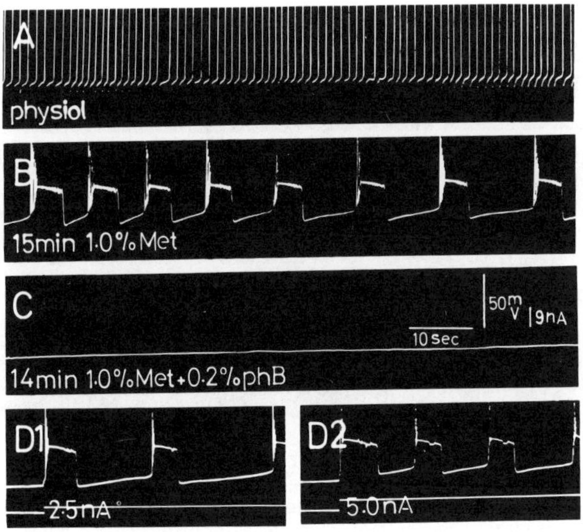

FIG. 4. Effects of phenobarbital on metrazolized plateau formation of the somatic biopotential. **A–C** and *upper traces of* **D1** and **D2:** Somatic biopotential. *Lower traces* of **D1** and **D2:** Intensity of the depolarizing current injected into the soma. **A:** Normal state. **B:** 15 min after the application of 10^{-2} kg/liter metrazol, abnormal plateau formations occurred repetitively. **C:** 14 min after 2×10^{-3} kg/liter phenobarbital in the presence of 10^{-2} kg/liter metrazol, the metrazolized neuromembrane hyperpolarized and altered to be in a silent state. **D:** Under the same conditions as **C,** depolarizing currents (**D1,** 2.5 nA; **D2,** 5.0 nA) were injected into the soma; abnormal plateau formation reappeared during the current injection which compensated for the hyperpolarizing effect of phenobarbital. (The somatic spikes in this figure have been retouched.) *Left vertical bar,* calibration of somatic biopotential (50 mV); *right vertical bar,* intensity of intracellularly injected current (9 nA) (for **D1** and **D2**); *horizontal bar,* time course (10 sec).

plateau formation. Application of 2×10^{-3} kg/liter phenobarbital in the presence of metrazol resulted in hyperpolarization of the neuromembrane and disappearance of the plateau formation. At this stage, the plateau formation reappeared with the depolarizing shift in the biopotential produced by intracellular injection of a current. Barbiturates, therefore, caused hyperpolarization of both the normal and the metrazolized neuromembrane. With adequate drug concentrations, a silent state of the neuromembrane occurred. Barbiturates, however, were not able to convert the metrazolized abnormal biopotential to the normal spike.

EFFECTS OF THEOPHYLLINE ON THE METRAZOLIZED PLATEAU FORMATION

Theophylline but not caffeine in concentrations of 5×10^{-4} to 10^{-3} kg/liter depolarized the neuron examined and increased the frequency of spike discharges. As shown in Fig. 5, theophylline at 10^{-3} kg/liter also

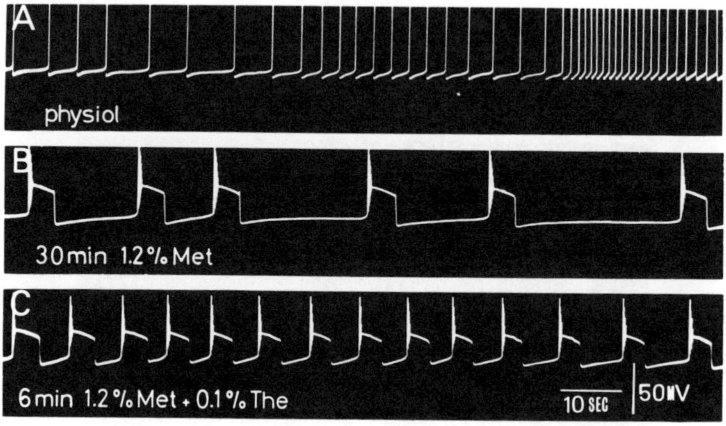

FIG. 5. Effects of theophylline on metrazolized plateau formation of the somatic biopotential. **A:** Normal state. **B:** 30 min after the application of 1.2×10^{-2} kg/liter metrazol. **C:** 6 min after 10^{-3} kg/liter theophylline in the presence of 1.2×10^{-2} kg/liter metrazol, the intervals between metrazolized plateau formations shortened with a depolarizing shift of the membrane potential. (The spikes in this figure have been retouched.) *Vertical bar,* calibration of somatic biopotential (50 mV); *horizontal bar,* time course (10 sec).

depolarized the metrazolized neuromembrane and resulted in shortening of the intervals between plateau formations. It has been reported (6) that the intracellular injection of a depolarizing current also resulted in shortening of the intervals between metrazolized plateau formations.

EFFECTS OF DIVALENT IONS ON METRAZOLIZED AND STRYCHNINIZED PLATEAU FORMATION

A lack of Ca^{2+} in the medium facilitated the effects of metrazol in producing the plateau formation, whereas an excess of Ca^{2+} prevented this plateau formation. Figure 6 shows such an experiment. In a Ca^{2+}-poor medium (1 mM, one-tenth of normal value), the neuromembrane depolarized and the frequency of spike discharges increased. Addition of 7.5×10^{-3} kg/liter metrazol to the Ca^{2+}-poor medium induced well-developed plateau formation. Metrazol in the same concentration with the Ca^{2+} concentration increased to the normal value (10 mM) converted the plateau to the slow depolarizing oscillation of the biopotential with grouped spike discharges. In the normal medium, 8.5×10^{-3} kg/liter metrazol was required to produce the plateau formation. When the Ca^{2+} concentration in the medium was increased to 30 mM (three times the normal value) in the presence of 8.5×10^{-3} kg/liter metrazol, the plateau formation of the somatic biopotential altered to a slow depolarizing oscillation with grouped spike discharges. In the Ca^{2+}-rich (30 mM) medium, 10^{-2} kg/liter metrazol was needed to produce the plateau formation. The effects of metrazol on the somatic bio-

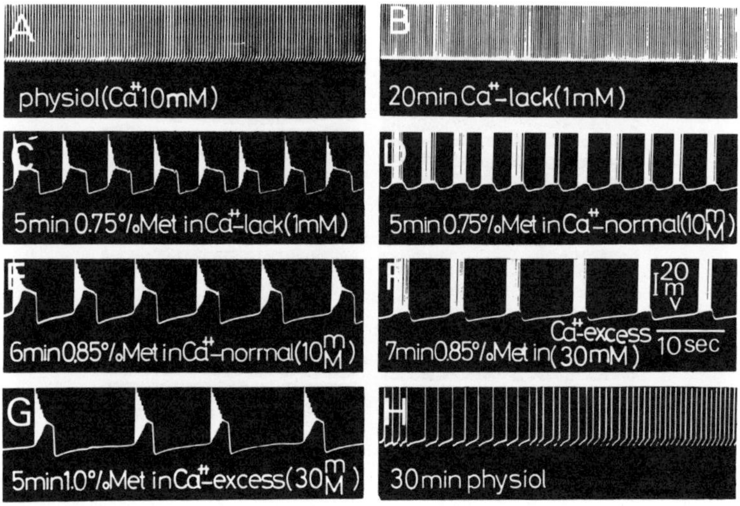

FIG. 6. Effects of changes in Ca^{2+} concentration of the medium on metrazolized plateau formation of the somatic biopotential. **A:** Normal state. **B:** 20 min after replacement with a Ca^{2+}-poor (1 mM) medium, the neuromembrane depolarized and the frequency of the spike discharges increased. **C:** 5 min after the application of 7.5×10^{-3} kg/liter metrazol in the Ca^{2+}-poor (1 mM) medium, plateau formation of the biopotential was well developed. **D:** 5 min after replacement with normal medium (Ca^{2+}, 10 mM) in the presence of 7.5×10^{-3} kg/liter metrazol, the biopotential showed slow depolarizing oscillations with grouped spike discharges. **E:** 6 min after 8.5×10^{-3} kg/liter metrazol in normal medium (Ca^{2+}, 10 mM), the plateau of the biopotential was again well developed. **F:** 7 min after replacement with a Ca^{2+}-rich (30 mM) medium in the presence of 8.5×10^{-3} kg/liter metrazol, the biopotential showed slow depolarizing oscillations with grouped spike discharges. **G:** 5 min after 10^{-2} kg/liter metrazol in the Ca^{2+}-rich (30 mM) medium, the plateau of the biopotential was well developed. **H:** 30 min after washing with the normal medium, the effects of metrazol had completely disappeared. (The spikes in this figure have been retouched.) *Vertical bar,* calibration of somatic biopotential (20 mV); *horizontal bar,* time course (10 sec).

potential completely disappeared after washing the ganglia with the physiological solution.

In another experiment, the critical concentrations of metrazol required for the production of the plateau formation of the biopotential in media with different concentrations of Ca^{2+} were determined as follows: 7.5×10^{-3} kg/liter metrazol in the Ca^{2+}-poor (1 mM) medium, 9×10^{-3} kg/liter metrazol in the normal medium (Ca^{2+} 10 mM), and 10^{-2} kg/liter in the Ca^{2+}-rich (30 mM) medium.

Changes in the Mg^{2+} concentration of the medium from zero to 33 mM (normal value: 11 mM) did not modify the above-mentioned effects of metrazol.

As shown in Fig. 7, when the abnormal biopotentials were caused by strychnine instead of metrazol, similar modifications were produced by changes in the Ca^{2+} concentration of the medium. In the Ca^{2+}-free medium,

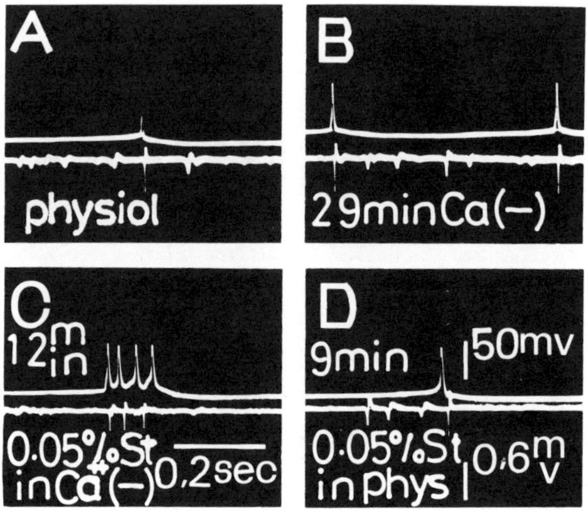

FIG. 7. Effects of changes in Ca^{2+} concentration of the medium on strychninized grouped spike discharges of the somatic biopotential. **A–D:** *Upper traces,* somatic biopotential; *lower traces,* axonic impulses recorded from the right pallial nerve. **A:** Normal state. **B:** 29 min after replacement with a Ca^{2+}-free medium. **C:** 12 min after the addition of 5×10^{-4} kg/liter strychnine to the Ca^{2+}-free medium, grouped spike discharges (four spikes) of the somatic membrane appeared, the axonic impulses not corresponding to the somatic spike discharges. **D:** 9 min after replacement with normal medium (Ca^{2+} 10 mM) in the presence of 5×10^{-4} kg/liter strychnine, a single spike, instead of the grouped spikes shown in **C,** appeared together with a related axonic impulse. *Upper vertical bar,* calibration of somatic biopotential (50 mV); *lower vertical bar,* calibration of axonic impulse (0.6 mV); *horizontal bar,* time course (0.2 sec).

the somatic membrane depolarized and the amplitude of the somatic spikes diminished. The addition of 5×10^{-4} kg/liter strychnine to the Ca^{2+}-free medium induced both grouped spike discharges of the somatic biopotential and multiple axonic impulses. There was, however, no correspondence between their appearance. Replacement with normal medium (Ca^{2+} 10 mM) in the presence of the same concentration of strychnine converted the grouped discharges of the somatic biopotential to a single spike with a related axonic impulse.

Other experiments examined the effects of a Ca^{2+} excess in the medium on the strychninized abnormal biopotential. Well-developed plateau formation obtained in, for example, 1.5×10^{-3} kg/liter strychnine in normal medium altered to double spikes after replacement with a Ca^{2+} excess (30 mM) in the same concentration of strychnine. This showed that the Ca^{2+} excess was antagonistic to such effects of strychnine. Changes in the Mg^{2+} concentration of the medium from zero to 33 mM (three times the normal value) did not affect the abnormal biopotentials induced by strychnine.

Barker and Gainer (4) reported the effects of divalent ions on physiologically occurring bursting pacemaker potential (BPP) activity in cell 11 of *Otala lactea* (in the activated condition) and R_{15} of *Aplysia californica* (13). Unfortunately, differences in their experimental protocol do not permit a strict comparison of their results with our present study; however, the results they obtained for the effects of divalent ions on the physiological appearance of BPP differ from our work on the abnormal biopotentials caused by convulsants, probably because of the following points. Both cell 11 of Otala and R_{15} of Aplysia failed to show the BPP in medium free of both Ca^{2+} and Mg^{2+}; and the BPP of R_{15} always developed well in the presence of more than 20 mM Ca^{2+} or Mg^{2+} in the medium.

EFFECTS OF TEMPERATURE CHANGE ON THE ABNORMAL BIOPOTENTIAL CAUSED BY CONVULSANTS

The temperature of the examined ganglia was controlled by thermomodules placed immediately under the experimental bath. The temperature of these thermomodules could be altered by passing a current. The temperature of the examined ganglia was monitored directly by a needle-type thermocouple with a telethermometer.

Experimental results for the metrazolized abnormal biopotential are summarized in Table 1. With the physiological solution, the neurons did not develop the abnormal biopotential in any of the temperature ranges examined, except for a case that showed slow depolarizing oscillation at 37°C, an unusually high temperature for the neuron. In the presence of metrazol, higher temperatures usually facilitated and lower temperatures

TABLE 1. *Effects of temperature change on the metrazolized abnormal biopotential*

Material no.	Medium (concn.:kg/liter)	Slow depolarizing oscillation	Plateau formation
1	Physiological solution	(−), until 30°C	
2	Physiological solution	(−), until 35°C	
3	Physiological solution	(−), until 34°C	
4	Physiological solution	(−), until 32°C	
5	Physiological solution	(+), at 37°C	(−), until 37°C
6	Physiological solution	(−), until 33°C	
7	4×10^{-3} Metrazol	(+), at 27°C	(−), until 33°C
8	5×10^{-3} Metrazol	(+), at 27°C	(+), at 32°C
9	5×10^{-3} Metrazol	(+), at 30°C	(−), until 32°C
10	6×10^{-3} Metrazol	(+), at 30°C	(−), until 32°C
11	7.5×10^{-3} Metrazol	(+), at 22°C	(−), until 33°C
12	8.5×10^{-3} Metrazol	(+), at 27°C	(+), at 32°C
13	9×10^{-3} Metrazol	(+), at 35°C	(−), until 35°C
14	10^{-2} Metrazol	(+), at 10°C	(+), at 22.5°C
15	10^{-2} Metrazol	(+), at 19°C	(+), at 35°C
16	1.2×10^{-2} Metrazol	(+), at 10°C	(+), at 19°C
17	1.2×10^{-2} Metrazol	(+), at 10°C	(+), at 19°C
18	1.3×10^{-2} Metrazol	(+), at 14°C	(+), at 17°C

usually prevented the occurrence of the metrazolized somatic abnormal biopotential.

Figure 8 shows an example of the modification of the metrazolized somatic plateau formation caused by gradual temperature change. Initially, the normal temperature was lowered in the presence of 1.3×10^{-2} kg/liter metrazol. Between 24 and 20°C, the plateau of the somatic biopotential was well developed, and axonic impulses showed no correspondence with the somatic local oscillation during its plateau phase. At 15°C, however, the plateau altered to a slow depolarizing oscillation, the somatic spikes being accompanied by related axonic impulses. At 10°C, a single spike appeared

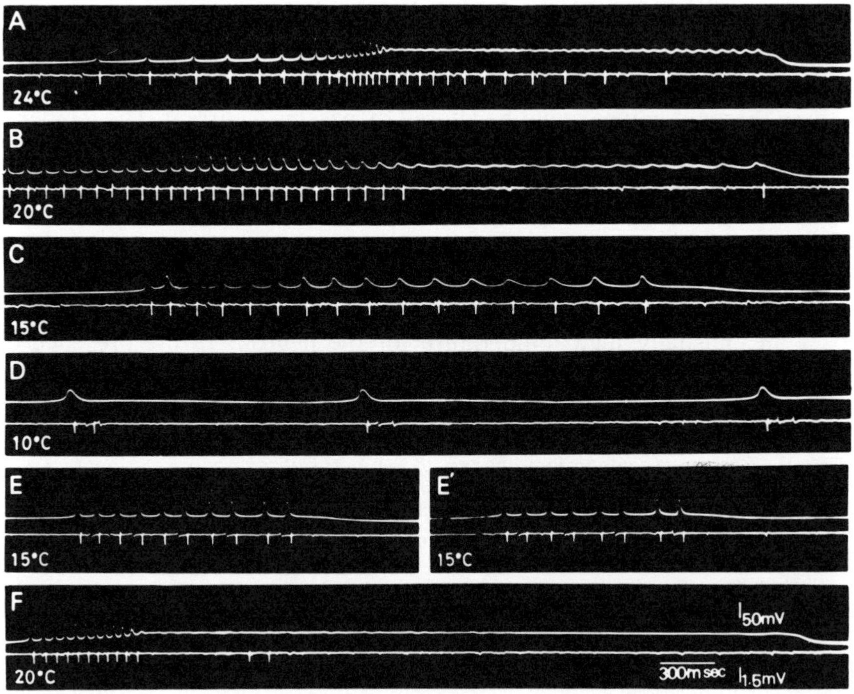

FIG. 8. Effects of temperature change on metrazolized plateau formation of the somatic biopotential. The experiment was performed in the presence of 1.3×10^{-2} kg/liter metrazol. Recordings were made from **A** to **F** in order, while the temperature of the dissected ganglia was gradually altered. **A–F:** *Upper traces,* somatic biopotential; *lower traces,* axonic impulses recorded from the right pallial nerve. **A:** At 24°C, the plateau of the somatic biopotential was well developed in the presence of metrazol (1.3×10^{-2} kg/liter). **B:** At 20°C. **C:** At 15°C, the plateau shown in **A** and **B** altered to a slow depolarizing oscillation. **D:** At 10°C, a single spike appeared even in the presence of metrazol in the same concentration. **E** and **E′**: At 15°C, the slow depolarizing oscillation reappeared. **F:** At 20°C, the plateau of the somatic biopotential was again well developed. (The axonic impulses in this figure have been retouched.) *Upper vertical bar,* calibration of somatic biopotential (50 mV); *lower vertical bar,* calibration of axonic impulse (1.5 mV); *horizontal bar,* time course (300 msec).

instead of the grouped spike discharges. The reverse situation was then observed by gradually increasing the temperature of the ganglia. At 15°C, the slow depolarizing oscillation of the somatic biopotential with grouped spike discharges appeared. At 20°C, the plateau formation was again well developed.

The strychninized somatic abnormal biopotential was similarly subjected to temperature changes, resulting in modifications generally similar to those of the metrazolized one. In an experiment shown in Fig. 9, the temperature of the ganglia was altered gradually from the lower extreme toward higher temperatures in the presence of 10^{-3} kg/liter strychnine. At 10°C, a single somatic spike occurred with a related axonic impulse. At 23.5°C, double spikes appeared. At 36°C, a shortened depolarizing oscillation with multiple (four) spike discharges appeared.

In contrast to these results for the effects of temperature on the abnormal

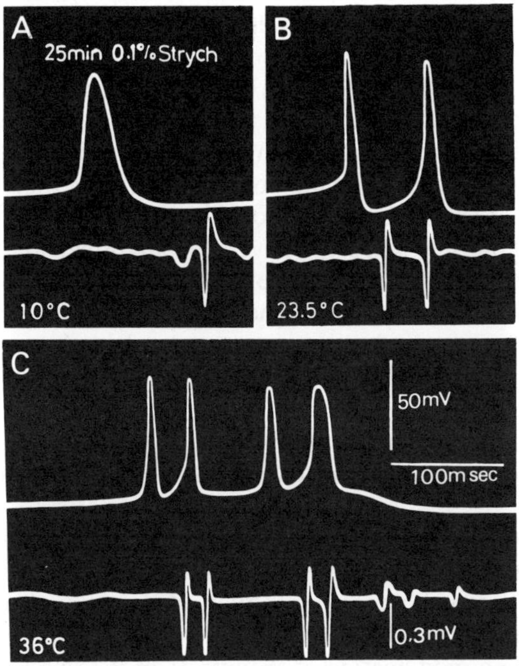

FIG. 9. Effects of temperature change on the strychninized grouped spike discharges of the somatic biopotential. The experiment was performed in the presence of 10^{-3} kg/liter strychnine. Recordings were made from **A** to **C** in order, while the temperature of the material was gradually altered. **A–C:** *Upper traces,* somatic biopotential; *lower traces,* axonic impulses recorded from the right pallial nerve. **A:** At 10°C. **B:** At 23.5°C, somatic double spikes appeared. **C:** At 36°C, somatic grouped spike discharges (four spikes) appeared. Note that the conduction velocity of the axonic impulse was slower at the lower temperature. *Upper vertical bar,* calibration of somatic biopotential (50 mV); *lower vertical bar,* calibration of axonic impulse (0.3 mV); *horizontal bar,* time course (100 msec).

biopotential caused by convulsants are the results of Barker and Gainer (4) who used cell 11 of Otala and R_{15} of Aplysia to study the effects of temperature on the physiological appearance of the amplitude of the BPP. Their results indicated that, for both neurons, the relationships between temperature and the amplitude of the BPP took the form of a parabolic curve; that is, in both the lower (near 10°C) and the higher (near 30°C) temperatures, these neurons lost their BPP activity, while the largest BPP occurred around 20°C.

CONCLUSIONS

A neuron was identified in the suboesophageal ganglia of *Helix pomatia* by means of simultaneous recordings of the intracellular somatic biopotential and of the axonic impulses from the right pallial nerve. Modifications of the convulsant-induced abnormal somatic biopotentials were examined using barbiturates, theophylline, divalent ion (Ca^{2+} and Mg^{2+}) concentration changes, and temperature changes.

Barbiturates (amobarbital, secobarbital, and phenobarbital) in high concentrations hyperpolarized both the normal and the metrazolized somatic membrane of the neuron examined. When metrazolized plateau formations occurred repeatedly, the application of a barbiturate prolonged the intervals of plateau formations with a hyperpolarizing shift of the biopotential. This addition of barbiturates failed to convert the abnormal somatic biopotential to a normal one. In the silent state of neuronal excitability obtained in the presence of both barbiturate and metrazol, the plateau formation of the biopotential reappeared after the intracellular injection of a depolarizing current to compensate for the hyperpolarizing effect of the barbiturate.

Theophylline, on the other hand, depolarized both the normal and the metrazolized membrane. In the presence of metrazol, theophylline shortened the intervals of the abnormal plateau formations with a depolarizing shift of the biopotential.

A lack of Ca^{2+} in the medium facilitated both the metrazolized and the strychninized abnormal biopotentials; an excess was antagonistic to the production of abnormal biopotentials by these two convulsants. A change in the Mg^{2+} concentration of the medium from zero to three times the normal value produced no modification of the effects of the two convulsants.

Higher temperatures (e.g., 30°C) facilitated and lower temperatures (e.g., 10°C) prevented the abnormal somatic biopotential caused by the two convulsants.

ACKNOWLEDGMENTS

This work was performed in the Institute of Neurophysiology and Psychophysiology, Department of Cellular Neurophysiology, C.N.R.S., Marseilles. The author wishes to express his thanks to Dr. A. Arvanitaki-Chalazonitis and Dr. N. Chalazonitis for their direction.

REFERENCES

1. Ayala, G. F., Dichter, M., Gumnit, R. J., Matsumoto, H., and Spencer, W. A. (1973): *Brain Res.*, 52:1–17.
2. Barker, J. L. (1975): *Brain Res.*, 92:35–55.
3. Barker, J. L. (1975): *Brain Res.*, 93:77–90.
4. Barker, J. L., and Gainer, H. (1975): *Brain Res.*, 84:479–500.
5. Cardot, H. (1933): *Ext. Ann. Physiol.*, 9: No. 4.
6. Chalazonitis, N., and Takeuchi, H. (1968): *C. R. Soc. Biol. (Paris)*, 162:1552–1554.
7. Creutzfeldt, O. D., Watanabe, S., and Lux, H. D. (1966): *Electroencephalogr. Clin. Neurophysiol.*, 20:19–37.
8. David, J. R., Wilson, W. A., and Escueta, A. V. (1974): *Brain Res.*, 67:549–554.
9. Dichter, M. A., and Spencer, W. A. (1969): *J. Neurophysiol.*, 32:649–662.
10. Ducreux, C., Takeuchi, H., and Chalazonitis, N. (1972): *C. R. Acad. Sci. (Paris)*, 275: 1907–1909.
11. Faugier-Grimaud, S. (1974): *Brain Res.*, 69:354–360.
12. Faugier, S., and Willows, A. O. D. (1973): *Brain Res.*, 52:243–260.
13. Frazier, W. T., Kandel, E. R., Kupfermann, I., Waziri, R., and Coggeshall, R. E. (1967): *J. Neurophysiol.*, 30:1288–1351.
14. Klee, M. R., Faber, D. S., and Heiss, W. D. (1973): *Science*, 179:1133–1136.
15. Matsumoto, H., and Ajmone-Marsan, C. (1964): *Exp. Neurol.*, 9:286–304.
16. Partridge, L. D., Jr. (1975): *Brain Res.*, 94:161–166.
17. Prince, D. A. (1969): *Electroencephalogr. Clin. Neurophysiol.*, 26:476–487.
18. Prince, D. A., and Futamachi, K. J. (1970): *Electroencephalogr. Clin. Neurophysiol.*, 29:496–510.
19. Sato, M., Austin, G. M., and Yai, H. (1967): *Nature*, 215:1506–1508.
20. Speckmann, E. J., and Caspers, H. (1973): *Epilepsia*, 14:397–408.
21. Takeuchi, H. (1968): *C. R. Soc. Biol. (Paris)*, 162:488–490.
22. Washizu, Y. (1965): *Comp. Biochem. Physiol.*, 15:535–545.
23. Washizu, Y., Bonewell, G. W., and Terzuolo, C. A. (1961): *Science*, 133:333–334.
24. Williamson, T. L., and Crill, W. E. (1976): *Brain Res.*, 116:217–229.
25. Williamson, T. L., and Crill, W. E. (1976): *Brain Res.*, 116:231–249.

Abnormal Neuronal Discharges, edited by N. Chalazonitis and M. Boisson. Raven Press, New York © 1978.

Effects of Pentylenetetrazol on Isolated Snail and Mammalian Neurons

E.-J. Speckmann and H. Caspers

Institute of Physiology, University of Münster, Germany

In mammalian neurons, convulsant drugs, e.g., penicillin and pentylenetetrazol (PTZ), induce a characteristic activity pattern commonly called paroxysmal depolarization shift (PDS) (1,29,32). A considerable amount of experimental work has already been devoted to the basic mechanisms underlying this special activity type. Apart from extraneuronal influences, such as potassium accumulation in the extracellular space (see refs. 13 and 24), a summation of excitatory postsynaptic potentials (EPSPs) or changes in the "intrinsic" properties of the neuronal membrane or both have been considered as elementary processes (2,3,5,6,8–10,18,19,26–28).

The present chapter aims to contribute additional data on the mechanisms of a PDS. To decide whether a synaptic input represents a prerequisite for the generation of a PDS, single neurons of *Helix pomatia* were treated with PTZ after complete isolation from synaptic inputs. The results of these investigations are described in the first sections and are simultaneously compared with the findings obtained in cortical neurons of the cat. Furthermore, the experiments performed on isolated snail neurons were supplemented by investigations on single ganglion cells cultured from the dorsal root of the rat. Finally, an attempt was made to contribute to the question of whether and to what extent changes of the so-called early outward current are involved in producing paroxysmal activity in snail neurons, since this current has often been regarded as essential in regulating burst activity (7,11,14,22).

METHODS

Experiments in *Helix pomatia*

The experiments were carried out in the identified buccal neurons B1 through B3 (30) as well as in neurons of the parietal and visceral ganglia of hibernating and nonhibernating snails (*Helix pomatia*). The tissue surrounding the neurons was destroyed and disintegrated as far as possible by means of microneedles and/or by a 30 min incubation in pronase Ringer's solution (20 mg pronase/ml Ringer's solution). Changes of the neuronal membrane properties have not been observed after these procedures. This method of preparation allowed us to impale the neurons by two separated microelectrodes. The disintegration of the surrounding tissue has the further advantage that extracellular potassium accumulation is considerably reduced (23). The technique for neuronal isolation has been described elsewhere (31).

The bath fluid was snail saline solution (Meng's solution I) (20) buffered by Tris-chloride, the pH ranging from 7.2 to 7.3. The experimental chamber was continuously perfused, the temperature of the fluid being kept constant at 20 ($\pm$1)°C. PTZ was admixed to the saline solution to give concentrations between 30 and 70 mmoles/liter.

Two separated glass microelectrodes were inserted into the neurons for intracellular recording and current injections. The voltage and current electrodes were filled with 3 moles/liter KCl or 1 mole/liter K-citrate and 0.5 mole/liter K-sulfate or 3 moles/liter Tris-propionate, respectively. To reduce the bath potential below 2 mV, the recording microelectrode and the bath fluid were connected to the amplifier by Ringer agar-AgAgCl-bridges. For potential recording, current injection, and voltage clamp measurements, conventional electronic techniques were applied.

Experiments in the Cat

Intracellular recordings were performed on pyramidal and nonpyramidal tract cells in the gyrus sigmoideus anterior and posterior. Seizure activity was elicited by intravenous administration of PTZ. Further technical details have been described elsewhere (32,33).

Experiments in Cultured Dorsal Root Ganglion Cells of the Rat

The explants were prepared from fetuses of 20 days' gestational age (4). The tissue was superfused with Grey's balanced salt solution with the temperature, pH, and osmolarity kept at normal levels. Intracellular recordings were performed between the second and third week after the ganglia had been explanted. PTZ was added to the control solution to give concentrations between 5 and 20 mmoles/liter. For intracellular recording and stimulation, conventional techniques were applied.

RESULTS AND DISCUSSION

Actions of PTZ on Resting Membrane Potential and on Discharge Pattern of Helix Neurons

In a first series of experiments, the actions of PTZ on single snail neurons were studied after the units had been isolated from synaptic inputs (31).

These investigations showed that the changes of the resting membrane potential (RMP) and of the resulting discharge pattern depended on the concentration of PTZ in the bath fluid.

With low PTZ concentrations (i.e., 10 to 20 mmoles/liter bath fluid) most often a small depolarization developed (Fig. 1A). Only a few neurons showed a transient hyperpolarization after PTZ had been administered (see ref. 25). A transient hyperpolarization of a similar amount occurred more frequently if the PTZ concentration was increased slowly (see Fig. 4B). During the depolarization, primarily silent neurons started to discharge and active neurons increased their discharge frequency. In both instances, the discharge pattern rapidly changed to short burst activity, which remained constant throughout the PTZ application (see refs. 6 and 15–17). These events were accompanied by alterations of the action potentials (see below).

After PTZ had been administered in medium concentrations (i.e., 30 to 60 mmoles PTZ/liter bath fluid), the changes in RMP and in discharge pattern at first resembled those described for low concentrations. Within 10 min, however, typical PDS developed (Fig. 1B). As shown by the tracings in Fig. 1B2, these PDS were characterized by an initial high frequency burst of action potentials, a sustained depolarization, and sinusoidal oscillations which finally turned over to a fast spike-like deflection. This sharp deflection was always followed by a rapid re- and hyperpolarization. In an

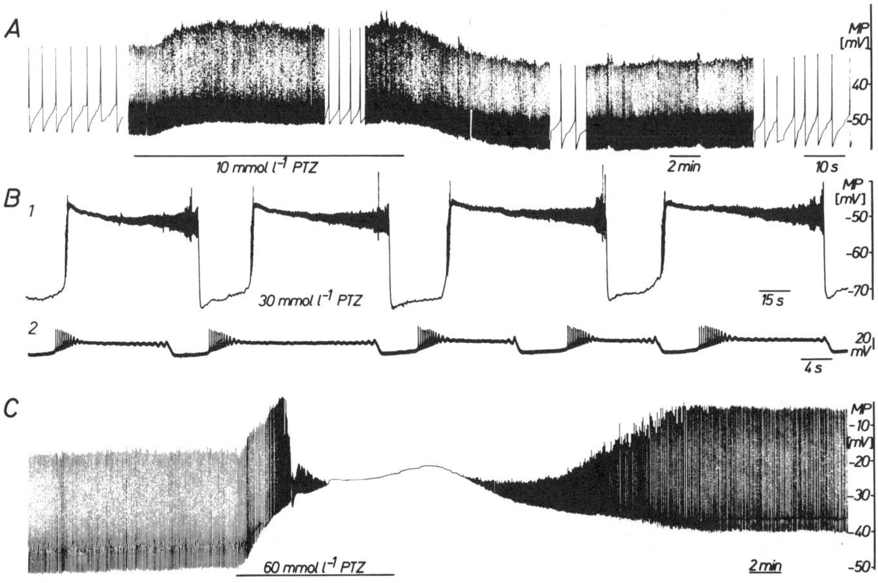

FIG. 1. Actions of PTZ in different concentrations on single neurons of *Helix pomatia*. **A, B1, B2,** and **C:** Neuron B3 of the buccal ganglion. Inkwriter tracings. B2: PDS occurring in an isolated neuron of the visceral ganglion after PTZ (30 mmoles/liter) had been applied. Oscilloscope tracings. (Experiments in collaboration with G. G. Küther.)

early stage, the PDS were comparable to the *ondes triangulaires* and *ondes carées* described by Chalazonitis and Takeuchi (6). With ongoing PTZ application, the PDS increased in duration and finally lasted up to some minutes. These findings were confirmed by experiments of Faugier-Grimaud (12) and Williamson and Crill (34) in other snail species.

It has often been reported (see refs. 11,16, and 17) that the occurrence of burst activity and of PDS is accompanied by the induction or accentuation of an anomalous rectification of the membrane. Only in some instances in the present study was the development of an anomalous rectification associated with burst and PDS activity found. Figure 2 shows a typical experiment. When the RMP as a function of time was superimposed on the I-V curves, it became apparent that the PDS originated from the lower

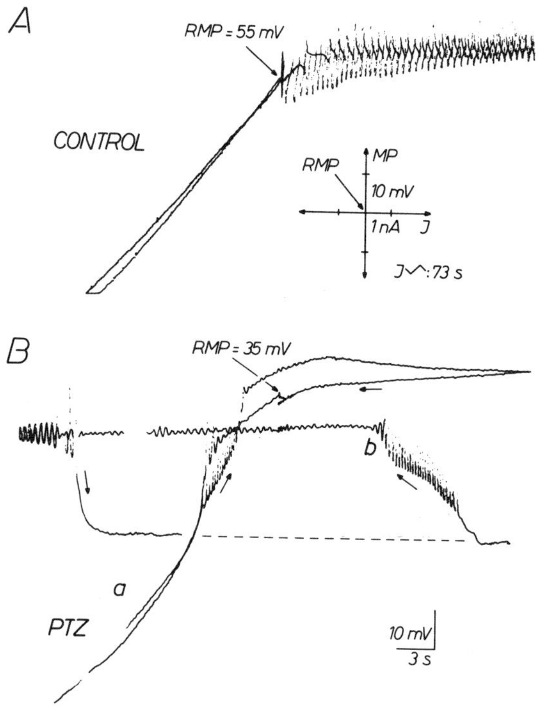

FIG. 2. Membrane resistance of neuron B3 of the buccal ganglion of *Helix pomatia* before **(A)** and after **(B)** PTZ (50 mmoles/liter) had been administered and its relationship to paroxysmal activity. The membrane resistance is displayed as current voltage relation. Triangular hyper- and depolarizing currents were injected through a separate microelectrode starting from the RMP. Duration of the whole current cycle, 73 sec. The directions of the current-induced RMP shifts are marked by arrows. After PTZ application, an anomalous rectification developed. The RMP as a function of time **(b)** is superimposed on the I-V curves **(a).** The sweep direction is indicated by arrows. The paroxysmal depolarization shift **(B)** started from the lower rectification point. (Experiments in collaboration with G. G. Küther.)

rectification point. Summarizing these findings, one can state that the appearance of an anomalous rectification was not a prerequisite for the occurrence of paroxysmal activity but facilitated the development of bistable states of the RMP.

In Fig. 3, the PDS led from a pyramidal tract cell of the cat and from isolated Helix neurons are compared. At first, the curves show a basic similarity of the paroxysmal discharges, but there are also some differences. In Helix, the depolarization remained stable throughout the whole PDS, except for the small sinusoidal oscillations. In the cat, on the other hand, the depolarization shifts of the pyramidal tract cells were superimposed with considerable fluctuations associated with fast spike-like deflections. These waves appeared with maximum probability 80 to 120 msec after the onset of the PDS. There are reasons to assume that they represent postsynaptic potentials elicited by reverberating excitation processes in thalamocortical circuits. Apparently, they modulate a basic form of PDS which is essentially identical with that in Helix neurons.

In snail neurons, PDS could be evoked only if PTZ in concentrations of 30 to 60 mmoles/liter had been administered. With high concentrations (i.e., 60 and more mmoles PTZ/liter bath fluid), a steep depolarization occurred until spike generation was blocked (Fig. 1C). Washing the preparation in normal bath fluid led to a reversal of the PTZ effects. The neuronal inactiva-

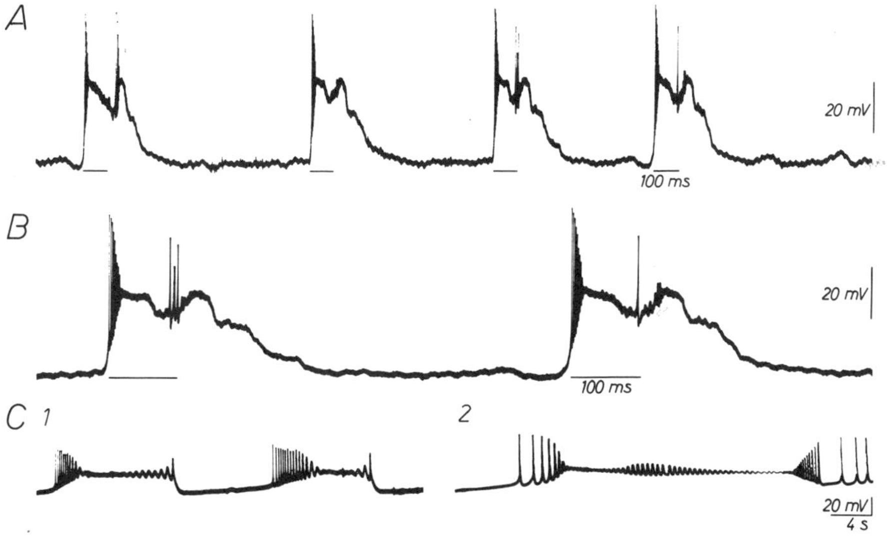

FIG. 3. Comparison of PDS elicited by PTZ in a pyramidal tract cell of the cat cortex **(A,B)** and in synaptically isolated neurons of the visceral ganglion of *Helix pomatia* **(C1, C2).** The PDS in the cat are superimposed by depolarizing deflections and spikelike discharges occurring with maximum probability 100 msec after the onset of the PDS (horizontal bars in **A** and **B**).

tion elicited by PTZ in a high concentration is in accordance with observations in mammals. In these animals, single unit activity, as well as the EEG waves, disappeared if the PTZ administration surpassed a critical extent. In summary, the observations made on neurons isolated from synaptic inputs suggest that the generation of PDS is, in principle, independent from synaptic activities and represents an "intrinsic" membrane mechanism.

PTZ-Induced Changes of Action Potentials in Helix Neurons

It has already been mentioned that the generation of action potentials (AP) was altered after PTZ administration. Therefore, the changes in AP and their relationship to shifts of the RMP were studied in more detail under the influence of increasing PTZ concentrations. To this purpose, AP were elicited by injection of depolarizing current pulses between burst discharges and PDS. A typical experiment is presented in Fig. 4A. The graphical

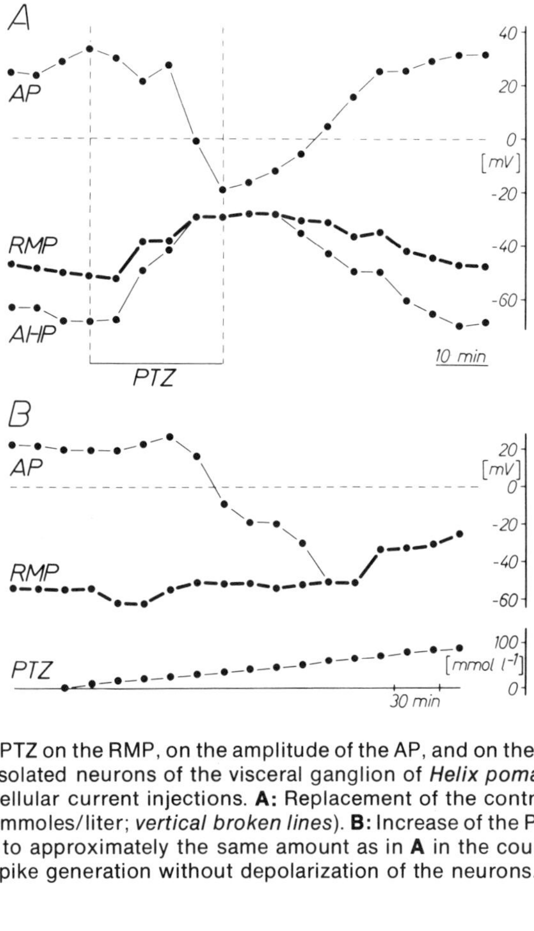

FIG. 4. Effects of PTZ on the RMP, on the amplitude of the AP, and on the afterhyperpolarization (AHP) in isolated neurons of the visceral ganglion of *Helix pomatia.* The AP were elicited by intracellular current injections. **A:** Replacement of the control bath fluid by a PTZ solution (70 mmoles/liter; *vertical broken lines*). **B:** Increase of the PTZ concentration in the bath fluid to approximately the same amount as in **A** in the course of about 4 hr. PTZ depressed spike generation without depolarization of the neurons.

evaluation indicates that the AP began to decrease in amplitude within a few minutes after PTZ had been applied. This diminution was often interrupted by a transient reincrease (Fig. 4A). The decrease in amplitude coincided with a reduction of the afterhyperpolarization (see also Fig. 1A) and an increase in the width of the AP. These changes ran in parallel to a progressive lowering of the RMP. Therefore, the question arises of whether the alterations of AP can be explained by the concomitant variation of the RMP alone. An answer was given by experiments in the course of which the concentration of PTZ was increased slowly to the same values as in the case of Fig. 4A. The graphical evaluation of such an experiment is shown in Fig. 4B. It reveals that the RMP failed to decrease when the PTZ concentration was raised to 60 to 70 mmoles/liter in the course of about 4 hr. AP, however, regularly showed the same changes as when the concentration was increased at a higher rate. The findings indicate that PTZ caused a complete inactivation of the spike generator, even if the RMP was unchanged. These results correspond with the reduction of the early inward and late outward currents found by Klee et al. (17) in Aplysia neurons after PTZ treatment.

Effects of PTZ on Cultured Dorsal Root Ganglion Cells of the Rat

The observations in neurons of invertebrates raise the question of whether the described reactions of RMP and AP can be elicited in isolated neurons of mammals as well. To give a first answer, dorsal root ganglion cells of the rat were investigated in tissue culture (4). This experimental procedure guaranteed a controlled isolation of the units.

Administration of PTZ in concentrations of 5 to 20 mmoles/liter evoked a sequence of RMP fluctuations. At first, a hyperpolarization appeared (Fig. 5A). It was reduced in amplitude when the preexisting RMP increased and failed to occur at RMP values of about −80 mV (Fig. 5B). Furthermore, it was associated with a decrease in membrane resistance (Fig. 5C). These findings suggest that the hyperpolarization is due to an increase in the permeability of the membrane for potassium and/or chloride ions (see ref. 25). When the administration of PTZ was continued, a depolarization developed (Fig. 5A). During this RMP shift, the membrane resistance was found to rise beyond initial values (Fig. 5C). The described sequence of RMP shifts proved independent of the actual PTZ concentration. PDS-like RMP fluctuations were not observed during the phase of PTZ administration or in the washing period.

Together with the RMP, AP of the cultured neurons evoked by outward current pulses were examined during PTZ administration. They exhibited the same impairments as described for snail neurons. As shown in Fig. 5B, the amplitude, slope of rise and fall, and hyperpolarizing afterpotentials were progressively diminished following PTZ administration. In some neu-

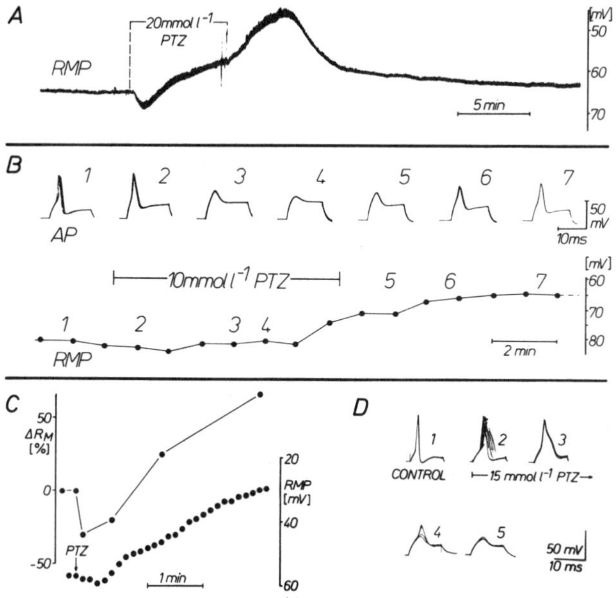

FIG. 5. Actions of PTZ on RMP **(A–C)**, on AP **(B, D)**, and on membrane resistance (R_M) **(C)** of cultured dorsal root ganglion cells of the rat. **A:** RMP was traced by an inkwriter. **B, D:** AP were elicited by intracellular current injection using a bridge circuit. AP in **B** are related to the RMP by corresponding numbers. **C:** Changes in R_M (ΔR_M) and in RMP were evoked by application of 20 mmoles PTZ/liter bath fluid. (Experiments in collaboration with D. Bingmann and F. Pietruschka.)

rons, the AP transiently developed a pronounced shoulder (Fig. 5D). The inactivation of the spike generator proved to be independent from the PTZ-induced RMP changes (Fig. 5B).

In summary, the experiments showed that single ganglion cells cultured from the dorsal root of the rat were unable to develop PDS, whereas the responses of the RMP and of the AP to PTZ resembled those in the snail neurons.

Early Outward Current in Relation to PTZ-Induced RMP Changes in Helix Neurons

In a final series of experiments, an attempt was made to determine which ionic mechanisms are underlying the PTZ-induced RMP changes in Helix neurons. The investigations focused on the so-called early outward current (I_A). Since this current has been found to counteract neuronal depolarizations, its activation or inactivation could be important for the genesis of PDS (7,11,14,22).

The voltage clamp recordings in Fig. 6A and B show that, besides the conventional early inward and late outward currents, an early outward cur-

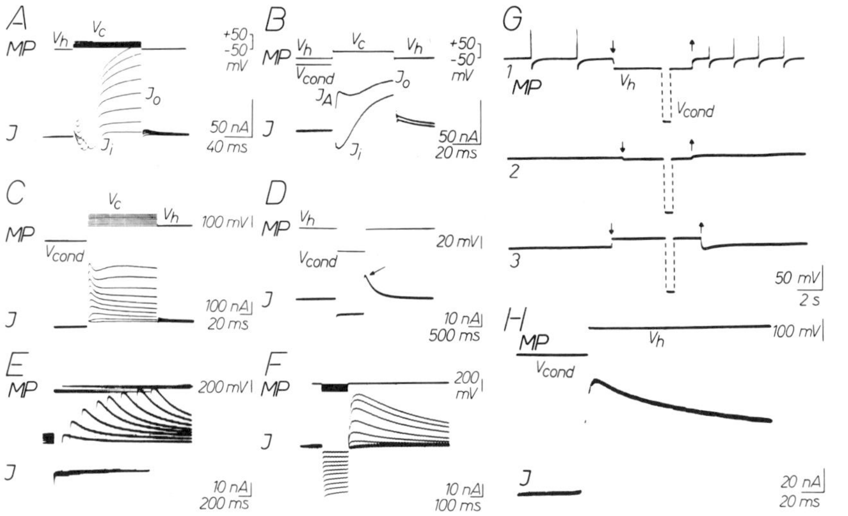

FIG. 6. Activation of the early outward current (I_A) during persistent and intermittent voltage clamp measurements. Neurons of the visceral ganglion of *Helix pomatia.* **A:** Ordinary inward (I_i) and outward (I_o) currents evoked by changing the membrane potential (MP) from the holding potential (V_h) to various command potentials (V_c). **B:** Appearance of the I_A when the V_c is preceded by a hyperpolarizing conditioning pulse (V_{cond}). **C:** Superimposed currents evoked by varying the V_c with the V_{cond} kept constant. **D:** Activation of the I_A (*arrow*) by applying only the V_{cond}. **E, F:** Dependency of the I_A from the duration **(E)** and the amplitude **(F)** of the V_{cond}. Experiment as in **D. G:** Intermittent voltage clamp at normal (2) and artificially decreased (1) and increased (3) MP. The MP was shifted by current injection. The period of voltage clamp is marked by arrows. **H:** Superimposition of the I_A obtained from the voltage clamp measurements in **G.** (Experiments in collaboration with F. Zimpel.)

rent became visible also in the investigated neurons. A prerequisite for the appearance of the I_A was a hyperpolarizing conditioning pulse preceding the command pulse. This finding is illustrated by the tracings in Fig. 6B. In a first sweep, the command pulse started from the holding potential near the normal RMP; and in a second sweep, a conditioning pulse was applied additionally. A net I_A became apparent only in the latter case. The activation characteristics of the I_A in these neurons showed similar current voltage relationships as found in other cells (Fig. 6C).

The particular aim of the present experiments was to examine the I_A in relation to the different PTZ-induced neuronal activities. Therefore, it had at first to be clarified what sequence of clamp pulses are applicable (a) to avoid a disturbance of the repeated occurrence of PDS, and (b) to measure the I_A independently from the actual unclamped RMP. Previous investigations had shown that the spontaneous activity of the neurons was disturbed by intermittent voltage clamp measurements to a minimum, if the command pulses were kept subthreshold for the activation of the early inward and late outward currents. For this reason, the I_A was elicited by switching the conditioning pulse directly to the holding potential (Fig. 6D). Thus, however, the amplitude of the I_A became comparatively small. The conditioning pulses were therefore adjusted to give the maximum I_A (Fig. 6E and F). As a rule, a pulse of 1 sec in duration and of 100 mV in amplitude proved most suitable. An independency of the I_A from the actual unclamped RMP could be achieved by clamping the membrane at a constant holding potential for about 3 to 4 sec before the conditioning pulse was applied. This finding is illustrated by an experiment without PTZ treatment in Fig. 6G and H. The superimposition of the current recordings in Fig. 6H shows that amplitude and time course of the I_A remained unchanged whether the clamp procedure started from depolarized, normal, or hyperpolarized RMP levels (Fig. 6G).

The described voltage clamp procedure was at first applied to measure the I_A during spontaneously occurring PDS elicited by PTZ in medium concentrations. As shown in Fig. 7A, with the appearance of the first PDS, the mean amplitude of the I_A was enhanced for about 20 min and returned thereafter to initial values. Besides such long-lasting fluctuations, variations in amplitude appeared which were strongly associated with the different phases of PDS activity. Each neuronal depolarization coincided with a depression of the I_A and each hyperpolarization with an increase of this current. This strong relationship between the activation of the I_A and the different phases of the PDS was found in several experiments for some hours. With the progressive depolarizations elicited by PTZ in high concentrations, the I_A showed a monodirectional decrease in amplitude (Fig. 7B). After reapplication of the control solution, the I_A increased rapidly and often reached the initial amount within 1 to 2 min after washing had started. In contrast to the I_A, the RMP returned only slowly to initial levels. The described findings are in general accordance with observations of Williamson and Crill (35).

As a cause for the reduction of the I_A during neuronal depolarization, a shift in the potassium equilibrium potential (E_{K^+}) must be considered (21). Paroxysmal neuronal activity could cause a K^+ accumulation in the extracellular space and therewith a reduction of E_{K^+}. However, from the preparation techniques applied, this possibility is not likely. Furthermore, the E_{K^+} could be decreased by a diminution of the K^+ activity in the intracellular space due to the decrease of the membrane resistance after PTZ treatment (36). Also, this interpretation is unlikely because the paroxysmal depolarizations were followed by hyperpolarizations which reached the control E_{K^+} and which were associated with an increase of the I_A beyond initial levels.

In summary, the changes of the measured I_A proved independent from actual fluctuations of the RMP and of the E_{K^+}. Therefore, the conclusion is drawn that variations in the amplitude of the I_A reflect changes in the activations of this current. Correspondingly, it may be assumed that periodic

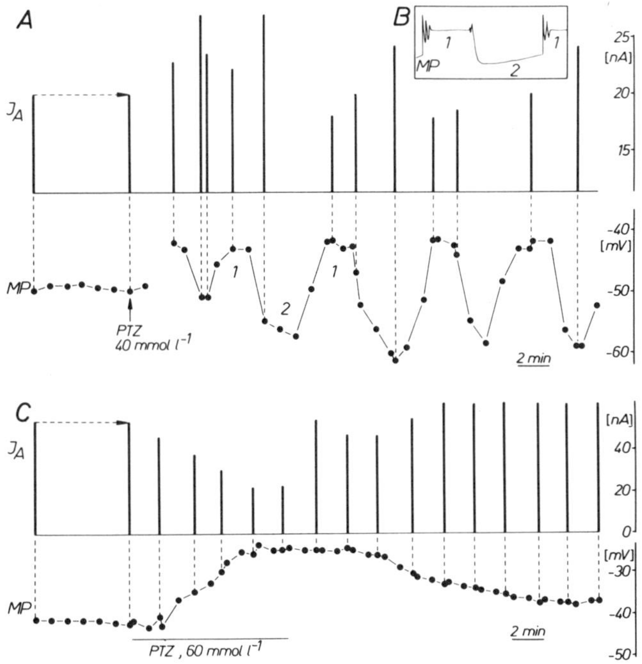

FIG. 7. Early outward current (I_A) in relation to PDS **(A)** and to a progressive neuronal depolarization **(C)** evoked by PTZ. Neurons of the visceral ganglion of *Helix pomatia*. **A:** Fluctuations of the I_A during repeated PDS after administration of 40 mmoles PTZ/liter bath fluid. The current measurements were performed as shown in Fig. 6G. Corresponding unclamped membrane potential (MP) changes, as shown by the schematic drawing in the inset **B,** are related to the spontaneous MP changes in the voltage clamp experiment in **A** by numbers. **C:** Depression of the I_A during a progressive diminution of the MP after 60 mmoles PTZ/liter bath fluid was applied. Voltage clamp procedure as in Fig. 6G. (Experiments in collaboration with F. Zimpel.)

activations and inactivations of the I_A and of inward currents (9) contribute essentially to the RMP fluctuations elicited by PTZ.

REFERENCES

1. Ajmone-Marsan, C. (1969): In: *Basic Mechanisms of the Epilepsies,* edited by H. H. Jasper, A. A. Ward, and A. Pope, pp. 299–319. Little Brown, Boston.
2. Ayala, G. F., Matsumoto, H., and Gumnit, R. J. (1970): *J. Neurophysiol.,* 33:73–85.
3. Ayala, G. F., and Vasconetto, C. (1972): *Electroencephalogr. Clin. Neurophysiol.,* 33: 96–98.
4. Bingmann, D., Speckmann, E.-J., and Pietruschka, F. (1977): *Neurosci. Lett.,* 4:73–76.
5. Calvin, W. H., Ojemann, G. A., and Ward, A. A., Jr. (1973): *Electroencephalogr. Clin. Neurophysiol.,* 34:337–351.
6. Chalazonitis, N., and Takeuchi, H. (1969): *C. R. Soc. Biol.,* 162:1552–1556.
7. Connor, J. A., and Stevens, C. F. (1971): *J. Physiol. (Lond.),* 213:21–30.
8. Dichter, M., and Spencer, W. A. (1969): *J. Neurophysiol.,* 32:663–687.
9. Ducreux, C., and Gola, M. (1975): *Pfluegers Arch.,* 361:43–53.

10. Ducreux, C., Takeuchi, H., and Chalazonitis, N. (1972): *C. R. Acad. Soc. (Paris)*, 275: 1907–1909.
11. Faber, D. S., and Klee, M. R. (1972): *Nature* [*New Biol.*]. 240:29–31.
12. Faugier-Grimaud, S. (1974): *Brain Res.*, 69:354–360.
13. Fischer, R. S., Pedley, T. A., Moody, W. J., and Prince, D. A. (1976): *Arch. Neurol.*, 33:76–83.
14. Gola, M., and Romey, G. (1971): *Pfluegers Arch.*, 327:105–131.
15. Johnson, W. L., and O'Leary, J. L. (1965): *Arch. Neurol.*, 12:113–127.
16. Klee, M. R. (1976): In: *Neurobiology of Invertebrates. Gastropoda Brain,* edited by J. Salanki, pp. 267–286. Publishing House of the Hungarian Academy of Sciences, Budapest.
17. Klee, M. R., Faber, D. S., and Heiss, W. D. (1973): *Science,* 179:1133–1136.
18. Matsumoto, H., and Ajmone-Marsan, C. (1964): *Exp. Neurol.*, 9:286–304.
19. Matsumoto, H., Ayala, G. F., and Gumnit, R. J. (1969): *J. Neurophysiol.*, 32:688–703.
20. Meng, K. (1960): *Zool. Jb.*, 68:539–566.
21. Neher, E. (1971): *J. Gen. Physiol.*, 58:36–53.
22. Neher, E., and Lux, H. D. (1971): *Pfluegers Arch.*, 322:35–38.
23. Nicholls, J. G., and Baylor, D. A. (1968): *Proc. Int. Union of Physiol. Sci.*, 6:117.
24. Pollen, D. A., and Trachtenberg, M. C. (1970): *Science,* 167:1252–1253.
25. Prichard, J. W. (1971): *Brain Res.*, 27:414–417.
26. Prince, D. A. (1967): *Neurology,* 23:83–84.
27. Prince, D. A. (1968): *Exptl. Neurol.*, 21:467–485.
28. Prince, D. A. (1969): *Electroenceph. Clin. Neurophysiol.*, 26:476–487.
29. Prince, D. A. (1974): In: *Handbook of Electroencephalography and Clinical Neurophysiology,* Vol. 2C, edited by A. Rémond, pp. 56–70. Elsevier Publishing Company, Amsterdam.
30. Schulze, H., Speckmann, E.-J., Kuhlmann, D., and Caspers, H. (1975): *Neuroscience Letters,* 1:277–281.
31. Speckmann, E.-J., and Caspers, H. (1973): *Epilepsia,* 14:397–408.
32. Speckmann, E.-J., Caspers, H., and Janzen, R. W. (1972): In: *Synchronization of EEG Activity in Epilepsies,* edited by H. Petsche and M. A. B. Brazier, pp. 93–111. Springer, New York.
33. Speckmann, E.-J., Caspers, H., and Janzen, R. W. C. (1977): In: *Architectonics of the Cerebral Cortex, IBRO Monograph Series, Vol. 3,* edited by M. A. B. Brazier and H. Petsche, pp. 191–209. Raven Press, New York.
34. Williamson, T. L., and Crill, W. E. (1976): *Brain Res.*, 116:217–229.
35. Williamson, T. L., and Crill, W. E. (1976): *Brain Res.*, 116:231–249.
36. Zidek, W., Speckmann, E.-J., and Caspers, H. (1976): *Pfluegers Arch.*, 362:R32.

Abnormal Neuronal Discharges, edited by N. Chalazonitis and M. Boisson. Raven Press, New York © 1978.

On the Behavior of Snail Neurons in the Presence of Cocaine

E. Lábos and E. Láng

Semmelweis University Medical School, First Department of Anatomy, Budapest, Hungary

During the past decade, investigations of the mechanisms underlying drug-induced central excitations and convulsions have been extended to lower animals as affording preparations suitable to perform experiments more easily clarifying cellular mechanisms. The knowledge acquired investigating lower organisms could be applied (although with obvious limitations) to interpret phenomena observed in higher animals.

The oscillations and paroxysmal depolarization shifts observed by Chalazonitis and Takeuchi (3) on snail giant neurons initiated detailed analytical works testing a wide variety of drugs on similar preparations contributing to the understanding of convulsions (see, e.g., ref. 7).

The actions of cocaine are classified as follows: (a) local anesthetic effect, (b) central excitation or convulsions, and (c) sensitization to catecholamines. Its local anesthetic effect has been demonstrated on molluscan or other invertebrate preparations (5,9,10). Explicitly excitatory responses also could be elicited, e.g., in larval molluscan neuromuscular system (7,8). However, effects related to the catecholamines were demonstrated in vertebrates (2,4).

Our aim was to describe some of those phenomena evoked by cocaine, which could not be regarded as results of its local anesthetic component of action. At the same time, the preparation utilized in the experiments (giant neurons of *Helix pomatia*) permits direct comparisons with other drug-induced excitations.

METHODS

The experiments were carried out partly during the spring on freshly collected *Helix pomatia* land snails and also with animals previously hibernated during winter. The subesophageal ganglion complex was isolated and kept in physiological solution at room temperature (17 to 25°C). The composition of the physiological solution was as follows: 51.7 mM NaCl, 4.7 mM KCl, 8 mM $MgCl_2$, and 10 mM $CaCl_2$.

In some experiments, the concentration of $CaCl_2$ was decreased or increased to 1 or 30 mM, respectively; this change was compensated by an equivalent amount of NaCl to keep the osmolarity at the level of the original solution. The pH of the solution was not controlled by buffers or $NaHCO_3$ and was, consequently, slightly acidic. Cocaine HCl was applied dropping its stock solution of 1 mg/ml into the bathing fluid of 3 ml volume. This method of application was mechanically nondisturbing; however, the appearance of effects was delayed by the diffusion time. The final concentrations of the cocaine in the different experiments were in the range of 0.3 to 1.5 mM/liter. Connective tissue layers covering the neurons were carefully removed. The recording was carried out with 2.5 mM KCl-filled micropipettes of 3 to 15 megohms resistance and with high impedence input units. The depolarization of the somatic membrane was realized by a bridge circuit through the recording pipette. Low intensities of injecting currents were applied in square-wave form at low frequency to avoid effects originating from stimulus repetitions. The identification of neurons was based on the topography and the characteristics of firing or individual spikes.

RESULTS

A map of some well-known identifiable neurons is demonstrated in Fig. 1. The effects of cocaine were tested on neurons most often localized in the right parietal and visceral ganglia of the subesophageal ganglion complex.

It has been generally observed that the neurons with either pacemaker-generated or synaptically evoked spontaneous activity showed a transient increase of firing rate followed by a decreased rate of discharges. During the drug effect, the shape of the action potential changes: (a) peak-to-peak amplitude of spikes gradually decreased; (b) duration of spikes measured

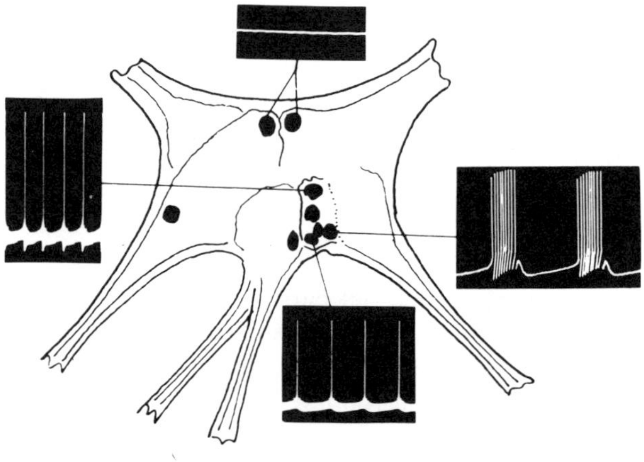

FIG. 1. Map of certain reference neurons. Dorsal surface of Helix subesophageal ganglion complex. The caudal part of this diagram is redrawn to show the localization of neurons in the following figures.

at the half-amplitude increased significantly; (c) at the final stage of effect, the soma failed to discharge. This slowing down of somatic potential generation could be observed in the phase of increased firing rate. The resting membrane potential did not respond to the drug application. However, sometimes a slight decrease in the resting membrane potential could be observed.

The interval histograms of neuronal spike discharges after the washing out of the drug usually were not identical to those of the control period.

Appearance of Spike Doublets and Triplets

The phase of either increased or decreased firing rate was often accompanied by the appearance of characteristic groups, consisting of two or three spikes and also of subthreshold potentials (Figs. 2C, E; 5B–D; 6D, F, G; 7). This phenomenon has been observed in different identifiable cells. However, the time intervals between the components of doublets were more or less

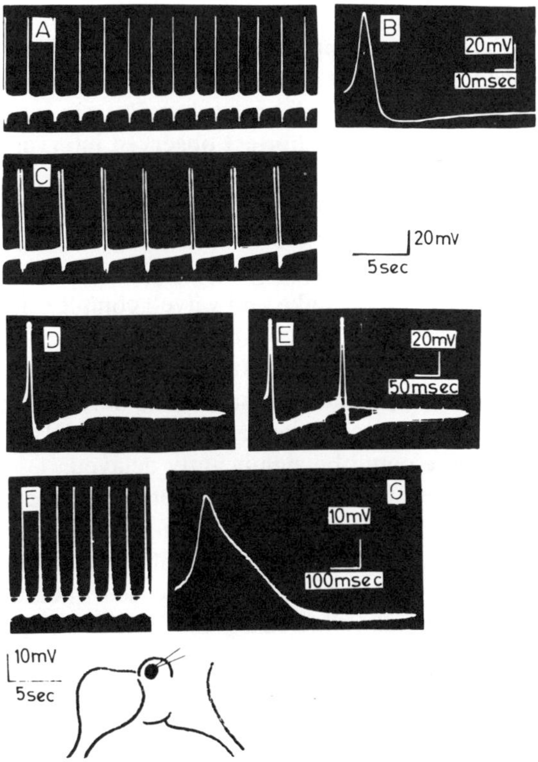

FIG. 2. Pacemaker neuron. Effect of 400 μg/ml cocaine, 10 mM $CaCl_2$ in the physiological solution. **A–B:** Control; **C–E:** between 10 and 25 min; **F–G:** after 30 min.

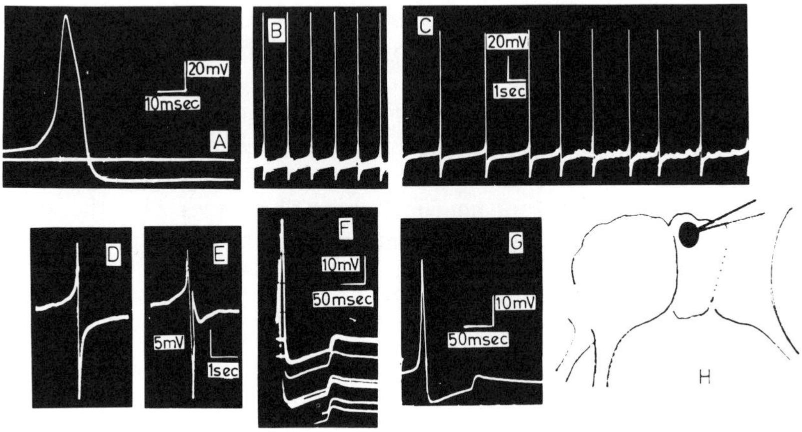

FIG. 3. Pacemaker neuron. Effect of about 150 μg/ml cocaine in physiological solution of 10 mM $CaCl_2$. **A–B:** Control; **C:** sample between 10 and 20 min; **D:** control; **E–G:** samples after 20 min; **H:** localization of the cell.

different in various cases. Figure 4 shows how intervals measured at successive doublets varied in a sample of a short period. This nonmonotonic increase of intervals was usually seen. At the onset of the sample, intervals of approximately 200 msec form the majority. Greater intervals of 250 to 400 msec occur more often. The shortest observed interval was about 100 msec (Fig. 7).

In extreme cases, instead of doublets of spikes only complexes of spikes and waves appeared (Fig. 3); triplets of spikes together with subthreshold potentials formed the complexes (Fig. 7). It is remarkable in the case demonstrated in Fig. 3 when a "spike and wave" complex appeared that the intervals between them were fairly constant (~100 msec).

In other cases, however, this temporal relationship was not so regular. Figure 7 shows that the appearance of second or third spikes could be preceded by the invasion of one or more generator potentials, and the appearance of the second spike could be significantly delayed.

The intervals measured between the doublets (or triplets) were usually greater than the average interval in the control period (Figs. 2C, 5D, 6F). This clearly suggests that the second spikes of the doublets influence the length of the following interval. Even in those cases when doublets are not generated, only single spikes followed by subthreshold depolarization, the interspike intervals tend to be separated into classes. Such a classification of intervals is demonstrated in Fig. 12. In the control period, only one type of spike appears (Fig. 12C, D, H). However, when the drug effect has already been developed, two types of spikes could be seen: spikes B followed by subthreshold potentials at regular intervals (Fig. 12E, G), and spikes A which were not followed. The four classes of intervals determined by these

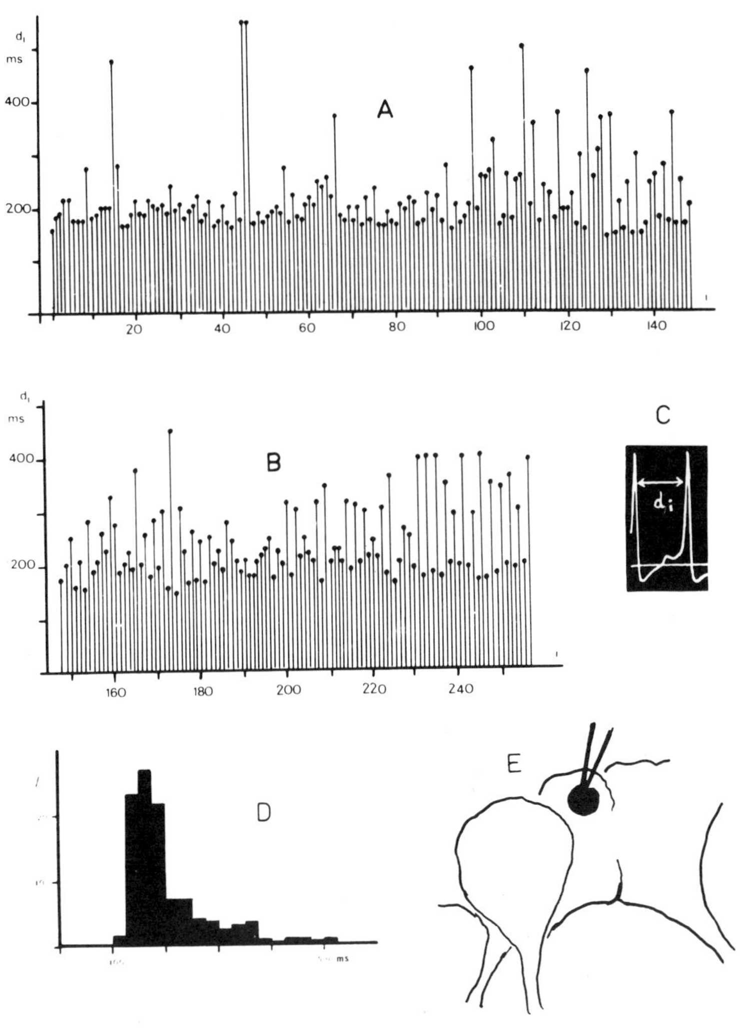

FIG. 4. Pacemaker neuron. **A–B:** The d_i interspike intervals are plotted against their sequence number; **C:** definition of d_i; **D:** distribution of intervals; **E:** localization of the neuron. The long intervals between doublets were not included in the graph.

two kinds of events (AA, AB, BA, or BB intervals) are distributed in different ways.

Depolarization Shifts

In our experiments, paroxysmal depolarization shifts comparable in duration or extent to those observed by Chalazonitis and Takeuchi (3) in the presence of metrazol were not observed at any applied concentration of

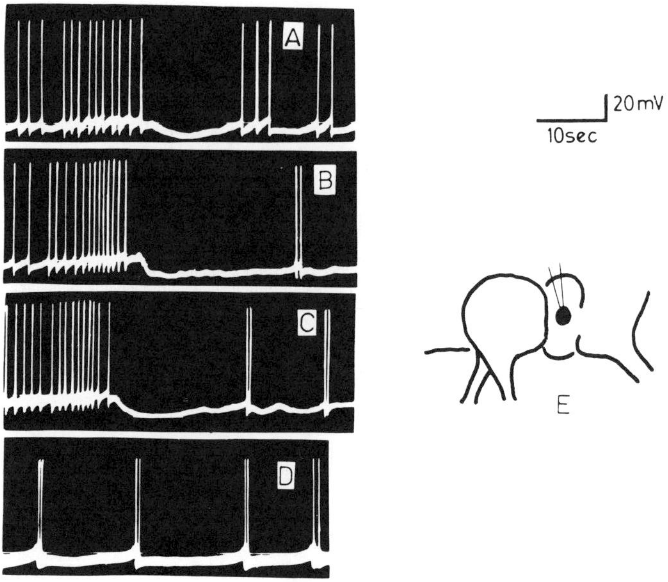

Fig. 5. Neuron with synaptic inhibitions. Effect of about 1 mg/ml cocaine, 10 mM $CaCl_2$ in the bathing fluid. **A:** Control; **B–D:** samples after 10 min; **E:** localization.

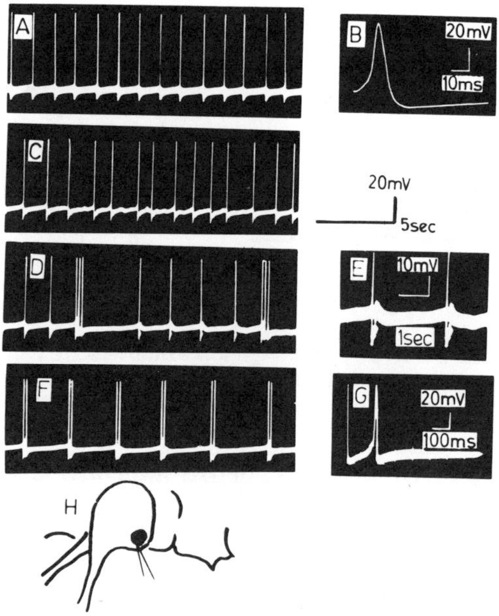

FIG. 6. Neuron in the visceral ganglion. Effect of about 500 μg/ml cocaine, 10 mM $CaCl_2$. **A–B:** Control; **C–G:** successive samples at the development of the effect after 10 min of the drug application. Note in **E** the appearance of triplets. **H:** Localization of the neuron.

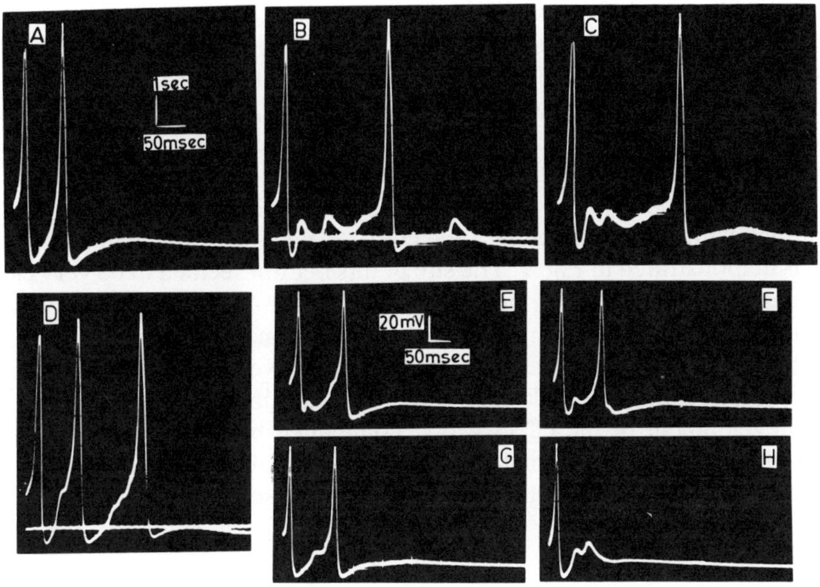

FIG. 7. Complexes consisting of spikes and subthreshold synaptic and axonal potentials to demonstrate various types.

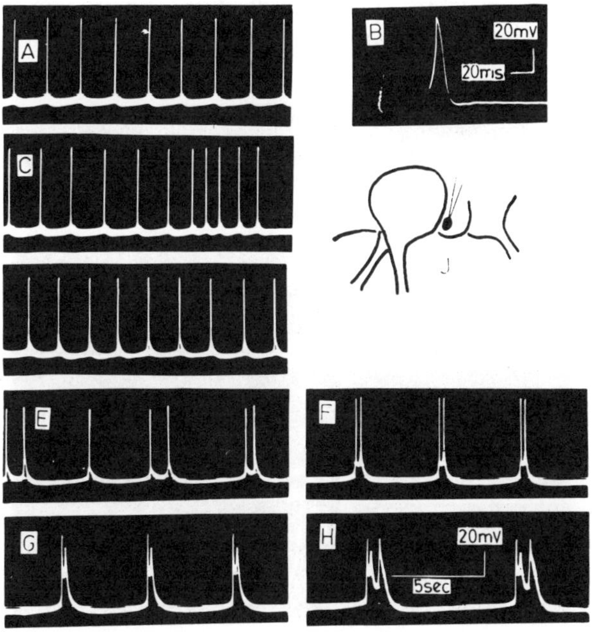

FIG. 8. Depolarization shift. **A–B:** Recordings before application; **J:** localization of the neuron; **C–H:** successive samples of activity between 10 and 30 min after drug application; 1 mg/ml cocaine, 10 mM $CaCl_2$.

$CaCl_2$ in the physiological solution. Nevertheless, two identifiable neurons did react to the drug by characteristic longer-lasting depolarizations. The first of these neurons is identifiable by its localization and on the basis of its characteristic slow generator potential (Fig. 8A, J). The effect of cocaine was manifested in the generation of doublets of spikes significantly different from those already described. These doublets are "sitting" on the plateau of a depolarization shift (Fig. 8F). In the later phase of the effect, slow oscillations with a rather peculiar time course developed. A similar phenomenon is demonstrated in Fig. 9. The site of this neuron seemed to be very close (or identical) to that of the neuron in Fig. 8. The second example of such reactions manifested in long-lasting depolarization shifts is that of the Br-like cell of Helix (Fig. 10). A prolongation of active bursts or even the total disappearance of hyperpolarized intervals was observable (Fig. 10C, I). The development of the effect was accompanied by a gradual decrease of the spike amplitude. The final stage of the neuronal behavior is reminiscent of slow nonsinusoidal oscillations and abortive discharges.

Changes in the Excitability of the Soma Membrane

To obtain information concerning the excitability of the somatic membrane, responses to standard depolarizing current injections were compared at the different phases of the drug effect. Figure 11H and J demonstrates a case when the excitability of the membrane was definitively

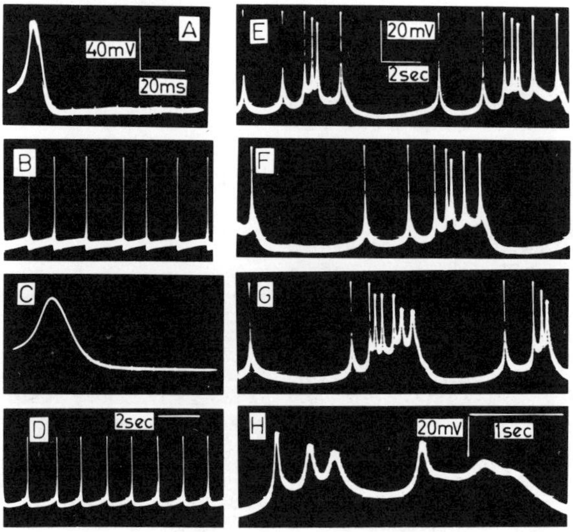

FIG. 9. Depolarization shift. Composition of the bath fluid: 10 mM $CaCl_2$ in the physiological solution, 120 μg/ml cocaine. **A–B:** Control; **C–H:** 18, 20, 30, 32, 40, and 50 min after application, respectively.

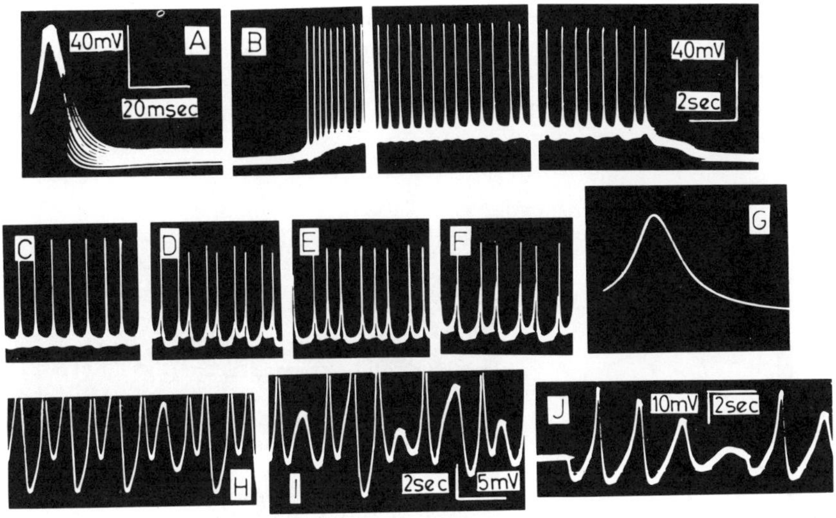

FIG. 10. Neuron with slow waves (1). Effect of 160 μg/ml cocaine, 32 mM $CaCl_2$ in the physiological solution. **A–B:** Controls; **C–I:** 5, 7, 15, 20, 50, 70, and 80 min after application; **J:** 100 min after washing.

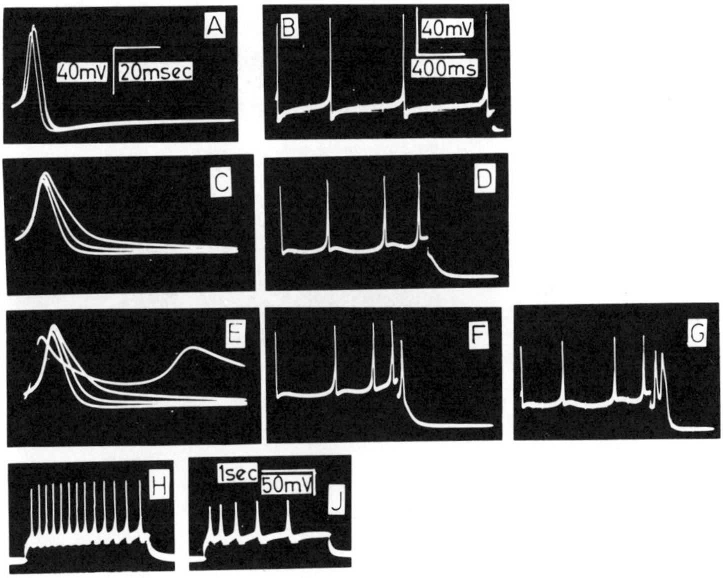

FIG. 11. Responses to soma depolarizations. **A–B:** Controls; **C** and **E:** 10 to 15 min; **D, F,** and **G:** 30 min after application of 120 μg/ml cocaine; 10 mM $CaCl_2$; **H:** control of another cell; **J:** 30 min after applying 160 μg/ml cocaine, 10 mM $CaCl_2$.

decreased, as could be expected on the basis of the local anesthetic effect of cocaine. However, responses seen in Fig. 11C, F, and G clearly show that in a certain phase of the cocaine effect, spikes separated by shorter intervals (as in the control responses) or even unusual slow waves (Fig. 11G) could be evoked.

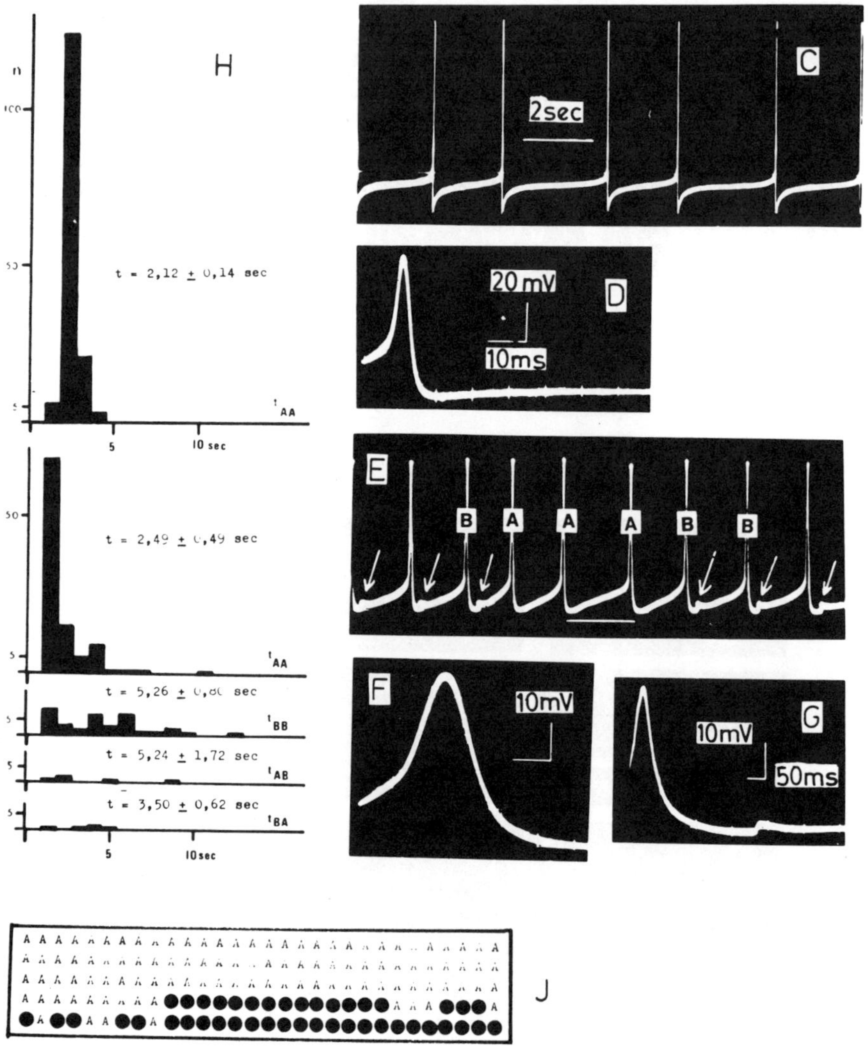

FIG. 12. Complexes of spike and subthreshold potential 1 mM $CaCl_2$ and 240 μg/ml cocaine in the physiological solution. **A–B:** Two different types of spikes, **B** followed by afterpotential as arrows indicate; **C–D:** control; **E–G:** samples between 17 and 40 min after drug application. **J:** types of successive spikes (*filled circle,* spike type **B**). **H:** Five interval histograms are demonstrated from top to bottom: control t_{AA}, t_{AA}, t_{BB}, t_{AB}, or t_{BA} classes of intervals after drug application.

DISCUSSION

The results of our studies can be summarized in the demonstration of "excitatory events" evoked by cocaine on Helix neuronal activity. Pharmacological textbooks usually regard cocaine as a local anesthetic but always describe its well-known central excitatory effects. Phenomena which could be regarded as "excitatory" were manifested in very short bursts (doublets) or relatively short "subparoxysmal" depolarization shifts. A plausible reason for the failure of cocaine to exhibit a more prominent excitation could be the following. Those variations regarded as local anesthetic effects (5) might be realized through effectively opposite variations of neuronal parameters than those which would result in increased rate of firing or appearance of depolarization shifts. At the same time, at least part of the extra spikes was evoked by synaptic activation similar to the effects of convulsant drugs (11). The lack of dramatic differences at lowered or increased $CaCl_2$ concentrations could also be interpreted by opposite complicated influences of Ca ions to the components of cocaine effect.

SUMMARY

The behavior of identified giant neurons in the subesophageal ganglia of *Helix pomatia* was observed by intracellular recording before and after application of cocaine. The neurons responded to the drug by transient increase of their spontaneous firing rate followed by a slowing down. At this final phase or even sooner, the time course of the action potentials changed the local anesthetic component of cocaine action. The transient excitatory phase often appeared in the form of doublets or triplets of spikes and subthreshold potential changes. Paroxysmal depolarization shifts were also observed in the responses of certain neurons to the drug; however, these never reached the degree observed in the case of other drugs. Neuronal responses to somatic depolarization evoked by current injection also showed the signs of excitatory influence of cocaine, e.g., by an increase of the number of responses.

The observed phenomena could represent both the local anesthetic and central excitatory effect of cocaine described for higher animals.

REFERENCES

1. Arvanitaki, A., and Chalazonitis, N. (1961): *Bull. Inst. Oceanogr. Monaco,* 1224.
2. Campos, H. A., and Urquilla, P. R. (1969): *J. Physiol.,* 200:311–320.
3. Chalazonitis, N., and Takeuchi, H. (1968): *C. R. Soc. Biol.,* 162:1552.
4. Fröhlich, A., and Loewi, O. (1910): *Arch. Exp. Pathol. Pharmacol.,* 62:129–169.
5. Just, W. W., and Hoyer, J. (1976): *Experientia,* 33:70–71.
6. Klee, M. R. (1976): In: *Neurobiology of Invertebrates. Gastropoda Brain,* edited by J. Salanki. Akademiai Kiado, Budapest.
7. Lábos, E. (1970): *Ann. Biol. Tihany,* 37:55–71.

8. Lábos, E. (1973): In: *Neurobiology of Invertebrates,* edited by J. Salanki. Akademiai Kiado, Budapest.
9. Láng, E., and Lábos, E. (1971): *37th Annual Meeting of Hungarian Physiological Society.*
10. Parker, G. H. (1910): *J. Exp. Zool.,* 8:1–43.
11. Takeuchi, H. (1971): *Acta Med. Okayama,* 25(6):615–626.

Abnormal Neuronal Discharges, edited by
N. Chalazonitis and M. Boisson. Raven
Press, New York © 1978.

Modulation of Endogenous Discharge in Neuron R_{15} Through Specific Receptors for Several Neurotransmitters

David O. Carpenter, Michael J. McCreery,
Catherine M. Woodbury, and Paul J. Yarowsky

*Neurobiology Department, Armed Forces Radiobiology Research Institute,
Bethesda, Maryland 20014*

For most nerve cells, the occurrence of paroxysmal depolarizing shifts (PDSs) is an abnormal discharge pattern; there are some neurons in both invertebrates and vertebrates, however, for which such a discharge pattern is a normal feature. In this chapter as well as in several that follow, an attempt has been made to study some of the characteristics of epileptiform discharges by studying the normal bursting pacemaker discharge of an identified invertebrate neuron.

Of all the identified neurons from a variety of invertebrates, none has captured quite the attention as cell R_{15} (16) of Aplysia. This cell, called by Strumwasser (32,33) the "parabolic burster," is characterized by a discharge pattern consisting of alternating periods of spiking and silence. This discharge pattern is truly an endogenous pacemaker. Alving (1) found that while all synaptic inputs to Aplysia neurons occur in the neuropile and not on the cell body, pacemaker generation occurs in the soma. She has illustrated an experiment on cell R_{15} (Fig. 4 in ref. 1) in which she tied a ligature between soma and axon-neuropile. In this particular experiment, the cell body with the ligated axon floated free of the ganglion to the opposite corner of the chamber, where Alving penetrated the totally isolated cell and recorded its characteristic bursting pattern.

The ionic mechanisms underlying this discharge have been analyzed in some detail (see ref. 10). The action potentials are primarily due to a Na^+ conductance but with a component due to Ca^{2+} (13). The slow wave underlying the bursting pattern can be separated from the spikes and appears to result from a not totally understood time and voltage variation of Na^+ and K^+ conductances (10,19,20). Although this neuron usually discharges in a bursting pattern, it may also be silent or discharge with a regular beating pattern. The discharge pattern varies with temperature and membrane resistance (9,10). Wilson and Wachtel (38) in R_{15} and Smith et al. (31) in the homologous cell in the land snail Helix have found a negative resistance region in the current-voltage relationship which correlated with the bursting pattern and which they suggest may be required for bursting. The presence of the bursting discharge is also dependent on divalent cation concentrations (3) and is influenced by some peptides, such as vasopressin and oxytocin (4). Strumwasser (32) has reported that cell R_{15} exhibits both a circadian and a lunar rhythm in the endogenous discharge. Although it is not certain that this is truly endogenous rather than an effect mediated through the eye (5), the idea that a single neuron could demonstrate such rhythms when isolated from the animal is provocative and has led some to believe that this cell is the "soul of the sea slug."

Although the pacemaker discharge is a regular feature of this cell, it is modified by synaptic inputs. R_{15} is excited by the identified cholinergic interneuron L_{10} (21). It also receives a large monosynaptic excitatory postsynaptic potential (EPSP) from an axon in the right connective nerve (16). Both of these inputs are blocked by hexamethonium, which appears to be a quite specific inhibitor of depolarizing Na^+-dependent acetylcholine (ACh) responses (28). When the various nerves to and from the abdominal ganglion are stimulated, mixed excitation and inhibition is produced in R_{15} (16). This cell also receives a synaptically mediated "inhibition of long duration," which appears to result from a burst of impulses from an interneuron making inhibitory synapses on R_{15} and which may depress the discharge for periods of many seconds to several minutes (16,35,36). Tauc (35) has noted that this long inhibition may be due to a prolonged action of a neurotransmitter, and has specifically suggested that dopamine (DA) might cause such an effect.

Our laboratory has recently been concerned with the study of receptors for neurotransmitter substances on Aplysia neurons, using ionophoretic drug application (14,27,29,34,39,40). We have described and characterized receptors for a variety of substances, including ACh, DA, octopamine (Oct), phenylethanolamine (Pnol), histamine, γ-aminobutyric acid (GABA), glutamic acid (Glu), and aspartic acid (Asp). Individual neurons may demonstrate one of at least three different ionic responses to each of these substances, due to Na^+, Cl, and K conductance increases, respectively. While receptors for each of these substances are specific, some are localized only in the neuropile, while others are present also on the cell body. In this chapter, we analyze the receptors on R_{15} in an effort to describe the kinds of neurotransmitters which may act on this cell as well as to determine the constancy of the type of response from one preparation to the next.

METHOD

R_{15} was identified by color, size, and location, and penetrated with one or two independent glass pipettes filled with K acetate (1 M). The preparation was constantly perfused with seawater containing three times the normal Mg^{2+} concentration in order to suppress natural synaptic transmission and to assure that the response to ionophoretic drug application was direct and not due to excitation of another neuron making contact with R_{15}. Putative neurotransmitters were applied through 1-, 3-, or 5-barrel ionophoretic electrodes using an ionophoretic control module designed to pass a predetermined charge over a variable time (37). This unit controls for electroosmosis to a greater degree than a constant current ionophoresis, but sequential drug application may be of different durations if the electrode resistance changes. All other procedures were as previously described (14).

RESULTS

Receptors on R_{15}

The effects of activation of several different receptors for putative neurotransmitters on the bursting pattern of cell R_{15} are shown in Fig. 1. We have found by transmitter ionophoresis specific receptors for GABA, ACh, DA, and Glu. Both GABA and ACh are depolarizing and excitatory, whereas the responses to DA and Glu are hyperpolarizing.

The brief responses to ACh and Glu consistently caused minimal disruption of the endogenous bursting pattern when elicited in the middle of the interburst hyperpolarization. These responses have a rapid peak onset and are relatively brief in duration. Certainly the ACh response, if of sufficient amplitude, would initiate or increase discharge, and the Glu response, if of sufficient size, could depress or stop discharge. At magnitudes similar to natural inputs, however, these receptors mediated only brief effects usually lasting not more than a few seconds. The responses to GABA and DA were considerably more prolonged and consistently produced alterations in the bursting pattern, even at threshold intensities. GABA consistently shortened the interburst intervals for at least one (and often several) cycle, while DA usually caused a pause in discharge lasting 30 to 60 sec.

R_{15} also has receptors for at least several other substances, which we have not yet studied in detail. Serotonin produces a depolarization, whereas both Oct and Pnol are hyperpolarizing. It is not yet clear whether the Oct and Pnol responses are mediated through the same or distinct receptors, since we have not found the localization of receptors by ionophoresis. Distinct receptors exist for these substances. The Oct receptor is most sensitive to Oct but somewhat sensitive to both norepinephrine and Pnol (12). The Pnol receptor is not sensitive to any other known substance (27). We have never found any response to bath-applied histamine or Asp.

Localization of Receptors

We have several impressions as to the localization of receptors, although there is considerable variability from preparation to preparation. The re-

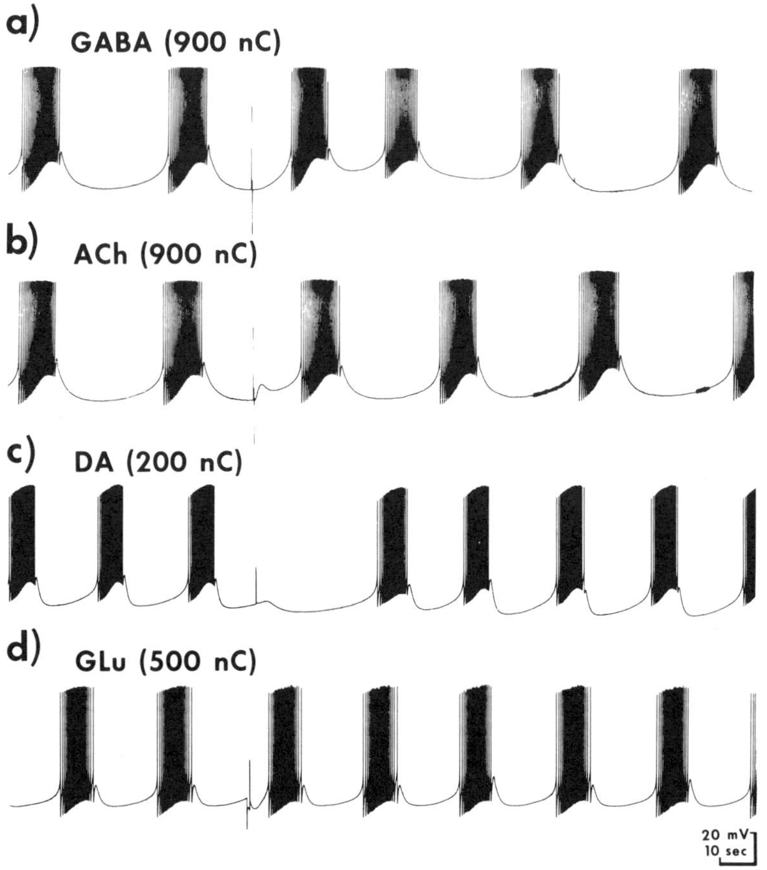

FIG. 1. Effects of GABA, ACh, DA, and Glu responses on endogenous discharge patterns in R_{15}. All records are from different preparations, except **a** and **b,** which are from an experiment utilizing a three-barreled electrode. In **c** and **d,** drugs were applied through a five-barreled electrode. In both cases, the barrels were filled with DA, Oct, GABA, Glu, and Asp, and there was no response to any other substance at that site other than that shown.

ceptors in the neuropile are quite difficult to find, and those receptors for one substance are not always in the same location in every cell. Most often we have obtained responses around the cell between 6 and 12 o'clock. The receptors are very discrete; and furthermore, those for one substance are not localized at the same site as those for others. In no experiment on DA or Glu responses did we ever find responses to ACh, GABA, 5-hydroxytryptamine (5-HT), Oct, or Pnol elicitable from the same location of the ionophoretic electrode. Often ACh and GABA responses could be obtained at one ionophoretic position. These observations support the supposition that the receptor localization reflects synapses which utilize that transmitter.

Ionic Basis of Excitatory Responses

The depolarizing responses to ACh and GABA differ in ionic basis and time course. Figure 2 shows the responses elicited by application of ACh and GABA by ionophoresis to the neuropile, which is the site of all synapses in this preparation. Aplysia neurons are all responsive to ACh applied to the soma (the recording site), although on different neurons ACh causes one of three different ionic responses. In R_{15}, however, we have found no receptors to any other substance on the soma. The ACh response in R_{15} is identical, whether elicited from soma or neuropile, and consists of a relatively rapid depolarization peaking at 1 to 2 sec. (Fig. 2a). The response is associated with an increased membrane conductance (Fig. 2b) and is totally abolished by replacement of external Na^+ with $Tris^+$, or mannitol plus high Mg^{2+}, all of which are effectively impermeable. These observations suggest that this response results from a relatively pure increase in Na^+ conductance, as has been suggested by Blankenship et al. (6).

The GABA response differs in that it is considerably longer in both latency and duration and is insensitive to removal of external Na^+ (Fig. 2a). In addition, this response is never associated with an increase in membrane conductance. In many experiments, no conductance change can be recorded, even (as in Fig. 2) when application of ACh at the same site in the neuropile results in a clear conductance increase. In other experiments (as in Fig. 3), the GABA response is associated with a clear decrease in membrane con-

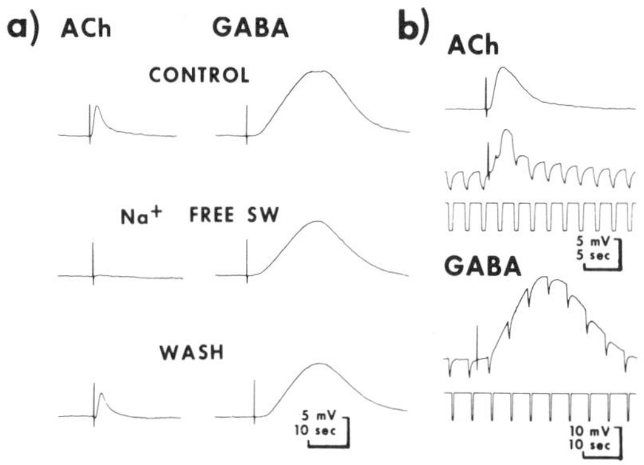

FIG. 2. Responses of R_{15} to ionophoretic application of ACh and GABA applied at a localized site in the neuropile. The ionophoretic electrode had three barrels, with the center barrel filled with distilled water to prevent coupling between ACh and GABA barrels. Ionophoretic charge was 300 nC for both substances. Membrane potential in all traces was maintained at −68 mV by application of hyperpolarizing current through a second intracellular microelectrode. Na^+ substitution was made with $Tris^+$ chloride seawater with the same osmolarity as control. **b:** Membrane conductance was measured by applying constant current pulses (2 namp) through the second intracellular electrode. Note that the ACh response is recorded at a faster sweep than elsewhere. There is a clear conductance increase to ACh but no apparent change to GABA.

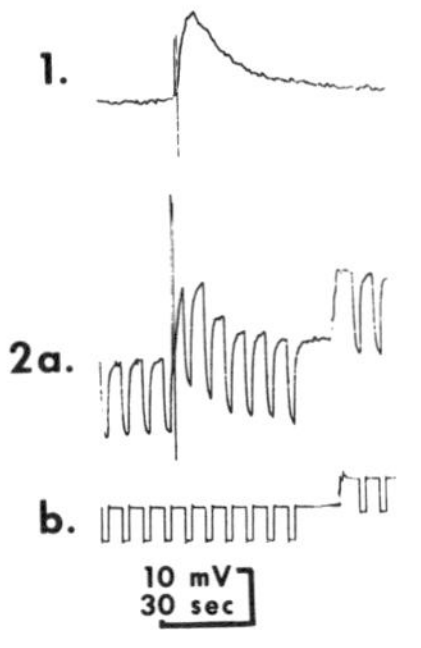

FIG. 3. Response to GABA from a different experiment on R_{15}. Ionophoretic charge was 200 nC. The response was associated with a 30% increase in membrane resistance **(2a).** The last two pulses in **2a** were taken after depolarization of the cell to the lead of the peak of the GABA response. The fact that these pulses are the same amplitude as those in the control indicate that the apparent decrease in conductance at the peak of the response does not result from membrane anomalous rectification. The current pulses **(2b)** were 1 namp.

ductance which is not an artifact of the nonlinear current-voltage characteristics of this cell. The response in Fig. 2 is recorded at a much slower time base and with considerably longer constant current pulses than that of Fig. 1. Usually the conductance change could be recorded only when the ionophoretic electrode was located relatively close to the cell body.

This response could be due to a conductance decrease to either Cl^- or K^+, since the equilibrium potential for both of these ions is negative relative to the resting potential. Yarowsky and Carpenter (40) have shown that the response is independent of Cl^- concentration and that its reversal potential varies with the K^+ concentration gradient. They conclude that GABA acts as an excitatory substance at these receptors by transiently decreasing K^+ conductance.

Ionic Basis of Inhibitory Receptors

Figure 3 illustrates the response to Glu applied to the neuropile. Glu causes a rapid hyperpolarization peaking at about 2 sec. This response reverses when external Cl^- is replaced by an impermeant anion, such as acetate, and is associated with an increase in membrane conductance. The reversal potential for the response in this neuron is −67 mV, which is within the known range of normal values for the equilibrium potential for Cl^-. These results suggest that Glu causes a specific increase in membrane Cl^- conductance.

In vertebrate systems, Asp is usually thought to mimic Glu as an excitatory neurotransmitter. In Aplysia, however, there are specific receptors for Asp, and both Asp and Glu may be either excitatory or inhibitory (39). In R_{15}, Asp has no effect on membrane potential or conductance when bath-perfused in concentrations up to 10^{-3} M. In lobster neuromuscular junction, there is an interaction of Asp with Glu which potentiates the action of Glu; this has been interpreted by Shank and Freeman (30) as an effect of Asp which increases the affinity of Glu for its receptor.

We have found a similar effect of Asp on the Glu receptor in R_{15} (Fig. 4). Asp alone at from 1 to 3 pulses of 1,000 nC each had no effect on potential

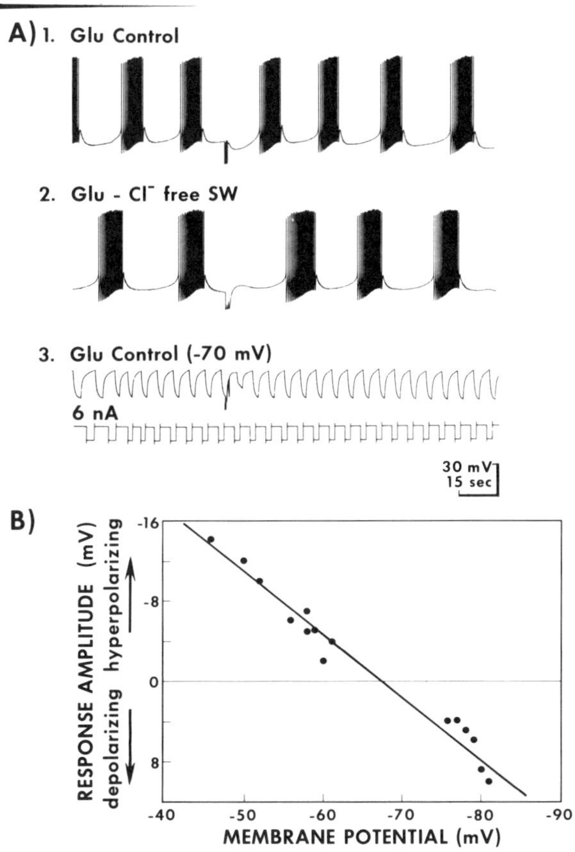

FIG. 4. Response to Glu on R_{15}. Ionophoretic charge was 3,000 nC in **A1–2** and 1,000 in **A3** and **B. A1:** At resting potential; illustrates the transient hyperpolarization caused by Glu and its effect on the spontaneous discharge pattern. **A2:** After a 10 min exposure to acetate-substituted (Cl^--free) seawater. Membrane potential was adjusted by DC current so that the peak of the interburst hyperpolarization is at the same level (−62 mV) as in the control. The response is now depolarizing, indicating that it is dependent on the Cl^- concentration gradient. **A3:** Increase in membrane conductance is shown while the cell was maintained at −70 mV, which is near the reversal potential for the response. Thus there is little or no voltage change, but a clear conductance increase. This record was fortuitously obtained after loss and reinsertion of one of the recording electrodes while the cell was in a state of partial injury, during which time the spontaneous discharge ceased. Under all other circumstances, the cell discharged at this potential. **B:** Plot of the response amplitude as a function of membrane potential.

or the burst pattern. Glu at 1,000 nC caused a small hyperpolarization when applied by itself (Fig. 4A2) but caused progressively greater responses when preceded by increasing pulses of Asp. When Glu preceded Asp, there was no interaction. Thus this response to Glu can be modulated by the concentration of Asp.

A second type of inhibition is shown in Fig. 5, elicited by application of

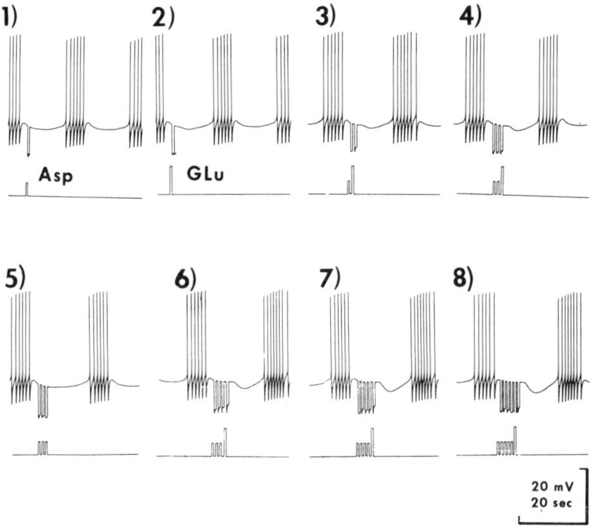

FIG. 5. Facilitation of R_{15} Glu response by Asp. Asp and Glu were each applied in pulses of 1,000 nC as indicated in the lower records of each trace, where Asp pulses are small and Glu large. Asp alone had no effect (1,5), but when Asp pulses were applied before Glu pulses, the Glu response was facilitated (3,4,6–8).

DA. This slow hyperpolarizing response peaks at approximately 20 sec, producing an interruption in the bursting pattern when the cell is active. When the cell is hyperpolarized to stop discharge (Fig. 5A–2), it is apparent that the voltage change lasts for nearly 3 min. The response is associated with an increased conductance (Fig. 5A–3), and the response is not inverted or diminished when the impermeant acetate is substituted for Cl^- (Fig. 5A–4). Hyperpolarizing the cell by current injection diminishes the amplitude of the response, but it is not reversed even at membrane potentials as low as −80 mV. These facts indicate that the response is not due to a conductance change to Cl^- but rather to an ion with an equilibrium potential at or beyond −80 mV. It is known that the equilibrium potential for K^+ is at about this level; this response can thus be attributed to an increase in K^+ conductance.

DISCUSSION

Location of Receptors and Their Role in Synaptic Transmission

All of the receptors for putative neurotransmitters which we have found on R_{15} are located exclusively in the neuropile, except for those to ACh. However, it is for ACh that there is the strongest evidence for a natural synaptic input. The interneuron L_{10} makes a monosynaptic excitatory connection on R_{15} (21), which shows ionic and pharmacologic identity with the response to ionophoretic ACh. Furthermore, L_{10} has been shown to contain ACh (22) and choline acetyltransferase, the enzyme responsible for ACh synthesis (18). The large EPSP which can be elicited by stimulation of the right connective nerve also shows a similar pharmacology, especially in sensitivity to concentrations of hexamethonium which have not been shown to affect responses to any other transmitter (28). These observations are perhaps the strongest indication that the ionic response elicited from receptors for ACh on the soma, which does not have synaptic connections, is indicative of the nature of the cholinergic input in the neuropile. A similar conclusion has been drawn by Gerschenfeld and Paupardin-Tritsch (17) in their study of 5-HT receptors on snail and Aplysia neurons.

The receptors for GABA, Glu, and DA are localized exclusively to the neuropile. Furthermore, they appear to be at different sites, since we have rarely seen responses to more than one substance ionophoresed at a specific location from a multiple-barreled electrode containing each of the other compounds. Interestingly, we often find no response to ACh at a location that gives a clear response to GABA, Glu, or DA, suggesting that not all of the neuropile is sensitive.

There are several reasons to suspect that the receptors we have isolated are functional at synapses on R_{15}: (a) It is known that R_{15} receives a variety of excitatory and inhibitory inputs (16). (b) The very specific localization of receptors for each of these substances and, indeed, the observation that usually the receptors for different substances are located at different sites argue strongly that the receptors are not simply nonspecific reactive groups. (c) The receptors for ACh, DA, Glu, and GABA always give rise to the same ionic and charge change in different preparations. (d) Although the blind probing of the neuropile with the large ionophoretic electrode presents technical difficulties, it is our strong impression that every R_{15} has each of these receptors. (e) Each of these substances is present in Aplysia nervous tissue (11,22,41) in a considerably greater concentration than in nonnervous tissue.

Unfortunately, pharmacologic tools are of limited value in identifying the actions of a specific neurotransmitter. Hexamethonium does appear to be a specific antagonist for the fast, Na^+-dependent response to ACh (28). Ergotamine and related compounds have been found to be relatively specific for DA, blocking both excitatory and inhibitory responses (2). However, these substances have not been tested on responses to Pnol or Oct. Curare has been found to block all fast Na^+ and Cl^- responses (14), including both the ACh and Glu responses on R_{15}. Picrotoxin and bicuculline, traditionally thought to be GABA receptor blockers, block Cl^- responses to several neurotransmitters, including the Glu response on R_{15}; but they do not affect the depolarizing GABA response (40). Penicillin also blocks all Cl^- responses (Pellmar and Wilson, *personal communication*). Thus there is no known agent that is specific for the Cl^- Glu, Na^+ Da, or conductance decrease GABA response.

Is Transmitter Specificity a Pre- or a Postsynaptic Property?

Our electrophysiologic studies on R_{15} and other identified neurons in Aplysia have indicated that the profile of receptors for and ionic responses to a variety of neurotransmitters is constant and can be used as an added method of identification of single neurons. Moreover, it appears that most single neurons have specific receptors for a surprisingly large number of putative neurotransmitters, although receptors for each substance may be very localized. These observations suggest a great specificity of the postsynaptic membrane.

It is usually believed that much of transmitter specificity is determined by presynaptic nerve terminals which contain and release a single neurotransmitter. Brownstein et al. (7), however, demonstrated that single Aplysia neurons contain considerable amounts of more than one putative neurotransmitter. This may not be surprising for such amino acids as Glu and Asp, which are present in all cells in relatively high concentrations (41), but it is unexpected that histamine, Oct, 5-HT (7), and DA (Kebabian, Swann, and Carpenter, *unpublished observations*) should coexist with ACh in neurons such as R_2. Dales' principle, as extended by others, has maintained that (a) a given neuron makes and releases only one neurotransmitter, and (b) all postsynaptic actions of a single neuron are identical, i.e., either excitation or inhibition but not both in different neurons (15). The second suggestion has convincingly been shown to be invalid by the demonstration that L_{10} may cause a cholinergic EPSP on R_{15} but a cholinergic inhibitory postsynaptic potential (IPSP) on L_{2-6} (21). There is no evidence that one Aplysia neuron releases more than one transmitter, although there is also no evidence that it releases only one. Some neurons, such as R_2 and L_{10} for ACh and C_1 for 5-HT, contain much higher concentrations of one transmitter than others, and thus the relatively low concentrations of other active substances may be only biologic noise. However, in both vertebrates (23) and Aplysia (25), there is morphologic evidence for more than one type of vesicle in a presynaptic terminal, and recently there is convincing immunocytochemical demonstration of substance P in a few large vesicles in a terminal containing many other small, unreactive vesicles (26). It is therefore increasingly likely that more than one substance may be released. Whether or not there is any response on the postsynaptic membrane is determined by the receptors present, and this may be the origin of the specificity. It is not inconceivable that the presynaptic neuron might release a veritable gemish, but that the postsynaptic neuron would respond to only one substance.

Relative Effects on Firing Patterns in R_{15} by Different Neurotransmitters

The four types of ionic responses described here fall into two distinctly different time courses of effect. Table 1 gives average values for the time to peak of the various ionic responses to ACh, DA, GABA, Glu, and Asp in unselected Aplysia neurons, including R_{15}. It is apparent that the fast Na^+ conductance increase responses and the Cl^- responses are both quite rapid, peaking at less than 2.5 sec, whereas the K^+ conductance increase responses to all these substances (including the DA response in R_{15}) is much slower, peaking at about 15 sec. The slow GABA depolarizing conductance-decrease response is only rarely found in other neurons, but it also peaks at about 10 sec in most cells (40).

As indicated in Fig. 6, these inputs may have very different effects on the endogenous bursting discharge pattern of R_{15}. Unless they are very large or are applied near the end of the interburst hyperpolarization, neither the Na^+ nor the Cl^- response alters the time of initiation of the next burst. Thus these events may add or subtract with other synaptic inputs but do not alone reset the bursting rhythm as do the GABA and DA responses. The DA response is very slow and may be the receptor mediating "inhibition of long duration" in R_{15}, as originally described by Tauc (35). As discussed below, however, responses of slow time course may be elicited by different neurotransmitters. Although we have not yet localized receptors on R_{15} by iono-phoretic application, there are responses to bath application on R_{15} of Oct and Pnol which look very similar to those of DA.

The effect of the GABA response may be much greater than is apparent from only the magnitude of the voltage change. Carew and Kandel (8) have shown that conductance decrease responses augment responsiveness to other synaptic inputs secondary to the increase in membrane resistance and may also result in an increased electrotonic coupling of cells. Thus this type of response may influence excitability to a considerably greater degree than a comparable conductance increase response, even if prolonged.

TABLE I.

MEAN TIME-TO-PEAK (IN SEC) OF RESPONSES OF APLYSIA NEURONS TO DIFFERENT PUTATIVE NEUROTRANSMITTERS

	IONIC REPONSE				
	Fast ↑G_{Na^+}	Slow ↑G_{Na^+}	↑G_{Cl^-}	↑G_{K^+}	↓G_{K^+}
Ach	1.8	–	2.4	12.2	–
DA	2.3	–	2.5	15.1	–
GABA	2.0	12.3	2.4	13.1	9.8
Glu	2.3	–	2.5	15.2	–
Asp	2.2	–	2.5	–	–

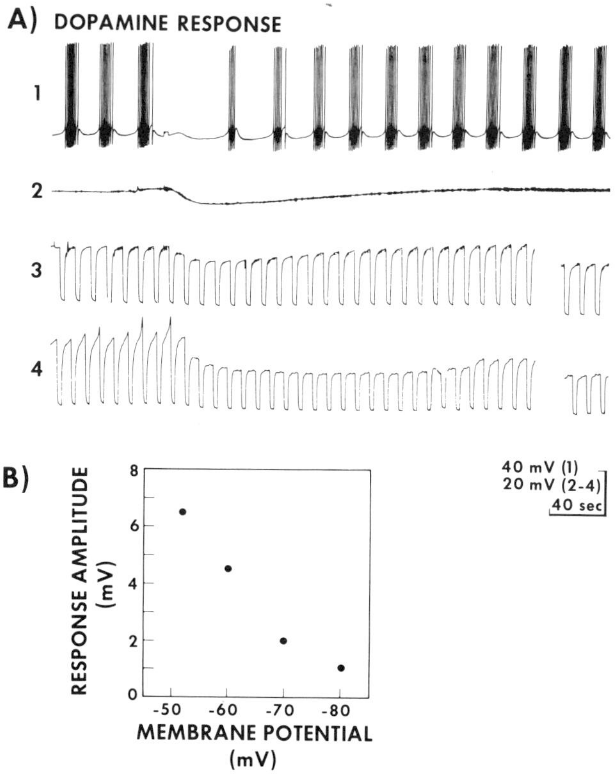

FIG. 6. Response of R_{15} to ionophoretic DA. Ionophoretic charge in each case is 1,000 nC. **A1:** Normal resting potential (peak of interburst hyperpolarization at −56 mV). **A2–4:** Hyperpolarization of the cell to −60 mV to prevent discharge. **A2,3:** Cell in normal seawater. **A3,4:** Membrane resistance was measured with constant current pulses of 6 nA. The pulses at the far right were taken with the cell hyperpolarized to the level of the peak of the response and show that the apparent decrease in resistance after DA is not due to anomalous rectification. **A4:** 45 min after beginning perfusion with acetate (Cl^--free) seawater. The response was not inverted and is actually somewhat increased in both duration and amplitude, probably secondary to the increase in membrane resistance in the Cl^--free solution. **B:** Response amplitude plotted as a function of membrane potential. It was not possible to reverse the response, but the extrapolated reversal potential is about −80 mV, known to be the equilibrium potential for K^+. The deviation from linearity at hyperpolarized levels may reflect the difficulty of altering potential at a distinct site when the current is applied in the soma.

The interaction we found between Asp and Glu at the Glu receptor is very similar to that described by Shank and Freeman (30) at lobster neuromuscular junction. While Asp caused very little effect when applied alone, it dramatically increased the responsiveness of the membrane to Glu. The response in lobster and crayfish is essentially a pure Na^+ conductance increase (24), whereas the response in R_{15} is to Cl^-. Shank and Freeman (30) have suggested that Asp acts to increase the affinity of the receptor for

Glu. Thus this is possibly another type of synaptic modulation and provides further reason to suspect that more than one active substance might be released from a single terminal.

On the Organization of Transmitter Receptors

We have previously suggested (14,29,34) that the receptor complexes for neurotransmitters on Aplysia are composed of two functional and structural entities: the receptor itself, which encodes transmitter specificity and binds the transmitter, and the ionophore, which results in the voltage conductance change. We have also suggested that a given receptor may be coupled to any of at least three ionophores, mediating Na^-, Cl^-, or K^+ conductances. In fact, it appears likely that there are more than three different ionophores, since a slower, curare-insensitive Na^+ conductance increase response and both depolarizing and hyperpolarizing conductance decrease responses have been found for several transmitters (17). The evidence in favor of this hypothesis is principally that at least the three common conductance increase responses have been found for each of nine different putative neurotransmitters, and, as indicated in Table 1, the time course of the responses are characteristic of the ionic response rather than the transmitter causing it.

Several features concerning the receptors on R_{15} are relevant to this hypothesis. In the central nervous system, Glu is considered an excitatory transmitter and GABA an inhibitory transmitter. These effects are reversed in R_{15}. Our hypothesis predicts that any transmitter could be excitatory or inhibitory, depending on to which ionophore the receptor was coupled. The failure of this GABA response to be blocked by bicuculline and picrotoxin, while all Cl^- responses to a variety of receptors are blocked, suggests a different site of action for these drugs than is generally appreciated. Finally, the demonstration of interaction of Asp with a Glu receptor mediating a Cl^- conductance increase is consistent with the view that the Glu receptor in R_{15} is similar or identical to that at the lobster neuromuscular junction where the response is a Na^+ conductance increase.

SUMMARY

Neuron R_{15} has receptors for several putative neurotransmitters which are discrete, localized, and specific. These receptors mediate responses that have different ionic bases, time courses, and effects on the endogenous discharge. ACh excites with a fast Na^+ conductance increase, whereas Glu inhibits with a fast Cl^- conductance increase. Both of these responses cause minimal disruption of the bursting discharge. DA causes a hyperpolarizing K^+ conductance increase which peaks at about 15 sec and may last up to

60 sec. GABA causes a depolarizing conductance decrease which also has a slow rise and long duration of effect. These responses both cause prolonged changes in the discharge pattern. It appears likely that ACh and GABA function as excitatory transmitters, and Glu and DA function as inhibitory transmitters on neuron R_{15}.

REFERENCES

1. Alving, B. O. (1968): *J. Gen. Physiol.,* 51:29–45.
2. Ascher, P., and Kehoe, J. S. (1975): *Handbook of Psychopharmacology, Vol. 4,* edited by L. L. Iverson, S. D. Iverson, and S. H. Snyder, pp. 265–310. Plenum Press, New York.
3. Barker, J. L., and Gainer, H. (1975): *Brain Res.,* 84:479–500.
4. Barker, J. L., Ifshin, M. S., and Gainer, H. (1975): *Brain Res.,* 84:501–513.
5. Beiswanger, C. M., and Jacklet, J. W. (1975): *J. Comp. Physiol.,* 103:19–37.
6. Blankenship, J. E., Wachtel, H., and Kandel, E. R. (1971): *J. Neurophysiol.,* 34:76–92.
7. Brownstein, M. J., Saavedra, J. M., Axelrod, J., Zeman, G. H., and Carpenter, D. O. (1974): *Proc. Natl. Acad. Sci. USA,* 71:4662–4665.
8. Carew, T. J., and Kandel, E. R. (1976): *Science,* 192:150–153.
9. Carpenter, D. O. (1967): *J. Gen. Physiol.,* 50:1469–1484.
10. Carpenter, D. O. (1973): In: *Neurobiology of Invertebrates, Tihany, 1971,* edited by J. Salánki, pp. 35–58. Akadémiai Kiadó, Budapest.
11. Carpenter, D., Breese, G., Schanberg, S., and Kopin, I. (1971): *Int. J. Neurosci.,* 2:49–56.
12. Carpenter, D. O., and Gaubatz, G. L. (1974): *Nature,* 252:483–485.
13. Carpenter, D., and Gunn, R. (1970): *J. Cell. Physiol.,* 75:121–128.
14. Carpenter, D. O., Swann, J. W., and Yarowsky, P. G. (1977): *J. Neurobiol.,* 8:119–132.
15. Eccles, J. C. (1957): *The Physiology of Nerve Cells.* The Johns Hopkins Press, Baltimore.
16. Frazier, W. T., Kandel, E. R., Kupfermann, I., Waziri, R., and Coggeshall, R. E. (1967): *J. Neurophysiol.,* 30:1288–1351.
17. Gerschenfeld, H. M., and Paupardin-Tritsch, D. (1974): *J. Physiol. (Lond.),* 243:457–481.
18. Giller, E., Jr., and Schwartz, J. H. (1971): *J. Neurophysiol.,* 34:93–107.
19. Gulrajani, R. M., Roberge, F. A., and Mathieu, P. A. (1977): *Biol. Cybern.,* 1:1–14.
20. Junge, D., and Stephens, C. I. (1973): *J. Physiol. (Lond.),* 235:155–181.
21. Kandel, E. R., Frazier, W. T., and Coggeshall, R. E. (1967): *Science,* 155:346–349.
22. McCaman, R. E., Weinreich, D., and Borgs, H. (1973): *J. Neurochem.,* 21:473–476.
23. Morest, D. K., Kiang, N. Y. S., Kane, E. C., Guiness, J. J., and Godfrey, D. A. (1973): In: *Basic Mechanisms of Hearing,* edited by A. Miller, pp. 479–504. Academic Press, New York.
24. Onodera, K., and Takeuchi, A. (1975): *J. Physiol. (Lond.),* 252:295–318.
25. Orkand, P. M., and Orkand, R. K. (1975): *J. Neurobiol.,* 6:531–548.
26. Pickel, V. M., Reis, D. J., and Leeman, S. E. (1977): *Brain Res.,* 122:534–540.
27. Saavedra, J. M., Ribas, J., Swann, J., and Carpenter, D. O. (1977): *Science,* 195:1004–1006.
28. Schlapfer, W. T., Woodson, P. B. J., Tremblay, J. P., and Barondes, S. H. (1974): *Brain Res.,* 76:267–280.
29. Shain, W., Greene, L. A., Carpenter, D. O., Sytowski, A. J., and Vogel, Z. (1974): *Brain Res.,* 72:225–240.
30. Shank, R. P., and Freeman, A. R. (1975): *J. Neurobiol.,* 6:289–303.
31. Smith, T. G., Jr., Barker, J. L., and Gainer, H. (1975): *Nature,* 253:450–452.
32. Strumwasser, F. (1965): In: *Circadian Clocks,* edited by J. Aschoff, pp. 442–462. North-Holland Publishing, Amsterdam.
33. Strumwasser, F. (1967): In: *Invertebrate Nervous System,* edited by C. A. G. Wiersma, pp. 291–319. University of Chicago Press, Chicago.
34. Swann, J. W., and Carpenter, D. O. (1975): *Nature,* 258:751–754.
35. Tauc, L. (1967): *Physiol. Rev.,* 47:521–593.
36. Waziri, R., and Kandel, E. R. (1969): *J. Neurophysiol.,* 32:520–539.

37. Willis, J. A., Myers, P. R., and Carpenter, D. O. (1978): *J. Electrophysiol. Tech.*, 6:34–41.
38. Wilson, W. A., and Wachtel, H. (1974): *Science*, 186:932–934.
39. Yarowsky, P. J., and Carpenter, D. O. (1976): *Science*, 192:807–809.
40. Yarowsky, P. J., and Carpenter, D. O. (1977): *Life Sci.*, 20:1441–1448.
41. Zeman, G. H., and Carpenter, D. O. (1975): *Comp. Biochem. Physiol.*, 52C:23–26.

Abnormal Neuronal Discharges, edited by
N. Chalazonitis and M. Boisson. Raven
Press, New York © 1978.

Endogenous and Synaptic Factors Affecting the Bursting of Double Spiking Molluscan Neurosecretory Neurons

P. R. Benjamin

Ethology and Neurophysiology Group, School of Biological Sciences, University of Sussex, Falmer, Brighton, England

The yellow cells of the pulmonate snail *Lymnaea stagnalis* are neurosecretory neurons with a well-defined set of morphological (15) and electrophysiological characteristics (2). They burst endogenously, but the bursts are unusual in that they contain single spikes, doublets, and other additional spike groupings (2). The mechanism by which bursts are formed appears similar to that described by Calvin (3–5) as extra spiking. The bursting activity of yellow cells is modulated by an excitatory synaptic input which is synchronous on the 10 to 15 cells of the visceral ganglion. Apart from the pioneering work of Chalazonitis (7), little has been published concerning the short-term modulatory effects of excitatory synaptic inputs on bursting neurons, although long-term effects have been studied by Parnas and Strumwasser (11). More is known about the effects of inhibitory inputs on endogenous bursters (12).

The object of this chapter is to describe some of the basic features of yellow cell endogenous activity and then to discuss the effects of excitatory inputs on the occurrence and timing of bursts. It will be shown that low amplitude excitatory postsynaptic potentials (EPSPs) can initiate bursts at any part of the endogenous interburst interval and, if occurring at low frequency, continuously reset the burst period. At frequencies of input approaching that of endogenous bursting, every yellow cell burst is initiated by an EPSP so that bursting activity is entirely timed by the synaptic input.

METHODS

Yellow cells were identified and recorded by intracellular techniques in isolated brains of *Lymnaea stagnalis*, as described previously (2). About 25 yellow cells occur in the central nervous system of *Lymnaea* in the locations described by Swindale and Benjamin (15). Visceral ganglion yellow cells were mainly recorded in the present experiments. A cluster of 8 to 10 yellow cells occurs near the viscero-right parietal connective (15), and other yellow cell somata occur near the origin of the visceral nerves and next to the visceral-left parietal connective.

One or two (independently manipulated) microelectrodes were placed in the cell bodies of yellow cells. One was used for passing current and the other for recording membrane potentials. Currents could also be passed through the recording electrodes via a bridge circuit. Pairs of yellow cells were also simultaneously recorded in some experiments. Current injection was monitored by an I-V convertor placed between the indifferent electrode and earth. Electrodes were in the 60 to 80 megohms range and filled by the glass fiber method with K_2SO_4.

Isolated brains were maintained during experiments in blood previously collected from the experimental animal or in a phosphate buffered saline (18) or in a Hepes buffered saline (pH 7.9) of the following composition (mM): Na^+ 59.4, K^+ 2.0, Ca^{2+} 4.0, Mg^{2+} 2.0, Cl^- 38.0, and HPO_4^- 0.1. Glucose (15 mg/liter) was added to this solution. Experiments were carried out at 20°C.

RESULTS

Patterns of Activity

Although a few yellow cells with high membrane potentials (> −55mV) were silent, most neurons were active when penetrated. The type of activity recorded was not the same in all cells (Fig. 1) but depended on membrane potential (Fig. 2A). There were two states of single spiking, one at high levels of membrane potential (Figs. 1A and 2A) and the other at low levels (Figs. 1E and 2A), while bursting activity occurred at intermediate levels (Figs. 1B and 2A). This is illustrated in Fig. 1G, in which a spontaneous slow fall in membrane potential occurred, converting the firing pattern from low frequency single spiking, through double spiking to bursting, and finally to high frequency single spiking.

A crucial feature of burst formation appeared to be the afterdepolarizing hump which followed single spikes at low frequency (Figs. 1A and 1G). As the membrane potential decreased, a second spike arose from this afterdepolarization to form a doublet. A triplet was formed by the addition of another spike to the doublet by the same mechanism, and a continuation of this process led to the formation of larger bursts. A similar method of burst formation has been found by Calvin (3–5) in vertebrate neurons.

With the formation of bursts from doublets in a slowly depolarizing yellow cell, the interspike interval shortened and the amplitude of the second spike became reduced compared with the first (Fig. 1G). In some cells, the reduction of spike amplitude and interspike interval continued for the addition of each spike up to a maximum of about eight spikes in a burst (Figs. 1C and 3A). In other cells, addition of extra spikes beyond three led to clearly

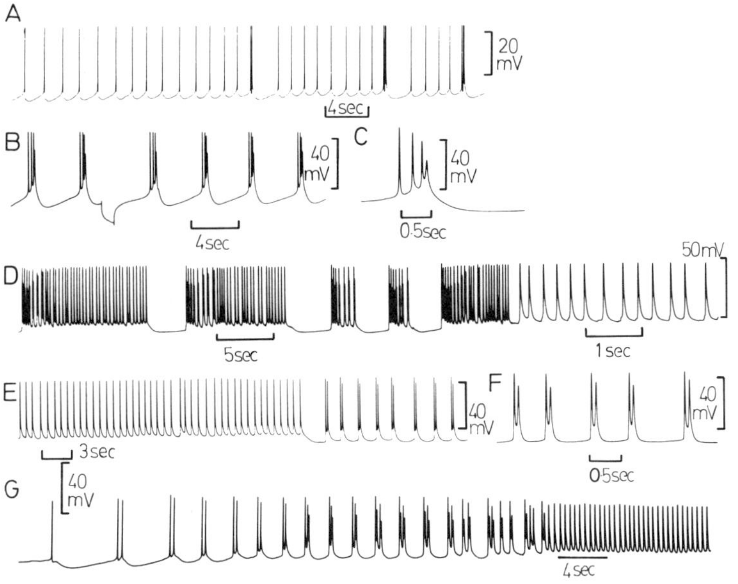

FIG. 1. Patterns of activity recorded in yellow cells. **A:** Low frequency single spiking pattern. It is just possible to see the depolarizing afterpotential occurring after each single spike. One doublet and two triplets also occur. The afterhyperpolarization following these extra spike groupings is larger than that occurring after single spikes. **B:** Burst formation in yellow cells with a small number of spikes per burst. Application of a hyperpolarizing pulse delays the onset of one of the bursts and resets the phase of the subsequent endogenous rhythm. **C:** Structure of a short burst on a fast time base. **D:** A neuron firing in very long bursts of variable duration which consist of doublet spikes but also many single spikes. This cell is near the level of membrane potential where high frequency single spiking occurs. **E:** High frequency single spiking cell. Note that the pronounced spike afterhyperpolarizations seen in **A** are absent. At the end of the trace, the cell has been tonically hyperpolarized and is doublet spiking. **F:** Doublet spikes with minimum interspike interval. **G:** Cell undergoing a long period of slow depolarization. The slow spontaneous depolarization shows the sequence of changes that occur at different levels of membrane potential in yellow cells. From left to right the sequence is: single spike with afterdepolarization, doublet with afterdepolarization, triplets, bursts with two doublets, doublets with minimum interspike interval, and second spike amplitude, high frequency single spikes.

discernible burst substructures when longer intervals than that between a doublet caused a recovery in spike amplitude and the formation of further doublets (Figs. 1D and 1G). It was interesting that the minimum interspike interval of a doublet occurred when bursting neurons were depolarized close to the level sufficient to produce high frequency single spiking (Fig. 1F) or when high frequency single spiking yellow cells were slightly hyperpolarized (Fig. 1E). Pure doublet spiking was seen in these circumstances with the interspike interval at a minimum of about 100 msec (about three times the duration of the first spike of the doublet).

If already bursting neurons were depolarized, then the frequency of burst-

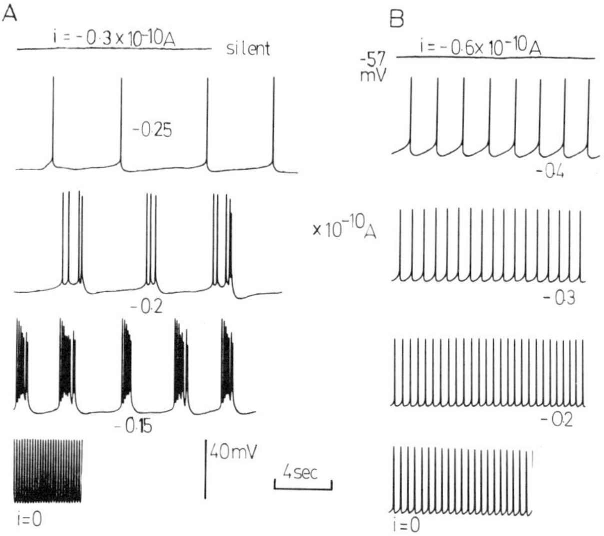

FIG. 2. Comparison of effects of different levels of tonically injected hyperpolarizing currents on firing in a bursty **(A)** and a nonbursty **(B)** yellow cell. Both of these cells were firing high frequency single spikes at penetration (*bottom traces*). They were hyperpolarized to silence by tonic hyperpolarizing currents passed through a second current injecting electrode. The activity of each cell was recorded at four different levels of tonic hyperpolarization. **A:** Bursty cell passed through sequence of activity from more to less hyperpolarized (top to bottom), which consisted of low frequency single spiking, low frequency bursting with a small number of spikes per burst, higher frequency bursting with increased number of spikes per burst, high frequency single spiking. **B:** Nonbursty cell increased frequency as it became more depolarized (top to bottom), but no change in type of firing is seen. Note that the values for injected current shown between spike traces are $\times 10^{-10}$ amperes. The levels of injected current required to change the cell from silence to high frequency single spiking are extremely small.

ing and the number of spikes per burst rose (Fig. 2A) in a manner characteristic of other molluscan bursting neurons (6,12). In bursts longer than about eight spikes, the spike frequency rose to a peak in the first part of the burst and then gradually decreased again (Fig. 3A, 10 and 12 spike bursts, Figs. 5C, 6C). The very long bursts seen in some yellow cells showed a gradual slowdown in frequency after the maximum had been reached early in the burst (Figs. 1D and 6B).

The interburst silent period was characterized by a rapid postburst hyperpolarization which occurred after the last spike of the burst (Figs. 1B and 3), followed by a gradual depolarization which eventually led to the onset of the next burst.

Not all yellow cells would burst. Many cells recorded in winter (snails collected from ponds from October to February) showed single spiking activity which could not be converted to bursting by membrane potential manipulations (compare the bursting yellow cell of Fig. 2A with the non-

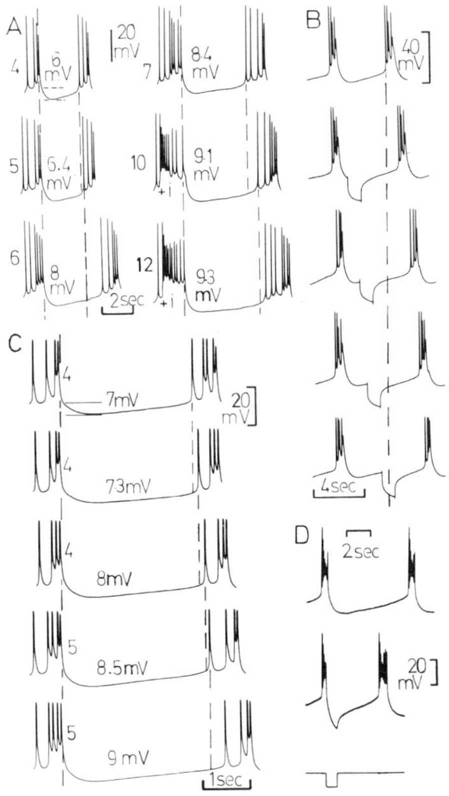

FIG. 3. Effects of endogenous factors and applied hyperpolarizing pulses on yellow cell bursting. **A:** Relationship between number of spikes in a burst, the amplitude of the post-burst hyperpolarization, and the duration of the interburst interval. Spikes in bursts indicated by numbers on left. Four, five, six, and seven spike bursts were spontaneously produced, whereas 10 and 12 spike bursts were depolarized by square current pulses applied during the burst. Size of postburst hyperpolarization indicated between bursts in millivolts. With increasing number of spikes, size of postburst hyperpolarization rises, as does the interburst interval. **B:** Effects of hyperpolarizing current pulses applied at different points in the interburst interval of a yellow cell. Gradual increase in delaying effects of pulse is obvious as it is applied later and later in the interburst interval compared with control burst (*top trace*). **C:** Structure of end of burst subgrouping of spikes affects the size of postburst hyperpolarization (values given in millivolts). Number of spikes in burst shown to left of each burst. The closer together the last three spikes of top three traces, the larger the postburst hyperpolarization and interburst interval. Adding of another spike (*bottom two traces*) produces even greater effect. **D:** Reducing number of spikes in a burst by application of hyperpolarizing pulse advances the onset of the next burst compared with the control (top trace).

bursting yellow cell of Fig. 2B). Such seasonal variations in "burstiness" were reported by Barker and Gainer (1) in the terrestrial snail *Otala*.

Endogenous Activity in Yellow Cells

The spiking activity already discussed occurred in yellow cells in the absence of synaptic input or where the synaptic input was too sporadic to be responsible for the observed activity. Three other pieces of evidence support the notion that yellow cell activity can be endogenously produced. First, the phase of bursting activity could be reset by the application of hyperpolarizing square pulses of current (Fig. 1B). Second, the interburst interval is related to the number of spikes in the previous burst (Fig. 3A). Third, the frequency of firing is related to the level of applied polarization (Fig. 2). This last point was mentioned in the preceding section and will not be discussed further.

Application of Hyperpolarizing Pulses

If a hyperpolarizing square current pulse was applied in the interburst interval of a yellow cell, then it invariably caused a delay in the onset of the next burst (Fig. 3B). How much the burst was delayed depended on the time at which it was applied. If it was applied just after the end of a burst, then its effect was less than if it was applied in the middle of the interburst interval. Its effect was maximal if it was timed to coincide with a burst (Fig. 3B). The overall effect was to cause a phase delay with the maximum effect if the hyperpolarizing pulse was applied at the end of the burst period (measured from the beginning of a burst). This is because bursts subsequent to the pulse occur at the same frequency as previously but are timed in relation to the burst following the hyperpolarization (Fig. 1B).

Numbers of Spikes in the Burst and Extra Spiking at the End of the Burst

The number of spikes in the bursts of a particular yellow cell varied within certain limits, if its activity was monitored for several minutes. The cell shown in Fig. 3A ranged from a minimum of four to a maximum of seven spikes per burst. There appeared to be a direct correlation between the number of spikes in the burst, the amplitude of the postburst hyperpolarization, and the duration of the interburst interval following the burst (Fig. 3A). If additional spikes were added to the burst by application of depolarizing pulses, then this increased the following interburst interval (Fig. 3A). Conversely, application of hyperpolarizing square pulses of current, which reduced the number of spikes in a burst, reduced the following interburst interval (Fig. 3D).

Even with a constant number of spikes, the duration of the interburst interval could vary; one factor which could affect this was the frequency of spikes in the last part of the previous burst. The neuron in Fig. 3C had a high frequency subgrouping at the end of its bursts, but the number of spikes and their interspike intervals varied within the subgrouping in different bursts over a period of 1 min or so. It was seen that even with the same total number of spikes in a burst, the occurrence of subgroupings of spikes with the largest number of spikes at a minimum aggregate interspike interval produced the largest postburst hyperpolarization and the longest subsequent interburst interval (Fig. 3C).

Synaptic Modulation of Yellow Cell Bursting

Identification of Excitatory Synaptic Input

Yellow cells in the visceral ganglion of Lymnaea received a synchronous synaptic input (Fig. 6A) which was identified as input 3 (18) by recording yellow cells simultaneously with other identified *Lymnaea* neurons, particularly the right pedal giant cell (Fig. 4), which were known to have this input. Input 3 was always excitatory when recorded in yellow cells. It was usually a compound potential formed of summating unitary EPSPs (Fig. 6A). Its duration could be as short as 1 sec or more commonly from 2 to 3 sec in a particular snail; it was never larger than a few millivolts, even when yellow cells were hyperpolarized to the point where no spiking occurred (Fig. 6A). EPSP frequency varied widely in different preparations. It could be sporadic, or it could occur at regular intervals of 30 to 40 sec or at a frequency which approached that of endogenous burst activity (every 6 to 20 sec). There were important differences noted in the effects of EPSP input at this higher frequency range compared with sporadic or low frequency input.

Low Frequency EPSP Input

EPSPs could evoke bursts at any part of the interburst interval of an endogenously active yellow cell (Fig. 4). The duration of the evoked burst was usually far greater than that of the EPSP, especially if the EPSP occurred late in the interburst interval (Fig. 4B and C). The number of spikes and the duration of bursts evoked by the EPSPs were greater if the EPSP arrived late in the interburst interval compared with if they arrived early (compare the evoked burst of Fig. 4C, top trace, with that of Fig. 4A, top trace). The effect of EPSPs adding extra bursts to endogenously active cells was to delay the onset of the following endogenous burst (Fig. 4). The size of the delay was much greater if the synaptically evoked burst occurred

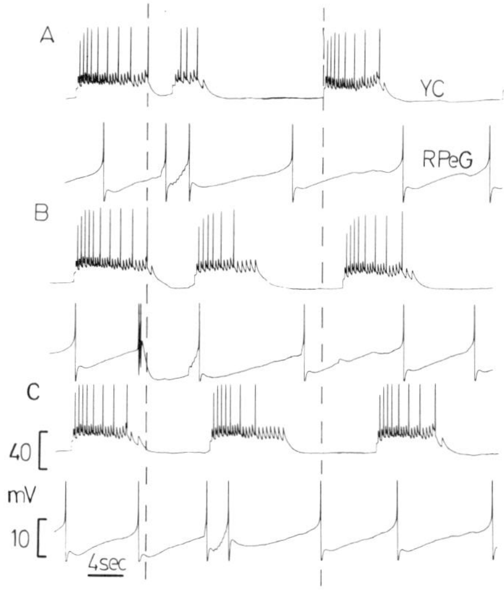

FIG. 4. Yellow cell (YC) bursts evoked by excitatory synaptic inputs arriving in the inter-burst interval of a YC. The YC is the top trace of each pair of recordings in **A–C;** the bottom trace of each pair is the right pedal giant neuron (RPeG) recorded simultaneously with the YC. Input 3 is an excitatory synaptic input which is identical on both cells (not shown), so the bottom trace of each pair of recordings acts as a monitor for EPSP input which occurs on both cells but which is masked by the occurrence of the EPSP evoked burst in the yellow cell traces. However, the onset of the EPSP input can be seen just prior to the synaptically evoked yellow cell bursts (bursts between the vertical dashed lines of all three traces). EPSP input occurred irregularly at a frequency of less than once every 30 sec in the cells; these records have been selected from a long series of recordings to illustrate the effect of synaptic input arriving at different parts of the endogenous interburst interval. Endogenous bursts in the three yellow cell traces in **A–C** have been arranged to the left of the figure so that the end of the bursts terminate at the same time (*left vertical dashed line*). The effect of an EPSP burst occurring at three different points in the interburst interval can be seen to delay the occurrence of the next endogenous burst (bursts to the right of the right-hand vertical dashed line). The delay in the occurrence of the endogenous burst was greatest in **C,** less in **B,** and least in **A.** Although the duration of EPSP in each trace varied little, there was a difference in the duration of the burst evoked by the EPSP in **A–C.** The synaptically evoked burst was longest in **C,** of intermediate duration in **B,** and shortest in **A.** Note that the duration of the EPSP as judged by its occurrence on the RPeG in **B** and **C** was much less than that of the evoked burst.

at the end of the interburst interval (Fig. 4C) compared with if it arrived just after a burst (Fig. 4A).

High Frequency EPSP Input Determines Burst Period

If the interval between EPSPs approached the endogenous burst period, then the timing of yellow cell bursts was governed by the occurrence of the EPSP input. Thus each burst was evoked by an EPSP and few or no bursts were endogenously timed (Fig. 5). The maximum frequency at which yellow

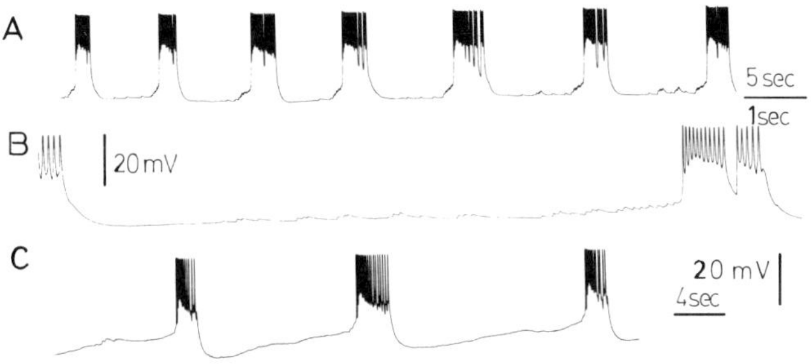

FIG. 5. Effect of high frequency EPSP input on yellow cell bursting. **A:** Every burst in this trace was evoked by an EPSP input; some small unitary EPSPs occur in the interburst interval of the yellow cell but are apparently not large enough to evoke a burst. **B:** Same yellow cell on a faster time base showing one interburst interval and EPSP input. **C:** Lower frequency input on another yellow cell does not always give rise to one for one EPSPs to bursts. The first burst on the left of the trace was endogenously timed but next two traces were synaptically evoked.

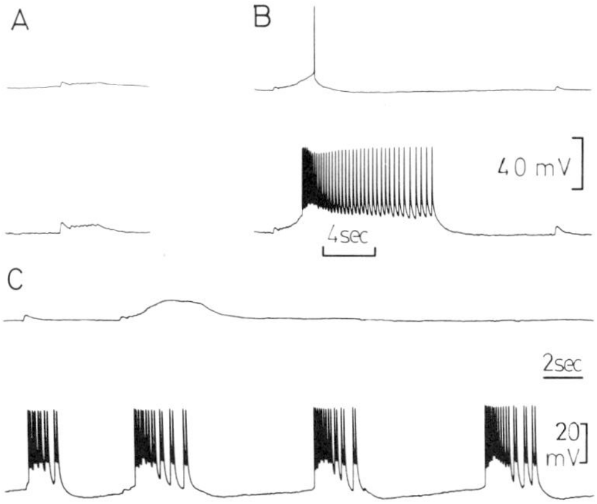

FIG. 6. Synchronous excitatory synaptic input in visceral ganglion yellow cells and the effects of this input on a cell prevented from firing by tonic hyperpolarizing currents. Same pair of yellow cells recorded in **A–C.** Yellow cell in top trace in **A–C** was silent and nonbursty (cells recorded in snail from October). Yellow cell in bottom traces of **A–C** was bursty and showed synaptically and nonsynaptically evoked bursts **(C). A:** Cell in bottom trace hyperpolarized to prevent spike activity. Cell in top trace at resting membrane potential. The EPSP (input 3) is synchronous on both cells. It is a compound potential in which individual summating potentials can be seen. The potential was presumably formed by a burst of spikes in the unidentified presynaptic cell. **B:** Cell in bottom trace was less hyperpolarized than in **A** but was silent unless the EPSP occurred. The synaptic input evokes a burst that far exceeds the duration of the EPSP (monitored in top trace). Membrane potential before and directly after the burst was about the same, which suggests that the underlying tonic level of membrane potential is at about the equilibrium potential for the postburst hyperpolarization. **C:** Cell of bottom trace less depolarized than in **A** or **B.** In this case, synaptically evoked bursts can be seen (first two bursts from left-hand side) but the last two bursts to the right are endogenously timed.

cells have been seen to follow EPSP inputs was once per 6 sec, but reasonable following occurred at less than half this frequency (Fig. 5C).

EPSPs Evoke Bursts in the Absence of Endogenous Membrane Potential Oscillations

When yellow cells were hyperpolarized by tonic hyperpolarizing currents, they ceased to generate spikes at between −50 and −55 mV. At this level of membrane potential, EPSPs evoked bursts of spikes which could far outlast the duration of the EPSP (Fig. 6B). It shows that the slow oscillatory waves, which in their depolarizing phase evoke spikes in endogenously active cells, are not necessary for burst generation in yellow cells (see also ref. 17).

DISCUSSION

Extra Spiking and Yellow Cell Bursting

Pacemaker burst discharges have been described in a number of molluscan preparations (8,13). Bursts of spikes alternate with postburst hyperpolarizations and a silent period. Within the burst, interspike intervals gradually decrease to a minimum and then increase again toward the end of the burst. These cells also have a single spiking or beating mode of firing (12), which can be converted to bursting by application of tonic hyperpolarizing currents (12). The endogenous activity of yellow cells in *Lymnaea* has some features similar to endogenously active cells in other gastropods, but the variety of activity seen in yellow cells is greater than that of other bursters, such as R_{15} of Aplysia. The greater variety of firing patterns recorded in yellow cells appears to be related to the extra spiking ability of these neurons.

The extra spike capability of yellow cells seems to be essential for bursting. Cells that will not form doublets at the appropriate level of membrane potential will not burst either (Fig. 2B). Extra spikes are formed on the afterdepolarizing hump of spikes in yellow cells. Sequences of spikes can be produced by this mechanism. The number of spikes in a burst is governed by the underlying tonic level of membrane potential; this is also the case in vertebrate neurons with extra spikes (3–5). Thompson and Smith (17) have recently shown that postspike depolarizations are important for burst formations in other well-known molluscan bursters, although no doublets or other extra spikes were seen in these cells.

Although sequences of spikes can be produced by spike afterdepolarizations, an additional mechanism is required to terminate the burst. In yellow cells, as in other bursting neurons, like R_{15} in *Aplysia* (9), burst termination is followed by a large hyperpolarization which resembles, in an enlarged form, the afterhyperpolarization seen in many neurons after single spikes. Like other molluscan bursting neurons, the amplitude of the postburst hyperpolarization and interburst interval of yellow cells is related to the number of spikes in the preceding burst (12,13). The frequency structure of the terminal part of the yellow cell burst can also influence the interburst interval. The higher the frequency of spikes in the last part of the burst, the longer the subsequent interburst interval. In other bursting neurons, the postburst hyperpolarization, forming the first part of the interburst interval, is caused by a delayed increase in K^+ conductance (9) stimulated by an inward flux of Ca^{2+} (10). There is a quantitative relationship between the number of spikes in the burst, the increment in intracellular Ca^{2+} produced by the spikes, and the amplitude of the postburst hyperpolarization (16). If similar mechanisms are involved in producing the postburst hyperpolarization of yellow cells, then it would explain the relationship obtained between the number of spikes in the burst and the postburst interval. Endogenous or evoked variation in spike number would influence the level of intracellular Ca^{2+} and thus the amplitude of the postburst hyperpolarization. The effect of end of burst frequency on the amplitude of the afterhyperpolarization could also be explained if intracellular Ca^{2+} concentrations were raised more rapidly by summation of Ca^{2+} influxes due to individual spikes (14).

Synaptic Modulation of Endogenous Bursting

Short-term modulatory effects of excitatory synaptic inputs on *Aplysia* Br bursters were reported by Chalazonitis (7). In yellow cells of *Lymnaea*, EPSPs evoked bursts at any point in the interburst interval of endogenously bursting neurons. The duration of the evoked burst was greatest when the EPSP arrived late in the interburst interval. Chalazonitis (7) also showed that EPSPs could evoke bursts in the interburst interval in Aplysia bursters; and like yellow cells, the occurrence of the next endogenous burst was delayed. Higher frequency input entirely determined the occurrence of bursts in yellow cells so that few endogenously timed bursts were seen. For this to occur, the frequency of synaptic input must be greater than the burst frequency which can be produced by the endogenous mechanism.

ACKNOWLEDGMENTS

I wish to thank the U.K. Science Research Council for financial support, and Malcolm Rose for his comments on this manuscript.

REFERENCES

1. Barker, J. L., and Gainer, H. (1973): *Nature,* 254:462–464.
2. Benjamin, P. R., and Swindale, N. V. (1975): *Nature,* 258:622–623.

3. Calvin, W. H. (1974): *Brain Res.*, 69:341–346.
4. Calvin, W. H., and Loeser, J. D. (1975): *Exp. Neurol.*, 48:406–426.
5. Calvin, W. H., and Sypert, G. W. (1976): *J. Neurophysiol.*, 39:420–434.
6. Carpenter, D. O. (1973): In: *Neurobiology of Invertebrates. Mechanisms of Rhythm Regulation,* edited by J. Salánki, pp. 35–59. Adadémiai Kiadó, Budapest.
7. Chalazonitis, N. (1967): In: *Neurobiology of Invertebrates,* edited by J. Salánki, pp. 201–226. Akadémiai Kiadó, Budapest.
8. Gainer, H. (1972): *Brain Res.*, 39:403–418.
9. Junge, D., and Stephens, C. L. (1973): *J. Physiol.*, 235:155–181.
10. Meech, R. W. (1974): *J. Physiol.*, 237:259–277.
11. Parnas, I., and Strumwasser, F. (1974): *J. Neurophysiol.*, 37:594–608.
12. Pinsker, H. M., and Kandel, E. R. (1977): *Brain Res.*, 125:51–64.
13. Strumwasser, F. (1967): In: *Invertebrate Nervous Systems,* edited by C. A. G. Wiersma, pp. 291–319. University of Chicago Press, Chicago.
14. Stinnakre, J., and Tauc, L. (1973): *Nature* [*New Biol.*], 242:113–115.
15. Swindale, N. V., and Benjamin, P. R. (1976): *Philos. Trans. R. Soc. Lond.* [*Biol.*], 274: 169–202.
16. Thomas, M. V., and Gorman, A. L. V. (1977): *Science,* 196:531–533.
17. Thompson, S. H., and Smith, S. J. (1976): *J. Neurophysiol.*, 39:153–161.
18. Winlow, W., and Benjamin, P. R. (1976): In: *Neurobiology of Invertebrates, Gastropoda Brain,* edited by J. Salánki, pp. 41–59. Akadémiai Kiadó, Budapest.

Abnormal Neuronal Discharges, edited by N. Chalazonitis and M. Boisson. Raven Press, New York © 1978.

Induction of Excitability in an Electrically Stable Membrane Following Chemical Interference with Sulfhydryl and Disulfide Groups at the Cell Surface

Conchita Zuazaga de Ortiz and José del Castillo

Laboratory of Neurobiology and Department of Pharmacology, Medical Sciences Campus, University of Puerto Rico, San Juan, Puerto Rico 00901

Crustacean muscle fibers kept in normal saline solution, such as van Harreveld's Ringer (24), are usually electrically inexcitable or give only graded responses to depolarizing current pulses (8,11). However, Fatt and Ginsborg (7) demonstrated that in the presence of high external concentrations of the divalent cations Ca^{2+}, Sr^{2+}, and Ba^{2+}, long, overshooting action potentials are generated by crustacean muscle. In addition, Grundfest and his co-workers (2,18,19,23) showed that electrical excitability is induced in the same tissue by several organic molecules, such as cocaine, procaine, methylene and sevron blues, acridine and neutral reds, tetraethylammonium, *d*-tubocurarine, and caffeine.

As emphasized by Reuben and Grundfest (19), the conversion from inexcitability, or graded responsiveness, to all-or-nothing activity by chemical agents can be accounted for in terms of changes in ionic conductances of the membrane and their voltage-dependent properties. However, one of the most interesting and elusive aspects of the problem is to define the nature of the chemical modifications of the membrane components responsible for the altered electrical properties. This problem is difficult to investigate using the above-mentioned compounds since they are capable of interacting with membrane constituents, particularly proteins, in many unforseen ways.

In the present chapter, we describe a simple procedure that converts a completely inexcitable muscle membrane into one that produces all-or-nothing action potentials with little accommodation. In addition, the chemical changes involved are relatively well defined.

In the course of an investigation on the effects of protein reagents on the glutaminergic receptors of the ventroabdominal flexor muscles of the freshwater shrimps *Atyas occidentalis* and *Xiphocaris elongatus*, we observed that exposure of the preparations to N-ethylmaleimide (NEM) induces the appearance of electrical excitability. Small depolarizations elicit long-lasting repetitive phenomena: volleys of action potentials that overshoot the zero baseline, subthreshold potential oscillations, and slow waves of depolarization with superimposed high frequency bursts. We refer to these phenomena as the NEM-effect, which thus far has only been reported in brief (6). The work presented herein focuses mainly on the chemical aspects of this effect.

METHODS

Preparations

Most experiments were performed on the ventroabdominal flexor muscles of *Atyas occidentalis* collected in streams near El Yunque rain forest; a few were conducted on the same muscles of *Xiphocaris elongatus*.

The specimens were decapitated and the tails cut out. Preliminary dissection was carried out on a cork platform under incident light. As soon as the flexor muscles were exposed, they were transferred to a Petri dish provided with a layer of Sylgard, fixed by insect pins, and the preparation was further cleansed.

Electrical Recording

Experiments were conducted on the same Petri dishes on which the final dissection of the muscle was carried out. The muscle was viewed by transillumination with a stereomicroscope at a magnification of 40×. The solution in the dish was connected by an agar bridge to a smaller chamber with an Ag-AgCl electrode connected to ground through a 10 KΩ resistor. The transmembrane potential was recorded with conventional glass microelectrodes filled with either KCl or K citrate connected to a CRO via a high input-resistance preamplifier. A digital meter connected to the output of the amplifier allowed continuous monitoring of the membrane potential. The potential across the muscle membrane was changed by injecting current through a second microelectrode inserted at a distance of less than 100 μm from the recording one. The applied current was monitored as the voltage drop across the 10 KΩ resistance and displayed in the second beam of the CRO.

Saline Solutions

The normal physiological saline, 524 mOs, was prepared according to van Harreveld (24) and had the following ionic composition (mM): NaCl 205, $CaCl_2$ 13.63, KCl 5.40, $MgCl_2$ 2.41, $NaHCO_3$ 2.02. It was then buffered to pH 7.3 to 7.4 with Tris.

The following modified saline solutions were also used: (a) Na-free, with the same millimolar amounts of $CaCl_2$, KCl, $MgCl_2$, and 258.6 mM Tris base (510 mOs), (b) Ca-free, with the same millimolar amounts of NaCl, KCl, $MgCl_2$, $NaHCO_3$, and 23.7 mM Tris base (469 mOs), (c) Na- and Ca-free, with the same millimolar amounts of KCl, $MgCl_2$, and 279.6 mM Tris base (522 mOs). Solutions a, b, and c were brought to pH 7.3 to 7.4 with HCl. (For the various reagents used and their abbreviations, see footnote).

RESULTS

The NEM Effect

Figure 1 illustrates the main action of NEM. Record A shows 11 catelectrotonic potentials elicited by injecting outward current pulses, about 490 msec in duration, into a muscle fiber of *Atyas* immersed in normal saline. Even though the fiber was depolarized by up to 70 mV (on a resting potential of 73 mV) there are no traces of any active response, nor any indication of changes in the passive electrical properties of the membrane.

In contrast to record A, record B shows rhythmic spontaneous firing of action potentials in a fiber of the same muscle after exposure to NEM. No current was injected in this instance; the action potentials were elicited by the depolarization caused by the insertion of the recording microelectrode. Spikes of just over 60 mV in amplitude were produced at a frequency of about 8.5 Hz. Action potentials similar to those of Fig. 1B are often recorded for periods of several minutes.

Concentrations of NEM as low as 1 mM acting on the muscle for periods of about 10 min are sufficient to cause the above changes. Higher concentrations (2 mM) take shorter times to act, but no effort has been made to quantitate such observations. The continued presence of NEM in the bath is not necessary for excitability. In fact, we have routinely washed out the NEM solution before recording.

It is doubtful whether after treatment with NEM the fibers can exhibit spontaneous electrical activity in the absence of some depolarization. In general, if the resting potential is high (60 mV or more), activity is seen only if a depolarizing current is injected in the fiber.

The NEM effect is transitory; typically, electrical excitability appears after 5 or 10 min exposure to NEM and lasts approximately 20 to 60 min. Thereafter, the fibers gradually lose their ability to generate action potentials and become inexcitable, just as before treatment with NEM. This occurs even if NEM is allowed to remain in the bath.

Reagents and Abbreviations: NEM, Eastman; dithiothreitol (DTT), Sigma; 5,5′-dithiobis-(2-nitrobenzoic acid) (DTNB), Sigma; *o*-iodosobenzoic acid (o-IB), Sigma; *p*-chloromercuribenzoic acid (PCMB), Sigma; *p*-chloromercuriphenyl sulfonic acid (PCMBS), Sigma; phenylmercuric acetate (PM), Sigma; methylmercuric chloride (MM), Alfa Inorganics; 6,6′-dithiodinicotinic acid (CPDS), Aldrich; N-phenylmaleimide (NPM), Nutritional Biochemicals Co.; iodoacetic acid, Sigma; cysteine, Sigma; tetrodotoxin (TTX), Calbiochem.

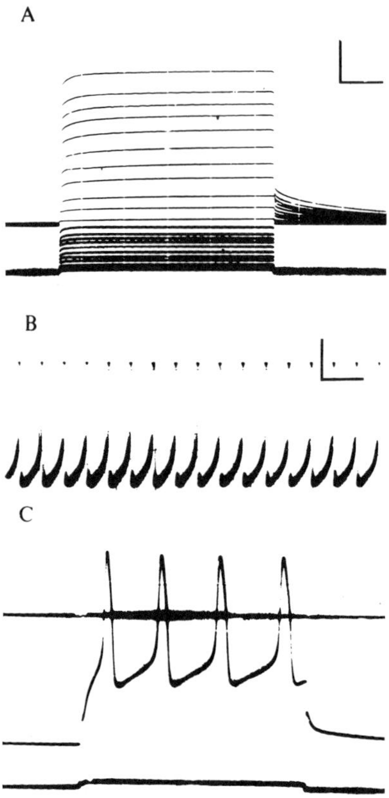

FIG. 1. Repetitive spike activity induced by NEM. **A:** Inexcitability in normal Ringer. **B:** Spontaneous firing of action potentials after exposure to NEM. **C:** Rhythmic generation of action potentials by a small depolarization after NEM treatment. Lower tracings in all records show the depolarizing current and upper tracings are membrane potential. Calibration: vertical, 20 mV and 2×10^{-6} A; horizontal, 100 msec (except **B,** 200 msec).

Electrophysiological Observations

Passive Membrane Properties

The resting potential (V_m) of *Atyas* muscle fibers does not change significantly after exposure to NEM. In one series of measurements, the mean V_m was −72.1 (range, 68 to 81 mV). After exposure to 2 mM NEM for 10 min, the average V_m was −70.8 (range, 68 to 74 mV).

The input resistance of the muscle fibers (R_{in}) also showed no appreciable changes following treatment with NEM. In a series of measurements before NEM, the average R_{in} was 128.9 kΩ; after treatment with 2 mM NEM, the average R_{in} was 126.0 ± 1.76 kΩ(SE; $n = 10$).

As shown in Fig. 2, the current-voltage curves obtained by injecting current in the muscle fibers were not affected by NEM. The relationship between current and voltage was linear in the hyperpolarizing direction; no rectification was observed even when the muscle fibers were hyperpolarized by almost 60 mV. A very small delayed rectification was seen when the membrane was depolarized by more than about 30 mV.

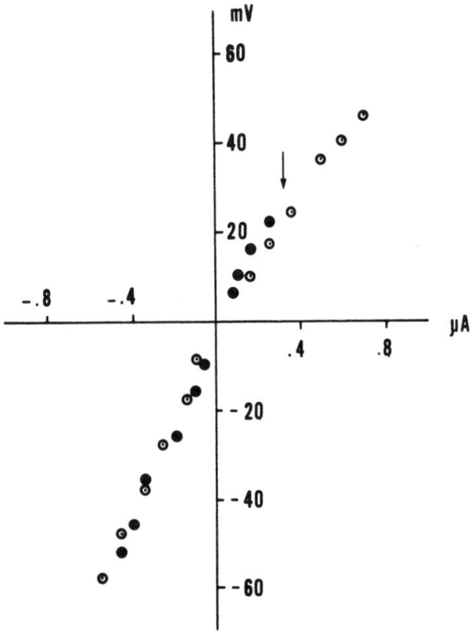

FIG. 2. Current-voltage relationship before (*dotted circles*) and after (*filled circles*) treatment with NEM (2 mM). *Arrow,* threshold for spike initiation.

The current-voltage relationship after treatment with NEM is also shown in Fig. 2. No significant change can be observed. However, no depolarizations greater than about 20 mV could be applied as the threshold for electrical excitation was reached.

It can be concluded, therefore, that NEM exerts no significant effect on the main passive properties of the *Atyas* muscle fibers.

Action Potentials

The action potentials generated by *Atyas* muscle fibers after exposure to NEM vary considerably in amplitude, shape, and duration. The largest spikes measured about 90 mV. On the other hand, we have often observed small (less than 1 mV) oscillatory changes of membrane potential.

In some fibers, the threshold is very low, only a few millivolts above the resting potential level. These are the fibers that exhibit repetitive activity as soon as they are impaled, without need to inject current. Others remain silent after insertion of the microelectrode, although repetitive activity can be elicited by a weak depolarization. In many instances, however, the rhythmic generation of action potentials requires a depolarization of between 20 and 30 mV (see Fig. 1C).

The reasons for such differences are not clear; a possibility that we have not been able to prove is that the changes induced by NEM may not occur uniformly over the surface of the muscle fiber but are limited to discrete areas of the membrane. If the whole membrane changes uniformly, a lower threshold of excitation would be expected. If the excitable areas are far from the microelectrodes, however, a higher apparent threshold would be observed.

Ionic Basis of the Action Potentials

Experiments were done with modified saline solutions in an attempt to determine which ions carry the currents responsible for the action potentials. The effects of TTX and Mn^{2+} ions were also examined.

Our results indicate that both Na^+ and Ca^{2+} ions carry the inward membrane currents that generate the action potentials. Active responses of the muscle membrane could not be suppressed by excluding either Na^+ or Ca^{2+} ions from the external solution, as illustrated in columns 1 and 2 of Fig. 3. As shown in column 3 of Fig. 3, however, electrical activity is completely although reversibly blocked when the muscle was immersed in a Na^+-Ca^{2+}-free solution.

TTX at a concentration of 50 nM did not exert appreciable effects on the action potentials, as shown in column 1 of Fig. 4. Control experiments showed that this compound, at a concentration of 50 nM from the same stock solution, completely suppressed the electrical activity of frog muscle.

Mn^{2+} at a concentration of 50 nM, however, suppresses reversibly the action potentials elicited by NEM. This is shown in column 2 of Fig. 4.

It can be concluded, therefore, that the action potentials generated after exposure to NEM are due to an inward movement of both Na^+ and Ca^{2+} ions, flowing presumably through common channels, not sensitive to TTX, but which are blocked by Mn^{2+}.

Repetitive Activity of the Membrane

The action potentials generated by membrane depolarization after exposure to NEM are in general repetitive, firing at a frequency that is a function of the applied current. In Fig. 5A the fiber had a V_m of 53 mV after insertion and in the absence of applied current, it did not fire. In Fig. 5B, a current pulse of less than 10^{-7} A with a duration of about 0.4 sec was injected and a single action potential was fired. By gradually increasing the strength of the current, the number of action potentials produced during the pulse increased (Fig. 5C–E).

In general, rhythmic activity is made up of axon-like action potentials. More rarely, the oscillatory response of the membrane to depolarizing pulses is composed of small cardiac-like action potentials. When the "cardiac-like" action potentials are well developed, i.e., with an amplitude of more than about 30 mV, the membrane shows little tendency to oscillate and only one action potential is elicited by the applied depolarizing pulse (see Fig. 6).

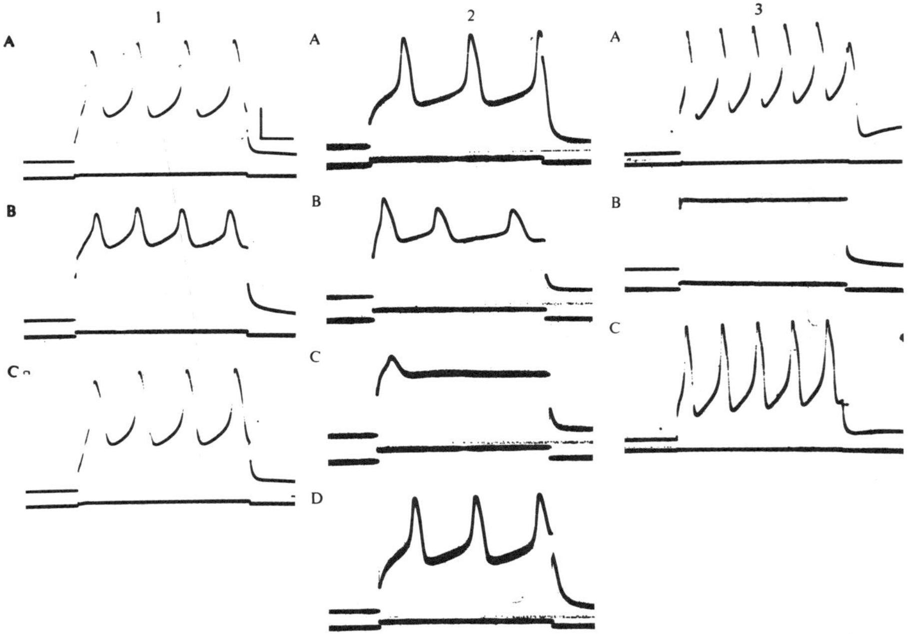

FIG. 3. Column 1: Effect of complete replacement of Na^+ with Tris. *A,* action potentials in normal Ringer; *B:* response after 2 min in Na^+-free solution; *C:* recovery 2 min after return to normal Ringer. **Column 2:** Effect of complete replacement of Ca^{2+} with Tris. *A,* normal Ringer; *B* and *C:* typical responses elicited in Ca^{2+}-free Ringer; *D,* return to normal Ringer. **Column 3:** Complete replacement of Na^+ and Ca^{2+} with Tris. *A:* normal Ringer; *B,* lack of activity in Na^+- and Ca^{2+}-free Ringer; *C,* return to normal Ringer. Calibration: vertical, 20 mV and 2×10^{-6} A; horizontal, 100 msec.

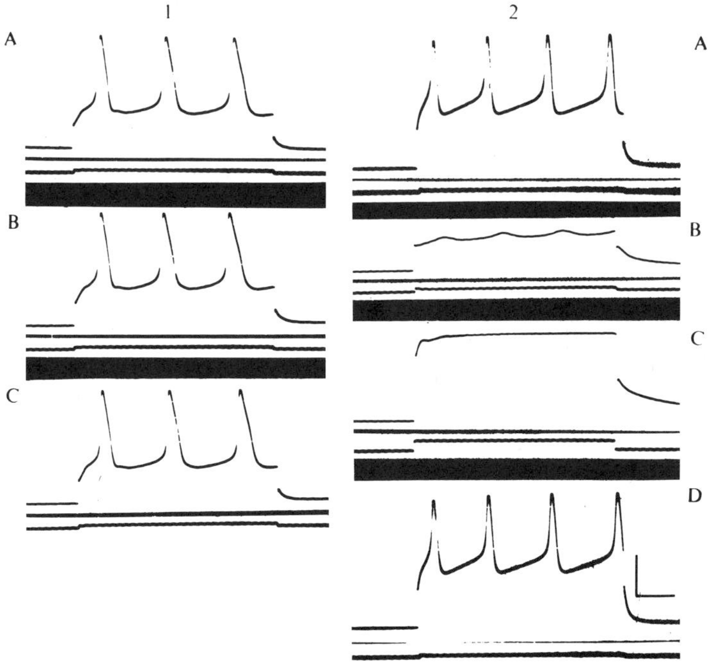

FIG. 4. Column 1: Inability of 50 nM TTX to block spikes. *A*, normal Ringer; *B*, response after 15 min in TTX; *C:* return to normal Ringer. **Column 2:** Suppression of spikes by 50 mM $MnCl_2$. *A*, normal Ringer; *B*, response after 2 min in $MnCl_2$; *C*, response after 3 min in $MnCl_2$; *D*, return to normal Ringer. Calibration: vertical, 20 mV and 2×10^{-6} A; horizontal, 100 msec.

Initiation of Electrical Activity After Exposure to NEM

The NEM effect has been monitored continuously in several preparations. A typical series of observations is summarized in Fig. 7. All the records were obtained from the same muscle fiber before and after exposure to 1 mM NEM. Before NEM was applied, the muscle membrane showed an almost completely passive electrical behavior. Only with depolarizations of 70 mV or more a small hump or graded response appeared at the beginning of the catelectrotonic potential. Figure 7B shows the first sign of electrical activity 2 min after exposure to NEM. Depolarizations of 40 mV or more gave rise to a downward notch which, in the top four pulses, was followed by a short oscillation.

Similar small negative responses have been observed at times in untreated fibers, but only temporarily. In Fig. 7C, taken 2 min after 7B, two new features are present. The negative notch has vanished in the largest depolarizations (60 to 70 mV), and the rising phase of the potential shows a short plateau followed by a small positive deflection. Concurrently, a new type of

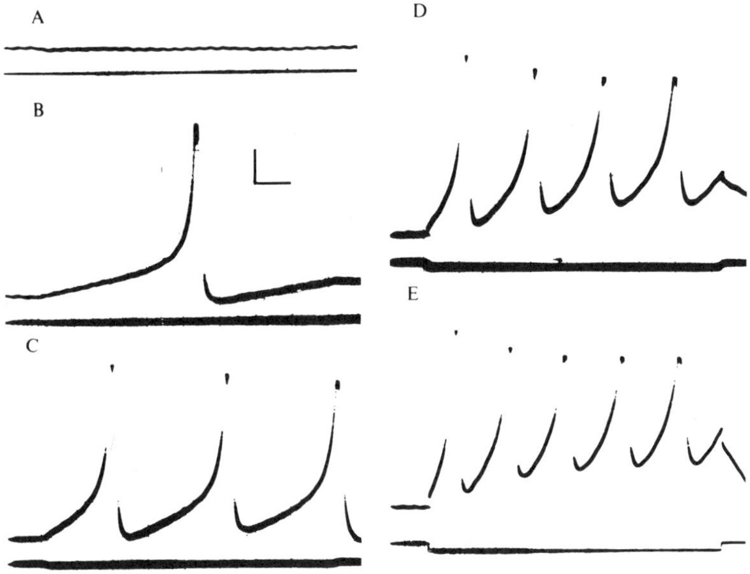

FIG. 5. Frequency of firing as a function of the applied depolarizing current. **A:** Lack of spikes in the absence of applied current. **B–E:** Rate of firing increases with the intensity of applied current. Calibration: vertical, 10 mV and 10^{-6} A; horizontal, 50 msec.

response appears; as seen in record C, a depolarization of about 40 mV now gives rise to a depolarizing response: a sudden step or shift to a new level.

In Fig. 7D, taken 4.5 min after exposure to NEM, the most obvious change is an increase in the amplitude of the response and a decrease in its length. In addition, a second similar response begins to appear, thus initiating oscillatory membrane behavior (see Fig. 7E, 5 min after NEM, and 7F, 7 min after NEM).

Chemical Observations

Chemical Basis of the NEM Effect

At the concentrations and periods employed, NEM is most likely to exert its excitability-inducing effect by combining covalently with SH groups (9). The effects of other maleimide derivatives have also been tested in this respect; NPM was also found to be effective; the substitution of an aliphatic ethyl group by an aromatic group did not make any apparent difference.

As already mentioned, the NEM effect lasts at most about 1 hr; after that time, the muscle fibers rapidly lose their ability to generate action potentials. In principle, this could be explained in two ways: (a) The bonds between

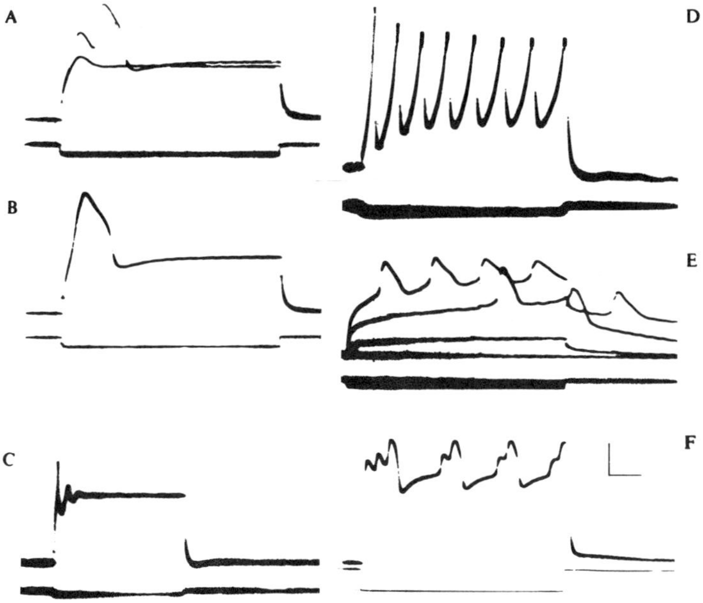

FIG. 6. Different types of responses elicited by depolarizing currents after exposure to NEM. **A, B:** Single cardiac-like action potentials. **C:** Damped oscillatory activity. **D:** Axon-like action potentials. **E:** Oscillatory activity made up by small cardiac-like action potentials. **F:** Repetitive activity made up by small double and one triple action potentials. Calibration: vertical, 20 mV **(A–F)** and 2×10^{-6} A **(A,B)**, 10^{-6} A **(C–E)**, 5×10^{-7} A **(F)**; horizontal, 50 msec **(A,B,E)**, 200 msec **(C)**, 100 msec **(D,F)**.

NEM and the SH groups are somehow broken, which is most unlikely because of their covalent nature; or (b) the NEM molecules, still attached to the S atom, undergo a change that alters their reactivity or other chemical properties.

If explanation (a) were true, one should expect that a second exposure of the preparation to NEM would restore excitability. This has been attempted many times with consistently negative results. It is likely, therefore, that the initial bonding of the SH groups to NEM maintains its integrity even after the effect vanishes. Therefore, the loss of electrical excitability cannot be attributed to a dissociation of the thioether bond.

The Appearance of New SH Groups Restores the NEM Effect

We found that after the NEM effect disappears, treatment of the muscle with DTT (1 mM, 10 min) before being exposed for a second time to NEM restores electrical excitability, although usually for a shorter time. Nevertheless, a second treatment with DTT, followed by new exposure to NEM, fails to induce electrical excitability a third time.

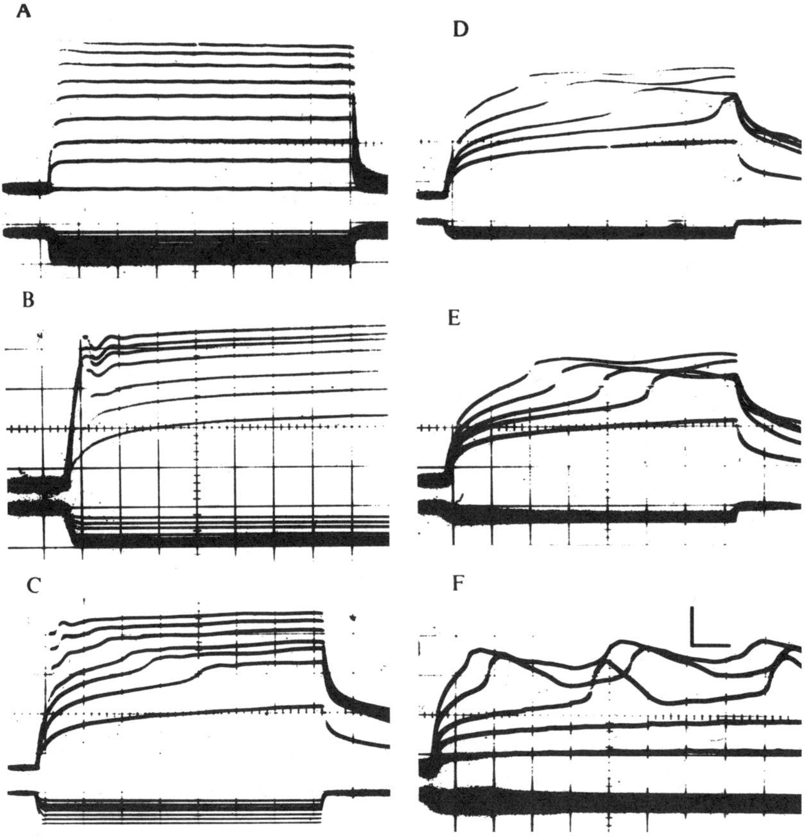

FIG. 7. Initiation of electrical activity under the influence of NEM. **A:** Very small graded responses to depolarizing currents before NEM treatment. **B–F:** Initiation and development of spike activity after exposure to NEM (1 mM) (see text). Calibration: vertical, 20 mV **(A–F)** and 2×10^{-6} A **(A)**, 10^{-6} A **(B–F)**; horizontal, 20 msec **(A,F)**, 5 msec **(B)**, 10 msec **(C–E)**.

This observation is interesting, as it proves that (a) the effect of DTT is to provide new SH groups by reduction of disulfide (S-S) bonds, and (b) the number of potentially available S-S groups is limited. This is also a rather convincing demonstration that free SH groups are needed for NEM to induce electrical excitability.

Potentiating Effect of Reducing Agents on the NEM Effect

It was also interesting to see that the duration of the NEM effect is dependent on the number of free SH groups present at the time the preparation was exposed to NEM. Indeed, if fresh preparations are treated with DTT

(1 mM, 10 min) before exposing them to NEM, the NEM effect usually lasts much longer (up to 3 hr in some instances).

Lack of Direct Effects of Reducing and Oxidizing Reagents

We have tested the effects of reducing agents (DTT, 2-ME) and of oxidizing agents (DTNB, o-IB) on the electrical properties of our preparations. The only effect of DTT has already been mentioned: pretreatment of the preparation results in a lengthening of the duration of the NEM effect, but neither reducing nor oxidizing agents alone induce excitability.

It was also interesting to see that treatment of the preparations with oxidizing agents does not prevent the NEM effect. This indicates that the SH groups, which on combining with NEM are responsible for the appearance of electrical excitability, are not close enough for the reaction $SH + SH \rightarrow S - S + H_2$ to proceed.

Effects of Heavy Metal Ions and Organic Mercurials

A simple interpretation of the excitability-inducing properties of NEM is to assume that they are the result of blocking or binding of free SH groups in the cell. It was reasonable to think that the same effect might be produced by other reagents that combine with SH groups, such as heavy metal ions, Hg^{2+} in particular, and organic mercurials, of the general type $R\text{-}Hg^+$, where R is either an aromatic or an aliphatic group.

The following compounds of this group were tested at the concentrations and time indicated: Hg^{2+} ions, 2 mM, 5 min; PCMB, 2 mM (saturated), 5 min; PCMBS, 2 mM (saturated), 5 min; MM, 2 mM, 5 min; PM, 2 mM (saturated), 5 min. The results of these experiments were negative as excitability did not develop in any of the preparations treated. It can be concluded, therefore, that a simple block of free SH groups is not sufficient to induce the NEM effect.

One aspect of these results was extremely informative; namely, both Hg^{2+} ions and organic mercurials tested as described above prevented the subsequent effect of NEM.

Location of the SH Groups Involved in the NEM Effect

The fact that the NEM effect is prevented by Hg^{2+} ions and organic mercurials is important also because it offers the possibility of obtaining information on the location of the SH groups. In fact, Hg^{2+} ions and the various organic mercurials differ markedly in their capacity to penetrate the cell. Hg^{2+} ions permeate rapidly, PCMB penetrates very slowly, and PCMBS penetrates little if at all. As a consequence, if only Hg^{2+} had prevented the effect of NEM, one could not exclude the possibility that such an effect might originate in reactions taking place in the cytoplasm. However, the fact that treatment with PCMBS for only 5 min also prevents the NEM effect can be regarded as evidence that the SH groups, which on combining with NEM give rise to electrical excitability, are situated on the surface of the cell.

We attempted to reverse the blocking action of Hg^{2+} ions and mercurials using thiols of different molecular weights and different permeant properties. We found that the blocking effect of organic mercurials can easily be reversed by treatment with cysteine (5 mM, 5 min, pH 7.85), a relatively small molecule (MW 121.2). However, we were unable to reverse the block by using either albumin or thiolated gelatin. These results, in conjunction with the blocking effect of PCMBS, support the view that the SH groups responsible for the NEM effect are situated on the cell surface membrane but are not directly accessible to the extracellular solution.

Other Compounds Tested

We have also tested several other compounds that combine with SH groups. Of these reagents, the only one that induced excitability, for brief periods of time, was maleate (HOOC—CH = CH—COOH) and only at a concentration of 10 mM and a period of exposure of 1 hr. Iodoacetate was inactive in either inducing excitability or preventing the occurrence of the NEM effect. It can be concluded, therefore, that the SH groups responsible for induction of excitability are not freely reacting ones. CPDS, a compound known to react reversibly with peripheral SH groups of Ehrlich ascites cells without entering the cell (10), did not induce excitability and did not prevent the effect of NEM in our preparations. These results suggest that CPDS is unable to combine with the SH groups involved in the NEM effect.

DISCUSSION

Both Bornstein-Schur (1) and Lorente de Nó (16) have reported that heavy metal ions block excitability in peripheral nerve, an effect described as irreversible since attempts to restore function by washing with Ringer's solution were unsuccessful. However, one of us, in collaboration with H. J. Hufschmidt (4,5), found that frog medullated nerve fibers blocked by low concentrations of heavy metal ions (i.e., 10^{-5} M Ag^{+}) became excitable within seconds after immersion in Ringer containing 1 mM cysteine. It was suggested, on those grounds, that excitability and conduction might depend on the integrity of free SH groups of proteins at the surface of the nerve fibers (5). Subsequent studies by a number of investigators have provided evidence indicating that membrane proteins are indeed responsible for gating the ionic currents that form the basis of both chemical and electrical excitability. In particular, those involved in the operation of voltage-dependent channels appear to be SH-rich proteins that can be blocked with appropriate reagents.

Smith (22) later repeated the experiments of del Castillo and Hufschmidt (5) and confirmed the idea that free SH groups are responsible for maintaining electrical excitability in frog nerve fibers. In addition, Smith (22) found that 1 mM NEM blocks excitation in lobster axons and 2 mM blocks the frog sciatic nerve.

Other workers have studied the effect of SH reagents on excitability. Huneeus-Cox et al. (12) showed that 1 to 2 mM NEM blocked the giant axon of the squid, applied either externally or by internal perfusion; Shrivastav et al. (21) found that MM blocks the action potentials in squid axons; Marquis and Mautner (17) demonstrated that the conduction block induced by thiol reagents on the squid axons is potentiated by repeated electrical stimulation; and Keana and Stämpfli (13), working on voltaged clamped nodes of Ranvier of the frog, observed that brief treatment with 41 mM NEM decreases both the early inward Na current and the delayed K current.

More recently, Shrager (20) has shown that exposure to NEM results in a specific reduction in sodium conductance in crayfish axons, while the resting potential, the delayed rise in potassium conductance, and the selectivity of the Na channels are unaffected. Experiments with structural analogs of NEM supported the idea that binding of SH groups is the cause of sodium inactivation. Shrager further observed that PCMBS, a compound that also reacts selectively with SH groups, decreases the sodium currents only to 50% of the control values (0.2 mM PCMBS, 25 min); attempts to suppress these currents completely by prolonged exposure to PCMBS resulted in rapid increase in leaking conductance.

In a sense, the experiments described in this chapter are in disagreement with the observations summarized above showing that compounds that bind SH groups block excitability. The opposite seems to occur in crustacean muscle, in which excitability does not appear unless some SH groups are bound by NEM.

Such a discrepancy may only be apparent. One can argue that blocking free SH groups may prevent a protein from participating in a reaction or undergoing a conformational change that in some instances may lead to excitation, whereas in others, it may block it. Indeed, a process that prevents the activation of early Na^+ currents may be similar to another that prevents an early activation of a potassium conductance, which, by opposing the effects of Na activation, will suppress excitability.

More interesting from a chemical viewpoint is the fact that the effects of NEM and organic mercurials are not fully equivalent. This is clear from our own observations and is in agreement with Shrager's (20) results with PCMBS. This compound does not fully block sodium currents in crayfish axons, nor does it induce excitability in crustacean muscle, where it prevents the NEM effect.

This indicates that the effects of binding free SH groups do not depend only on the fact that the —CH_2—SH side chains are blocked and therefore unable to function in the usual way. It rather suggests that the characteristics of the thioethers (of the general type: protein —CH_2—S—R) formed as a result of the various reactions may be of importance in determining the properties of the modified proteins.

NEM possess two salient features: the two carbonyl groups and the tertiary nitrogen. Both, as emphasized by Shrager (20), are shared by some local anesthetics, such as procaine, a compound that causes slow inactivation of Na^+ channels in frog node (14,15) and also induces the conversion of graded to all-or-none responsiveness in lobster muscle fibers (19).

It is plausible, therefore, that the new properties induced in a membrane protein by NEM are the result of the formation of thioethers that possess carbonyl groups and a tertiary nitrogen. Of these two features, the carbonyl groups appear to be more important, as shown by the fact that maleate also induces excitability. It is not unlikely that, following the binding of maleate to a SH group, this compound may adopt a molecular configuration resembling that of maleimide, at least in regard to the position of the carbonyl groups (see Fig. 8B).

It is proposed as a working hypothesis that the slow inactivation of sodium currents in crayfish axons and the induction of excitability in crustacean muscle depend on the conversion of —CH_2—SH side chains to thioethers similar to that shown in Fig. 8B. Such a thioether would be a completely different side chain having a new reactivity and functional properties. The new carbonyl groups may react with neighboring amino groups; in fact, the possibility of alcylation of amino groups by carbonyl groups is well documented (e.g., see ref. 3).

The new NEM or succinic side chains may lead to the formation of bonds between different regions of a protein or between different protein

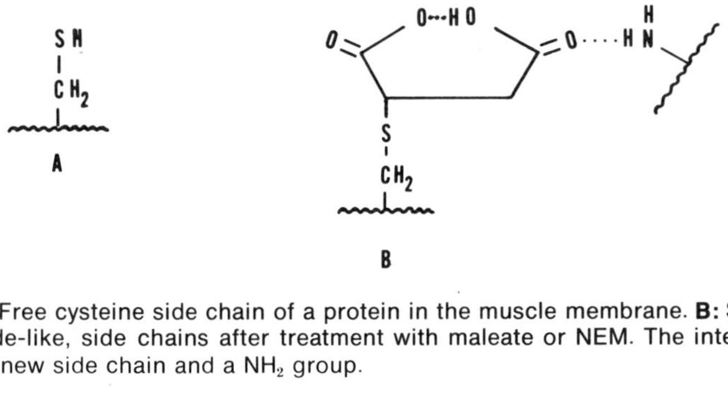

FIG. 8. A: Free cysteine side chain of a protein in the muscle membrane. **B:** Succinic, or succinimide-like, side chains after treatment with maleate or NEM. The interaction between the new side chain and a NH_2 group.

units that may prevent the occurrence of conformational changes. The above-mentioned alcylation of amino groups by carbonyl groups is only one possibility. Hydrogen bonds may also be established through the latter groups, and this would explain why the Na inactivation induced by NEM in crayfish axons (20) can be reversed by long hyperpolarizing prepulses that create sufficient stress to break the H-bonds.

ACKNOWLEDGMENTS

This work was supported by USPHS grants NS-07464, NS-14938, and GM 05784 MRC. (Contribution No. 77 of the Laboratory of Neurobiology).

Note Added in Proof. The hypothesis that the carbonyl groups of the thioether side chains are responsible for the induction of excitability has recently received support from the observation that 4-cyclopentene-1,3-dione, a compound similar to maleimide but which has a methyl group instead of the nitrogen and also combines with free SH groups [Toro-Goyco E., Zuazaga de Ortiz C., and del Castillo J. (1978): *Experientia,* in press], induces excitability in *Atyas* muscle (Zuazaga de Ortiz C. and del Castillo J. (1978): *Experientia,* in press).

REFERENCES

1. Bornstein-Schur, E. (1935): *Quoted in ref. 5.*
2. Chiarandini, D. J., Brandt, P. W., and Reuben, J. P. (1967): *J. Gen. Physiol.,* 50:2501.
3. Cohen, L. A. (1968): *Ann. Rev. Biochem.,* 37:695–726.
4. del Castillo, J., and Hufschmidt, H. J. (1950): *Z. Angewandte Chemie,* 62:168.
5. del Castillo, J., and Hufschmidt, H. J. (1951): *Nature,* 167:146–148.
6. del Castillo, J., and Lara-Estrella, L. O. (1973): *Fed. Proc.,* 32:772.
7. Fatt, P., and Ginsborg, B. L. (1958): *J. Physiol.,* 142:516–543.
8. Fatt, P., and Katz, B. (1953): *J. Physiol.,* 120:171–204.
9. Friedman, E., Marrian, D. H., and Simon-Reuss, I. (1949): *Br. J. Pharmacol.,* 4:105–108.
10. Grassetti, D. R., Murray, J. F., and Ruan, H. T. (1969): *Biochem. Pharmacol.,* 18:603–611.
11. Hodgkin, A. L., and Rushton, W. A. (1946): *Proc. R. Soc.,* 133:444–478.
12. Huneeus-Cox, F., Fernández, H. L., and Smith, B. H. (1966): *Biophys. J.,* 6:675–689.
13. Keana, J. F. W., and Stämpfli, R. (1974): *Biochim. Biophys. Acta,* 373:18–33.
14. Khodorov, B. I., Shishkova, L. D., and Peganov, E. M. (1974): *Bull. Exp. Biol. Med.,* 77:226–229.
15. Khodorov, B., Shishkova, L., Peganov, E., and Revenko, S. (1976): *Biochim. Biophys. Acta,* 433:409–435.
16. Lorente de Nó, R. (1947): *A Study of Nerve Physiology,* Studies from the Rockefeller Institute, vol. 132, pp. 179–185.
17. Marquis, J. K., and Mautner, H. G. (1974): *J. Memb. Biol.,* 15:249–260.
18. Ozeki, M., Freeman, A. R., and Grundfest, H. (1966): *J. Gen. Physiol.,* 49:1319–1334.
19. Reuben, J. P., and Grundfest, H. (1960): *Biol. Bull.,* 119:335.
20. Shrager, P. (1977): *J. Gen. Physiol.,* 69:183–202.
21. Shrivastav, B. B., Brodwick, M. S., and Narahashi, T. (1976): *Life Sci.,* 18:1077–1082.
22. Smith, H. M. (1958): *J. Cell. Comp. Physiol.,* 51:161–171.
23. Werman, R., and Grundfest, H. (1961): *J. Gen. Physiol.,* 44:997–1027.
24. van Harreveld, A. (1936): *Proc. Soc. Exp. Biol. Med.,* 34:428–432.

IONIC MECHANISMS IN ABNORMAL NEUROMEMBRANES

Abnormal Neuronal Discharges, edited by
N. Chalazonitis and M. Boisson. Raven
Press, New York © 1978.

Chemical Stimulants and Real-Time Spectrum Analyzer Used for Studying Properties of Membrane Excitable Sites

Ichiji Tasaki

Laboratory of Neurobiology, National Institute of Mental Health, National Institutes of Health, Bethesda, Maryland 20014

In 1936, Arvanitaki (2) discovered "oscillatory local responses" in crab axons. This discovery preceded Hodgkin's well-known work on subthreshold responses published in 1938 (9). These oscillatory local responses are directly related to Monnier's "pararesonance" in electrical stimulation discussed in 1933 (10). In 1936, Fessard (5) published his work on rhythmical activity of nerve fibers. Later, Arvanitaki (3) showed that, in rhythmical activity, full-sized action potentials are released at the peaks of oscillatory local responses.

Recently, using squid giant axons under intracellular perfusion, my collaborators and I have reexamined Arvanitaki's observations. (Matsumoto, Tasaki, and Inoué, *in preparation*). The results obtained are as follows: (a) Lowering of the external divalent cation concentration, application of various oxidizing reagents, and internal perfusion with an alkaline solution bring about oscillatory changes of the membrane potential; the frequency of this oscillation is close to that described by Arvanitaki (3). (b) When the impedance of the axon membrane is determined as a function of frequency under these conditions, it shows a sharp maximum at the frequency, which corresponds to Monnier's (10) pararesonance. (c) Through a simple Fourier transform, mathematical expressions describing the frequency-dependence of the membrane impedance and the phase-shift can be described from the time-course of a damped oscillation observed following the delivery of a subthreshold pulse of electric current. Through these mathematical analyses, we arrived at the notion that the axon membrane can be regarded as an assembly of minute oscillators that have a power spectrum of the Lorentzian type with its peak-frequency located at the point of pararesonance. Following these preliminary studies, we tested whether or not such oscillators can be demonstrated by the use of a commercially available real-time spectrum analyzer. The results of our test clearly indicate that such minute oscillators do exist.

In this chapter, various properties of the excitable sites (ionic channels) in the squid axon membrane, as shown by the use of chemical stimulants and a real-time spectrum analyzer, are described. The importance of periodic processes in the nerve membrane is repeatedly demonstrated in our investigations. Dr. Arvanitaki, as well as Professors Monnier and Fessard, emphasized the importance of periodic phenomena in the nerve membrane a long time before the advent of modern techniques for studying electrophysiological properties of the nerve membrane.

METHODS

Squid giant axons (available in the Marine Biological Laboratory in Woods Hole, Massachusetts) were used in all experiments described herein. After removing the major portion of the small nerve fibers, we mounted the giant axon horizontally in a Lucite chamber filled with artificial seawater (ASW). The concentrations (expressed in mM) of the salts in ASW were: 423 NaCl, 9 KCl, 9.3 $CaCl_2$, 23 $MgCl_2$, 25.5 $MgSO_4$, and a small amount of tris(hydromethyl)aminomethane-HCl buffer (pH 8.0 ± 0.1). The ASW in the chamber was grounded with a large agar-filled salt-bridge connected to an Ag-AgCl-agar (KCl) electrode. A glass pipette electrode of the Ag-AgCl-agar (KCl) type, approximately 100 μm in outside diameter, was introduced longitudinally along the axis of the axon. The DC resistance of the glass-pipette electrode was between 0.6 and 1.0 megohms. (In a series of experiments, platinized platinum electrodes were used internally and externally; the "signals" observed with these metal electrodes were essentially the same as those obtained with the nonpolarizable electrodes of the Ag-AgCl type.)

The internal recording electrode was connected to a Bak unity-gain electrometer stage (Electronics for Life Science, Inc.), and the potential difference across the axon membrane was measured with a Tektronix 502 oscilloscope. The output of the electrometer stage was led to a low-level preamplifier (Tektronix type 122) with a gain of approximately 100. The output of this AC-coupled preamplifier was connected to a Mini-Ubiquitos spectrum analyzer (Nicolet Scientific Corp.). The output of the spectrum analyzer was led further to a Tektronix 7704 oscilloscope, and the spectra on the oscillograph screen were photographed with a Polaroid camera. Because the amplitude of the signals developed by the axon membrane varied over a wide range (1 to 100 μV peak-to-peak), the vertical (amplitude) axis of the spectral record was on a logarithmic scale. The horizontal axis was always linear. The device described above for recording small oscillatory signals was calibrated with both sinusoidally varying voltage waves and random voltage noise.

To lower the external divalent cation concentration, the method of mixing ASW with an istotonic (530 mM) NaCl solution was used. Various chemical stimulants were dissolved in ASW and applied externally to the axons in the nerve chamber. When an oscillograph record of a propagated action potential was needed, a brief electric shock was delivered near the end of the axon after disconnecting the spectrum analyzer from the output of the low-level preamplifier. All the experiments were carried out at room temperature (19 to 21°C).

RESULTS

Using the experimental setup described under Methods, attempts were made to record electric signals developed by giant axons immersed in normal seawater. By comparing the noise spectrum observed with the two recording electrodes placed outside the axon with that obtained after introduction of the small electrode into the axon, it was sometimes possible to demonstrate very small signals of the axonal origin located at a frequency of approximately 150 Hz. However, the amplitude of these signals was usually so close to the level of the noise generated by the recording device that it was not possible to determine accurately the spectrum of such signals.

The amplitude of the signals at about 150 Hz was found to increase gradually when the external divalent cation concentration was lowered by a factor of 3 or 4. During the period of this gradual increase in the signal amplitude, there was little or no change in the peak frequency of the signal. When the signal amplitude was between a few and about 30 μV peak-to-peak, the signal spectrum appeared to be roughly Lorentzian (see the examples shown in Fig. 1, bottom). With signals in this range of amplitude, it is difficult to recognize their periodicity on the screen of an ordinary oscilloscope. These signals are designated in this chapter as "periodic miniature responses."

When the signal amplitude exceeded about 50 μV, there was a rapid increase in the amplitude associated with a reduction in the band width of the spectrum. The signal was soon found to be transformed into oscillatory local responses with a more or less well-defined frequency (as examined with an ordinary cathode-ray oscilloscope).

In recent years, tetrodotoxin (TTX) and tetraethylammonium (TEA) have been used to characterize the properties of excitable membrane sites. For this reason, the effects of TTX and TEA on periodic miniature responses were examined using axons immersed in a medium with a reduced divalent cation concentration.

In all nine axons examined, the periodic miniature responses were found to be promptly and completely suppressed after addition of TTX to the external medium at a concentration of about 5×10^{-8} M (see Fig. 1, left). The noise spectrum obtained from an axon under the action of TTX was very similar to that taken with the intracellular electrode withdrawn from the interior of the axon and kept in external seawater.

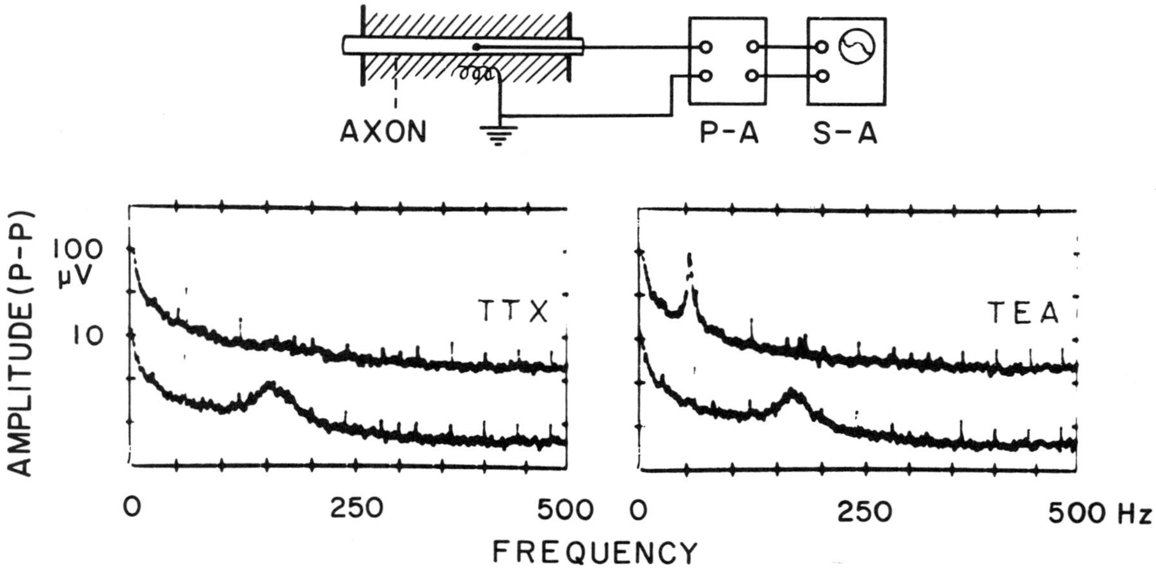

FIG. 1. Top: Schematic diagram illustrating the experimental setup used to detect oscillatory miniature responses generated by giant nerve fibers of the squid. AXON, Loligo nerve fiber; P-A, amplifier with a high input impedance; S-A, real-time frequency analyzer. **Bottom left:** Spectral records showing suppression of oscillatory miniature responses by TTX. The voltage scale on the vertical axis applies only to the bottom trace taken immediately before application of TTX. The upper trace, taken 1 min after application of TTX, was displaced upward by one division; note the disappearance of the signal at about 150 Hz. The sharp vertical lines, which appear at multiples of 60 Hz, are artifacts generated by the power source. **Bottom right:** Spectral records showing a large shift of the peak frequency of oscillatory miniature responses brought about by intracellular injection of TEA. Note that the signal recorded before TEA injection had a peak frequency of about 160 Hz and that the frequency shifted to about 54 Hz after injection.

The effect of TEA was examined using giant axons into which both the injection pipette and the recording electrode were introduced initially. At the time when a signal at about 150 Hz appeared under the influence of a low divalent cation medium, TEA was injected uniformly into the entire 28 mm length of the axon by the procedure described by Tasaki and Hagiwara (14). This procedure brought about an immediate lowering of the peak-frequency of the signal (see Fig. 1, right). The amplitude of the signal was usually enhanced to some extent by TEA injection. With the injection fluid containing 160 mM TEA, the peak-frequency observed after injection was between 30 and 36 Hz (six axons). With 500 mM TEA solution, the peak-frequency observed after injection was between 13 and 17 Hz (three axons) . These observations indicated that the excitable membrane sites involved in generation of these miniature responses are sensitive to both TTX and TEA.

It is of some interest to test the effect of TTX following injection of TEA into the axon. Figure 2 shows an example of such tests; the first (left) record was taken from an axon immersed in a medium with a reduced divalent cation concentration. The action potential record in the inset was taken immediately after the signal spectrum was photographed. The next (middle) signal spectrum was taken about 1 min after injection of TEA. The inset in the middle shows that the action potential duration was markedly prolonged by TEA injection. The last (right) record was taken immediately after addition of TTX to the external medium. The miniature responses were completely suppressed and the action potential amplitude was diminished by TTX.

It was frequently noted that periodic miniature responses are more sensitive to TTX than are full-sized action potentials. At the moment when the miniature responses were completely suppressed by addition of TTX to the medium, it was found that the action potential amplitude was almost normal. In TEA-treated axons, the action potential duration was shortened by TTX before there was a significant decrease in the action potential amplitude. Eventually, TTX completely suppressed propagated action potentials.

Next, to elucidate the mechanism of generation of miniature responses, the potential difference across the axon membrane was measured during the course of development of responses following application of various chemical stimulants. Figure 3 shows an example of such measurements. Initially, the axon under study was kept in normal ASW. After establishing that both the resting and the action potential of the axon remained normal and stable, a spectrum of the noise developed by the recording system was photographed (see the smooth spectrum, Fig. 3, middle). The divalent cation concentrations in the medium were then reduced to one-quarter of the normal values by mixing ASW with an isotonic NaCl solution. The potential difference across the axon membrane (at rest) was found to be

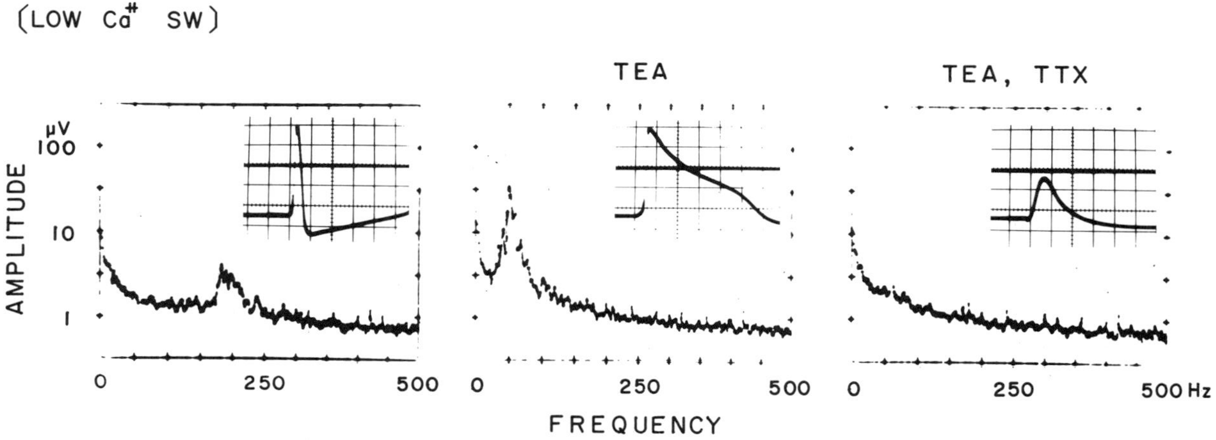

FIG. 2. Effects of TEA and TTX applied successively to a squid giant axon developing periodic miniature responses. **Left:** Spectrum of periodic miniature responses obtained approximately 5 min after introduction of a medium with low divalent cation concentrations. **Middle:** Spectrum obtained about 1 min after injection of TEA. **Right:** Spectrum taken 1 min after addition of TTX to the medium. The oscillograph records in the insets were taken approximately 30 sec after the corresponding spectra were photographed.

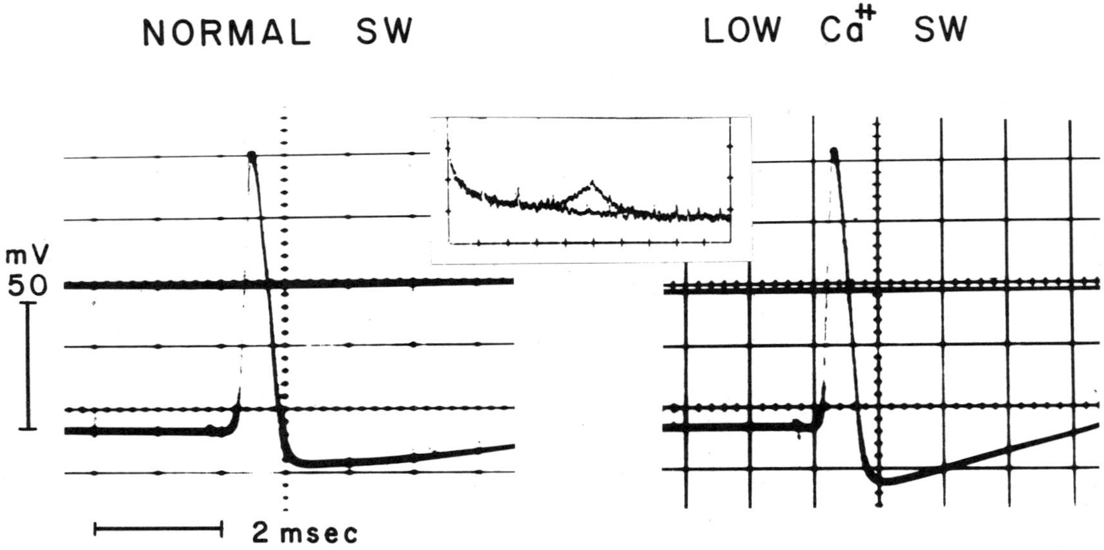

FIG. 3. Records of the resting and action potentials taken before **(left)** and after **(right)** lowering the divalent cation concentrations in the external medium. The spectrum of periodic miniature responses in the middle (taken in a medium with low divalent cation concentrations) were superimposed on a spectrum taken from the same axon while it was still in normal seawater. Note that generation of miniature responses was not associated with a detectable fall in the resting membrane potential.

increased slightly by this procedure (the axon interior becoming 1 to 2 mV more negative). Soon, development of miniature responses became apparent on the screen of the spectrum analyzer (see the spectral record in the middle of Fig. 3). The oscillograph record (Fig. 3, right) was taken immediately after the signal spectrum in the middle was photographed. In all axons thus examined, it was found that there is no fall in the resting membrane potential at the time when periodic miniature responses were evoked by a reduction of the external divalent cation concentrations.

Similar observations were made using 4-dimethylaminopyridine and scorpion venoms as chemical stimulants. The results of these observations are published elsewhere.

DISCUSSION

The experimental results described above may be summarized as follows: (a) Various chemical stimulants applied to squid giant axons are capable of generating periodic responses (rhythmical activity) with extremely small amplitudes (a few microvolts peak-to-peak across the axon membrane). (b) By lowering the external divalent cation concentration, periodic miniature responses of about 30 μV in amplitude can be generated without being accompanied by a fall in the potential difference across the axon membrane. (c) Periodic miniature responses are very sensitive to both TTX and TEA. (d) The frequency of these miniature responses is close to Monnier's (10) frequency of pararesonance and to Arvanitaki's (2,3) frequency of oscillatory local potentials.

There is little doubt that these miniature responses represent repetitive excitation (rhythmical activity) localized at a small number of excitable sites (or ionic channels) in the axon membrane. Thus we see that the unitary processes taking place at the individual membrane sites involved are essentially periodic.

It is evident from result (b) above that miniature responses are not triggered by membrane depolarization. It is also clear that there is little or no electric interaction between the membrane sites generating these miniature responses.

Many investigators regard excitable membrane sites as consisting of two independent, spatially separate ion channels (or ion pores), namely, of Na- and K-channels (see ref. 12). Several authors cautioned against wholehearted acceptance of this simplistic view. Frankenhaeuser (7), for example, stated: "No finding directly decides whether sodium and potassium are moving through the same or through different sites. Calcium clearly affects both the sodium and potassium systems." In contrast, Mullins (11) assumes the existence of only one kind of ion pore. He explains the process of excitation by assuming that the Na-selective pores are converted, during voltage-clamp, into K-selective pores. Some time ago, I (13) expressed the view that the membrane macromolecules (proteins) have properties of cation-exchanges, and nerve excitation is an electrochemical manifestation of a macromolecular phase-transition associated with a change in the ratio of di- and univalent cations at the ion-exchange sites. The process of generation of miniature responses is now considered on the basis of these different viewpoints as to the structure of the excitable membrane sites.

The macromolecular theory mentioned above was advanced from the outset to interpret the excitation of the nerve membrane by lowering of the external divalent cation concentration in the medium. The effect of other chemical stimulants may be treated on the same basis. Oxidizing reagents, for example, are expected to increase the net negative charges of membrane macromolecules and to change the cation selectivities. The rate of conformational changes and of relaxation processes, which can be affected by TTX and TEA, are assumed to determine the magnitude and repetition frequency of these miniature responses.

Ion pores of particular dimensions have been postulated to explain the effects of TEA and TTX (1,8). The finding that TTX readily suppresses miniature responses may be taken, according to this postulate, as an indication of involvement of Na-channels in generation of these responses. The observations made with TEA seem to suggest that K-channels are also involved in these processes. It is very difficult, if not impossible, to explain the effects of TTX and TEA on the assumption that Na- and K-channels are spatially separate and independent of each other. Mullin's (11) hypothesis (postulating only one kind of ion pore) appears to be more adequate than the Armstrong-Hille hypothesis for explaining the effects of TEA and TTX shown in Fig. 2.

Finally, it may be pointed out that there is a close relationship between the membrane noise in the squid axon membrane (4,6) and the miniature responses described herein. I have discussed this relationship in some detail in papers that have appeared in the *Japanese Journal of Physiology* (Vol. 27, p. 643 and Vol. 28, p. 89).

SUMMARY

By using a real-time spectrum analyzer in combination with various chemical stimulants, electrical signs of periodic responses localized at a small number of excitable sites in the squid axon membrane are demonstrated. The amplitude of these responses was 1 to 30 μV peak-to-peak, and their frequency was close to that of Monnier's (11) pararesonance and to that of Arvanitaki's (2,3) oscillatory local responses. Experimental evidence is presented, indicating that these responses were not triggered by membrane depolarization. Both TTX and TEA were found to exert strong effects on these responses. These findings are interpreted as being consistent with the macromolecular hypothesis of nerve excitation.

REFERENCES

1. Armstrong, C. M., and Binstock, L. (1965): *J. Gen. Physiol.,* 48:859–872.
2. Arvanitaki, A. (1936): *J. Physiol. Pathol. Gen.,* 34:1182–1197.
3. Arvanitaki, A. (1939): *Arch. Int. Physiol.,* 49:209–256.
4. Conti, F., DeFelice, L. J., and Wanke, E. (1975): *J. Physiol.,* 248:45–82.
5. Fessard, A. (1936): *L'Activité Rhythmique des Nerfs Isolé,* Paris, Hermann & Cie.
6. Fishman, H. M., Poussart, D. J. M., and Moore, L. E. (1975): *J. Membr. Biol.,* 24:281–304.
7. Frankenhaeuser, B. (1938): In: *Progress in Biophysics and Molecular Biology,* edited by J. A. V. Bultler and D. Noble, pp. 97–105. Pergamon Press, Oxford.
8. Hille, B. (1968): *J. Gen. Physiol.,* 51:199–219.
9. Hodgkin, A. L. (1938): *Proc. R. Soc. Lond.* [*Biol.*], 126:87–121.
10. Monnier, A. M. (1933): *C. R. Soc. Biol.,* 114:1295–1297.
11. Mullins, L. J. (1968): *J. Gen. Physiol.,* 52(3):550–553.
12. Narahashi, T., and Moore, J. W. (1968): *J. Gen. Physiol.,* 52(3):553–555.
13. Tasaki, I. (1968): *Nerve Excitation: A Macromolecular Approach.* Charles C Thomas, Springfield, Ill.
14. Tasaki, I., and Hagiwara, S. (1957): *J. Gen. Physiol.,* 40:859–885.

Abnormal Neuronal Discharges, edited by N. Chalazonitis and M. Boisson. Raven Press, New York © 1978.

A Model for the Production of Slow Potential Waves and Associated Spiking in Molluscan Neurons

M. Gola

Institute of Neurophysiology and Psychophysiology, C.N.R.S., Marseilles, France

Endogenously active neurons fire action potentials because they have pacemaker properties rather than synaptic activation. Among these neurons, bursting neurons fire groups of spikes separated by regular, long silent periods. Such pacemaker neurons have been identified first in the mollusc Aplysia (2) and then in a number of phyla and have been particularly studied in molluscan neuronal systems. The bursts of spikes are mediated by a slow, subthreshold membrane potential oscillation. Using the voltage-clamp method, it has been shown by several authors (11,15,17,36) that in molluscan bursting neurons, the slow potential oscillations are due to slow conductance changes involving Na^+, Ca^{2+}, and K^+ ions.

Bursting activity can also be elicited by a variety of physical or chemical means in many neurons that do not burst spontaneously (1,3,4,6,7). In some instances, aspects of these burstings are similar to those seen in spontaneously bursting neurons, particularly upon addition of the convulsant drug pentylenetetrazol (PTZ) (6) or upon moderate warming of cell L_{11} of Aplysia (1). The membranes of such induced bursting neurons have several common properties with normally bursting neurons, as revealed in voltage-clamp conditions (9,10,16), which led several authors to explain the various bursting activities by a common model (17,36,40). In terms of this model, in addition to the fast conductance changes that underlie single spikes, the membrane of bursters shows (a) a steady state negative conductance that provides positive feedback for the bursting rhythm, and (b) slow conductance changes which account for the potential trajectory in the silent period.

Similar membrane properties exist in cells displaying slow potential waves without associated spiking, such as in L_{11} cell of Aplysia at high temperature (16), in TTX-treated R_{15} cell of Aplysia, and in certain Ba-treated molluscan cells (19), and in cells displaying regular high frequency discharges or paroxysmal depolarization shifts (PDSs) (9,10). In light of the widespread occurrence of slow potential wave-generating systems, it appears that the membranes of molluscan pacemaker neurons possess in common (a) a spike-generating mechanism similar to that described in a number of excitable cells and characterized by fast kinetic parameters, and (b) a modulating system characterized by slow kinetic parameters and resulting in various electrical behaviors, among which the bursting activity is only a particular case, even if physiologically important.

Interferences between the two systems (fast and slow) must exist; but because of the great difference in the magnitudes of their time constants, they can be conveniently studied separately. The present chapter describes slow wave-generating systems of several normally or experimentally in-

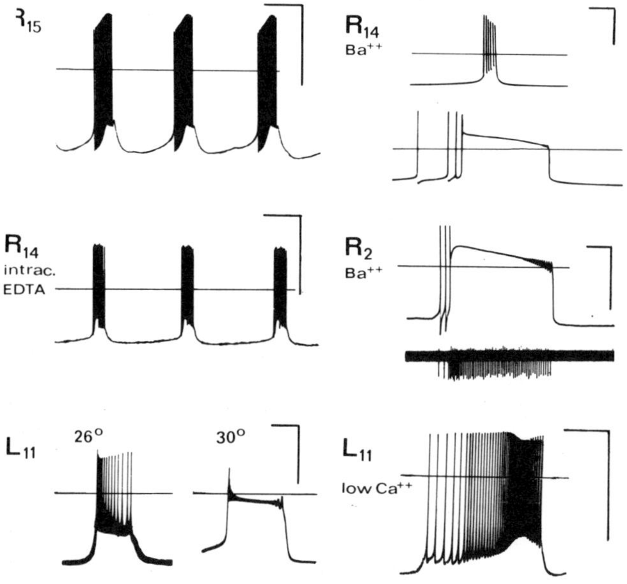

FIG. 1. Spontaneous or experimentally induced burst-generating neurons in Aplysia. [The cells are identified according to the nomenclature of Frazier et al. ref. 14a]. **Left:** R_{15} cell (*Aplysia californica*) is the well-known bursting neuron of Aplysia, first identified in *Aplysia fasciata* (2). R_{14} (*Aplysia californica*) is a silent neurosecretory neuron; the regular bursts were observed after a 3 hr iontophoretic injection of EDTA (or EGTA) in order to reduce the intracellular free calcium concentration. L_{11} (*Aplysia fasciata*) is regularly firing at room temperature. Upon warming, it generates bursts of spikes (26°C) or long plateaus (30°C). **Right:** R_{14} cell of *Aplysia rosea. Upper row,* burst of spikes some minutes after substituting Ba^{2+} for Ca^{2+} (10 mM) in artificial seawater; *lower row,* spontaneous burst-plateau pattern after 30 min Ba-Ca substitution. The pattern is reproduced regularly with a 3 to 4 min period. R_2 cell of *Aplysia californica.* In this cell, barium is also effective in inducing burst-plateau patterns, although at a higher concentration (Ba^{2+} = 20 mM). The recording below the intracellular trace (from the right connective which contains the axon of R_2) shows that the soma plateau induces a high frequency discharge in the axon. L_{11} (*Aplysia californica*) generating long bursts of spikes following the reduction of extracellular calcium concentration from 10 to 2 mM. Vertical calibrations, 50 mV for all recordings; horizontal calibrations, 10 sec for all recordings except Ba^{2+}-treated R_2 cell and L_{11} at low Ca^{2+}, 2 sec.

duced burst-generating neurons of Aplysia. Some of these neurons are illustrated in Fig. 1. Much attention has been paid to the slow potential waves induced by barium because their analysis in voltage-clamp conditions provide all the data necessary for the reconstruction of the potential changes.

Finally, starting from a brief analytical study of the model set up for the Ba-induced slow potential waves, a generalized model is presented with the aim of predicting the slow potential behavior of several burst-generating neurons.

TRIGGERING MECHANISM OF SLOW POTENTIAL WAVES

In the past two decades, it has been possible to evaluate the properties of the ionic currents in the somata of silent and pacemaker neurons of molluscs. On the basis of voltage-clamp data, Connor and Stevens (8) have been able to predict correctly the firing of normally silent neurons in response to a constant stimulus current. However, the mechanism responsible for the slow cyclic potential changes in bursting neurons remained unknown until 1973, when voltage-clamp studies revealed the existence of a persistent inward current. This inward current, carried by Na^+ and/or Ca^{2+} ions, resulted in an N-shaped, steady state current-voltage relationship, i.e., with a region of negative slope conductance (Fig. 2a). The inherent regenerative nature of such negative conductance has been utilized as the basis for models of bursting activity, analogous to the Trautwein (38) model originally proposed for the sinus pacemaker of the heart. These models imply the existence of a high resting Na^+ conductance and of time-dependent fluctuations in K^+ conductance (5,35) which have been effectively observed in bursting neurons by Junge and Stephens (24).

The persistent or slowly inactivating inward current of bursting neurons (12,29) well accounts for the depolarizing phase of the slow wave which triggered the burst of spikes; because of the positive feedback induced by the negative conductance, the depolarization must grow exponentially. The "explosive" aspect of the depolarization is well evidenced when the spikes are suppressed (by TTX or in Na-free saline) (30).

Similar long-lasting inward currents are observed in several experimentally induced bursting neurons, as well as in neurons generating recurrent long plateaus. There can be some differences in the ionic nature and kinetic parameters of these inward currents, but in all cases, the instability induced by the negative conductance acts to depolarize the membrane toward an excited state (defined as the null current point of the current-voltage relationship) which may lead to firing. The straight correlation between the existence of a region of negative conductance and bursting activity has been outlined by several authors (9,10,15,36,40) and is well illustrated in cell L_{11} or Gen of Aplysia. At room temperature, this neuron fires at a relatively high frequency. When temperature is raised

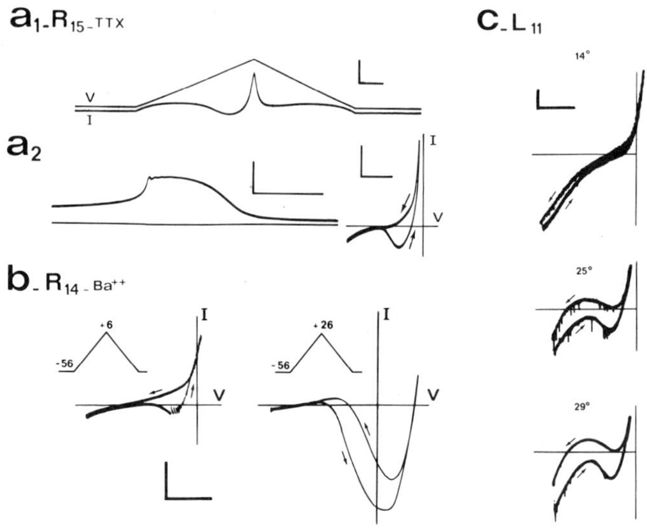

FIG. 2. N-shaped current-voltage relationships in burst-generating neurons. The current-voltage curves are obtained in voltage-clamp conditions with slow ramp pulses. **a**$_1$: TTX-treated R_{15} neuron of *Aplysia fasciata. Upper trace,* symmetrical voltage ramp pulse; *lower trace,* resulting transmembrane current. The amplitude and slope of the ramp are, respectively, 56 mV and 2.4 mV/sec. **a**$_2$: Spontaneous slow potential wave in TTX-treated R_{15} neuron of *Aplysia fasciata* and current-voltage curves obtained as described in **a**$_1$. Outward currents are upward, and positive potentials are on the right. Horizontal and vertical axes are, respectively, zero current and zero potential baselines. The small arrows indicate the direction of potential changes. Upon slow depolarization (2.2 mV/sec), the I-V curve is N-shaped with a null-current point near −15 mV, whereas the N-shape disappears upon slow repolarization. **b:** Current-voltage curves of a R_{14} neuron of *Aplysia rosea* before (*left*) and 30 min after (*right*) substitution of Ba^{2+} for Ca^{2+} in artificial seawater. The voltage programs are shown in insets. **c:** Current-voltage curves in L_{11} neuron of *Aplysia fasciata* at three different temperatures. The N-shaped curves are obvious above 20°C when the cell generates either bursts of spikes or long plateaus. (Note that the hysteresis almost disappears at low temperature.) Calibrations: voltage, 25 mV in all records; current, 20 nA in all records (40 nA in **a**$_1$); time scale (for **a**$_1$ and **a**$_2$), 5 sec.

above 25°C, it generates high frequency bursts of spikes separated by long silent periods (see Fig. 1). At higher temperatures (around 30°C), the spikes are considerably reduced and very often disappear and the potential alternatively passes from a hyperpolarized (−40 to −50 mV) to a depolarized level (−10 to −20 mV). The current-voltage relationships obtained with the slow ramp method show a progressive increase in the magnitude of the negative slope when temperature is increased, with an optimum at about 30°C (Fig. 2c). This cell is of interest because it offers the opportunity to analyze the membrane properties underlying very different electrical behaviors (tonic firing, bursting activity, and long plateau production) induced in an easily controlled and reversible way. For instance, the null-current points of the I-V curves obtained with slow depolarizing ramps are near −15 mV whether the cell generates bursts of spikes (at 25°C) or long plateaus (at 30°C). Since the level of the null-current point is a measure of the depolarizing tendency, it is clear that the production of bursts or pure plateaus depends on the interplay between the slow wave-generating mechanism and the spike-forming system which must be depressed by excessive warming.

Using the slow ramp method, a steady state negative conductance caused by a persistent inward current has also been observed in the following burst-generating neurons: R_{14} neuron of Aplysia in the presence of 10^{-4} dopamine (18), R_{14} neuron intracellulary injected with EGTA, Aplysia and Helix neurons bathed in saline containing Ba^{2+} instead of Ca^{2+} (19), R_{14} and R_2 neurons of Aplysia in the presence of veratridine.

The characteristics of dopamine-treated R_{14} (which is much like the R_{15} bursting neuron of Aplysia) are described in the following chapter by Boisson and Gola (*this volume*). The recurrent burst-plateau patterns induced by replacing Ca^{2+} with Ba^{2+} (see cell R_2 and R_{14} bathed in Ba^{2+} saline, illustrated in Fig. 1) are of interest because (a) the long potential waves appearing here as plateaus are generally devoid of spikes, and (b) the persistent inward current is much larger (up to 100 nA) than that in bursting neurons and thus can be quantitatively studied (Fig. 2b).

The depolarizing effects of veratridine on axons are well documented (39) and have been attributed to a noninactivating sodium current. The same depolarizing effect has been observed in molluscan cells (28); but, particularly in R_2 and R_{14} cells of Aplysia, veratridine is able to induce recurrent slow waves with associated high frequency bursts. These slow waves are characterized by a fast depolarizing phase followed by a very slow repolarization lasting 40 to 60 sec.

The convulsant PTZ induces the regular production of PDSs in most molluscan nerve cells and particularly in normally firing cells (6). As in bursting cells, there is a close correlation between the ability of the treated cells to generate PDS and the existence of a quasisteady state negative conductance (9,10); both are favored by increasing temperature (10).

Finally, in all cells generating slow potential waves (with or without associated spiking), a common membrane property exists (the negative slope conductance), which triggers the passage from a resting or hyperpolarized state to an excited or depolarized state. This triggering mechanism [in the sense introduced by Franck (14)] is, for the present, due to similar ionic processes, that is, a persistent activation of an inward current carried by Na^+ and/or Ca^{2+} (or Ba^{2+}). However, other ionic mechanisms—more particularly, a depolarizing potassium inactivation—could be considered. It is worth noting that the triggering process implies the activation of an inward current, which is a voltage-dependent process, rather than a high resting Na^+ or Ca^{2+} conductance (as in the Trautwein model), which is voltage independent.

Some differences can be noted in the magnitude and time course of the activated inward current: in R_{15} bursting neurons of Aplysia, the inward current develops slowly (compared to the spike time constant), and its amplitude reaches no more than 20 nA (17). In PTZ-treated neurons, the persistent inward current is also very small, but it develops more rapidly. In Ba^{2+}- or veratridine-treated neurons, the inward current reaches its maximal amplitude in a few milliseconds. However, the last two cases do not show clearly if separate channels exist (as in bursting neurons) for the quickly inactivated inward current and the persistent inward current or if the same channel undergoes incomplete inactivation.

RECOVERY PROCESSES

The persistent activation of an inward current resulting in a quasisteady state negative slope conductance is an essential condition for slow potential wave generation. It is well known, however, that such a negative conductance is able to produce oscillations only under suitable additional conditions, which imply that the load line intersects the current-voltage relationship only in the negative slope region. Here, the load line of the system is a I = constant straight line (generally the I = 0 axis under stimulus-free conditions), which actually intersects the positive slope limbs of the N-shaped I-V curve in one or two stable points so that the system exhibits monostable or bistable behavior. Such behaviors have been effectively described in several excitable membranes (20), and particularly in non-classic R_{15} neurons of Aplysia (17). They have been explained on the basis of the properties of the N-shaped current-voltage relationships which always have at least one stable state where the system stabilizes following the cessation of a stimulus. In the case of the current-voltage curves described in the preceding section, this stable state corresponds to the excited state leading to continuous firing or permanent plateau (as in veratrinized axons). Thus some recovery process must exist and must result in the temporary disappearance of the stable excited state. Since the latter was created by the persistent inward current, the recovery process can be due to slow disappearance of that inward current or to slow development of an outward current, or to both phenomena. The first demonstration of the existence of slow recovery processes in bursting neurons was obtained by the ramp voltage-clamp method (16). When the ramp was applied to the membrane from a hyperpolarized state, the current-voltage relationship was N-shaped, as previously described. When the potential was initially held at a low level (near 0 mV), however, the N-shape disappeared and the current-voltage relationship was lifted (Fig. 2). It has been shown that the slow changes in the current-voltage relationship were brought about by a slowly developing outward current and partly by a slow decline in the inward current (17,36). As a result, the excited state was changed to a new stable resting state (near −45 mV), inducing the repolarization of the cell. These two mechanisms appear to be very general since they have been observed in a variety of burst-generating cells. For instance, in Ba-treated neurons, the main role in the recovery process is played by a very slow inactivation of the inward current ($\tau \simeq 20$ sec) and to a lesser extent by a slowly developing outward current (19). In warmed L_{11} cells generating bursts or plateaus, the recovery process implies mainly a slowly developing outward current (16), whereas in veratrinized cells, it implies the slow disappearance of the inward current. The slow outward currents in three different slow wave-generating cells (R_{15}, warmed L_{11}, and Ba-treated R_{14}) are illustrated in Fig. 3a–c.

The slow outward current is generally carried by K^+ ions, and the increase in potassium conductance is in some way related to the influx of

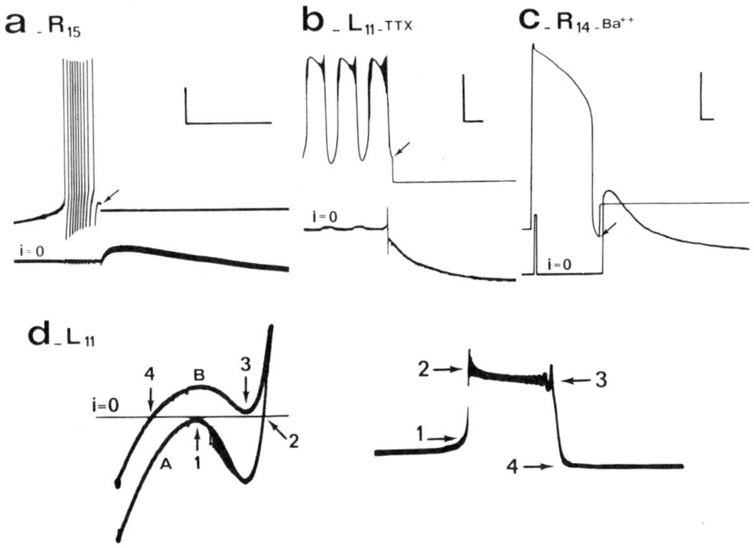

FIG. 3. Slow current changes in burst-generating neurons. Current changes with slow kinetics were observed after passage to voltage-clamp conditions (at the point indicated by the small arrow) just after a burst or plateau. In the three cells illustrated in **a, b,** and **c,** the current changes correspond to a decrease in outward current. **a:** R_{15} cell of *Aplysia depilans.* The holding potential is at −45 mV and the outward current declines with a time constant of 12 to 15 sec. **b:** L_{11} cell of *Aplysia fasciata* bathed in seawater containing TTX 10^{-5} M in order to reduce the spikes. The tail current upon institution of the voltage-clamp ($V = -65$ mV) has an exponential decay with a time constant of 58 sec. **c:** R_{14} cell of *Aplysia rosea* bathed in artificial seawater containing Ba^{2+} instead of Ca^{2+}. The time constant of the tail current decrease is 50 sec. **d:** Interpretation of the square-shaped waves, generated in a warmed (29°C) L_{11} *Aplysia fasciata* neuron, based on the current-voltage curves obtained with the slow ramp method. The I-V curve shifts from A to B following depolarization or from B to A following hyperpolarization, respectively, due to a decrease or an increase in the slow outward current. The points 1 to 4 on the I-V plane refer to corresponding points on the square wave. Calibrations: voltage, 20 mV in all records; current, 5 nA in all records (8 nA in **a**); time, 10 sec in all records.

Ca^{2+} ions during the burst (22,23,37). There is some controversy concerning the way in which Ca^{2+} ions interfere with the potassium conductance (26,31–33), but this mechanism clearly illustrates the fact that a delayed counteraction exists between the trigger and recovery processes. In the case of Ba^{2+}-treated neurons, a similar coupling between the influx of Ba^{2+} ions and a slow outward current has been suggested (19).

The slow disappearance of the inward current has generally been attributed to a slow inactivation of the corresponding conductance or to a decrease in the electromotive force due to accumulation of the invading ions in the intracellular space. In both cases (voltage-dependent inactivation and accumulation), however, the larger the change due to the triggering system, the larger the consecutive slow recovery process.

Finally, as a result of the recovery processes, the cell repolarizes and slowly recovers its ability to generate a depolarizing transition to the excited state. In such two-process systems with kinetic coupling, as depicted by Franck (14), the slow potential waves underlying the bursts of spikes arise from successive self-triggering and recovery (Fig. 3d).

TRIGGER AND RECOVERY PROCESSES IN BARIUM-TREATED BURST-GENERATING CELLS

The convulsive effects of Ba^{2+} ions on various excitable membranes are well documented. Recently, it has been shown that the substitution of Ba^{2+} for Ca^{2+} resulted, in some molluscan cells, in the production of regular patterns consisting of bursts of spikes followed by a long plateau. During the plateau (close to +20 to +30 mV), the membrane resistance collapses and the membrane becomes highly permeable to Ba^{2+} ions (19). The voltage-clamp method shows that the main effects of barium were (a) to reduce the delayed potassium conductance (21), and (b) to induce a slowly declining inward current mainly carried by Ba^{2+} ions.

The Ba-induced slow waves are a convenient model for the study of slow potential wave genesis, since the various processes described in the preceding sections are present and can be easily analyzed.

Triggering System of Ba-Induced Slow Waves

After the Ba-Ca substitution, the current-voltage relationship obtained with the slow ramp method exhibits a well-pronounced N-shape with a null-current point near +30 mV (see Fig. 2b). The N-shape results from the fast activation ($\tau < 10$ msec) of an inwardly directed Ba^{2+} current which can be blocked by Co^{2+}, Mn^{2+}, or La^{3+} ions. Since the delayed outward current is strongly reduced by barium, a voltage-clamp step of short duration (less than 100 msec) produces an inward current which remains partly activated at the end of the step and which is well evidenced when the step is preceded by a conditioning pulse in order to inactivate the delayed outward current or when the saline contains TEA. The threshold of the Ba current activation is close to −30 mV. The Ba-current is maximum at +10 mV and its reversal potential is almost +40 mV (with $[Ba^{2+}] = 10$ mM). The limiting slope of the Ba current-voltage curve leads to a maximum conductance of 3.5 nA/mV (Fig. 4a). The activation-voltage curve ($\bar{a}_{Ba}$ versus V) is shown in Fig. 4b.

Recovery Processes of Ba-Induced Slow Waves

Using long command potential steps, it is clear that the net current does not remain steadily inward; it slowly decreases with a time constant of several seconds ($\tau \simeq 20$ sec at positive potentials). The disappearance of the net inward current is mainly due to a slow decrease in the Ba current and to a lesser extent to a slowly developing outward current (19). The reduction of the Ba current during long-term depolarizations was attributed to a voltage- and time-dependent inactivation rather than to accumulation of Ba^{2+} inside the cell, even if the latter phenomenon must play a role in the overall decrease in Ba current. The very slow kinetic of the Ba current inactivation accounts for the long duration of the depolarized plateau; conversely, the very long time of recovery from inactivation (τ up to 220 sec

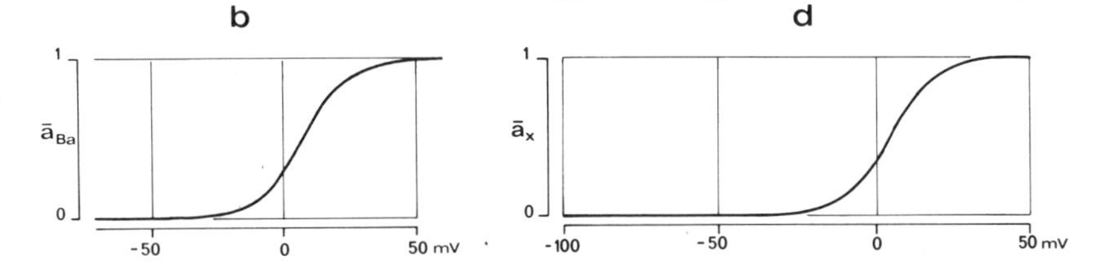

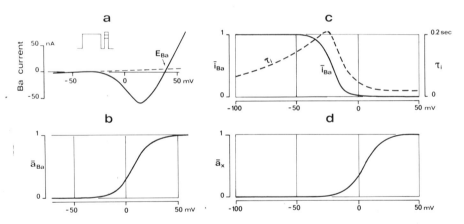

FIG. 4. Characteristics of the trigger and recovery processes of Ba-treated R_{14} neuron of Aplysia generating regular burst-plateau patterns. **a:** Inward Ba current remaining after 100 msec depolarization from a holding potential of −50 mV. The test pulse was preceded by a conditioning pulse (60 mV, 500 msec) in order to inactivate the delayed outward current. The Ba current is maximal at +10 mV and reverses direction close to +40 mV (small vertical arrow). **b:** Activation-voltage curve of the Ba current obtained by dividing the I_{Ba} (V) curve of a by the quantity (V-E_{Ba}). The resulting curve $\bar{a}_{Ba}$(V) is normalized to 1 for maximal values observed at positive potentials. **c:** Steady state inactivation [$\bar{i}_{Ba}$(V)] and inactivation time constant [τ_i(V)] curves of the slowly inactivated Ba current. **d:** Steady state activation curve [$\bar{a}_x$(V)] of the slowly developing outward current. The time constant of the a_x process is almost independent of voltage and is approximately 50 sec.

at −50 mV) accounts for the long silent period separating two successive bursts. The time constants involved here are three to four orders of magnitude larger than those involved in the spike-forming system.

The properties of the Ba current inactivation [steady state inactivation curve $\bar{i}_{Ba}(V)$, kinetics of development and removal of inactivation $\tau_i(V)$] are diagramed in Fig. 4c.

The second component of the recovery process, namely, a slowly developing outward current, was observed with the following experimental procedure: The voltage-clamp was instituted in the course of, or following, a spontaneous slow wave; the potential was held constant at a level more negative than the threshold of the Ba current (−30 mV) (see Fig. 3c). Under these conditions, the tail current displayed a fast declining phase of inward current as expected from the fast turning-off of the Ba current. This phase was followed by a slower phase ($\tau \simeq 50$ sec) of outward current for potentials more positive than −60 mV or inward current for more negative potentials. The level of the reversal potential ($\simeq -60$ mV) shows that the slow component of the tail current is always outward in the potential region swept by a slow wave and that it is probably carried by K^+ ions. The activation-voltage curve of the slow outward current $\bar{a}_x(V)$ is shown in Fig. 4d.

The role of the slow outward current in the shape and duration of the slow wave is probably small since its corresponding maximum conductance reaches only 0.2 nA/mV (at the end of a wave) instead of 3.5 nA/mV for the slowly inactivated Ba conductance, but it controls the potential trajectory in the interwave phase and, consequently, the period of the bursts.

COMPUTER SIMULATION OF THE BARIUM-INDUCED BURST-PLATEAU PATTERNS

The following model has been put forward in order to (a) describe the endogenous burst-plateau patterns of Ba-treated neurons, (b) investigate the effects of the main parameters on the behavior of the model and particularly the interferences between the trigger and recovery processes, and (c) explore the interplay between the spike-forming system and the slow wave-generating system.

The slow wave-generating system is composed of the slowly inactivated Ba current (SI_{Ba}) and the slowly developing outward current (SI_x) (S denotes slow system). The spike-forming system is mainly derived from the Connor and Stevens model (8), except that the parameters have been modified to fit our results and the fact that no transient outward current (g_A or g_{Ks}) was incorporated; it is composed of a classic, quickly inactivated inward current (I_{in}) and a delayed potassium current (I_K).

The membrane behavior under an applied current (I) is governed by the equation:

$$I = I_C + I_L + I_{in} + I_K + SI_{Ba} + SI_x$$

where I_C and I_L are the capacitive and leakage currents, respectively:

$$I_C = C\frac{dV}{dt}$$

Each individual ionic current I_j (j = Ba, Na, or K) obeys the following relationship:

$$I_j = g_j\,(V, t)\,(V - E_j)$$

where g_j is the instantaneous conductance of the channel j and E_j is the reversal potential of the corresponding ion. When describing the kinetics of each ionic current, we used the formalism of the Hodgkin-Huxley model: the instantaneous conductance of the species j is governed by two voltage- and time-dependent parameters describing the activating ("a" process) or inactivating ("i" process) properties of the channel

$$g_j = \bar{g}_j\, a_j^n\,(V, t)\, i_j\,(V, t)$$

where $\bar{g}_j$ is the maximal value of g_j, and a_j and i_j are dimensionless parameters ranging between zero and one.

The time dependence of the a and i variables arises from the following first-order equations:

$$\frac{da}{dt} = f_a\,(V, a) = \alpha_a(V)\,(1 - a) - \beta_a(V)\,a$$

and

$$\frac{di}{dt} = f_i\,(V, i) = \alpha_i(V)\,(1 - i) - \beta_i(V)\,i$$

The rate constants αs and βs are voltage dependent so that the kinetic behavior and the steady state levels (indicated by a bar) of the "a" and "i" processes are:

$$\tau = 1/(\alpha + \beta) \qquad \bar{a} = \alpha_a/(\alpha_a + \beta_a) \qquad \bar{\imath} = \alpha_i/(\alpha_i + \beta_i)$$

As shown in Fig. 4, the "a" functions are S-shaped and "i" functions are Z-shaped. The functions $\alpha(V)$ and $\beta(V)$ for the various ionic currents have been described previously (19). Some of the parameters used in the standard model are listed in Table 1.

TABLE 1. *Parameters used in standard model*

I_j	j	$\bar{a}$(V)	$\bar{\imath}$(V)	n	E_j (mV)	g_j (nA/mV)
I_{in}	$Na^+ + Ca^{2+}$	0–1	0–1	3	+40–+50	6
I_K	K^+	0–1	0–1	1	−70	5–10
SI_{Ba}	Ba^{2+}	0–1	0–1	1	+40	3.5
SI_x	K^+(?)	0–1	1	1	−60	0.5
I_L	$Cl^- + K^+ + Na^+$	1	1	—	−30–−35	0.1

The model was assessed in two steps, first with the slow wave-generating system alone, then with the whole model. The simplified model (not including the spike elements) generates regular long plateaus lasting 14 sec and having the same configuration as the plateaus induced by Ba^{2+} in R_{14} cell of Aplysia (Fig. 5a). A number of tests have been performed with the simplified model, and the results satisfactorily reproduce those obtained in Ba-treated cells: increase (respective decrease) of the period of plateaus by a constant inward (respective outward) current, cyclic changes in input resistance tested by regular current pulses (Fig. 5c), slow component in the tail current following institution of voltage-clamp conditions during the plateau.

The model also allows the study of the effects of the two recovery processes on the production of regular slow waves. For instance, replacement of the voltage-dependent slow outward current by a constant conductance ($Sg_x = 0.1$ nA/mV) does not significantly alter the shape and duration of the

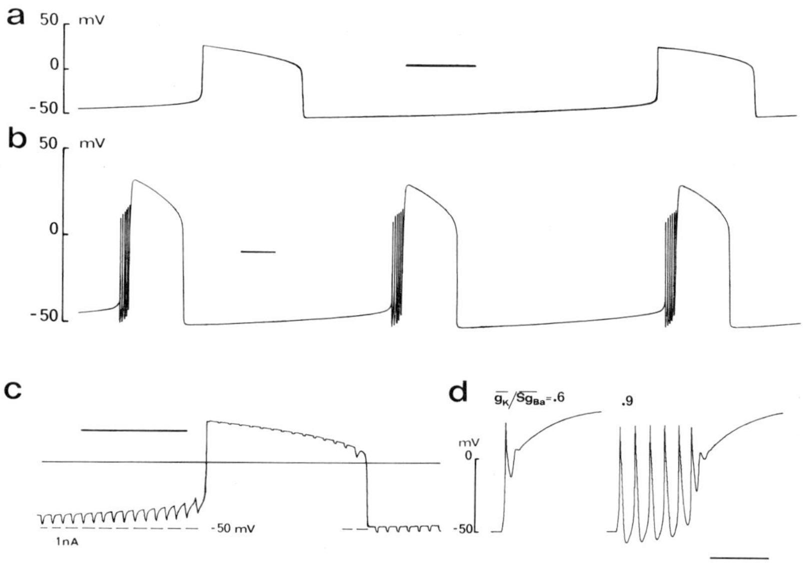

FIG. 5. Reconstruction of Ba-induced burst-plateau patterns. **a:** Long (14 sec) square-shaped potential waves regularly generated by the simplified model (not including the spike elements). **b:** Incorporation of the spike elements (as described in the text) results in burst-plateau patterns similar to those observed in Ba-treated R_{14} cell. **c:** Cyclic changes in input resistance (tested by regular current pulses; 1/sec, 200 msec). The input resistance is higher just before the plateau; it collapses on the plateau and recovers slowly in the interwave period. **d:** Influence of the spike parameters on the plateau formation. By increasing the delayed potassium conductance (g_K), the formation of the plateau is delayed and the burst becomes longer. The rounded time course following the burst due to inactivation of g_K is absent when the spike is not incorporated in the model (see **a**). Horizontal bar, 10 sec in **a, b,** and **c; 1** sec in **d.**

wave; however, it does considerably increase the period, since the trajectory of the potential in the hyperpolarized phase is then solely controlled by the recovery time constant of the Ba current inactivation. Conversely, when suppressing the slow inactivation of the Ba current, periodic plateaus are no longer produced and the membrane remains depolarized on the excited stable state close to +30 mV. However, a short pulse of inward current induces a repolarizing response similar to those described by Grundfest (20) and recently observed in some Aplysia neurons (17). Without inactivation of the Ba current, recurrent long plateaus appear again if either Sg_{Ba} is reduced by 40% (equivalent to a fast inactivation of 40% of the Ba current) or Sg_x is doubled.

The complete model with the standard parameters for the spike- and slow wave-generating mechanisms generates regular burst-plateau patterns similar to those observed in Ba-treated R_{14} neurons (Fig. 5b). The transition from the burst to the plateau phase, however, was highly dependent on the ratio g_K/Sg_{Ba}, i.e., on the ratio of the delayed outward current to the slowly inactivated inward current; the higher the ratio g_K/Sg_{Ba}, the longer the duration of the burst (Fig. 5d). When this ratio is reduced to 0.6, the model generates single spikes immediately followed by a plateau; the pattern is much like the one obtained in warmed L_{11} cells of Aplysia and in certain PTZ-treated molluscan cells.

It is to be expected that further modifications of the steady state parameters and kinetic variables will give more indications about the interplays between the trigger and recovery processes on the one hand and between the fast and slow potential-generating mechanisms on the other. The great number of parameters of the whole model, as well as the long time necessary to establish the various programs and to solve the system by conventional methods, prohibits such an exploratory study. However, the qualitative behavior of the system can be related to the various parameters by methods using the phase plane or phase space analysis. Even if these methods give little information on the time course of the potential changes, they permit the definition of minimal conditions that must be fulfilled for an oscillatory solution. This approach has been used successfully by Fitzhugh (13) for the study of the Hodgkin-Huxley system and more recently by Plant and Kim (35). The following results (obtained with Dr. J. Argemi) come from a preliminary qualitative study of the slow wave-generating mechanism of Ba-treated neurons.

First we observed that the set of four differential nonlinear equations describing the slow system can be separated into two groups according to the magnitude of their time constant: the voltage and the triggering process (activation of the Ba current) vary much more rapidly than do the two recovery processes (the ratio of the rate constants of the fast phases to those of the slow phases amounts to 10^4) which allows the whole system to be studied in two steps:

1. When the triggering system is put into action, the voltage changes rapidly but the two recovery processes can be considered as being at rest since they do not have enough time to change significantly.

2. During the slow potential changes corresponding to the effects of the two recovery processes which are now operating, the triggering system adapts almost instantaneously to the voltage level and thus can be considered voltage dependent only.

The first case leads to the study of the (V, a_{Ba}) system [analogous to the reduced (V, m) system of Fitzhugh (13)] where i_{Ba} and a_x are given constants defined by the initial state of the membrane ($i_{Ba} \simeq 1$ and $a_x \simeq 0$ for a hyperpolarized cell). The whole system reduces to a set of two coupled differential equations:

$$\frac{dV}{dt} = F\,(V, a_{Ba})$$

$$\frac{da_{Ba}}{dt} = f(V, a_{Ba})$$

in which i_{Ba} and a_x are constants.

The solution of this reduced system can be easily found in the (V, a_{Ba}) plane where the singular points, corresponding to $\frac{dV}{dt} = \frac{da_{Ba}}{dt} = 0$ are either stable nodes (Fig. 6) or saddle points or saddle nodes. When the cell is

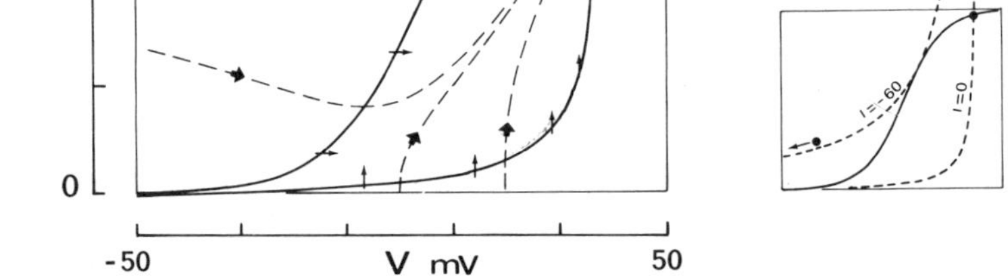

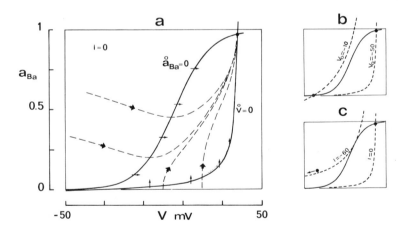

FIG. 6. Plots of the (V-a_{Ba}) reduced system showing the V-nullcline ($\mathring{V} = 0$) and the a_{Ba}-nullcline ($\mathring{a}_{Ba} = 0$). **a:** With the standard parameters, all state variables set at their resting values (−50 mV) and I = 0, the V-nullcline intersects the a_{Ba}-nullcline at +38 mV. The solutions of the (V-a_{Ba}) system are represented by trajectories marked with arrowheads. **b:** Same representation as in **a.** The two V-nullclines are for initial conditions corresponding to a cell either hyperpolarized ($V_0 = -50$ mV) or depolarized ($V_0 = -10$ mV). The singular points (***black dots***) are both stable points located at +38 mV for the cell initially hyperpolarized or −40 mV for the cell initially depolarized. **c:** Effect of current on the state of the system. With inward current, the stable point at +38 mV moves leftward and for current intensity larger than 60 nA, the stable point is located at negative potentials (out of scale).

initially hyperpolarized (all steady state parameters set to their hyperpolarized value), the system has one singular point (a stable node) which is at +38 mV; i.e., the membrane tends to reach this excited stable state (Fig. 6a). Conversely, when the cell has been initially depolarized (let us suppose at +38 mV), i.e., when the two recovery processes are fully developed, the singular point is a stable node located at −45 mV, and the cell tends to repolarize (Fig. 6b). This short analysis shows how the slow oscillations appear by the alternative occurrence of stable nodes, either depolarized or hyperpolarized. The same study permits the determination of the extreme values of the main parameters compatible with an oscillatory solution. For instance, with the standard parameters, the cell will remain depolarized if the slow inactivation of the Ba current is only acting on 35% of the Ba current. Figure 6c illustrates the fact that an injected inward current larger than 60 nA is necessary to cut off a plateau immediately after its formation.

The second case studied was the reduced system in which the a_{Ba} function adapts instantaneously to the potential level, i.e., the system describing the slow behavior of the membrane. The system to be solved then reduces to:

$$\frac{dV}{dt} = F(V, i_{Ba}, a_x)$$

$$\frac{di_{Ba}}{dt} = f_1 (V, i_{Ba})$$

$$\frac{da_x}{dt} = f_2 (V, a_x) \qquad \text{with} \qquad a_{Ba} = \bar{a}_{Ba} (V)$$

The solution of this reduced system is to be found in the (V, i_{Ba}, a_x) space. With standard parameters, the conditions $\frac{dV}{dt} = \frac{di_{Ba}}{dt} = \frac{da_x}{dt} = 0$ led to a folded surface, (Fig. 7a) on which the potential moves slowly, whereas outside the surface the changes in potential operate almost instantaneously and are then described by the (V, a_{Ba}) reduced system. A periodic solution representing recurrent slow potential waves appears as a limit cycle with slow kinetics for the paths located on the slow behavioral surface and fast kinetics for the paths outside this surface. Similar to the (V, a_{Ba}) reduced system, the extreme values of the main parameters compatible with an oscillatory solution are easily determined; for instance, if $I < -3$ nA, the shape of the slow behavioral surface is thoroughly changed, and it is obvious that a limit cycle can no longer exist (Fig. 7b). The slow behavioral surface resembles a surface used by Zeeman (41) for the Hodgkin-Huxley model and actually represents one of the seven figures of the catastrophe theory of Thom, the "cusp catastrophe."

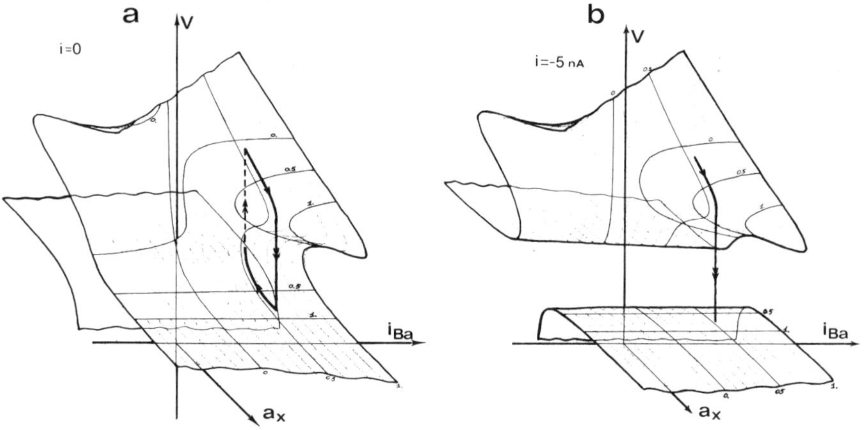

FIG. 7. Plots of the (V, i_{Ba}, a_x) reduced system for two different current intensities (0 and −5 nA). The solutions are represented by slow paths located on a folded surface or slow behavioral surface. When the paths reach the edge of the pleat, the paths fall (or jump up) to the bottom (or top) sheet of the surface. The slow potential changes are indicated by an arrowhead and the fast one by a double arrowhead. A stable periodic solution representing the square-shaped potential waves induced by barium exists with $I = 0$. With $I = -5$ nA, the shape of the slow behavioral surface is thoroughly changed and all the paths end on the bottom sheet. The cell remains silent with a high membrane potential level.

DISCUSSION

The occurrence of slow changes in specific ionic conductances of molluscan neuronal membranes is now firmly established. These slow changes imply mainly Na^+, K^+, and Ca^{2+} ions as the main charge carriers. For instance, slow inactivation of the Na-Ca system (25,27) and slowly developing K current (34) have been observed in several molluscan cells, either silent or regularly firing.

However, the most obvious and well-documented slow conductance changes are those occurring in a variety of pacemaker cells having in common the ability to display slow potential changes. This chapter shows that the basic membrane properties of these neurons are very similar, and that the slow potential production is accounted for by similar ionic mechanisms which have been conceptually grouped in two systems: a trigger system and a recovery system. [These terms have been introduced by Franck (14) in his description of two-process systems for biological oscillators.] If the production of bursts of spikes or long plateaus are underlain by similar mechanisms, they are also functionally equivalent since the plateaus initiated at or near the soma level induce a simultaneous spiking in the axon. Thus in both cases the output of the neuron is a regular bursting activity (see Fig. 1 where the axon of Ba-treated R_2 cell is firing at the time when a long plateau is generated in the soma).

In burst-generating cells, the interplay between the trigger and recovery processes results in a slow oscillatory behavior, but other types of behavior can be induced by small changes in the magnitude of each process. For instance, the regular firing of cell L_{11} of Aplysia at room temperature would result from a lack of recovery process so that the potential tends to stabilize on an excited stable state (near −20 mV) inducing a regular high frequency firing. Increasing the temperature increases the recovery process (a slowly developing outward current), and the slow waves appear. Conversely, a burst-generating cell can be stabilized if the recovery process overwhelms the trigger process, as in dopamine-treated R_{15} cell of Aplysia (18). Other behaviors induced by adjustments of the two processes include the working on monostable modes in which the cells respond by a long-lasting depolarization or hyperpolarization to a short stimulus, such as synaptic potentials (17).

The voltage-clamp analysis of Ba-treated neurons allows the description of all the processes implicated in the genesis of the burst-plateau patterns. However, it must be emphasized that the Ba model can be generalized to other burst-generating neurons only if additional specific characteristics, such as the Ca^{2+}-dependent activation of potassium conductance, are in-

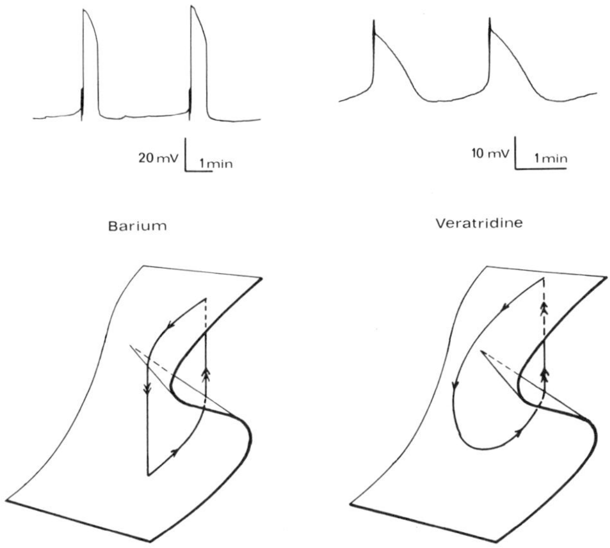

FIG. 8. Square- and saw-tooth-shaped potential waves induced by barium or veratridine. The slow behavioral surface of Ba-treated cells is reproduced schematically from Fig. 7 (with I = 0). The stable limit cycle is completely included in the folded part of the surface so that both depolarizing and repolarizing phases have fast kinetics. The slow behavioral surface of veratridine-treated cells is hypothetical. The periodic solution represented by the closed curve with arrowheads must pass only once over the pleat, so that only the depolarizing phase of the wave is fast.

cluded. Nevertheless, the study of the model with the mathematical methods of qualitative analysis could aid in the understanding of the slow potential mechanism in different neurons. One illuminating example is provided by the slow waves occasionally induced in Aplysia neurons by the alkaloid veratridine. The waves are characterized by a fast upstroke inducing a transient high frequency discharge followed by a very slow return to the original level, as in TTX-treated R_{15} cell. On the basis of the slow behavioral surface of the Ba model, it can be deduced that the potential trajectory must pass only once on the folded part of the surface (the depolarizing transition) instead of two, as in Ba-induced slow waves (Fig. 8).

Since the magnitude of the folded part of the surface is directly related to that of the trigger process, it follows that in veratridine-treated neurons generating slow potential waves, the trigger system must be reduced compared to the situation in Ba-treated cells and that the recovery process must be large enough so that the trajectory passes outside the folded part of the surface, as illustrated in Fig. 8.

Finally, it can be hoped that further study of the model and the incorporation of new data from various bursting neurons will help to answer the questions: What are the various mechanisms of burst production? Why are some factors effective in inducing bursting activity? Why are some cells more susceptible to generating slow potential waves than others?

ACKNOWLEDGMENTS

I am indebted to Dr. C. Ducreux and Dr. H. Chagneux for collaboration in the experimental part, and to Dr. J. Argemi, who introduced me to the idea of a dynamical space, for the opportunity, kindly given, to study unpublished results. I wish to thank Ms. M. André and Mr. G. Chatzopoulos, R. Fayolle and G. Jacquet, for invaluable technical assistance.

REFERENCES

1. Arvanitaki, A. (1962): *C. R. Acad. Soc.*, 255:1523–1525.
2. Arvanitaki, A., and Chalazonitis, N. (1968): In: *Neurobiology of Invertebrates*, edited by J. Salanki, pp. 169–199. Akademiai Kiado, Budapest.
3. Barker, J. L., and Gainer, H. (1974): *Brain Res.*, 65:518–520.
4. Barker, J. L., Ifshin, M. S., and Gainer, H. (1975): *Brain Res.*, 84:501–513.
5. Both, R., Finger, W., and Chaplain, R. A. (1976): *Biol. Cybern.*, 23:1–11.
6. Chalazonitis, N., and Takeuchi, H. (1968): *C. R. Soc. Biol.*, 162:1552–1556.
7. Chaplain, R. A. (1976): *Brain Res.*, 106:307–319.
8. Connor, J. A., and Stevens, C. F. (1971): *J. Physiol. (Lond.)*, 213:31–53.
9. David, R. J., Wilson, W. A., and Escueta, A. V. (1974): *Brain Res.*, 67:549–554.
10. Ducreux, C., and Gola, M. (1975): *Pfluegers Arch.*, 361:43–53.
11. Eckert, R., and Lux, H. D. (1975): *Brain Res.*, 83:486–489.
12. Eckert, R., and Lux, H. D. (1976): *J. Physiol. (Lond.)*, 254:129–151.
13. Fitzhugh, R. (1960): *J. Gen. Physiol.*, 43:867–896.
14. Franck, U. F. (1973): In: *Biological and Biochemical Oscillators*, edited by B. Chance, E. K. Pye, A. K. Ghosh, and B. Hess, pp. 7–30. Academic Press, New York.

14a. Frazier, W., Kandel E., Kupferman, I., Waziri, R., and Coggeshall, R. (1967): *J. Neurophysiol.,* 30:1288–1351.
15. Gola, M. (1974): *Pfluegers Arch.,* 352:17–36.
16. Gola, M. (1976): *Experientia,* 32:585–587.
17. Gola, M. (1976): In: *Neurobiology of Invertebrates: Gastropoda Brain,* edited by J. Salanki, pp. 381–424. Akademiai Kiado, Budapest.
18. Gola, M., and Boisson, M. (1974): *J. Physiol. (Paris),* 69:154A.
19. Gola, M., Ducreux, C., and Chagneux, H. (1977): *J. Physiol. (Paris),* 73:407–440.
20. Grundfest, H. (1972): In: *Perspectives in Membrane Biophysics,* edited by D. P. Agin, pp. 37–63. Gordon and Breach Science, New York.
21. Hagiwara, S., Fukuda, J., and Eaton, D. C. (1974): *J. Gen. Physiol.,* 63:564–578.
22. Heyer, C. B., and Lux, H. D. (1976): *J. Physiol. (Lond.),* 262:319–348.
23. Heyer, C. B., and Lux, H. D. (1976): *J. Physiol. (Lond.),* 262:349–382.
24. Junge, D., and Stephens, C. L. (1973): *J. Physiol. (Lond.),* 235:155–181.
25. Kostyuk, P. G., Krishtal, O. A., and Shakhovalov, A. (1977): *J. Physiol. (Lond.),* 270: 545–568.
26. Kostyuk, P. G., and Krishtal, O. A. (1977): *J. Physiol. (Lond.),* 270:569–580.
27. Kostyuk, P. G., Krishtal, O. A., and Pidoplichko, V. I. (1975): *Nature,* 257:691–693.
28. Leicht, R., Meves, H., and Wellhöner, H. H. (1971): *Pfluegers Arch.,* 323:50–62.
29. Lux, H. D., and Eckert, R. (1974): *Nature,* 250:574–576.
30. Mathieu, P. A., and Roberge, F. A. (1971): *Can. J. Physiol. Pharmacol.,* 49:787–795.
31. Meech, R. W. (1974): *Comp. Biochem. Physiol.,* 48A:387–395.
32. Meech, R. W. (1974): *Comp. Biochem. Physiol.,* 48A:397–402.
33. Meech, R. W., and Standen, N. B. (1975): *J. Physiol. (Lond.),* 249:211–239.
34. Partridge, L. D., and Stevens, C. F. (1976): *J. Physiol. (Lond.),* 256:315–332.
35. Plant, R. E., and Kim, M. (1976): *Biophys. J.,* 16:227–244.
36. Smith, T. G., Barker, J. L., and Gainer, H. (1975): *Nature,* 253:450–452.
37. Stinnakre, J., and Tauc, L. (1973): *Nature [New Biol.],* 242:113–115.
38. Trautwein, W. (1973): *Physiol. Rev.,* 53:793.
39. Ulbricht, W. (1969): *Ergebn. Physiol. Rev. Physiol,* 61:18–71.
40. Wilson, W. A., and Wachtel, H. (1974): *Science,* 186:932–934.
41. Zeeman, E. C. (1973): In: *Dynamical Systems,* edited by M. Peixoto, pp. 683–741. Academic Press, New York.

Abnormal Neuronal Discharges, edited by N. Chalazonitis and M. Boisson. Raven Press, New York © 1978.

Dual Effects of Catecholamines on Burst Production by Aplysia Neurons

M. Boisson and *M. Gola

*Centre Scientifique de Monaco, Principauté de Monaco; and *Institut de Neurophysiologie et Psychophysiologie, C.N.R.S., Marseilles, France*

One of the most original properties of molluscan neurons is their ability to convert from one type of activity to another in response to various physical or chemical alterations of environmental conditions. For instance, regularly firing or even silent neurons can be converted into oscillatory or bursting neurons; conversely, bursting neurons can be converted into silent neurons (8). Such changes of the electrical activity of molluscan cells have been observed following either slight changes in the ambient physical conditions [temperature (6), PO_2 (10), light (7)] or on addition of artificial or naturally occurring substances [metrazol or PTZ (9), cycloheximide (11), low Ca^{2+} (2), peptides (3,21)].

The type of conversion depends on the nature of the experimentally altered factor as well as on the properties of the target cell. For instance, moderate warming (from 20 to 25°C) converts the regularly firing L_{11} cell of Aplysia into a bursting cell (1) but not the other identified Aplysia neurons. Metrazol has been found particularly efficient in inducing high frequency discharges in normally regularly firing cells (13); visible light can convert firing cells into silent cells and vice versa (7).

In this chapter, we present two opposite conversions of electrical activity of Aplysia into a bursting cell (1) but not the other identified Aplysia neurons. same neurogenic substances (catecholamines). The conversions were: (a) a silent cell into a bursting cell (cell R_{14} of Aplysia) (4a,16), and (b) a bursting cell into a silent cell (cell R_{15} of Aplysia) (4,5,12).

The catecholamines were dopamine (DA), epinephrine, and norepinephrine. The overall effects of all were similar, but at different concentrations; the following results are restricted to the effects of DA.

STABILIZING EFFECT OF CATECHOLAMINES ON BURSTING NEURONS

The addition of DA (10^{-3} to 10^{-4} M) to artificial seawater induced two reversible effects in R_{15} bursting cells of Aplysia (4): (a) The membrane potential slowly increased (up to 15 mV) over the course of 30 to 100 sec, resulting in the cessation of the spike discharge (Fig. 1A). (b) The modal activity was thoroughly changed, as can be observed with a constant outward current injected 120 sec after the drug addition; the depolarization caused by the current led to a phasic type response, and no more bursts of spikes were observed, whatever the current intensity might be (Fig. $1A_2$). Thus in the presence of DA, R_{15} became hypoexcitable and behaved like a stable neuron.

The membrane characteristics of normal or DA-treated R_{15} were determined with slow ramp pulses applied to the membrane in voltage-clamp conditions (15). In a typical experiment, the membrane was previously hyperpolarized (at −70 to −90 mV) by a constant inward current; the current-voltage relationships were then obtained with a slow depolarizing ramp followed by a symmetrical repolarizing ramp. The amplitude of the command potential ramps was from 60 to 80 mV and the slope of the ramp pulses was from 1 to 5 mV/sec.

With this method, it has been possible (15) to demonstrate the existence

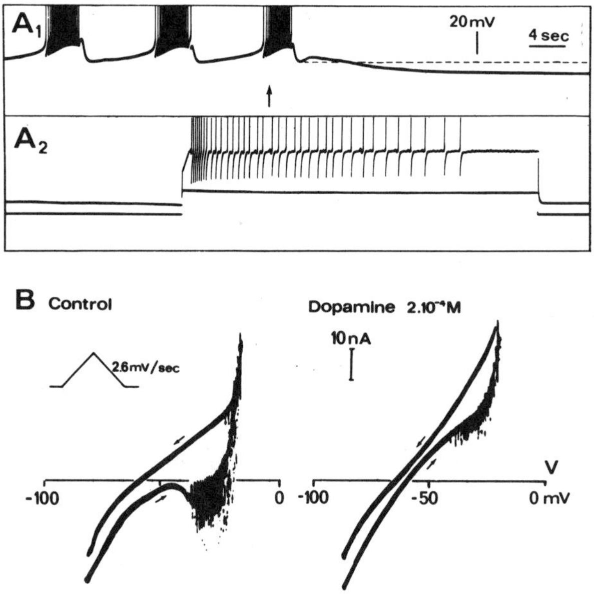

FIG. 1. Stabilization of R_{15} bursting neurons by catecholamines. **A_1**: Abolition of bursts and hyperpolarization of an R_{15} neuron induced by 5×10^{-4} M DA (added at arrow). **A_2**: Phasicotonic firing of a DA-treated R_{15} neuron in response to an outward current (*lower trace*). **B:** Current-voltage relationships in normal and DA-treated R_{15} neurons. The curves were obtained in voltage-clamp conditions with the slow ramp pulses shown in inset (slope, 2.6 mV/sec). Outward currents are upward and the horizontal line is the zero current level. Changes in potential (starting from the left) are indicated by the small arrows.

of two features essential for the genesis of the slow potential waves underlying the bursting activity of R_{15}. The first is a region of negative slope in the current-voltage relationship (I-V curve) obtained with the depolarizing pulse; the second is a hysteresis of the current, linked to the sweeping direction (Fig. 1B).

On each I-V curve, an operating point, defined by the intersection of the $I = 0$ axis, corresponds to the curve; for the membrane initially hyperpolarized, the operating point is at -16 ± 5 mV, and for the membrane initially depolarized this point is at -56 ± 4 mV. Thus membrane hyperpolarization (either passive or active) leads to depolarizing processes, whereas its depolarization induces processes causal to membrane potential increase. On that basis, the slow potential changes underlying bursting activity can be interpreted as due to the slow transition between the two current-voltage curves obtained from a depolarized state or from a hyperpolarized state (15).

The I-V curves of R_{15} were drawn as indicated above, during normal bursting activity and in the hyperpolarization phase following the addition of DA (0.5 to 1 μM/ml). The most immediate effects of DA were the important decrease and sometimes the disappearance of both the negative slope region of the current-voltage curves and of the current hysteresis (Fig. 1B).

Thus DA has suppressed the two conditions we considered essential for the genesis of the slow waves of potential. It is not surprising that the membrane characteristics of a DA-treated R_{15} are then like those of the R_2 stable neuron.

In addition, DA increases the membrane slope conductance between -65 and -45 mV. Normally, this conductance is very low (0.15 to 0.4 nA/mV), whereas for membrane polarizations exceeding 70 mV, the slope conductance increases up to 2 nA/mV. This change in membrane conductance is known as anomalous rectification (17); in R_{15}, the change of slope occurs at -65 mV. By adding DA, the conductance between -65 and -45 mV greatly increased to 0.7 to 1.5 nA/mV, while the conductance in hyperpolarization was not changed. Consequently, the anomalous rectification ratio, defined as the ratio of the conductance above and below -65 mV, fell from 6 to 10 to 1.5 to 2. In the presence of DA, the intersections of both I-V curves with the $I = 0$ axis were between -60 and -65 mV, i.e., more negative than the threshold of spikes (-40 mV); therefore, these points correspond to conditions of electrical stability.

It is worth noting that the same conversion of electrical activity and of the current-voltage relationships were observed during the hyperpolarization phase of the well-known inhibition of long duration (ILD) triggered by repetitive stimulation of the left connective (5). ILDs have been assumed to be mediated by DA, which increases the potassium conductance of the membrane (22).

OSCILLOGENIC EFFECT OF CATECHOLAMINES ON SILENT NEURONS

This transformation mainly involves a secretory group of nerve cells, the white cells, in which the R_{14} neuron (or A′) is the most representative cell.

When incubated in artificial seawater containing 1 μM/ml DA, R_{14} hyperpolarized by 18 mV in 10 to 20 sec (Fig. 2). Being normally silent, this neuron became hypoexcitable after such a hyperpolarization. In addition to its hyperpolarizing properties, however, DA (as well as the other catecholamines) had a more specific and remarkable action on R_{14}. In response to a suprathreshold depolarizing direct current, the DA-treated secretory neurons displayed a bursting activity (Fig. 2C) instead of the normal adaptive discharge (Fig. 2A) (4a).

The membrane characteristics of DA-treated R_{14} neurons were determined with the same voltage-clamp method as used for the study of the R_{15} bursting cells. Figure 3 shows that the two features we considered essential for the slow wave generations in R_{15} are then present, i.e., a quasisteady state negative slope conductance (when the ramp is depolarizing) and a current hysteresis. By comparing Fig. 3B and Fig 1B, it can be seen that the current-voltage relationships of DA-treated R_{14} look like those of R_{15} (except that the negative slope part of the current-voltage relationship of R_{14} was not entirely contained in the inward current region).

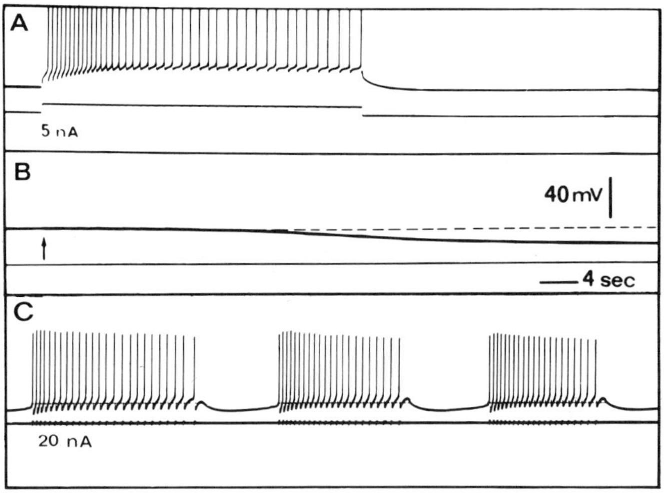

FIG. 2. Bursting activity induced by DA in an R_{14} neuron of Aplysia. **A:** Adaptative discharge induced by a step of outward current in an R_{14} cell bathed in normal saline. **B:** Addition of 2×10^{-4} M DA (*arrow*) hyperpolarizes the cell. **C:** DA-treated R_{14} neuron; regular bursting activity obtained with a constant outward current (in order to compensate the hyperpolarization induced by DA). The bursts of spikes are superimposed on slow potential oscillations, as in normally bursting neurons.

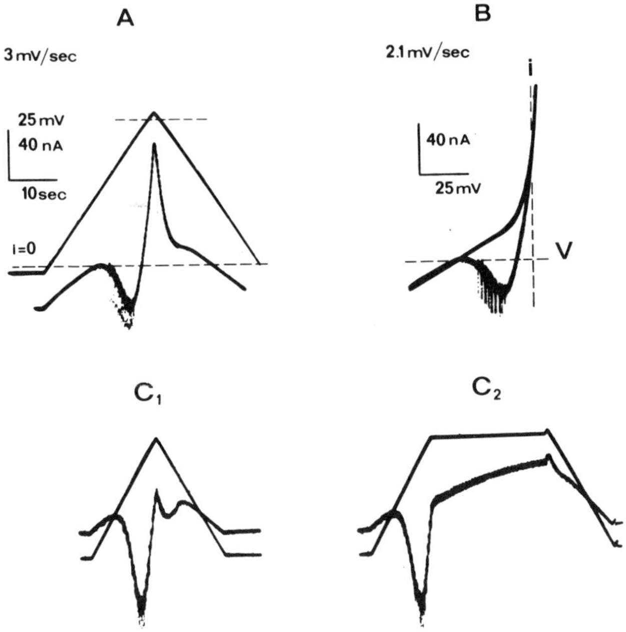

FIG. 3. N-shaped current-voltage relationships in DA-treated R_{14} neurons. **A:** Current-time record (*lower trace*) in response to slow depolarizing and repolarizing ramps (*upper trace*) applied in voltage-clamp conditions. The horizontal dashed lines are zero potential level (*top line*) and zero current level (indicated by i = 0). **B:** X-Y display of the current (*vertical*) and potential (*horizontal*) changes obtained as indicated in **A** (slope of the ramps, 2.1 mV/sec). The horizontal and vertical lines are, respectively, zero current and zero potential baselines. The current-voltage relationship is N-shaped with the depolarizing ramp and monotonous with the repolarizing ramp, as in R_{15} bursting neurons (compare with Fig. 1). **C:** Slow disappearance of the N-shape consecutive to depolarization. With the repolarizing ramps, the current trace is still N-shaped in $\mathbf{C}_1$ but not in $\mathbf{C}_2$, where a slow outward current develops during the depolarization (same scale as in **A**).

By comparing the current-voltage relationships before and after the addition of catecholamines, it appeared that the catecholamines induced an outward current whose intensity increased upon depolarization. As can be seen in Fig. 4A, the DA-induced current is not linearly related to the membrane potential, as would be the case if DA affected merely the resting conductances of the membrane; the curve of the DA-induced current versus potential shows a threshold at about −40 mV and has a shape similar to that of the delayed potassium current-voltage curve. This would suggest that in R_{14}, DA exerts a direct activating effect on the delayed potassium conductance (Fig. 4B). Such direct controls by catecholamines of the processes involved in the generation of action potentials are well known in cardiac muscles (20) but have been only recently observed in the nervous system (18).

The reason for the specificity of catecholamines on R_{14} are as yet unknown

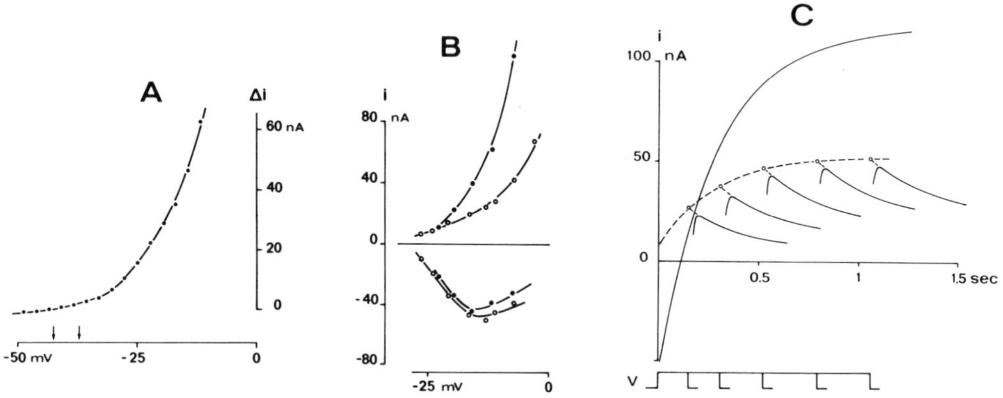

FIG. 4. A: DA-induced outward current in an R_{14} neuron, deduced by subtracting the I-V curves (obtained with slow depolarizing ramps) of DA-treated and normal R_{14} neurons. The two vertical arrows are the resting potentials in normal saline (*right-hand arrow*) and in the presence of DA (*left-hand arrow*). **B:** Persistent inward currents and delayed outward currents in normal R_{14} (*open circles*) and in the presence of DA (*filled circles*). The current was measured after 1 sec depolarization. DA affected mainly the outward delayed current, whereas the persistent inward current (Na) was only slightly reduced. **C:** Kinetics of the delayed outward current in R_{14}. *Upper curve,* delayed outward current (K-current) produced by a 1.2 sec step depolarization to 0 mV (holding potential, −40 mV); *Lower traces,* tail currents following pulses of increasing duration, as indicated in the bottom trace. The tail current slowly declines to zero with a time constant of 250 msec; *dashed curve,* outlines the time course of the increase of the initial tail current amplitude (*open circles*). The outward current grows exponentially with a time constant of 280 msec.

(and could eventually be related to its neurosecretory nature); but the overall result is a strong sharpening of the spike (duration reduced by 50 to 80%) due to both the hyperpolarization of the cell and the increase in the available potassium conductance (Fig. 5).

Other unusual membrane properties were observed in R_{14}; some of them must play a role in the production of bursts upon the addition of catecholamines: (a) The delayed potassium channel is weak compared to the situation in the other molluscan cells, and its time course is much slower (time constant of activation from 120 to 250 msec instead of 20 to 40 msec) (see Fig. 4C). It is activated at potentials more positive than −25 mV and its reversal potential is between −50 and −55 mV. All these features account for the long duration of the spike (up to 70 msec). (b) A persistent Na current is activated at potentials more positive than −40 mV. Thus the steady state I-V relationship has a negative slope (with a peak at −15 mV), which very likely supports the tonic firing of the cell upon constant current injection. (c) A slow outward current (time constant of 10 to 25 sec) develops when the cell is depolarized (see Fig. $3C_2$).

Another point must be emphasized concerning the mechanism of sustained slow oscillations and bursts production. As stated above, the nega-

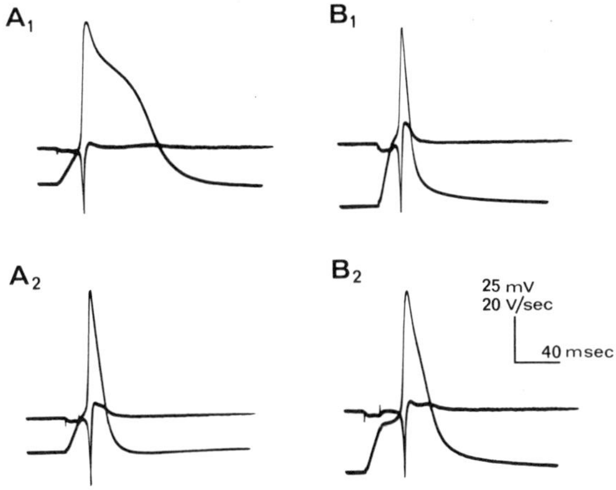

FIG. 5. Effects of DA on the action potentials of R_{14}. In each record, the second trace is the first derivative (inverted) of the potential. **A_1**, **B_2**: Cell in normal saline. **A_2**, **B_1**: Cell in the presence of DA. **A_1**, **A_2**: Initial potential corresponding to the resting potential in control saline. **B_1**, **B_2**: Initial potential corresponding to the resting potential in the presence of DA (hyperpolarized compared to series **A**). **A_1**: Typical long-lasting action potential of R_{14}. **B_1**: Following addition of 2×10^{-4} M DA, the cell is hyperpolarized and the spike is considerably sharpened with a fast repolarizing phase. **A_2**: Hyperpolarization induced by DA was compensated by a constant outward current, but the spike remained shorter than in control saline. **B_2**: After washing, the cell was maintained hyperpolarized at the level reached with DA (same level as in **B_1**). Nevertheless, the duration of the spike increased.

tive slope conductance exists before the addition of DA and is mainly due to a persistent Na current (plus a small Ca contribution) since it almost completely disappears with 10^{-5} M TTX and in O-Na^+ seawater. Thus the augmentation of the potassium conductance by DA must be a shunt for this inward current; indeed, the negative slope conductance region was reduced by DA. However, despite a well-pronounced quasisteady negative slope conductance, R_{14} does not work like a bursting cell but responds to a constant stimulus by a tonic discharge; despite the reduction of the negative slope by DA, however, the cell is then able to fire bursts of spikes (under the effect of a depolarizing current). This emphasizes that the slow potential wave generation requires a quantitative adjustment between the two features described in R_{15} (presented in more detail by Gola, elsewhere in this volume). This conclusion is reinforced by observing that higher concentrations of DA (or other catecholamines) failed to induce a bursting activity in R_{14} and in other related cells.

In conclusion, the stabilizing actions of catecholamines on bursting cells, as well as their oscillogenic effects on stable neurons, are operated by similar modifications affecting the potassium conductances of the membrane. Thus the slow potential oscillations underlying bursting activity can be generated only if the potassium conductance (and particularly its slow component) has an optimum amplitude; reducing (normal R_{14}) or increasing (DA-treated R_{15}) the potassium component of the ionic conductances results in the stabilization of the cell. Such results are in agreement with the present ideas on the ionic mechanisms underlying the slow potential generation in bursting cells and in other related cells: the slow potential waves arise from the depolarizing tendency induced by persistent Na^{+}-Ca^{2+} inward currents (14,15,19) and from the delayed counteraction played by a slow activation of a potassium conductance (15). If the second component of that system, the repolarizing potassium activation, is too large or too weak, then the cell will remain either hyperpolarized or depolarized.

REFERENCES

1. Arvanitaki, A. (1962): *C. R. Acad. Sci.,* 255:1523–1525.
2. Barker, J. L., and Gainer, H. (1974): *Brain Res.,* 65:516-520.
3. Barker, J. L., and Gainer, H. (1974): *Science,* 184:1371–1373.
4. Boisson, M., and Chalazonitis, N. (1972): *Comp. Biochem. Physiol.,* 41A:883–885.

4a. Boisson, M., and Chalazonitis, N. (1973): *C. R. Acad. Sci.,* 276:1025–1028.

5. Boisson, M., and Gola, M. (1976): *Comp. Biochem. Physiol.,* 54C:109–113.
6. Chalazonitis, N. (1961): *C. R. Acad. Sci.,* 255:1652–1653.
7. Chalazonitis, N. (1964): *Photochem. Photobiol.,* 3:539–549.
8. Chalazonitis, N. (1970): *J. Physiol.,* 73:441–452.
9. Chalazonitis, N., and Arvanitaki, A. (1973): In: *International Encyclopedia of Pharmacology and Therapeutics, Section 19: Anticonvulsant Drugs, Vol. 2,* edited by J. Mercier, pp. 401–424. Pergamon Press, Oxford.
10. Chalazonitis, N., Gola, M., and Arvanitaki, A. (1965): *C. R. Soc. Biol.,* 159:2451–2455.
11. Chalazonitis, N., and Morales, T. (1971): *C. R. Soc. Biol.,* 165:1923–1928.
12. Ducreux, C., and Chalazonitis, N. (1971): *C. R. Soc. Biol.,* 165:1350–1353.
13. Ducreux, C., and Gola, M. (1975): *Pfluegers Arch.,* 361:43–53.
14. Eckert, R., and Lux, H. D. (1975): *Brain Res.,* 83:486–489.
15. Gola, M. (1974): *Pfluegers Arch.,* 352:17–36.
16. Gola, M., and Boisson, M. (1974): *J. Physiol. (Paris),* 69:154A.
17. Kandel, E. R., and Tauc, L. (1966): *J. Physiol. (Lond.),* 183:287–304.
18. Koketsu, K., and Minota, S. (1975): *Experientia,* 31:822–823.
19. Smith, T. G., Barker, J. L., and Gainer, H. (1975): *Nature,* 253:450–452.
20. Trautwein, N. (1973): *Physiol. Rev.,* 53:798.
21. Treistman, S. N., and Levitan, I. B. (1974): *Nature,* 261:62–64.
22. Walker, R. J., Ralph, K. L., Woodruff, G. W., and Kerkut, G. A. (1971): *Comp. Gen. Pharmacol.,* 2:5–25.

Abnormal Neuronal Discharges, edited by
N. Chalazonitis and M. Boisson. Raven
Press, New York © 1978.

Pharmacological Alteration of the Ionic Currents Underlying Slow Potential Wave Generation in Molluscan Neurons

C. Ducreux

Department of Biophysics of Neuromembranes, Institute of Neurophysiology and Psychophysiology, C.N.R.S., Marseilles, France

Paroxysmal depolarization shifts (PDSs) in cortical neurons are induced by a number of drugs or even by alteration of the ionic content of the extracellular fluid (23,26). Similar activities had been observed in molluscan neurons (5); for instance, long plateaus lasting 1 to 60 sec with associated spiking are induced in these neurons by pentamethylenetetrazol (PTZ) or barium ions. By analogy, these patterns are also called PDSs but they are also described by the terms burst plateau pattern or square-shaped potential waves.

The study in voltage-clamp conditions of such neurons shows that slow current changes are occurring and that the I-V relation, obtained by the slow potential ramp method, exhibits a negative slope region due to a persistent inward current. Since this inward current is similar to the one observed in neurons exhibiting either spontaneous (R_{15}) (12) or induced slow waves, the PDSs can be considered as a particular case of slow potential wave. The computed simulation of the PDSs based on voltage-clamp data implies that a specific ionic conductance remains activated during the plateau (14). Our purpose in this chapter is to identify the ion(s) flowing through the membrane during the plateaus induced by either PTZ or barium ions substituting calcium ions.

It has been emphasized that a permanent or slowly inactivating inward current is a necessary but not sufficient condition for the cell to generate slow potential waves or PDSs; and conversely, its lack would indicate that the neurons are not able to generate slow waves. Thus in each case, to control the ability of neurons to produce burst or plateau patterns, the current-voltage relationship of the membrane of PTZ- or barium-treated neurons was determined with the slow potential ramp method. The experiments were performed in *Helix pomatia* and *Aplysia rosea* neurons; detailed techniques have already been described (7).

The role of calcium and sodium conductances was analyzed with the

TABLE 1. *Summary of experiments*

Ion	PDS inductor	Inhibitor	Changes in ionic concentration
Na^+	PTZ	TTX	$Na^+ = 0$
	Ba^{2+}	TTX	$Na^+ = 0$
Ca^{2+}	PTZ	Co^{2+}	$Ca^{2+} = 0$
	Ba^{2+}	Co^{2+}	$Ca^{2+} = 8$ or 10 mM
Mg^{2+}	PTZ	No test	$Mg^{2+} = 0$
	Ba^{2+}	No test	$Mg^{2+} = 0$

specific blockers Co^{2+} and tetrodotoxin (TTX) added to the PTZ or barium saline. Additional evidence of Na-Ca implication was obtained by modifications of the concentrations of calcium, sodium, or magnesium ions of the saline. Table 1 summarizes these different experiments.

RESULTS

Role of Sodium Ions in Plateau Formation

The PTZ-treated neurons of Helix (70 mM) displayed slow potential waves or PDSs 5 to 10 min after addition of the drug. Generally, the slow waves were observed at low temperature (less than 20°C) (8) or low concentration of PTZ (less than 70 mM), whereas the PDSs appearing here as long plateaus were observed with higher values of both parameters (25). In the plateau configuration, the membrane depolarized at nearly −10 mV during 0.5 to 10 sec. Under our experimental conditions (21°C, 70 mM), the PDSs appeared more frequently than the slow waves. In some experiments, the membrane remained depolarized in plateau as long as the drug was present, but PDSs could be established by injection of a constant inward current. When either slow waves or PDSs were reproduced regularly (constant amplitude, duration, and frequency), as shown in Fig. 1A1, TTX (3×10^{-5} M) was added to the PTZ saline. In less than 1 min, the plateau patterns disappeared and the neuron became silent. Figure 1B1 shows that an outward current could not trigger PDSs but only spikes probably produced by a calcium ion flow (11). The slow rising potential ramp method was used before and after addition of the drug (slope of the ramp, 6.3 mV/sec; holding potential, ~ −55 mV; amplitude, 50 mV). In normal PTZ saline (Fig. 1A2), the current-voltage curve displayed a negative slope in the range −40 to −30 mV; this negative slope was induced by an inward current of 5 nA (difference between the two null slope points); a steady state current-voltage curve with a negative slope induced by PTZ has been already described by David et al. (6), Johnston and Ayala (15), and Ducreux and Gola (8). After TTX addition, and with the same voltage-clamp

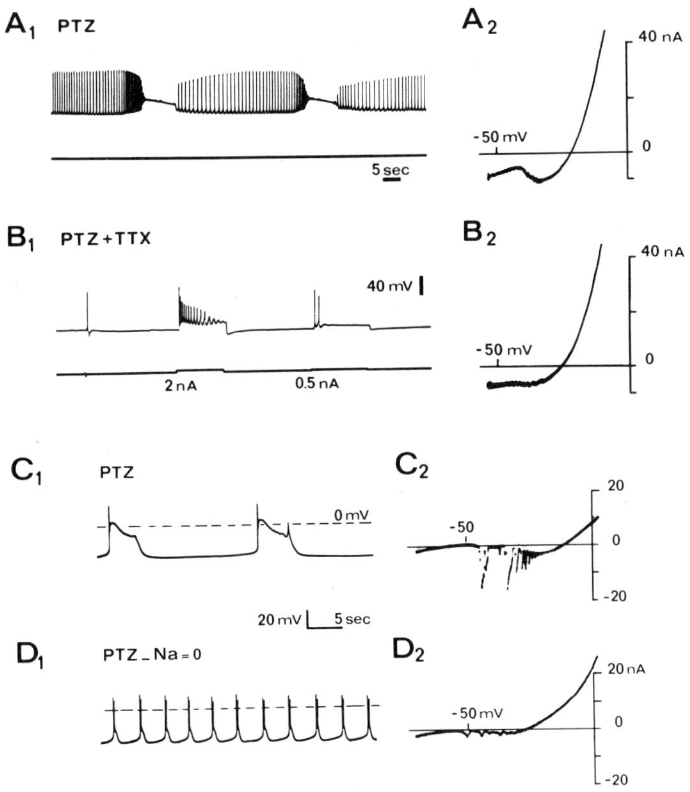

FIG. 1. Effects of TTX and Na-free saline on PTZ-treated neurons of Helix. $\mathbf{A}_1$: Periodic high frequency discharges and plateaus induced by PTZ in a regularly firing neuron of Helix. $\mathbf{A}_2$: Corresponding I-V curve obtained with a slow potential ramp (slope, 6.2 mV/s; amplitude, 45 mV). Inward currents are downward. *Horizontal line,* $i = 0$; *vertical line,* $V = 0$. The curve is N-shaped with a peak inward current of 5 nA. $\mathbf{B}_1$: Five minutes after TTX addition, the neuron is silent and a depolarizing current induces spikes without PDSs. $\mathbf{B}_2$: Corresponding I-V curve obtained with the same potential ramp as in $\mathbf{A}_2$. After TTX addition, the negative slope has disappeared. $\mathbf{C}_1$: Regular PDSs in PTZ-treated neurons of Helix (different from **A** and **B**). PDS duration, 5 sec; amplitude, 40 mV; spike overshoot, +20 mV; period, 23 sec. $\mathbf{C}_2$: Corresponding I-V curve exhibiting an inward current of 4 nA induced by a slow potential ramp (slope, 3.2 mV/sec; amplitude, 70 mV). $\mathbf{D}_1$: Same cell 6 min after Na removal; the PDSs are replaced by a regular burst of two spikes triggered by slow potential waves (duration, 0.7 sec; amplitude, 12 mV; spike overshoot, +15 mV; period, 4 sec). $\mathbf{D}_2$: The I-V curve remained slightly N-shaped (peak inward current of 0.5 nA). (Fig. 1C and D drawn from ref. 7.)

ramp (Fig. 1B2), the inward current disappeared and the I-V curve showed only a flattened region.

This TTX effect suggests an important role played by the sodium channel in the genesis of PTZ slow waves. To support this result, the sodium ions were removed from the PTZ saline when the PDSs showed a regular pattern. Similar experiments have been performed by Smith et al. (24) to determine the role of Na ions on the slow waves and the negative slope of the I-V curves. In Fig. 1C1, the PDSs were reproduced regularly (duration, 5 sec; amplitude, 40 mV); in Na-free PTZ saline (substituted by tris) (Fig. 1D1), the PDSs disappeared in 3 min, and there remained only a regular bursting activity of brief potential waves inducing doublets (duration, 0.7 sec; amplitude, 12 mV). The negative slope region induced by an inward current of 4 nA in normal PTZ (Fig. 1C2) is reduced to 0.5 nA after removal of Na^+ ions (Fig. 1D2).

These effects are fully reversible on Na introduction. The remaining slow waves, as well as the reduced inward current observed in Na-free PTZ saline, could be attributed to either residual extracellular sodium or a calcium component.

The Ba-treated neurons of Helix (8 mM) or Aplysia (10 mM) showed paroxysmal depolarizations characterized by a burst of spikes followed by a long plateau. The plateau level is about +20 mV for 10 to 60 sec duration in Aplysia neurons, and nearly +10 mV for 5 to 20 sec in Helix neurons (13). Generally, the pattern was reproduced regularly at intervals from 4 to 6 min in Aplysia neurons and from 5 to 20 sec in Helix neurons. Some neurons remained silent in Ba saline, but PDSs were induced by injecting a small outward current. Contrary to the PTZ effects, the membrane had never been observed to be blocked on the plateau level, and decreasing the temperature or the Ba concentration did not change the PDSs into slow waves.

The barium effects were stabilized 15 to 20 min after ionic substitution. The PDSs of an L_{11} Aplysia neuron are shown in Fig. 2A and B before and after TTX addition. In contrast to similar experiments with PTZ, the barium-induced PDSs remained without alteration upon TTX addition, whereas the spike amplitude was slightly decreased (about 5 mV). The increase in plateau duration can be accounted for by the low frequency of occurrence of the plateau. Similar results could be observed with R_{14} Aplysia neurons and with Helix neurons: the spike amplitude slightly decreased but the PDSs were not altered by TTX. The effect of TTX on the inward current was controlled before and after its addition with the slow ramp method. The ramp slope was from 2 to 10 mV/sec; a larger span than that for PTZ-treated neurons was used (about 80 mV, from −50 to +30 mV) because the inward current is larger than in PTZ-treated cells and develops from about −30 mV to positive potentials (+10 to +40 mV) (9,19). The peak inward current (near +5 mV) remained unchanged before and after TTX addition, whereas the hump near −20 mV, as well as the fast transients originating in the axon, was reduced by TTX (Fig. 2C and D).

To control this lack of effect of TTX on barium-treated neurons, the ganglion of Helix was immersed in Na-free Ba saline after recording regular PDSs in normal Ba saline (Fig. 2E). The membrane hyperpolarized slightly (about 5 mV) but the PDSs remained, although with a reduced frequency

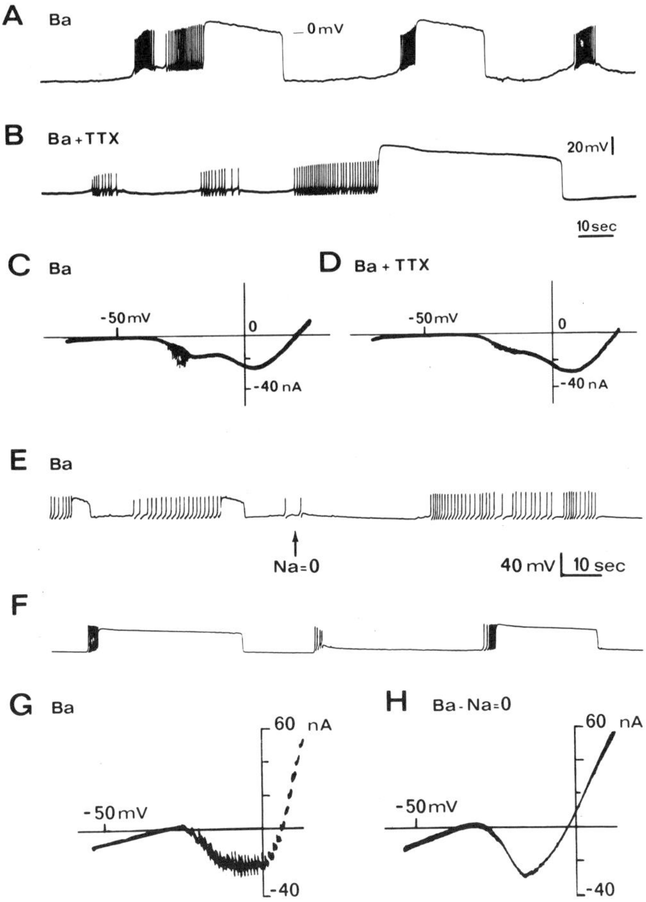

FIG. 2. Effects of TTX and Na-free saline on Ba-treated neurons. **A–D:** L_{11} neuron of Aplysia. **A:** Spontaneous PDSs after 60 min in Ba artificial seawater (ASW). The spike overshoot before the plateau is near +15 mV; PDS amplitude, 60 mV. **B:** Spontaneous activity 3 min after TTX addition. The PDSs are not blocked but the spike is slightly depressed. Overshoot near 0 mV; PDS amplitude, 50 mV. **C–D:** I-V curves obtained with the slow potential ramp method (slope, 10 mV/sec; amplitude, 95 mV) applied before **(C)** and after **(D)** TTX addition. The peak inward current (near +5 mV) is not altered by TTX which reduces the hump near −20 mV and the axonal fast transients. **E–G:** Helix neuron. **E:** Spontaneous PDSs induced by Ba (amplitude, 40 mV). Na ions are removed at arrow. **F:** Continuous recording of **E.** PDSs remain (same amplitude). **G–H:** I-V curves obtained with slow potential ramps (slope, 2.6 mV/sec; amplitude, 65 mV) before **(G)** and after **(H)** Na ion removal from the Ba saline, showing that the peak inward current of the N-shaped I-V curve is not reduced in Na-free Ba saline. (From ref. 7.)

and an increased duration (Fig. 2F). This effect was observed during 10 to 15 min but the neurons were not kept any longer in Na-free Ba saline to prevent deterioration of the cell. The same results were observed with R_{14} Aplysia neurons, but the membrane hyperpolarized about 15 mV and an outward current was necessary to trigger PDSs. In voltage-clamp conditions, the peak inward current exhibited by the I-V curves remained unchanged in the Na-free Ba saline, as shown in the recordings of Fig. 2 G–H.

In conclusion, sodium is necessary in the slow waves or PDSs induced by PTZ, whereas its role seems negligible in Ba-induced PDSs.

Role of Calcium Ions in Plateau Formation

The early inward current of the spike (11) and the slow decreasing inward current in R_{15} (12) are carried by both sodium and calcium ions. Also, during the plateaus induced by PTZ, an influx of Ca^{2+} ions cannot be excluded. On the other hand, in Ca-free Ba saline, the inward current is probably carried by Ba^{2+} ions flowing in the calcium channel. To control the part played by calcium, the cobalt ions, known as a calcium-conductance blocker, were used (1,10). A second set of experiments implies either the removal of calcium ions from the PTZ saline or the addition of calcium ions to the Ba saline.

After recording of the spontaneous PDSs induced by PTZ (in Fig. 3A1: duration, 5 sec; amplitude, 40 mV; period, 18 sec), cobalt (8 mM) was added to the PTZ saline. Fig. 3B1 shows that there is no important change in the PDS activity of the neuron, even after 10 min of Co^{2+} action, except a slight hyperpolarization (4 mV) and a consecutive increase in the period (30 sec) and the duration (6.5 sec) of PDSs. However, the spike overshoot was reduced by 15 mV. The I-V curves recorded in voltage-clamp conditions exhibited the same shape before and after cobalt addition (Fig. 3A2 and B2); in each case the negative slope was produced by an inward current of 3 nA.

The effects of Ca-free PTZ saline (substituted by tris) corroborated those of cobalt ions; in Fig. 3C1 PTZ induced pseudoplateaus of 1 sec duration with associated abortive spikes. Figure 3D1 shows the spontaneous activity of this neuron 15 min after removal of calcium ions: the resting potential decreased by about 2 mV, the spike amplitude was slightly decreased (about 8 mV), but the PDSs remained with similar duration. However, the plateau level originally at 0 mV (at the beginning of the plateau) was shifted to about −10 mV in Ca-free saline. The I-V curves exhibited a negative slope due to an inward current of 7 nA in Ca-free saline instead of 10 nA in control PTZ saline (Fig. 3C2 and D2). These results indicate that calcium ions are not essential for the production of plateaus by PTZ, even if they participate in their production.

In the experiments on Ba-induced PDSs, we cannot study directly the contribution of Ca ions since the Ba-induced slow waves are impeded by calcium. In most experiments, cobalt ions (10 mM) in artificial seawater were used as a conductance blocker; and in some experiments, lanthanum and manganese were used (at lower concentrations, 3 to 6 mM) giving

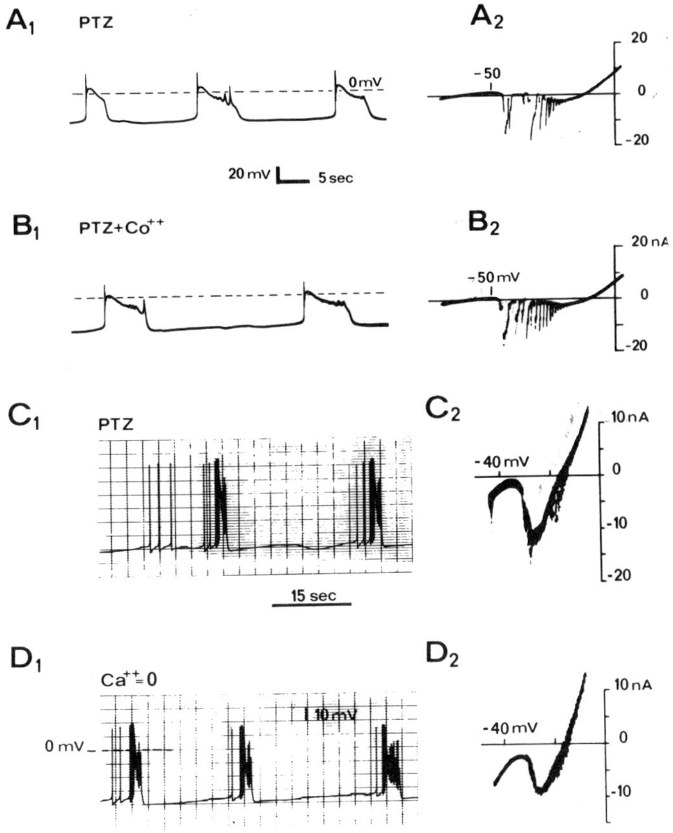

FIG. 3. Effects of cobalt and Ca-free saline on PTZ-treated Helix neurons. **A**$_1$: Spontaneous PDSs in the same neuron as illustrated in Fig. 1C and D (duration, 5 sec; amplitude, 40 mV; period, 18 sec; spike overshoot, +20 mV). **A**$_2$: Corresponding I-V curve. The peak inward current inducing the negative slope is 4 nA (ramp slope, 3.2 mV; amplitude, 70 mV). **B**$_1$: After 10 min of Ca^{2+} action, the PDSs persist without important changes (duration, 6.5 sec; amplitude, 40 mV; period, 30 sec; spike overshoot, +15 mV). **B**$_2$: I-V curve of cell illustrated in **B**$_1$ showing that its shape is unchanged. **C**$_1$: Spontaneous high frequency discharges in another PTZ-treated neuron. Spike overshoot, +25 mV. **C**$_2$: Corresponding I-V curve exhibiting a negative slope produced by an inward current of 10 nA (ramp slope, 3.2 mV/sec; amplitude, 40 mV). **D**$_1$: After 15 min in Ca-free PTZ saline, the spontaneous PDSs are still present with a spike overshoot at +17 mV. **D**$_2$: Corresponding I-V curve with an inward current slightly decreased (7 nA peak to peak).

similar results, except that the blocking effect of La^{3+} was incomplete and irreversible (14).

Figure 4A shows regular PDSs produced in Ba saline by a direct outward current of 3 nA in an R_{14} neuron of Aplysia. When the cobalt ions were added to saline, the membrane depolarized without spiking but no PDSs appeared, even with an outward current of 4 nA. The I-V curves obtained

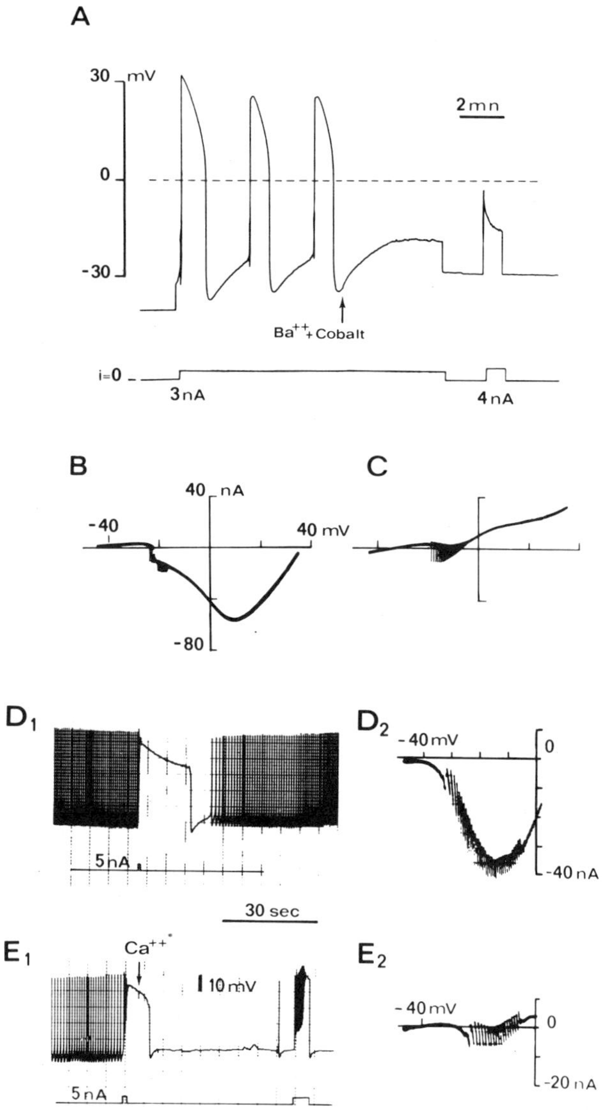

FIG. 4. Effects of cobalt and calcium on Ba-treated neurons. **A:** PDSs in Ba-treated R_{14} neuron of Aplysia triggered by an outward current of 3 nA. At arrow, cobalt ions (10 mM) are added to the ASW and the membrane depolarizes without generating PDSs. A current pulse of 4 nA fails to produce PDSs. **B–C:** I-V curves obtained with slow potential ramps (slope, 3.2 mV/sec; amplitude, 80 mV) before **(B)** and after **(C)** cobalt addition. The inward current of 55 nA producing an N-shaped I-V curve in **B** disappears completely in **C.** $\mathbf{D_1}$: Regularly firing neuron of Helix with PDSs induced by an outward current pulse of 5 nA. Mean duration of PDSs, 17 sec. $\mathbf{D_2}$: I-V relationship of cell illustrated in $\mathbf{D_1}$ showing a peak inward current of 35 nA (ramp slope, 2.1 mV/sec; amplitude, 50 mV). $\mathbf{E_1}$: Calcium ions (8 mM) are added at the beginning of a PDS (*arrow*); the membrane returns to the resting level 3 sec later, shortening the plateau duration to 8 sec (compare with $\mathbf{D_1}$); 48 sec later, a second pulse is unable to induce another PDS. $\mathbf{E_2}$: I-V relationship of cell illustrated in $\mathbf{E_1}$. The negative slope has completely disappeared and only a flattened region remains.

with a slow potential ramp exhibited a negative slope due to an inward current of 55 nA (peak to peak) which completely disappeared when cobalt was added (Fig. 4B and C). Since during the plateau the membrane conductance increases 60 times (13) and the cell becomes highly permeable to Ba^{2+} ions, it follows that (a) Ba ions are the main charge carrier during the plateau, (b) the Ba current is blocked by ions known as blockers of Ca^{2+} conductance, and (c) Ba ions probably flow in Ca channel whose conductance is enhanced by the Ba-Ca substitution.

The role of Ca^{2+} in Ba-induced PDSs was assessed by a second set of experiments in which Ca ions were added to the Ba Ringer. Figure 4D1 shows the normal PDS of a Helix neuron bathed in Ca-free Ba saline triggered by a current pulse of 5 nA. When several PDSs were recorded with a constant duration (about 17 sec), calcium ions were added to the Ba saline just at the beginning of a PDS (Fig. 4E1); in 3 sec, the potential returned to the resting level so that the PDS lasted only 8 sec. A second pulse applied 48 sec later induced spikes but no plateau. For Helix neurons, the calcium was added at its normal concentration of 8 mM, but lower concentrations (2 mM) were effective in blocking the PDSs. Similarly, 10 mM Ca^{2+} also blocked PDSs in R_{14} neurons of Aplysia. The inward current induced by the slow potential ramps in normal Ba saline (about 35 nA in Fig. 4D2) disappeared almost completely in the presence of Ca ions (Fig. 4E2). Therefore, Ca ions have a stabilizing effect on Ba-treated neurons by reducing and/or masking the persistent inward Ba current.

In conclusion, in Ba-treated neurons, the production of PDSs implies mainly a calcium channel permeable to Ba ions, whereas this channel plays a secondary role in PDS production induced by PTZ.

Role of Magnesium in Plateau Formation

It is likely that Mg ions do not play a direct role in plateau genesis; but, like other divalent cations, they can modify the state and kinetic parameters of ionic conductances (2) and thus interfere in the production of PDSs. To examine this possibility, Mg ions were removed from the PTZ or Ba saline when the PDSs were reproduced regularly.

In Fig. 5A1, the typical PTZ-induced PDSs lasted 7 sec and the plateau level was between −5 and −20 mV. In Mg-free saline (substituted by tris) the PDSs remained but the duration increased (11 sec), and the plateau level (Fig. 5B1) was between −15 and −25 mV. The I-V curves were similar in these two cases with an inward current of 10 nA (Fig. 5A2 and B2).

The Ba-induced PDSs of Fig. 5C1 lasted 16 sec and the top of the plateau was at +20 mV. When the Mg was removed, PDSs were still produced (by a pulse of outward current) but the same results as described in PTZ-treated cells were observed. The duration increased (20 sec), and the plateau level decreased (to +12 mV) (Fig. 5D1). These results are in agreement with those of Barker and Gainer (2), who show that the amplitude of bursting

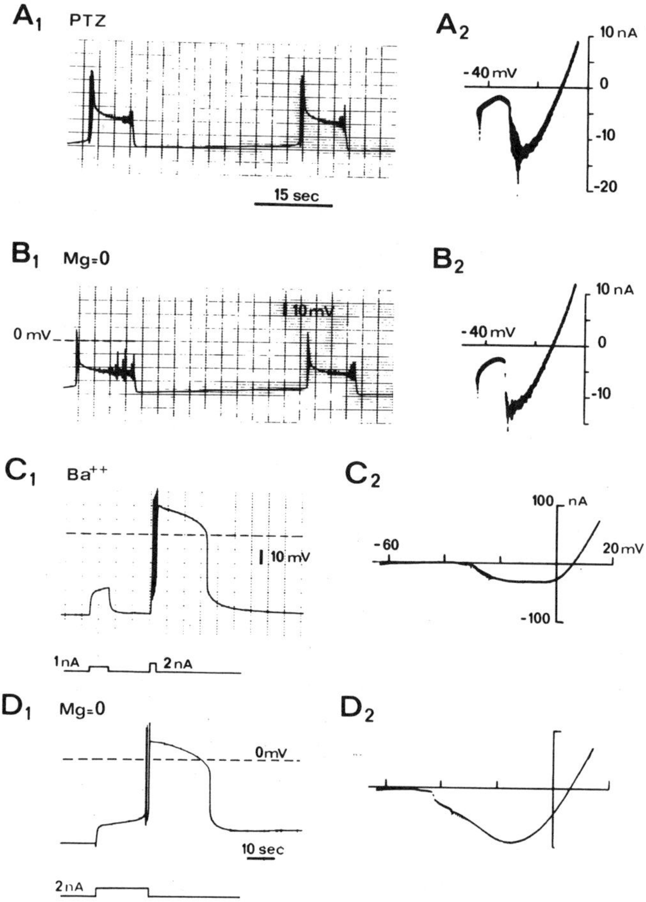

FIG. 5. Effect of Mg-free saline on PTZ- or Ba-treated neurons. **A–B:** PTZ-treated neuron of Helix. **A_1**: Spontaneous PDSs of an initially silent neuron (period, 40 sec; duration, 8 sec; plateau level between −5 and −20 mV; spike overshoot, +18 mV). **A_2**: N-shaped I-V curve corresponding to A_1 and exhibiting an inward current of 8 nA peak to peak (ramp slope, 3.2 mV/sec; amplitude, 40 mV). **B_1**: Spontaneous PDSs in Mg-free PTZ saline (period, 44 sec; mean duration, 10 sec; plateau level between −15 and −25 mV; spike overshoot, +8 mV). **B_2**: Corresponding I-V curve with the same inward current of 10 nA as in **A_2**. **C–D:** Ba-treated neuron of Aplysia. **C_1**: PDSs triggered by an outward current of 2 nA (duration, 16 sec; plateau level, +20 mV; spike overshoot, +32 mV). **C_2**: Corresponding I-V curve, N-shaped by an inward current of 30 nA (ramp slope, 7.6 mV/sec; amplitude, 80 mV). **D_1**: PDSs (triggered by an outward current) in Mg-free Ba saline (duration, 20 sec; plateau level, +12 mV; spike overshoot, +26 mV). **D_2**: Corresponding N-shaped I-V curve exhibiting an inward current of 90 nA.

pacemaker potential (BPP) in Aplysia neurons increases with Mg^{2+} concentration. With the slow potential ramp method, it appeared that the inward current was greatly increased by removing Mg^{2+} from the Ba saline. In Fig. 5C2 and D2, the inward current passed from 30 nA in the Ba saline to

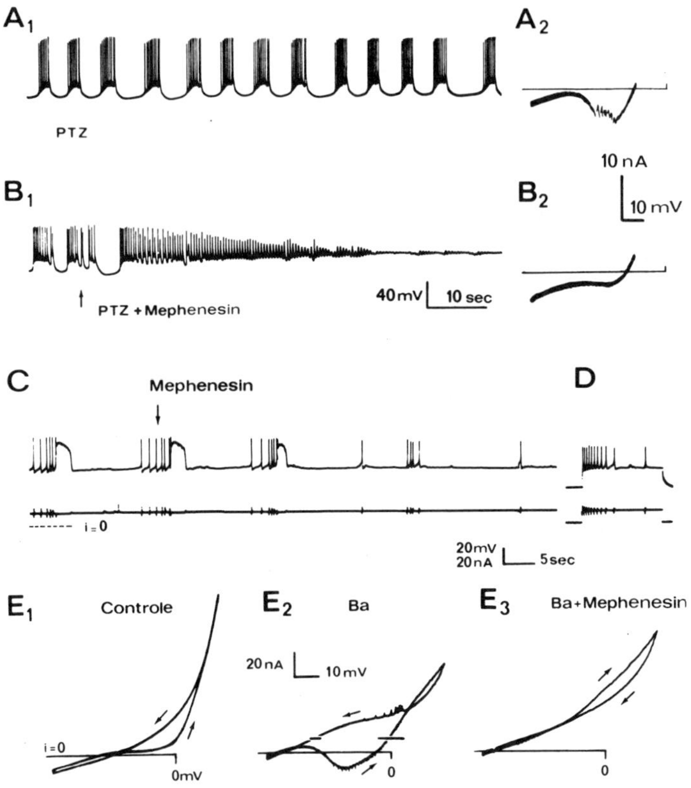

FIG. 6. Blocking effects of mephenesin on the PTZ- or Ba-treated neurons of Helix. **A_1**: Bursting activity induced by PTZ in a regularly firing neuron. **A_2**: I-V curve obtained just before **A_1** recording with a slow potential ramp (slope 3 mV/sec; amplitude, 50 mV). Horizontal line represents $i = 0$; inward currents are downward and the small bar at right represents $V = 0$. A peak inward current of 6 nA results in an N-shaped curve. **B_1**: Four minutes after recording **A,** mephenesin (added at arrow) blocks the slow waves and the spikes progressively disappear. **B_2**: Corresponding I-V curve without negative slope. **C:** Regular PDSs induced by a constant outward current of 17 nA (*lower trace*) in a Ba-treated neuron of Helix. Mephenesin added to the Ba saline (*arrow*) blocks the plateau in 30 sec and reduces the spike amplitude. **D:** Two minutes after adding mephenesin, the same depolarizing current fails to produce the burst-plateau pattern and the cell generates spikes of reduced amplitude. **E:** Same cell as illustrated in **D.** I-V curves obtained with slow symmetrical potential ramps (slope, ± 6 mV/sec; amplitude, 75 mV). The small arrows indicate the direction of sweeping, starting from the left. The cell was bathed in normal snail Ringer (**E_1**), in Ba saline (**E_2**), and in Ba saline with mephenesin (**E_3**). The N-shape of the I-V curve (corresponding to the depolarizing ramp) induced by Ba ions disappears with mephenesin. The small horizontal bars in **E_2** represent the level of the current used in **C** to trigger the PDSs. The intersection of this current level with the N-shaped I-V curve corresponds roughly to the plateau level of PDSs in **C,** whereas the intersection with the repolarizing I-V curve corresponds to the potential level between PDSs. (From ref. 7.)

90 nA after removal of Mg ions. This effect must be attributed to an increase in Ba conductance when the Mg concentration is lowered.

In conclusion, Mg ions exert a blocking effect on the ionic conductances implied in PDS genesis; and, conversely, lowering the Mg^{2+} concentration favors the production of PDSs by PTZ or Ba ions.

Mephenesin Effects on the PDSs Induced by PTZ or Barium

The anticonvulsant mephenesin is known to block both sodium and calcium early inward currents in molluscan cells, in contrast to specific blocking agents, such as TTX and cobalt ions, which act on only one of the two early currents (16). In this section, our purpose is to compare the action of mephenesin (5 mM) with that of TTX or cobalt ions on the PDSs and on the slow-decaying inward current. The same experimental methods were used; when the PDSs or slow waves (Fig. 6A1) were reproduced regularly, the mephenesin was added to the PTZ saline. Figure 6B1 shows the blocking effect of the drug on the slow waves; the spikes disappeared and the cell became silent. Inward or outward currents could not regenerate the slow waves or PDSs. In the same cell, the negative slope of the I-V curve was induced by an inward current of 6 nA (Fig. 6A2) in PTZ saline. When the slow waves disappeared, the I-V curve displayed only a flattened region (Fig. 6B2). In Ba-treated neurons, the mephenesin also blocked the PDSs (Fig. 6C) in less than 20 sec, and a depolarizing pulse could not restore PDS activity (Fig. 6D). The same blocking effect was observed in Aplysia neurons. The I-V curve exhibited in barium an inward current of 20 nA (Fig. 6E2) which completely disappeared when the mephenesin was added to saline (Fig. 6E3).

In conclusion, mephenesin is apt to block unspecifically (a) the PDSs induced by either PTZ or barium, (b) the underlying slow-decaying inward current carried by Na^+ or Ba^{2+}, and (c) the fast transient Na^+ or Ca^{2+} currents generating the spikes.

DISCUSSION

In this chapter, we compare the effects of Na^+ and Ca^{2+} specific blockers on the PDSs induced by either PTZ or barium with the aim of determining the active channel during the plateau. These results were controlled by the effects of various concentrations of these ions; the two methods led to similar conclusions. In PTZ saline, the TTX or Na-free saline blocks the PDSs and the negative slope of the I-V curves disappears, whereas with Co^{2+} or in Ca-free saline, the PDSs, as well as the negative slope resistance, persist. Thus during the plateau whose conductance is increased (4) or when the inward current develops, the role of main charge carrier is played by sodium ions flowing through the membrane; calcium ions play a secondary role.

In Ba saline, the PDSs remain in TTX or in Na-free saline; we conclude that Na ions play a secondary role in the genesis of PDSs. On the contrary, cobalt blocks the PDSs, indicating that the calcium channel is involved during the plateau. Since the plateau level varies with the barium concentration as expected for a barium electrode (13), and since the membrane conductance increases 60 times during the PDSs, it can be concluded that during the plateau, the barium ions flow through the membrane via the calcium channel.

Calcium added to the Ba saline blocks the PDSs; on the other hand, Barker and Gainer (2) reported that, when calcium and magnesium were replaced by strontium, the R_{14} neuron exhibits behavior that resembles that of the R_{15} bursting neuron. We have also observed similar activity in a few R_{14} neurons in Ca-free saline. Thus the decrease of the extracellular calcium ions favors the production of PDSs or slow waves. The following results confirm that these ions play an indirect but important role by controlling the state of the other conductances. Indeed, the PDSs induced by PTZ can be blocked when the calcium concentration is raised from 8 to 30 mM (H. Takeuchi, *personal communication*); and, conversely, a lack of calcium ions in the saline does not prevent the PDSs and even facilitates their formation.

Similar results have been reported by Klee et al. (17) in Aplysia neurons treated by strychnine. The decrease in calcium concentration induces long-lasting depolarizations, and the increase in calcium concentration reduces but does not block the amplitude of the waves. The above assumption about the role of calcium is supported by the results of Barker and Gainer (2) showing that seasonal and physiological changes are implicated in the slow wave formation. They reported that a neuron from active but not estivated snails generates BPP in normal Ringer and that BPPs appear in estivated snails only in Ca-free saline. A similar observation was made in the course of PTZ experiments. The PTZ is effective in inducing PDSs in active but not in hibernating snails, although we used the same saline throughout the year. These seasonal variations could be related to the annual cyclic changes in the overall ionic concentration of the hemolymph and particularly in Ca concentration (3,18,22), which decreases in active snails. An experimental background for this hypothesis can be found in the results of Meech and Strumwasser (21) and of Meech and Standen (20), showing that the increase in intracellular calcium concentration augments a Ca-dependent potassium conductance. Thus an increase in Ca^{2+} concentration, which occurs in nonactive animals, tends to stabilize the membrane and counteracts the epileptogenic effects of PTZ and Ba. When the Ca concentration of the hemolymph is lowered (as in active animals or in experiments with reduced Ca concentration), the Ca influx, as well as the Ca-dependent potassium activation, is reduced, and the membrane tends to remain on the depolarized plateau. It is worth noting that the extension of plateau duration is also produced by TEA, which blocks part of the potassium conductance (not illustrated).

These results show that the internal and/or external Ca ions play a regulating role in the ability of the cell to generate slow waves or PDSs. Because of influx of Ca ions during the bursts of spikes and the concomitant changes in intracellular Ca concentration, the regulating role of calcium is not restricted to a constant background level but varies periodically during successive PDSs, thus acting as a negative feedback to stabilize the membrane after each PDS.

SUMMARY

In Helix or Aplysia neurons, paroxysmal depolarizations were induced either by drugs, such as PTZ, or by the ionic substitution Ba-Ca. On the plateau, the membrane potential reached about −15 mV in PTZ-treated neurons and +25 mV in Ba-treated neurons; simultaneously, the membrane resistance decreased, indicating an increase of a specific ionic conductance. Our purpose was to determine the main ions generating the plateau in these two cases. The slow potential ramp method of clamping was used, and the treated neurons displayed a negative slope region in their I-V curves; this negative slope is due to a persistent inward current and indicates the ability of neurons to produce plateaus.

TTX or Na-free saline blocked paroxysmal depolarizations and the persistent inward current induced by PTZ but had no effect on those induced by Ba. Co^{2+} ions, which block Ca conductances, had no effects on the paroxysmal depolarizations nor on the persistent inward current induced by PTZ; but they did block those induced by Ba. The plateaus induced by PTZ persisted in Ca-free saline. Those induced by Ba disappeared when Ca was added to saline. Both types of paroxysmal depolarizations persisted in Mg-free saline, indicating that Mg ions play a secondary role. We conclude that during the plateau, the main role is played by Na ions in PTZ-treated neurons and by Ba ions flowing through the Ca channel in Ba-treated neurons. Despite their different ionic nature, both types of paroxysmal depolarizations were blocked by the anticonvulsant mephenesin. The role of Ca ions in the PDS genesis is discussed.

ACKNOWLEDGMENTS

I wish to thank Dr. M. Gola for his helpful criticism of this chapter. I am indebted to Ms. M. Andre, Mr. R. Fayolle, and G. Jacquet for technical assistance.

REFERENCES

1. Barker, J. L., and Gainer, H. (1975): *Brain Res.*, 84:461–477.
2. Barker, J. L., and Gainer, H. (1975): *Brain Res.*, 84:479–500.
3. Burton, R. F. (1971): *Comp. Biochem. Physiol.*, 39:267–275.
4. Chalazonitis, N., and Arvanitaki, A. (1973): In: *International Encyclopedia of Pharmacology and Therapeutics, Section 19, Vol. 2*, edited by J. Mercier, pp. 401–424. Pergamon Press, New York.
5. Chalazonitis, N., and Takeuchi, H. (1968): *C. R. Soc. Biol.*, 162:1552–1556.
6. David, R. J., Wilson, W. A., and Escueta, A. V. (1974): *Brain Res.*, 67:549–554.
7. Ducreux, C. (1978): *Comp. Biochem. Physiol.* 59:51–57.
8. Ducreux, C., and Gola, M. (1975): *Pfluegers Arch.*, 361:43–53.
9. Ducreux, C., and Gola, M. (1977): *Brain Res.*, 123:384–389.
10. Eckert, R., and Lux, H. D. (1976): *J. Physiol.* (*Lond.*), 254:129–151.
11. Geduldig, D., and Gruener, R. (1970): *J. Physiol.* (*Lond.*), 211:217–244.
12. Gola, M. (1974): *Pfluegers Arch.*, 352:17–36.
13. Gola, M., and Ducreux, C. (1977): *Experientia*, 33:328–329.
14. Gola, M., Ducreux, C., and Chagneux, H. (1977): *J. Physiol.* (*Paris*), 73:407–440.
15. Johnston, D., and Ayala, G. F. (1975): *Science*, 189:1009–1011.
16. Klee, M. R., and Faber, D. S. (1974): *Pfluegers Arch.*, 346:97–106.
17. Klee, M. R., Faber, D. S., and Heiss, W. D. (1973): *Science*, 179:1133–1136.
18. Kostyuk, P. G. (1968): In: *Neurobiology of Invertebrates*, edited by J. Salanki, pp. 145–167. Plenum Press, New York.
19. Magura, I. S. (1977): *J. Membr. Biol.*, 35:239–256.
20. Meech, R. W., and Standen, N. B. (1975): *J. Physiol.* (*Lond.*), 249:211–239.
21. Meech, R. W., and Strumwasser, F. (1970): *Fed. Proc.*, 29:834.
22. Meincke, K. F. (1975): *Comp. Biochem. Physiol.*, 52:135–140.
23. Mercier, J. (1953): Thesis. University of Marseilles.
24. Smith, T. G., Barker, J. L., and Gainer, H. (1975): *Nature*, 253:450–452.
25. Sugaya, A., Sugaya, E., and Tsujitani, M. (1973): *Jpn. J. Physiol.*, 23:261–274.
26. Woodbury, D. M. (1969): In: *Basic Mechanisms of the Epilepsies*, edited by H. H. Jasper, A. Ward, and A. Pope, pp. 647–681. Little Brown, Boston.

Abnormal Neuronal Discharges, edited by
N. Chalazonitis and M. Boisson. Raven
Press, New York © 1978.

Doublet Discharges and Bistable States Induced by Strychnine in a Neuronal Soma Membrane

M. R. Klee, *D. S. Faber, and **J. Hoyer

Max Planck Institute for Brain Research, Department of Neurobiology, Frankfurt am Main, Federal Republic of Germany

The general belief is that nearly all drugs that affect the function of a neuronal system do so by acting primarily at chemical excitable membrane patches (thus changing the efficacy of excitatory or inhibitory potentials) or by interfering with presynaptic biochemical processes, e.g., transmitter production, turnover, reuptake. An example of such action is the suppression of transmitter release at a cholinergic synapse owing to the action of ethanol (5,6); however, ethanol in similar concentrations has also a well-defined blocking action on the inward current system of the electrical excitable membrane in Aplysia neurons (2).

The blocking or reducing action of convulsants (e.g., strychnine) on postsynaptic potentials (PSPs) is well known. Surprisingly, some phenomena that had been explained as the consequence of their action on the PSPs can be evoked by strychnine as well as by pentylenetetrazol (PTZ) even in those preparations of invertebrate ganglia in which synaptic transmission has been blocked by either ligature or a complete isolation of cell somata from their presynaptic structures (1,8,17,24). This includes two phenomena that seem to be prerequisites of convulsive activities: (a) doublet or multiple discharges, and (b) long-lasting depolarizations, observed first in mammalian neurons and termed paroxysmal depolarizing shifts (PDS) (19), although they are better described as a bistable state of the membrane potential. Especially the latter phenomenon was believed to depend exclusively on an increased excitatory input to the neuron (see Prince and Schwartzkroin, and Andersen et al., *this volume*). The fact that the same phenomenon can be evoked by convulsants in the absence of synaptic input raises the question as to whether some convulsants act directly at membrane patches responsible for the generation of action potentials and/or slow changes of the soma membrane potential, probably in addition to their effects on synaptic processes.

Present address: **Research Institute on Alcoholism, Buffalo, New York; and **Institut für Allgemeine und Vergleichende Physiologie de Universität Wien, Schwarzpanierstrasse 17, A-1090 Vienna, Austria.*

The experiments and data presented here were restricted to the action of one convulsant, strychnine, and the emphasis was put on mechanisms that are terminating or gating the induced hyperexcitability rather than on processes that induce the abnormal neuronal activity.

Three points are discussed: (a) Which changes of the current-voltage relationship are combined with the generation of multiple discharge after strychnine application? (b) Which change of the I/V characteristic might be responsible for the preference for doublet discharges versus higher multiple discharges after strychnine application? (c) Which additional change in the I/V curve must be induced in order to evoke bistable states of the membrane potential?

METHODS

Experiments were done on neurons of the visceral ganglion of *Aplysia californica,* kindly supplied by Dr. R. C. Fay, Venice, California. The preparation was perfused by artificial seawater (ASW) to which drugs, such as strychnine sulfate (0.1 to 1.0 mM), cobalt chloride (30 mM), and mephenesin (5 to 15 mM), were added. The temperature was kept constant at 16°C. Cells were impaled with a set of double electrodes and voltage or current clamped by the two-needle method. The overall current was measured by an I/V converter. Clamp commands were either square or ramp pulses, the latter of different slopes. Outward current was measured in addition by an operational amplifier (Tektronix A502), set in an integrator configuration. Current and voltage were recorded on either film or paper by a penwriter (Brush 280).

RESULTS

Ten to twenty min after the application of strychnine in concentrations between 0.1 and 1.0 mM to the perfusing seawater, the neurons in the visceral ganglion of Aplysia start to generate doublet discharges. This abnormal discharge pattern can be induced in all cells, regardless of whether they show in the control a so-called bursting behavior (e.g., cells R_{15} and L_2 to L_6), a regular beating pacemaker activity (cells R_3 to R_{13}), no discharges (the silent cells R_2, R_{14}, and L_1) or a mixed synaptic plus pacemaker activity, which is characteristic for most of the neurons of this ganglion. The doublet discharge pattern is illustrated in Fig. 1B for the cell R_{15} and in Fig. 8B for the cell R_2 during the injection of a depolarizing current.

In some cells, especially in those with an excitatory synaptic input which at that time is not blocked by the action of strychnine (5), the numbers of spikes per burst might be greater than two; but the majority of cells are discharging with doublets.

As has been described elsewhere (13,14,17), voltage-clamp data have

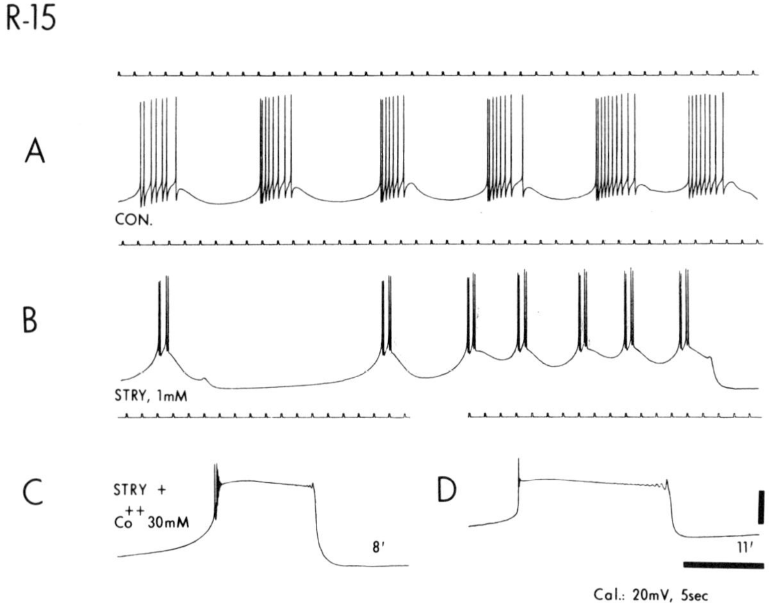

FIG. 1. Induced doublet discharge and bistable state in cell R_{15} by strychnine and strychnine plus cobalt. **A:** Control, bursting pattern of cell R_{15}. **B:** Twenty-five min after addition of 1 mM strychnine to the ASW, R_{15} discharges with doublets with nearly unchanged underlying oscillatory waves. **C:** After addition of 30 mM cobalt chloride to the strychnine-ASW, the cell starts to show bistable state phenomenon after 8 min **(C)** with spontaneous repolarizations.

shown that strychnine is reducing the inward as well as the outward current and is shifting the maximum of the inward current on the voltage axis to lower voltages, i.e., close to a reduced resting membrane potential, and the outward current to the right. A fast, K^+-dependent outward current (11,21, 25) present in some Aplysia neurons (16) is reduced in the same range as the delayed outward current. The effect of strychnine on the membrane resistance can be rather small, while the action potential is reduced and its duration prolonged. These effects of strychnine are illustrated in Fig. 2 for a cell of the group L_2 to L_6.

By using a method first invented by Rojas et al. (22), we checked whether the increase in duration of the action potential is related to an increased current flowing through the channels of the inward current system. As shown in Fig. 3 for a cell of the same group as that of Fig. 2 expressed by the current tails for the inward (downward) and the outward (upward) currents similar to the relationship in nerve fibers, the maximum inward current coincides with the spike peak, while the maximum of the outward current coincides with the second repolarizing phase of the action potential. Whereas the duration of the action potential is increased nearly threefold in Fig. 3C

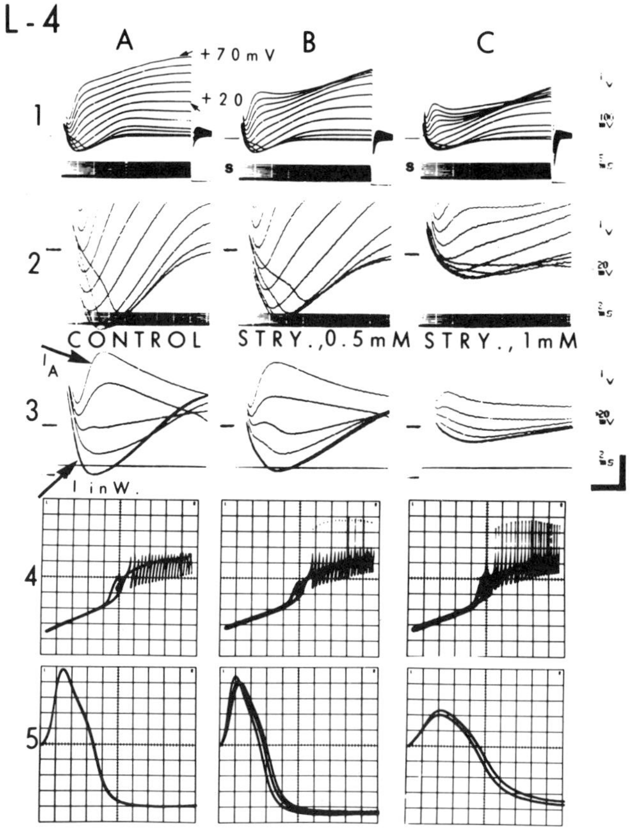

Cal.: 200mV, 2μA, 10msec(1); 200mV, 0.4μA, 4msec(2); 200mV, 0.3μA, 4msec(3); 20mV, 10nA/div.(4); 10mV, 5msec/div.(5).

FIG. 2. Effect of 0.5 and 1 mM strychnine on the inward and outward current of cell L_4. Stepwise reduction of the inward and outward current following strychnine application (*1* and *2*), combined with a reduction of the fast K^+-outward current (*3*). Almost no change in the slope of the membrane resistance, displayed on an X-Y scope (*4*), while the action potential is reduced especially in overshoot and prolonged in duration (*5*). Notice the increase in afterhyperpolarization in *B5.*

and the outward current is markedly reduced, the relationships between the different phases of the action potential and the peaks of the inward and outward currents remain essentially the same. Moreover, as can be seen in Fig. 3C, the inward current system does not display any sign of an incomplete inactivation, as seen by the absence of inward current, when the outward current reaches its maximum.

As can already be seen in Fig. 2C1, in some cells strychnine produces not only a reduction of the amplitude of the outward current but also a change in

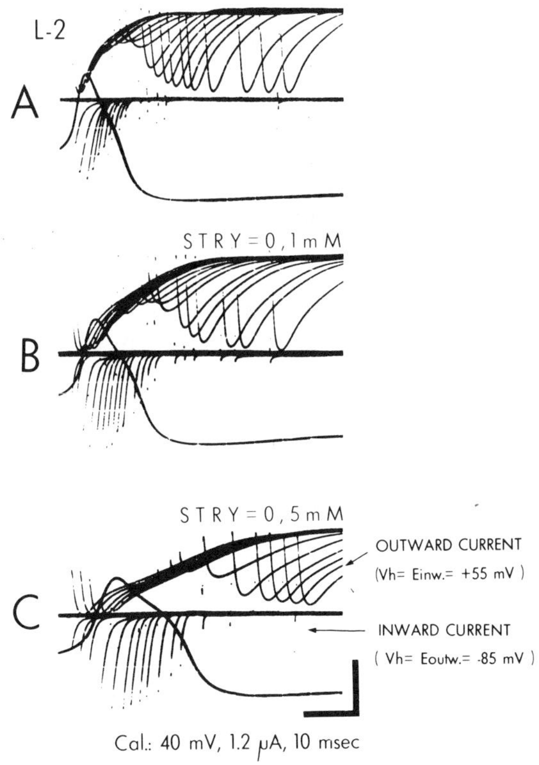

FIG. 3. Different time relationship between the inward and the outward current to different phases of the action potential. At different times during the course of the action potential, the membrane was clamped either to the equilibrium potential of the outward current (−85 mV) so that only the tail currents of the inward current is displayed (*lower signals*), or to the inward current (+55 mV) showing the magnitude and time course of the outward current (*upper signals*). Maximum inward current during the peak of the action potential in control, maximum outward current during the second repolarizing phase of the spike. Although the duration of the action potential increases with increasing concentration of strychnine (**B** and **C**), the time relationship remains the same. Complete shut-off of the inward current during maximum activation of the outward current.

its time course. During a depolarization of only 40 msec, one can observe that the outward current, measured at the end of the pulse, for a command to +20 mV is somewhat greater than the outward current flowing while the membrane is depolarized to +70 mV.

As has been described for neuron A of *Helix pomatia* (12,20) and for neurons of Tritonia (26), the delayed outward current can be separated into two components: a voltage-dependent, TEA-sensitive potassium current and an outward current which depends on the flow of calcium ions across the membrane but is depressed following the injection of calcium ions and is partially resistant to TEA injection (12). Because of its dependence on the flow of Ca ions, the current becomes smaller as the potential change approaches the Ca^{2+}-equilibrium potential. Therefore, the I/V curve of cells displaying this Ca-dependent potassium current ($I_{K(Ca)}$) is N-shaped at a voltage between +25 and +100 mV, showing a null point which will be related to $E_{Ca}{}^{2+}$ around +100 mV (20,26) or at higher values (12).

Some Aplysia neurons differ from Helix and Tritonia neurons in that N-shape, maximum of Ca^{2+}-dependent current, and null point are often reached only after depolarizations up to +250 to +300 mV. An example of this cell type is the regular burster cell R_{15}. As shown in Fig. 4A1, even during depolarization of the membrane up to +180 mV the outward current rises stepwise in a linear relationship to the imposed voltage change. The same holds for a neuron of the group RC (Fig. 6B). Other cells, such as the burster neurons of the group L_2 to L_6 or the pacemaker cells of the group R_3 to R_{13}, exhibit the N-shape at lower potentials, i.e., at +80 to +100 mV (Figs. 6A and 7A1).

The effect of strychnine is a shift of the N-shape and maximum of $I_{K(Ca)}$ to smaller positive membrane potentials. As shown in Fig. 4B1, 75 min after the application of 1 mM strychnine, the outward currents flowing during depolarizations to +40 to +100 mV show the characteristics of the $I_{K(Ca)}$. The same holds for the effect of strychnine shown in Fig. 2B1 and C1. Figure 6 shows the difference in the I/V characteristic of the outward current of two neurons of the visceral ganglion of Aplysia. The cell R_3, shown in Fig. 6A, displays a pronounced N-shaped I/V curve with a maximum of $I_{K(Ca)}$ around +100 mV. Fifty minutes after strychnine application, this maximum shifts on the voltage axis below +50 mV. The slope of the outward current below and above the maximum is still running parallel to the control values, as can be seen for the potentials from 0 to +50 mV and from +150 to +200 mV. The cell of the group RC, shown in Fig. 6B, shows a linear I/V relationship following depolarization up to +200 mV. After the addition of strychnine, the I/V curve displays an N-shape with a maximum of $I_{K(Ca)}$ around +100 mV. After adding 5 mM mephenesin to the strychnine ASW, the I/V curve again becomes linear with a slope different from that during control. The N-shape of the outward current reappears and remains after 90 min of washing.

Another characteristic of the $I_{K(Ca)}$ is its ability to facilitate during repetitive activation, which is essentially the opposite for the TEA-sensitive late outward current. The faster rise of the outward current to a second pulse has been described by Clusin et al. (3) for the electroreceptor of the skate, by Ulbricht and Wagner (27) for the Ranvier node of the frog after treatment with 4-AP, and by Heyer and Lux (12) for Helix neurons after application of TEA. This characteristic is shown in Fig. 4F–K. A series of 5 or 10 pulses was given at intervals of 40 or 20 msec. In control, starting with the second pulse, the outward current decreases continuously (Fig. 4F). Following the application of strychnine, the total outward current is reduced

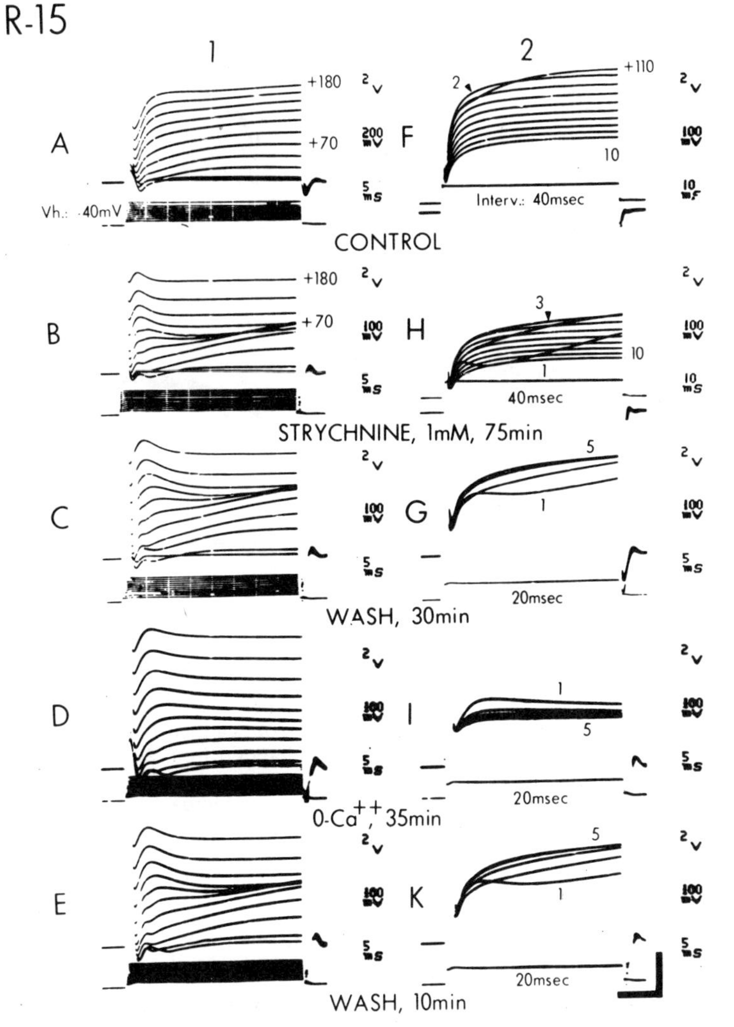

Cal.: A: 0 .4V, 4μA, 10msec; F-G: 0 .4V, 2μA, 20msec;
B-E: 0 .4V, 2μA, 10msec; H-K: 0 .4V, 2μA, 10msec.

FIG. 4. Relationship between changes in the outward current and the effect of repetitive activation of the outward current. Equal spacing of the outward current in cell R_{15} in control during depolarizations up to +180 mV **(A).** During repetitive activation, the outward current gradually declines from first to tenth pulse **(F).** Following strychnine application, the total outward current is reduced by 50% (see different sensitivity of the current trace in **A** versus **B–E**). Outward current behaves opposite to control during repetitive activation with the same frequency: increased and faster turn-on of the outward current up to the third pulse **(H).** No essential change after 30 min of washing in ASW (**C** and **G**). Replacing ASW by zero-Ca^{2+}-seawater reinduced a more equally spaced outward current, while the repetitive activity is blocked and facilitation disappeared (**D** and **I**). Same characteristics as in **B–H** and **C–G** after only 10 min washing in normal ASW (**E** and **K**).

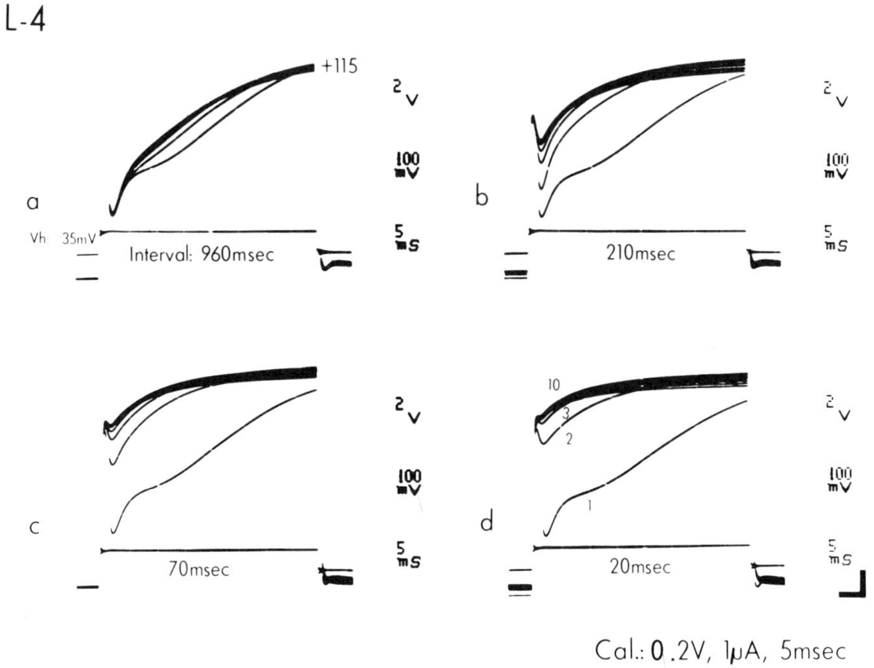

FIG. 5. Relationship between pulse interval and time course of activation of K^+ current during strychnine action. Repetitive activation of the outward current by a train 10 pulses. Time of repolarization (pulse interval) has been changed from 960 **(a)** to 20 msec **(d).** Outward current in response to first and second pulse in **d** increased by 87%. This increase in speed of the outward current decreases with increased interval (**d** versus **a**).

by about 60% (the sensitivity of the current trace in Fig. 4A is half of that in B to E). At the same time, the facilitating characteristic of the outward current becomes obvious at the same potential (+110 mV). In response to the same series of pulses, the outward current increases at least up to the fifth command. This phenomenon lasts also during washing together with the phenomenon of the N-shape (Fig. 4C and G). When the neuron is washed in zero calcium seawater, both phenomena disappear (Fig. 4D and I) and are restored after 10 min of washing in normal ASW (Fig. 4E and K). As shown for the cell L_4 in Fig. 5, the facilitating effect is stronger with short pulse intervals in the range of 20 to 100 msec (Fig. 5d and e) and nearly disappears with intervals of about a second (Fig. 5a).

During the experiments in cell R_{15}, we compared the changes of firing pattern with the described changes in the outward current. As soon as the doublet discharges appeared, the N-shape of the outward current was also present in the range of +20 to +100 mV. To test any relationship between doublet discharges and the change in the $I_{K(Ca)}$, we checked whether mephenesin, which has been shown to stop the doublet discharges (15) or is

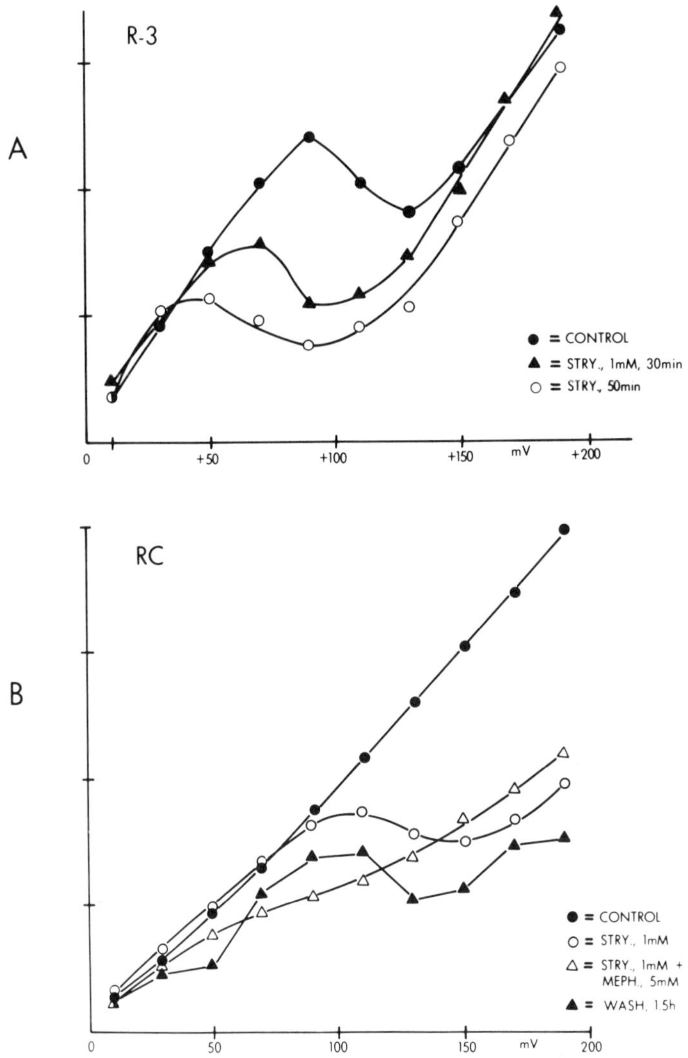

FIG. 6. Effect of strychnine on a neuron showing N-shape I/V characteristic **(A)** and on a cell with linear I/V curve **(B)**. **A:** Strychnine shifts the maximum $I_{K(Ca)}$ from approximately 100 mV to about +50 mV after 50 min. **B:** After strychnine, the membrane displays an N-shape with a maximum around +100 mV. After adding 5 mM mephenesin to the strychnine-ASW, the N-shape disappears and returns during washing. *Abscissa,* absolute membrane potential; *ordinate,* voltage output of the integrator (V × 220 nA × sec).

reducing or blocking the calcium and sodium inward current, as mephenesin and ethanol do in Aplysia (2), is also counteracting the $I_{K(Ca)}$. As shown in Fig. 7B1 and 2, 15 mM mephenesin completely blocks not only the inward current but also the N-shape phenomenon in the outward current. In con-

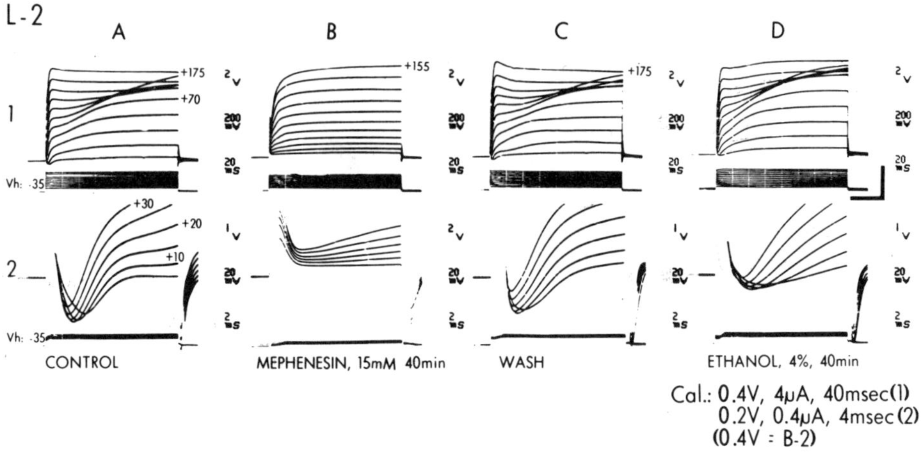

FIG. 7. Effect of mephenesin and ethanol on inward and outward currents. The cell shows an $I_{K(Ca)}$ in the range of +90 to +135 mV **(A1).** Adding 15 mM mephenesin blocks both the inward current **(B2)** as well as $I_{K(Ca)}$ **(B1).** Ethanol reduces the inward current **(D2)** but $I_{K(Ca)}$ is unchanged **(D1).**

trast, 4% ethanol reduces the inward current but does not interfere with the outward current (Fig. 7D1 and 2).

The time-locked occurrence of doublet discharges and a shift of the maximum of $I_{K(Ca)}$ to between near 0 and +50 mV tempts one to speculate whether the facilitating characteristic of this current might be an explanation for the preference of doublet discharges versus multiple discharges in these neurons. As shown in Fig. 5d measured after 10 msec, the outward current caused by the second pulse is 87% greater than the one following the first pulse. We also observed in R_{15} that during doublet discharges displayed in the phase-plane method, the voltage change (dV/dt) during the repolarizing phase was faster than in control. It seems plausible, therefore, that the increased activation of the outward current system during the second action potential is strong enough to prevent a third activation of the inward current system.

Aplysia neurons show differences from neurons of Helix and Archidoris in that both strychnine and PTZ (the latter in concentrations up to 70 mM) do not induce bistable states in the neurons (see refs. 8,9,24, and 28). However, if cobalt chloride (30 mM) was added to the strychnine-seawater solution, or the calcium concentration was reduced, bistable states could be observed (17,18), as shown in Fig. 1C and D and Fig. 8A.

Here, again, we were interested in the mechanism responsible for the termination of the long-lasting depolarization. Using the method of Gola (10), i.e., applying slow ramp voltage commands, we found that both depolarization and repolarization of the bistable state coincide in voltage with a negative conductance displayed in the current slope (Fig. 8D, E, and G)

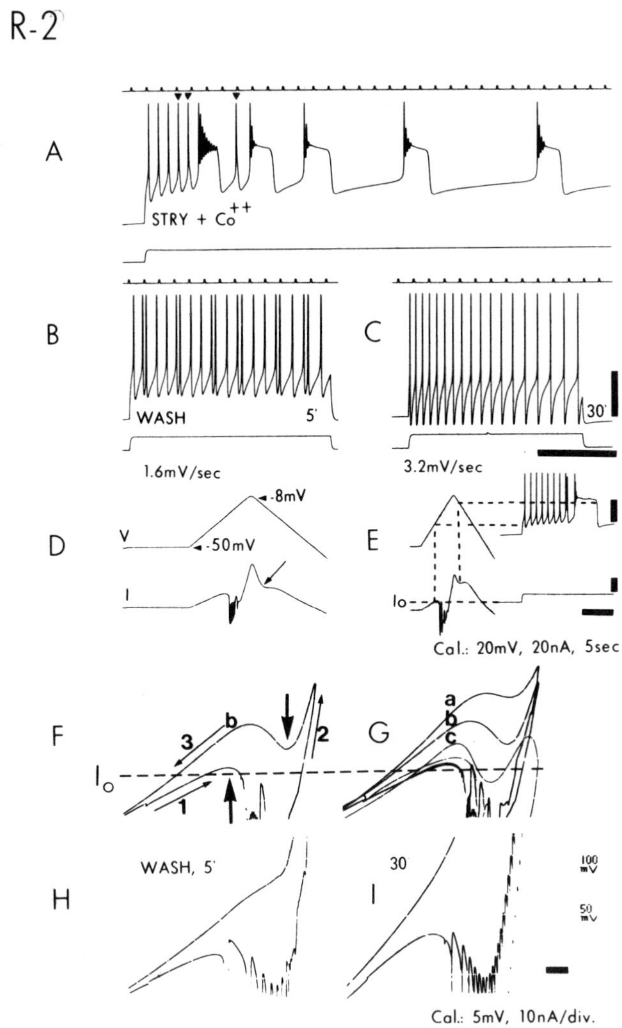

FIG. 8. Negative conductances during bistable state induced by strychnine plus cobalt treatment. **A–C:** Voltage recording. Cell R_2 shows during the injection of a depolarizing current action potentials with complete repolarization (*triangle*) mixed with incomplete repolarizations leading to bistable states. **B** and **C:** Recovery after 5 and 30 min, respectively. **D–I:** Voltage-clamp experiments during bistable states **(D–G)** and during washing **(H and I)**. Voltage commands: Ramps of different speed. Direction of the current signal on the X-Y scope is illustrated in **F** by arrows 1 to 3. During the bistable state, a second negative conductance can be seen during repolarization, most prominent in *c* of **G.** The voltage change is 8 mV/sec for the command in *c.* Complete recovery, i.e., disappearance of the negative conductance on repolarization in **I.**

(14). A relationship exists between the occurrence of the negative conductance during repolarization and the bistable state, both disappearing after 5 min of washing (Fig. 8H). These results agree with those described by Ducreux and Gola (4) for the characteristics of the I/V relationship of neurons of Helix during bistable states induced by PTZ.

DISCUSSION

As has already been discussed (13,14,17), the combined changes of both inward and outward currents in the I/V relationship could explain the induction of multiple discharges in nerve cells of Aplysia after strychnine application. It is doubtful that, in addition to the shift of the maximum inward current to lower voltages, the reduction of I_A is also essential for bursting (29). This current can be detected only in some of the Aplysia visceral ganglion cells (16), but the same type of multiple discharges occurs in all cells investigated. Even the tendency for bursting is not greater in the cells displaying I_A than in those in which this fast K^+-current cannot be detected. The question arises as to which mechanism causes the discharge pattern to consist mainly of doublets. As suggested above, the facilitating characteristic of $I_{K(Ca)}$ might be responsible for the phenomenon, i.e., the fact that in a given voltage range, which coincides with the voltage around the maximum of $I_{A(Ca)}$, the activation of the outward current is much faster in response to a second depolarization following the first pulse with an interval of 20 to 100 msec.

This phenomenon is not "induced" by strychnine; in Aplysia, it can always be elicited best in the range of the maximum $I_{K(Ca)}$, mainly in the range between +100 and +180 mV, and in the vertebrate Ranvier node at a depolarization of +147 mV (27). Therefore, the impression of a changed characteristic of the outward current is caused by the fact that during control the repetitive activation might have been checked in a voltage range less positive than the maximum of $I_{K(Ca)}$. If, owing to the effect of strychnine, the maximum of $I_{K(Ca)}$ is now shifted into the voltage chosen for control, the phenomenon will occur, as is shown for the cell R_{15} in Fig. 4, in which case the shift might be as much as 100 mV. The importance for this shift of the maximum of $I_{K(Ca)}$ might be that the voltage range approaches the voltage range of the peak of an action potential and is therefore important for the repolarization of the membrane following a spike.

As has been described, the faster activation of the outward current during repetitive activation has been observed by Heyer and Lux (12) following TEA treatment and by Ulbricht and Wagner (27) following 4-AP treatment. Does strychnine have a TEA-like action? This has been postulated recently by Shapiro (23) for the effects of strychnine on the outward current in the Ranvier node, in which doublets do not occur.

As has been shown (12), TEA does not shift the maximum of $I_{K(Ca)}$ to lower potentials in Helix neurons. Experiments on Aplysia neurons showed that TEA added in concentrations of 50 mM to the ASW not only did not change the voltage range of $I_{K(Ca)}$ but instead reduced and blocked nearly equally both components of the outward current. The facilitating effect on the speed of the outward current during repetitive activation, as well as the N-shape in the I/V curve, were reduced and blocked regardless of the duration of the depolarization. From our experience, only benzodiazepines can also induce doublet discharges (Hoyer et al., *this volume*).

The shift of the maximum $I_{K(Ca)}$ has been shown by Meech and Standen (20) following changes of the outside calcium concentration. As this concentration was not changed, one must speculate whether a change in the inside calcium concentration could be responsible. As has been shown by Heyer and Lux (12), increasing the calcium concentration inside the neuron by pressure injection of Ca^{2+} ions depressed rather than enhanced the outward current. Strychnine reduces both components of the fast inward current, i.e., the sodium and the calcium current components. This reduced permeability toward calcium might be the explanation for the data presented. Similarly, the action of mephenesin can be explained: By blocking the inward current completely, Ca^{2+} has no facilitating effect on the potassium current. Why ethanol, although it reduces the inward current substantially, has no effect on $I_{K(Ca)}$ cannot be answered. In conclusion, the effect of strychnine seems to be rather different from that of TEA on Aplysia neurons; more data about the effect of TEA on $I_{K(Ca)}$ are necessary.

That neither strychnine nor PTZ induces bistable states in Aplysia neurons when given alone suggests that the remaining outward current, probably $I_{K(Ca)}$, is able to guarantee a sufficient repolarization. Only when, in addition, $I_{K(Ca)}$ is blocked by cobalt or reduction of the calcium concentration of the ASW can bistable states be evoked by strychnine (18). The stable points of the membrane potential coincide with those being registered simultaneously in voltage recordings. These data (14) agree with those of Ducreux and Gola (4) from Helix neurons under PTZ. The data presented stress the importance of the movement of calcium ions across the membrane for the generation of phenomena related to abnormal "convulsive" activity of neurons.

ACKNOWLEDGMENTS

The authors would like to thank Dr. M. R. Park and Dr. W. Daunicht, Ch. Huber, M. Duesmann, E. Park, and H. Thomas for help during the experiments, for the preparation of the manuscript, and for photography.

REFERENCES

1. Arvanitaki, A., Chalazonitis, N., and Otsuka, M. (1956): *C. R. Acad. Sci.*, 243:307–309.
2. Bergmann, M. C., Klee, M. R., and Faber, D. S. (1974): *Pfluegers Arch.*, 348:139–153.

3. Clusin, W., Spray, D. C., and Bennett, M. V. L. (1975): *Nature,* 256:425–427.
4. Ducreux, C., and Gola, M. (1975): *Pfluegers Arch.,* 361:43–53.
5. Faber, D. S., and Klee, M. R. (1974): *Brain Res.,* 65:109–126.
6. Faber, D. S., and Klee, M. R. (1976): *Brain Res.,* 104:347–353.
7. Faber, D. S., and Klee, M. R. (1977): In: *Alcohol and Opiates,* edited by K. Blum, pp. 41–63. Academic Press, New York.
8. Faugier-Grimaud, S. (1974): *Brain Res.,* 69:354–360.
9. Faugier, S., and Willows, A. O. D. (1973): *Brain Res.,* 52:243–260.
10. Gola, M. (1974): *Pfluegers Arch.,* 352:17–36.
11. Gola, M., and Romey, G. (1971): *Pfluegers Arch.,* 327:105–131.
12. Heyer, C. B., and Lux, H. D. (1976): *J. Physiol.,* 262:349–382.
13. Klee, M. R. (1975): In: *Subcortical Mechanisms and Sensorimotor Activities,* edited by T. L. Frigyesi, pp. 89–120. Huber, Bern.
14. Klee, M. R. (1976): In: *Neurobiology of Invertebrates. Gastropoda Brain,* edited by J. Salánki, pp. 267–286. Akademiai Kiado, Budapest.
15. Klee, M. R., and Faber, D. S. (1974): *Pfluegers Arch.,* 346:97–106.
16. Klee, M. R., and Faber, D. S. (1974): In: *Central Rhythmic and Regulation,* edited by W. Umbach and H. P. Koepchen. Hippocrates Verlag, Stuttgart.
17. Klee, M. R., Faber, D. S., and Heiss, W.-D. (1973): *Science,* 179:1133–1136.
18. Klee, M. R., and Heiss, W.-D. (1969): *Electroencephalogr. Clin. Neurophysiol.,* 27:683–684.
19. Matsumoto, H. (1964): *Electroencephalogr. Clin. Neurophysiol.,* 17:249–307.
20. Meech, R. W., and Standen, N. B. (1975): *J. Physiol.,* 249:211–239.
21. Neher, E., and Lux, H. D. (1971): *Pfluegers Arch.,* 322:35–38.
22. Rojas, E., Bezanilla, F., and Taylor, R. E. (1970): *Nature,* 225:747–748.
23. Shapiro, B. I. (1972): *J. Gen. Physiol.,* 69:897–914.
24. Speckmann, E. J., and Caspers, H. (1973): *Epilepsia,* 14:357–408.
25. Stevens, C. F. (1969): In: *Basic Mechanisms of the Epilepsies,* edited by H. H. Jasper, A. Pope, and A. A. Ward, pp. 76–82. Little Brown, Boston.
26. Thompson, S. H. (1977): *J. Physiol.,* 265:465–488.
27. Ulbricht, W., and Wagner, H.-H. (1976): *Pfluegers Arch.,* 367:77–87.
28. Williamson, T. L., and Crill, W. E. (1976): *Brain Res.,* 116:217–229.
29. Williamson, T. L., and Crill, W. E. (1976): *Brain Res.,* 116:231–249.

Abnormal Neuronal Discharges, edited by N. Chalazonitis and M. Boisson. Raven Press, New York © 1978.

Changes in Ionic Currents Associated with Flurazepam-Induced Abnormal Discharges in Aplysia Neurons

J. Hoyer, *M. R. Park, and *M. R. Klee

*Institut für Allgemeine und Vergleichende Physiologie der Universität Wien, Vienna, Austria; and *Max-Planck-Institut für Hirnforschung, Neurobiologische Abteilung, Frankfurt am Main, Federal Republic of Germany*

Studies have been made in a number of laboratories in an attempt to determine the basic events of seizure discharge, as well as the possible mechanisms of action of anticonvulsant agents in the relatively simply structured and electrophysiologically accessible neurons of certain molluscs. Yet for every investigator of invertebrate preparations who seeks a more detailed understanding of mechanisms, the question is ever present whether his findings have a significance in regard to the clinic. The data may not fit with known facts of the mammalian system. For example, diphenylhydantoin is claimed to exhibit its anticonvulsant activity in reversing a pentylenetetrazol (PTZ)-induced negative resistance in *Aplysia* neurons (15), although it is known that diphenylhydantoin is not an antagonist of PTZ seizures in mammals (26). On the other hand, the actions of a substance may be manifold, providing no clue to which of the actions, if not all, are the consequent ones. Thus the barbiturates have been shown to hyperpolarize all types of oscillating cells, normal or abnormal (2), whereas a selective depression of excitatory postsynaptic potentials (EPSP) is claimed by Barker and Gainer (1) to be the mechanism of their action.

Similarly, divergent findings exist concerning the mode of action of the benzodiazepines. Studies to date have focused almost entirely on benzodiazepine effects on synaptic transmission. The benzodiazepines are accused of interfering in some way with a large number of known or putative transmitters. They are reported to act on GABA receptors, synergistic to GABA (9) as well as antagonistically (7,22). An action on glycine receptors is postulated (27) and also questioned (4). Interactions with serotonin, dopamine, acetylcholine, and norepinephrine are reported (see Costa and Greengard, ref. 3, for survey). Only little attention has been paid to the effects of the benzodiazepines on the nonsynaptic excitable membrane itself, this being a conceivable central point of action that could have at least some of the above-mentioned synaptic actions as a consequence.

Liebeswar (17) has already demonstrated a reduction of inward current in the action potential of papillary muscle fibers of the guinea pig heart. Besides the manifold results, the wide spectrum of benzodiazepine action in the clinic has also impeded the finding of clear answers, denying a clear conceptual framework for the investigator. The present study, therefore, is concerned with the elucidation of the actions of a benzodiazepine, flurazepam-HCl, on the nonsynaptic neuronal membrane. We believe that such groundwork is necessary as a first step in undertaking the study of the specific means by which the substance acts as an anticonvulsant.

Flurazepam-HCl was used in the present study because of its good water solubility compared to that of other benzodiazepines and because its actions are representative, in animal screening experiments, of the benzodiazepines (19). The electrophysiological effects of flurazepam-HCl were also found to be similar to those of other benzodiazepines in this preparation (12).

METHODS

Standard current and voltage-clamp techniques were used for recordings from neurons of the visceral ganglion of *Aplysia californica* obtained from Dr. R. C. Fay, Pacific Biomarine Supply Co., Venice, California. The nomenclature used is that of Frazier et al. (6). Flurazepam-HCl was added to artificial seawater in concentrations of 0.1 to 0.5 mM. The artificial seawater used was Instant Ocean at pH 7.8 to 8.0 which was aerated and cooled (16 ± 1°C) during perfusion. A complete report of this work, including a detailed description of the methods, is in preparation (14,16).

RESULTS

Discharge Pattern

Flurazepam-HCl changes the discharge pattern of all the cells investigated in this study. The nature of these changes follows the spontaneous firing pattern in control conditions with a more or less pronounced tendency to form bursts of multiple discharges in all types of cells. The silent cells, R_2 and the left pleural ganglion giant cell, are changed in their behavior into cells with slow oscillations of the resting membrane potential (Fig. 1), exhibiting a discharge pattern similar to bursting pacemaker cells. Cells with endogenous bursting behavior (e.g., L_3 in Fig. 1) have, under the influence of flurazepam-HCl, an increase in the duration and amplitude of their oscillations. The increased duration of the depolarizing wave triggers more action potentials per oscillation, except that after longer periods of perfusion the number of action potentials begins to decrease. At that time, the action potential is extremely prolonged. Regular cells (e.g., R_3 in Fig. 1) become irregular in their firing pattern. This type of cell is the least prone to

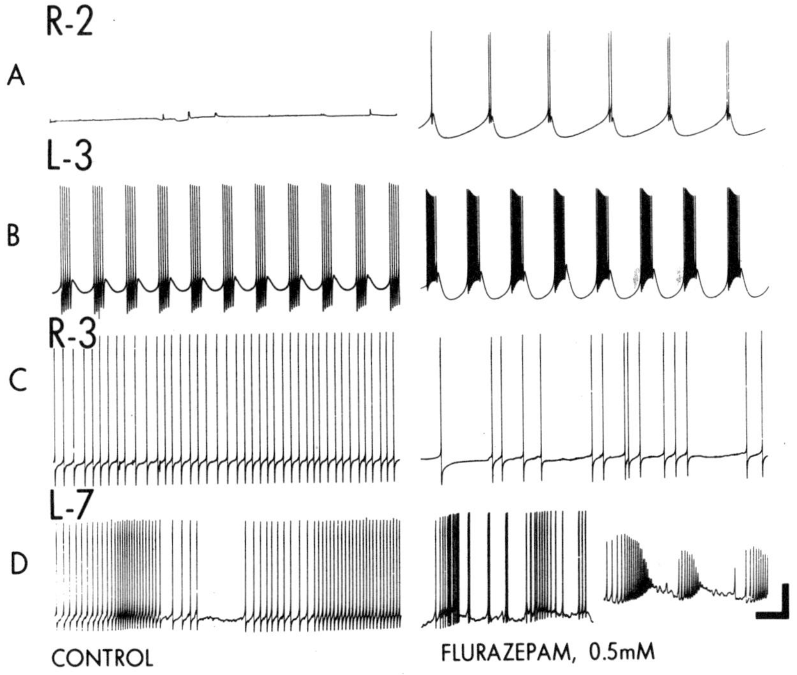

FIG. 1. Alterations in the four types of discharge patterns following flurazepam-HCl treatment. **A:** Silent cell R_2 begins spontaneous bursting. **B:** Oscillations in membrane potential of the spontaneously bursting cell L_3 are enhanced. **C:** Regular cell R_3 becomes irregular. **D:** Irregular cell L_7 remains irregular but with a tendency to form bursts (*lower right*). Cal: 20 mV, 15 sec.

exhibiting double or multiple discharges under the influence of flurazepam-HCl. Irregular cells (e.g., L_7 in Fig. 1) show a tendency to form bursts or multiple discharges. After a prolonged period of drug perfusion, a bursting pattern consisting of a series of action potentials reduced in amplitude and prolonged in duration together with abortive spikes remains (Fig. 1D, right trace).

Action Potential

In action potentials having in control conditions no discernible pause in repolarization, flurazepam-HCl reduces the rising and falling phases about equally. The result is a prolonged action potential of reduced amplitude. In those cells whose action potential in control has a biphasic repolarization, seen as a break and shoulder in the falling phase, the second part of the repolarization is more sensitive to reduction by flurazepam-HCl than the rising phase and first part of the falling phase. The effect of the drug on this second type of action potential results in a reduced amplitude, slower depolarizing phase, and slow partial repolarization, which is then followed by a plateau lasting for several hundred milliseconds, in some cases with oscillations superimposed on the plateau. The plateau caused by the strong antagonism of flurazepam-HCl to the second part of repolarization results in a figure reminiscent of paroxysmal depolarizing shifts.

Membrane Currents

Two sets of voltage-clamp data have been obtained from these neurons. The current-voltage relationship of the slow currents, which determines the resting potential of the cell and the degree to which it is stable, is conveniently studied after Gola (8), using slow de- and hyperpolarizing ramp voltage commands. Certain cells (e.g., R_2) develop with flurazepam-HCl a negative slope in their I-V curve (Fig. 2), which is described by several authors as an essential condition for endogenous oscillation (5,8,21). In cells already exhibiting this negative slope under control conditions, such as R_{15}, this property is accentuated by flurazepam-HCl.

The relatively more transient and much larger currents that flow during an action potential were studied under voltage-clamp conditions by the standard method of using depolarizing step commands of different amplitude from a holding potential chosen near the resting membrane potential, i.e., −40 to −50 mV. Both inward and outward currents are reduced by flurazepam-HCl. As previously reported (13), the inward current is shifted to more positive voltages. To distinguish the effect of flurazepam-HCl on the two different outward current systems (see Discussion), the I-V curves for outward currents were measured in normal and O-Ca^{2+} seawater. With flurazepam-HCl, there is a fast inactivation of the outward current (most prominent with strong depolarizations), which remains in O-Ca^{2+} solution (Fig. 4, 2A versus 2B). Using depolarizing voltage-clamp commands of longer duration (more than 100 msec) and high amplitude (up to approximately +150 mV), a nonlinearity in the current-voltage curve for the outward current can be observed (Fig. 4A). This nonlinear, N-shaped I-V characteristic is described for other invertebrate neurons (11,18,24), as well as for strands of guinea pig myometrium (25), as being dependent on an inward flux of Ca^{2+} ions and has been shown to vary in proportion to the extracellular Ca^{2+} concentration (18).

The voltage-clamp data are presented as I-V curves in Fig. 5. Measuring the outward current with respect to the time after the beginning of the depolarizing command, the amount of reduction of the outward current by flurazepam-HCl is seen to depend on the time chosen for analysis (Fig. 5A). The small depression after 6 msec compared to the strong depression after 36 msec reflects the fast inactivation of the outward current. To avoid divergent results by measuring at different times, the I-V curves in Fig. 5B and C are obtained by integration of outward current over the whole

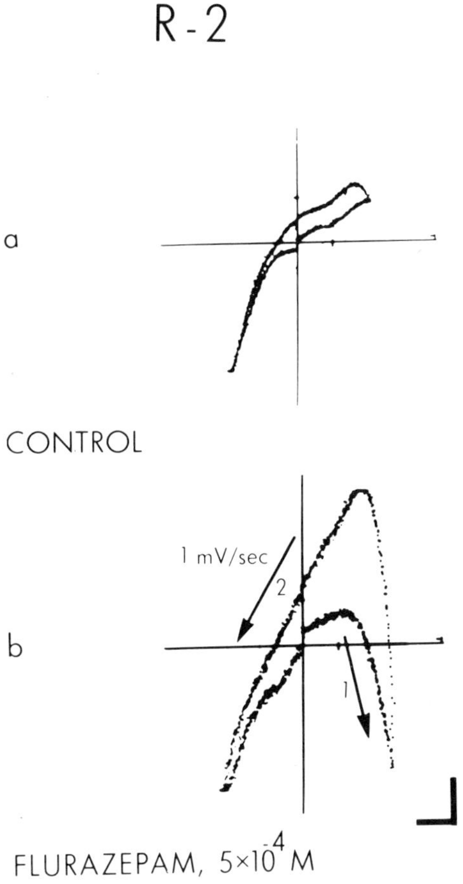

FIG. 2. Induction of a region of negative slope in the I-V curve for slow currents. Current (*ordinate,* inward is down) is a function of membrane potential (*abscissa,* depolarization to the right) in this voltage-clamp experiment on the giant cell R_2. Ramp voltage commands, first depolarizing, then hyperpolarizing (limbs *1* and *2* are so indicated for the flurazepam-HCl record, **b**) were used, from a holding potential of −35 mV. Cal: 1 mV, 10 nA.

time of the depolarizing command, 160 msec. Figure 5B demonstrates the I-V relationship under control conditions (which is N-shaped) and in O-Ca^{2+} together with the influence of flurazepam-HCl on the outward current under these conditions. The differences between outward current in control and in O-Ca^{2+} medium and the difference between the outward current reduced by flurazepam-HCl, again, in seawater of normal ionic composition and O-Ca^{2+}, are shown in Fig. 5C. The maximum of the Ca^{2+}-dependent outward current (the peak values) (Fig. 5C) is shifted in flurazepam-HCl solution from +130 to +50 mV. Likewise, the extrapolated

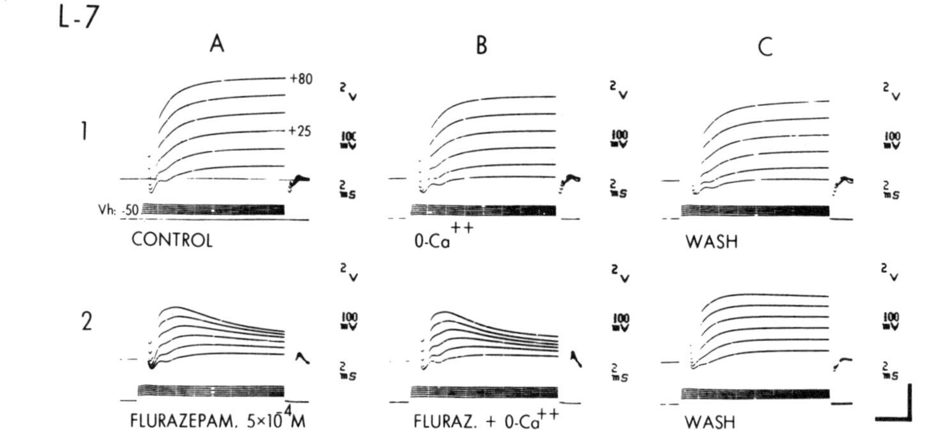

FIG. 3. Effects of flurazepam-HCl on the outward currents in voltage clamp. Voltage commands of +40, 60, 80, 100, 120, and 140 from a holding potential of −50 mV were used. The experiment proceeded from left to right and from top to bottom. The fast inactivation, within 3 msec, of the outward current caused by flurazepam-HCl is independent of any effect of Ca^{2+}. Cal: 0.4 V, 2 μA, 4 msec.

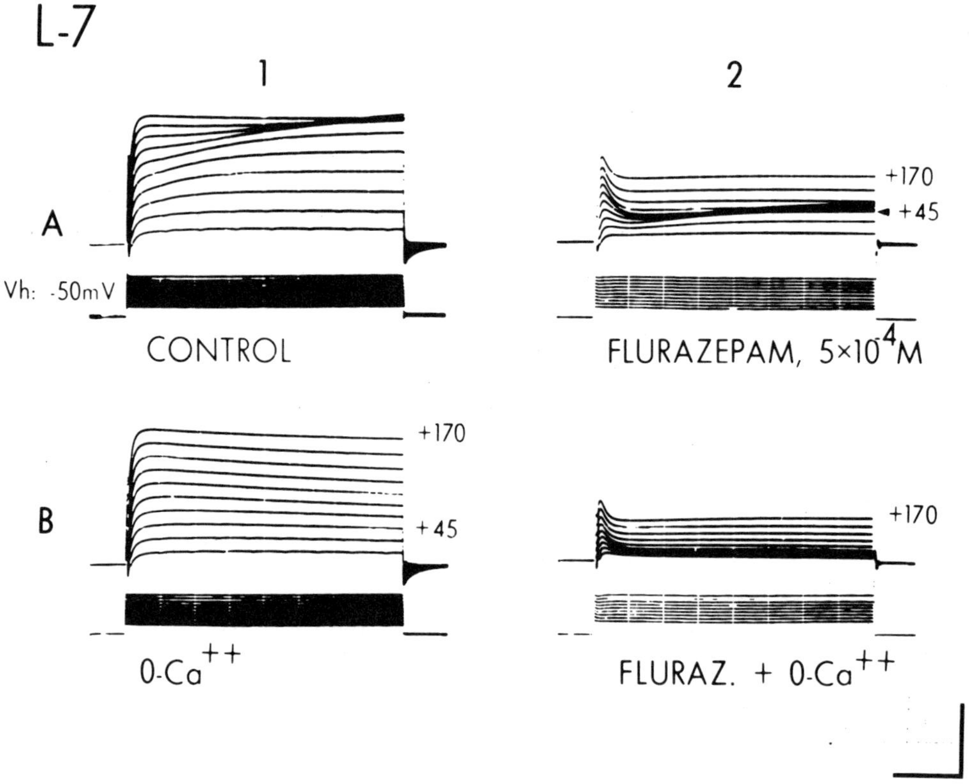

FIG. 4. Effects of flurazepam-HCl on the outward currents. Same cell as Fig. 3. Higher commands, 40, 60, 80, 100, 120, 140, 160, 180, 200, 220, reveal the Ca^{2+} dependence of the outward current as a crossing of the current traces (**A1** and **2**). This nonlinearity is absent in O-Ca^{2+} seawater (**B1** and **2**). Cal: 40 msec, 400 mV, 2 μA.

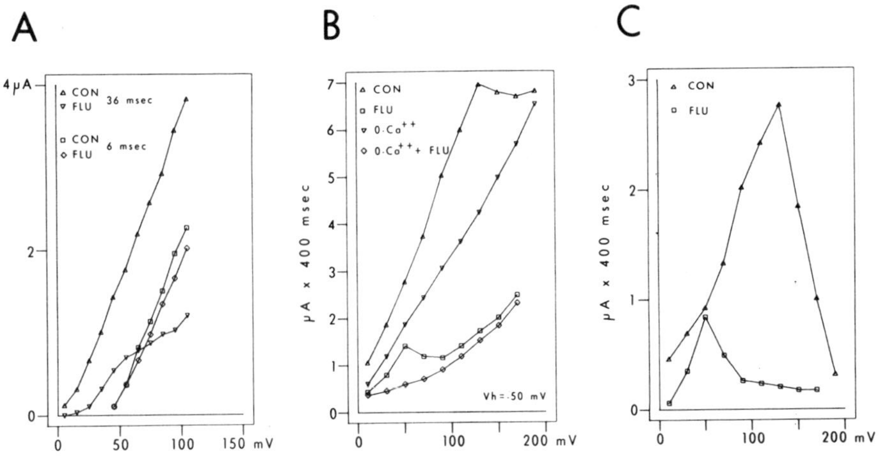

FIG. 5. Flurazepam-HCl effects on the two outward current systems. **A:** Amplitude of net outward current measured 6 msec after the beginning of a depolarizing step command is hardly affected by perfusion by flurazepam-HCl, 0.5 mM. The same measurement 36 msec later shows the sensitivity to flurazepam-HCl. The cell is an L_3. **B** and **C:** Made from the L_7 portrayed in Figs. 3 and 4. **B:** N-shaped I-V curve can be seen in seawater with normal ionic content with and without flurazepam-HCl (truncated in the former case). Its lack in the O-Ca^{2+} curves, with and without flurazepam-HCl, reflects the absence of a Ca^{2+}-mediated K^+ current. The difference between the curves of O-Ca^{2+} and normal ionic strength results in the plot of C, a representation of the I-V relationship of the Ca^{2+}-mediated K^+ current. Important in this figure is the shift of this current to lower voltages.

point at which the Ca^{2+}-mediated outward current is null shifts from +195 to +115 mV, also 80 mV. At +50 mV potential, the ratio between the voltage-dependent and the Ca^{2+}-mediated K^+ current changes from about 2:1 to 1:1.5 under the influence of flurazepam-HCl.

DISCUSSION

The discharge patterns and alterations of the action potential, which we have been able to relate to voltage-clamp-determined changes in transmembrane currents, are similar to phenomena observed in mammalian neurons. For example, a tendency to discharge in bursts or groups of action potentials, rather than to discharge continuously, is reported by Steiner and Hummel (23) in hippocampal and lateral geniculate cells of the cat. The occurrence of double and multiple discharges in single fibers of the optic tract of cats under the influence of diazepam is described by Heiss et al. (10). In *Aplysia* neurons, the tendency to form bursts can be accounted for by the appearance, attributable to flurazepam-HCl, of a region of negative slope on the I-V curve for slow currents.

The reduction of inward current we observed is in accordance with Liebeswar's (17) presumption of reduced inward current in guinea pig papillary muscles, which he based on data from action potential recordings.

The shift of maximal inward current to higher voltages (12) indicates an increase in firing threshold. The effects on the outward currents (reduction and fast inactivation) are an explanation of the increased duration of the action potentials. It is well established that the late outward current consists of two components (11,18): a Co^{2+}-resistant, tetraethylammonium (TEA)-sensitive, voltage-dependent current known from axonal membranes, while soma membranes exhibit, in addition, a TEA-resistant, Ca^{2+}-dependent outward current. The latter component is responsible for the N-shaped I-V curve for outward current in the range of +50 to +200 mV. By using O-Ca^{2+} seawater, the I-V plot for the Ca^{2+}-dependent outward current may be determined by taking the difference between the normal and O-Ca^{2+} curves. In our experiments with *Aplysia,* this shows a maximum around +130 mV and an extrapolated null crossing potential, an indication of the Ca^{2+} equilibrium potential, of +180 mV. Meech and Standen (18) have shown that a 10-fold change in the external Ca^{2+} concentration shifts the null point and peak potential by 30 mV. As shown in Fig. 5B and C, flurazepam-HCl has a dual action on the late outward current. The peak potential of the Ca^{2+}-dependent outward current is shifted to a lower voltage (+50 mV). Both the peak and extrapolated null point are shifted by about 80 mV. At the same time, the Co^{2+}-resistant outward current, isolated in O-Ca^{2+}, is reduced by two-thirds. Therefore, the action of flurazepam-HCl on the outward currents can be described as a TEA-like action which might be expressed by the fast inactivation of outward current at commands of higher voltage (Figs. 3 and 4) and the effect of a reduced outward Ca^{2+} concentration.

An effect of flurazepam-HCl on Ca^{2+} processes could explain the findings of other groups on synaptic mechanisms. Thus the diverse effects of benzodiazepines on the various transmitter systems would be plausible. An increase in the facilitation of an EPSP by flurazepam-HCl (14) matches the effects of low Ca^{2+}, high Mg^{2+}, or Co^{2+} ions on the same EPSP (20).

SUMMARY

Current and voltage-clamp techniques were used while recording from *Aplysia* neurons under the influence of 0.1 to 0.5 mM flurazepam-HCl, a benzodiazepine easily dissolved in water. Flurazepam-HCl causes changes in the action potential itself, as well as alterations in the pattern of discharge. The cells investigated change their pattern of spontaneous firing in a reproducible manner: silent cells into bursting, regular into irregular, irregular into irregular with burst tendency; bursting cells remain bursting but with increases in amplitude and duration of each period. For the action potential, several parameters are affected by flurazepam-HCl. There is an increase in action potential duration. The overshoot, the maximum rate of rise, the rate of repolarization, and the afterhyperpolarization are all reduced.

Voltage-clamp experiments corroborate these observations. Related to firing pattern, flurazepam induces a negative slope and a pronounced anomalous rectification in the I-V curve (this in cells which are changed from silent into bursting). Corresponding to action potential changes, flurazepam-HCl causes a decrease in both inward and outward currents. Additionally, using depolarizing step commands to the region of +150 mV for at least 100 msec, the N-shaped I-V curve for outward current is flattened. This reflects a reduction in total outward current and a shifting of the peak of Ca^{2+}-dependent K^{+} current to less positive voltages.

ACKNOWLEDGMENTS

This work was partly supported by a grant from the Österreichischen Fonds zur Föderung der wissenschaftlichen Forschung. Flurazepam-hydrochloride was a gift from Hoffmann-LaRoche.

The authors wish to thank M. Duessman, C. Huber, E. Park, G. Thalhammer, and H. Thomas for their assistance.

REFERENCES

1. Barker, J. L., and Gainer, H. (1973): *Science,* 182:720–722.
2. Chalazonitis, N. (1977): *This volume.*
3. Costa, E., and Greengard, P., Eds. (1975): *Advances in Biochemical Psychopharmacology, Vol. 14: Mechanism of Action of Benzodiazepines.* Raven Press, New York.
4. Curtis, D. R., Game, C. J. A., and Lodge, D. (1976): *Br. J. Pharmacol.,* 56:307–311.
5. David, R. J., Wilson, W. A., and Escueta, A. V. (1974): *Brain Res.,* 67:549–554.
6. Frazier, W. T., Kandel, E. R., Kupfermann, I., Waziri, R. E., and Coggeshall, R. E. (1976): *J. Neurophysiol.,* 30:1288–1351.
7. Gähwiler, B. H. (1976): *Brain Res.,* 107:176–179.
8. Gola, M. (1974): *Pfluegers Arch.,* 352:17–36.
9. Haefeli, W., Kulcsar, A., Möhler, H., Pieri, L., Polc, P., and Schaffner, R. (1975): In: *Advances in Biochemical Psychopharmacology, Vol. 14: Mechanism of Action of Benzodiazepines,* edited by E. Costa and P. Greengard, pp. 131–151. Raven Press, New York.
10. Heiss, W.-D., Heilig, P., and Hoyer, J. (1969): *Vision Res.,* 9:493–506.
11. Heyer, C. B., and Lux, H. D. (1976): *J. Physiol.,* 262:319–348.
12. Hoyer, J. (1977): *Pharmakopsychiatrie,* 10:271–280.
13. Hoyer, J., Klee, M. R., and Heiss, W.-D. (1976): In: *Neurobiology of Invertebrates. Gastropoda Brain,* edited by J. Salanki, pp. 253–265. Akadémiai Kiadó, Budapest.
14. Hoyer, J., Klee, M. R., Heiss, W.-D., and Park, M. R. *In preparation.*
15. Johnston, D., and Ayala, G. F. (1975): *Science,* 189:1009–1011.
16. Klee, M. R., Hoyer, J., Park, M. R., and Heiss, W.-D. *In preparation.*
17. Liebeswar, G. (1972): *Naunyn Schmiedebergs Arch. Pharmacol.,* 275:445–456.
18. Meech, R. W., and Standen, N. B. (1975): *J. Physiol.,* 249:211–239.
19. Randall, L. O., Schallek, W., Scheckel, C. L., Stefko, P. L., Banzinger, R. F., Pool, W., and Moe, R. A. (1969): *Arch. Int. Pharmacodyn. Ther.,* 178:216–241.
20. Schlapfer, W. T., Woodson, P. B. J., Tremblay, J. P., and Barondes, S. H. (1974): *Brain Res.,* 76:267–280.
21. Smith, T. G., Barker, J. L., and Gainer, H. (1975): *Nature,* 253:450–452.
22. Steiner, F. A., and Felix, D. (1976): *Nature,* 260:346–347.
23. Steiner, F. A., and Hummel, P. (1968): *Int. J. Neuropharmacol.,* 7:61–69.
24. Thompson, S. H. (1977): *J. Physiol.,* 265:465–488.
25. Vassort, G. (1975): *J. Physiol.,* 252:713–734.

26. Woodbury, D. M., and Esplin, D. W. (1959): *Proc. Assoc. Res. Nerv. Ment. Dis.*, 37:24–56.
27. Young, A. B., Zukin, S. R., and Snyder, S. H. (1974): *Proc. Natl. Acad. Sci. USA*, 71: 2246–2250.

Abnormal Neuronal Discharges, edited by
N. Chalazonitis and M. Boisson. Raven
Press, New York © 1978.

Unusual Properties of the Ca-K System Responsible for Prolonged Action Potentials in Neurons from the Snail *Helix Pomatia*

C. B. Heyer and H. D. Lux

Department of Neurophysiology, Max-Planck-Institute for Psychiatry, Munich, Federal Republic of Germany

The basic currency of long-distance information transfer in the nervous system is the classic all-or-none action potential. The frequency with which this event occurs marks the digital coding of such analog signals as receptor potentials and summations of graded postsynaptic potentials. Although it is true that information can be conveyed by the average frequency of firing, it is also likely that additional characteristics of spike trains will be important in determining postsynaptic responses (47). Specific temporal patterning is often obvious in naturally occurring discharges of action potentials (9,55). A gradual decrease in frequency (adaptation) characterizes many receptor processes, whereas some neurons typically fire in bursts of action potentials. That the postsynaptic response is sensitive to the fine structure of presynaptic activity has been shown in several preparations (12,48,52). Indeed, changes in activity patterns toward the grouping of spikes into bursts (for example, in epileptic discharges) may be a more powerful mechanism for increasing postsynaptic responses than simple alterations in average spike frequency.

A second potentially important element of variability involves parameters of the action potential itself. Changes in action potential duration can be expected to alter the efficacy of synaptic transmission in both electrical and chemical synapses (25). For chemical synapses, this is partly due to simple changes in activation of the relatively slowly rising calcium current mediating transmitter release (31). However, recent work (15,16,20,21,32,36) on bursting pacemaker neurons of *Helix pomatia* also suggests that spike elongation itself reflects unusual properties of conductances of both Ca and K ions.

Although the phenomenon of grouped discharges has received much attention, sufficient information on membrane conductance changes which could produce bursting behavior is not available from classic axon preparations. The squid giant axon, for example, has only a limited ability to fire repetitively, and then generally does so at rather constant frequencies (14,18,54). There is very little variability in action potential duration; even if the slow calcium current might also serve as a model for synaptic calcium currents, it is discouragingly small (4,5). In the study of presynaptic currents (31), the squid giant synapse has been most useful. Again, this is basically a relay synapse, and plasticity, although evident (22), is limited.

More recently, the soma membranes of bursting pacemaker molluscan neurons have been studied. Certain identifiable cells display relatively complex patterns in the generation of spikes and in the characteristics of action potentials during a burst (3,55). Within a single burst of spikes, there are progressive changes in the height (some increase) and duration (marked increase) of the action potentials. It is the purpose of this chapter to review more recent findings on bursting neurons in *Helix pomati* including also some results by Hofmeier and Lux (unpublished) which implicate a unique Ca-K system responsible for the production of patterned output and the progressive changes in action potential characteristics during such patterns.

METHODS

A primary tool for studying membrane currents is the voltage clamp. We have used a conventional two-electrode voltage clamp (electrodes C and D in Fig. 1) [details of the clamp circuit have been described (44)] to study membrane currents in the somata of neurons in the snail *Helix pomatia*. However, two properties of the soma preclude simple application of this technique. The first is a geometrical consideration. The axon extending from the soma provides a significant loading problem: it cannot be maintained isopotential with the soma membrane and therefore cannot be clamped with the soma (see Fig. 1) (16). Second, there may be significant nonuniformity in membrane properties. Variations in the proportion of Na and Ca in the action potential have been found between soma and axon membranes in these neurons (24,59). There is also evidence of membrane nonuniformity over the soma. For example, receptors for neurotransmitters are not distributed evenly on the soma surface (28,53). For these reasons, it is most useful to be able to isolate for study a small, more uniform patch of soma membrane. To do this, we used the patch-clamp technique (44). The data presented herein are for currents flowing across a small patch of membrane ($\leq$ 120 μm diameter, depending on the size of the patch electrode, electrode E in Fig. 1).

As we have shown previously (20,32,34), results from voltage-clamp and/or the better patch-clamp data are inadequate for determining changes in the flow of ions across these membranes. When there are large simultaneous inward and outward currents, as there are in many Helix neurons, the ability to record only changes in net current does not permit an increase in one to be differentiated from a decrease in the other. To make such a determination, a second and independent measure of at least one of the

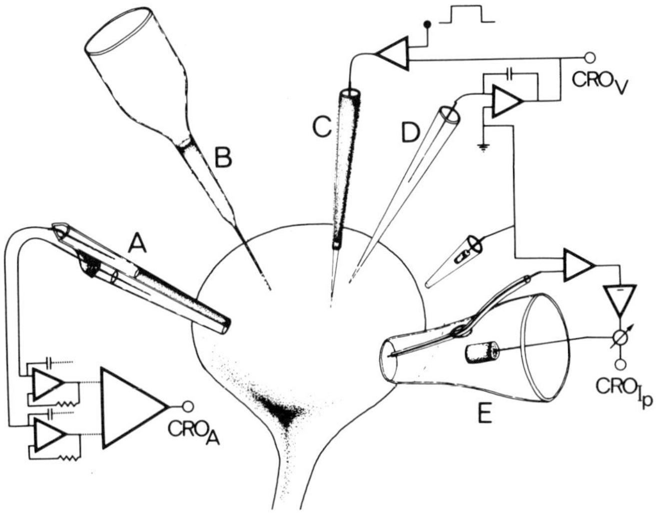

FIG. 1. Diagram of the experimental setup. On the surface of the neuron are two electrodes. **A:** Ion-sensitive electrode made from theta capillaries and consisting of an ion-sensitive side (upper barrel in figure, filled with an ion exchange resin selective for K or Ca ions), which records a signal proportional to the ion activity plus the extracellular potential, and a reference side (usually filled with NaCl). With appropriate circuitry, the reference signal is subtracted, leaving a potential representing the activity of selected ions (CRO_A). **B:** injection electrode for pressure (shown) or iontophoretic injection of one of a variety of substances. **C:** Shielded current injection electrode. **D:** Voltage-sensing electrode for a conventional intracellular voltage clamp (CRO_V). (See text for further details.) **E:** Patch-clamp electrode for measuring the current across an isolated patch of membrane (CRO_{IP}). In all experiments there were also either two or three intracellular electrodes.

mobile ions is necessary. We have employed several techniques for separately assaying inward or outward currents. Extracellular, double-barreled, ion-sensitive microelectrodes (e.g., electrode A in Fig. 1) were used in many experiments. These electrodes (made from theta capillaries) consist of an ion-sensitive side, whose response reflects both changes in the activity of certain ions at the tip and of the potential at that point, and a reference side for registering the potential. When the two signals are electrotonically subtracted, the remaining reading reflects the ion activity at the tip. Electrodes sensitive to changes in potassium (using the Corning resin 477317) (60) and calcium (using a Ca-sensitive resin) (46) were employed in different experiments. Response characteristics of these electrodes have been described (19,33,37,45).

Calcium-selective electrodes were used to measure decreases in extracellular Ca (Lux, *unpublished observations*) caused by Ca entering the cell (see Fig. 3B). The electrochemical potential which determines the responses of ion-sensitive electrodes is a logarithmic function of the change in ionic activity. The potential difference for equal concentration changes is smaller for ions of higher valency. This and the high concentration of Ca in the

Ringer solution (10 mmoles/1) reduces the resolution of the signal mediated by changes in Ca activity and estimates require many trials. The concentration of extracellular K (4.0 mM) is less than that of Ca; therefore, K-selective microelectrodes appear to be more sensitive to extracellular changes in K activity. The efflux of K was compared with total net currents under voltage clamp. A striking example of the application of this technique is shown in Fig. 2. There is clearly an increase in K efflux during stimulation, although the net voltage-clamp currents are inward. The error in determining absolute values for K efflux exceeds the variation expected for larger depolarizations. Therefore, we normally investigate changes in the ratio of K measurements and net current measurements. If there is no inward current, and for a single location of the K electrode, the ratio of measured K accumulation to net current should be the same for all stimulus paradigms. If there is an inward current of constant size, then the K/I ratio should not vary as a function of stimulus frequency. This is not true for many Helix neurons; the discrepancy, a deficit in net charge transfer, reflects changes in the magnitude of the inward current (32). Records in Fig. 3A show typical current and K efflux changes with repetitive stimulation. Although the net outward current declines with repetition, the K signal shows little decrease; a deficit of 12% for the second pulse and about 30% for the third pulse (both compared with pulse one) are calculated.

Changes in calcium activity can also be detected in the light output from the photoprotein aequorin. Therefore, we pressure injected neurons with aequorin and measured the light emission and net voltage-clamp currents under a variety of stimulus paradigms. [Techniques have been described (30,35). Pressure injection was accomplished through an additional elec-

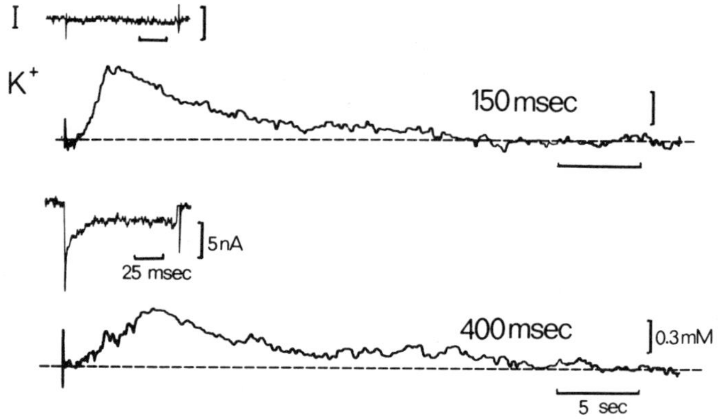

FIG. 2. Demonstration of voltage-dependent potassium efflux in the presence of net inward currents. Pulses (100 msec) were presented every 150 msec (*upper*) or 400 msec (*lower traces*); five trains of 10 pulses each were averaged. The net inward current (I) is much greater at the longer intervals; the K efflux (K^+) is not significantly different in the two trials.

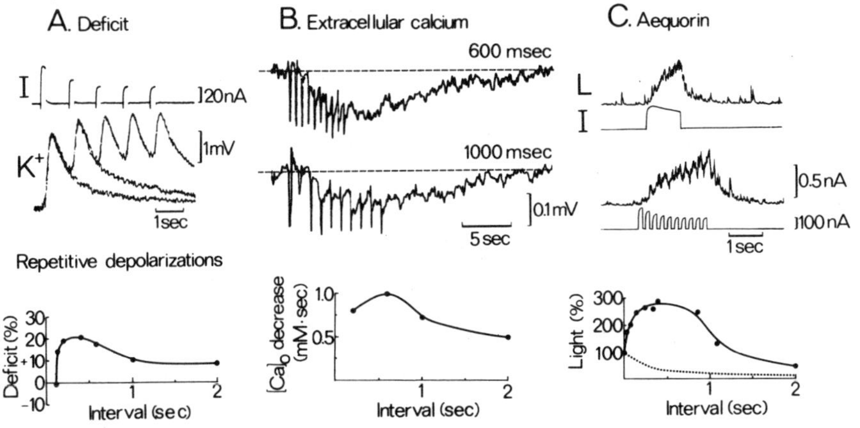

FIG. 3. Three methods for assaying a persistent slow inward Ca current in the presence of large outward currents. **A:** Deficit in net charge transfer is calculated by comparing the time integral of the net current (I) with the time integral of the increase in $[K]_o$. Compared with the first pulse, there are deficits of 12% for the second pulse and 30% for the third pulse in this example. When the sum of I and K for repetitive depolarizations is compared with I and K values for a continuous depolarization, the K/I ratio is greater for repeated pulses. The percentage increase in the deficit in net charge transfer is presented as a function of the interstimulus interval for repeated pulses (*lower graph*). **B:** Decrease in $[Ca]_o$ during repetitive depolarizations is measured for the sum of 10 pulses. Records of $[Ca]_o$ for pulses of 600 and 1,000 msec intervals are shown. The decrease in $[Ca]_o$ as a function of interstimulus interval is plotted. **C:** Outward current (I) and voltage-dependent increases in light emission (L) from neurons injected with the Ca-sensitive photoprotein aequorin in response to continuous (*upper traces*) depolarization are compared with the sum of repetitive (*middle traces*) depolarizations. *Lower graph,* increase in light produced by spaced depolarizations (expressed as a percentage of the light from a continuous depolarization) is presented as a function of stimulus interval. Dashed line indicates light output predicted by a simple model of constant Ca influx during the pulses and removal during interpulse intervals; data for B and C are from refs. 35, 36. (All voltage-clamp pulses are from −50 to 0 or +10 mV; the sum of responses to 10 100 msec pulses was used to determine frequency dependence in the three graphs. See text for further discussion.)

trode, electrode B, Fig. 1.] An example of the type of records obtained is seen in Fig. 3C.

A variety of other substances was injected either by pressure or iontophoretically (e.g., through electrode B, Fig. 1) in different experiments. These include [ethylenebis(oxyethylenenitrilo)] tetraacetic acid (EGTA; 250 mM adjusted to pH 7.5 with KOH), calcium chloride (1 to 100 mM; ± EGTA buffers; ± 66 to 100 mM KCl), tetraethylammonium chloride (TEA; 250 mM to 1 M; ± 66 mM KCl), and cesium chloride (100 mM). Details have been described (20,21).

BURSTING NEURONS

Many neurons in molluscan ganglia show distinctive patterns of spike activity. Perhaps the best known of these is R_{15}, the parabolic burster of

Aplysia (3,55). As the name implies, this neuron normally fires bursts of action potentials; when the intervals between spikes are plotted versus time, the result often describes a parabola. Burst generation is endogenous to the soma membrane and continues even after mechanical isolation of the soma from its axon and the remainder of the ganglion (1).

We have studied the R_{15} homolog in *Helix pomatia* (27,49), which we have designated the "fast burster" to differentiate it from a neuron usually located next to it and showing a slower burst pattern, the "slow burster." Behavior of the two neurons is summarized in Fig. 4. In Fig. 4A an intracellular record from a fast burster (upper trace) shows the pattern of bursts and deep postburst hyperpolarizations characteristic of this cell. The lower traces in Fig. 4A are the extracellular currents from this cell and the nearby slow burster monitored simultaneously by the patch-clamp electrode. The frequency of action potentials within a burst and the frequency of bursting are less for the slow burster. Within a burst, the action potentials become increasingly longer in the fast burster (up to 2 to 3 times the half-width of the first spike), and the interspike intervals undergo a characteristic modulation (superimposed upper traces, Fig. 4B). The variation in action potential duration is much less for the slow burster, and marked shoulders do not occur on the spikes (the lower trace in Fig. 4B shows the typical spread in action potential duration during a burst in the slow burster). During single voltage-clamp pulses, the net outward current decays. The decay is more

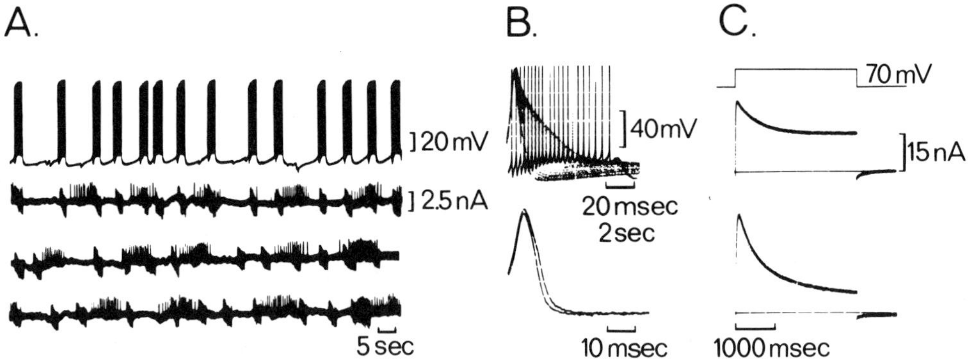

FIG. 4. Characteristics of the fast and slow bursters from *Helix pomatia.* **A:** Intracellular record from the fast burster (*upper trace*) and continuous extracellular fields (*lower traces*) recorded with the patch-clamp electrode simultaneously monitoring the activity of the fast and slow bursters in a single preparation. **B:** Typical increases in spike duration during a burst in the fast (*upper traces*) and slow (*lower traces*) bursters. A burst recorded at slower speed is superimposed for the fast burster showing the typical pattern of interspike intervals and postburst hyperpolarization; only the shortest and longest spikes are shown for the slow burster. **C:** Responses to sustained (3 sec) voltage-clamp pulses (*upper trace*) in the fast (*middle trace*) and slow (*lower trace*) bursters.

rapid and the relative decrease greater for the outward currents of the slow burster (Fig. 4C). Additional differences are discussed below.

Several membrane currents have been associated with the phenomenon of bursting, both as a normal behavioral pattern and as a response to drug applications or temperature changes (16,17,58,61). We review here some of the properties of currents in bursting pacemaker neurons of *Helix pomatia* which should contribute to the production of bursts and the variations of action potentials within bursts.

CURRENTS FOUND IN BURSTERS

Inward Currents

During single voltage-clamp pulses (e.g., from holding potentials of −50 to +10 mV), there is an initial inward current which rapidly turns into an outward current, the latter decaying with sustained depolarization. A complex pattern of inward and outward tail current follows repolarization. The short duration of the net inward current with large depolarizations does not reflect its true time course. With low levels of depolarization, a noninactivating inward current ($I_{in\ slow}$) has been identified in these neurons (15,16). Current and potassium records in Fig. 2 reveal the persistent net inward current. In this particular example, part of the potential-dependent outward current (see below) has been blocked by intracellular pressure injection of cesium ions. Similar records, however, are obtained in untreated cells. The net inward current does not inactivate completely during 100 msec pulses (from −50 to 0 mV in Fig. 2), despite the clear activation of outward K currents revealed by the K-electrode recording.

The presence of a persistent inward current during stronger depolarizations was first inferred from the deficit in net charge transfer calculated by comparing magnitudes of K efflux and net currents. An example of typical recordings is shown in Fig. 3A. With repetitive stimulation, the proportion of inward current (i.e., the deficit) increases (see details in Methods). The absence of a net charge transfer deficit in the presence of blockers of Ca currents (e.g., Co) and the persistence of a deficit with large depolarizations, reaching a minimum at membrane potentials of about +130 mV, suggested that $I_{in\ slow}$ was primarily carried by Ca ions. The primary role of Ca has been confirmed by demonstrations that the increase in intracellular Ca measured as light emission from cells injected with the photoprotein aequorin, and decreases in extracellular Ca measured with ion-selective electrodes, have time and voltage relationships similar to those of the deficit current (see Fig. 3B, C).

Two properties of the slow inward Ca current seem especially important for understanding the behavior of the fast burster. This current does not inactivate completely with sustained depolarization (15,16,20). Indeed, even with holding potentials equivalent to the most hyperpolarized levels of the interburst interval, there is a significant inward current (16). This would presumably provide a depolarizing drive throughout the burst.

With repetitive voltage-clamp depolarizations, a second property of the inward Ca current is revealed: the magnitude of the Ca current is a function of the stimulus interval. In Fig. 2, the net inward current for 400 msec stimulus intervals is more than four times larger than that for the shorter intervals. The K efflux is virtually identical for the two frequencies. Therefore, the increased inward current represents a true increase in the Ca current. Measurements of the deficit in net charge transfer show that total magnitude of the inward current increases with increasing intervals between the depolarizing pulses up to some intermediate value (e.g., up to about 400 msec intervals between pulses for 100 msec pulses from -50 to $+10$ mV) and then decreases with further interval increases (e.g., to the steady state for intervals exceeding 2 sec). The summed inward current for widely spaced depolarizations exceeds the current calculated for sustained depolarizations of the same total duration. This may be due in part to some inactivation of inward currents during continuous depolarization. However, the greater magnitude inward current at intermediate intervals of depolarization must thus reflect a facilitation of $I_{in\ slow}$. The typical frequency dependence of calculated deficits is shown in Fig. 3A.

Facilitation of an inward Ca current is also obvious from aequorin data (see Fig. 3C). Records of decreasing extracellular Ca reveal a similar pattern. In Fig. 3B the measurement of $[Ca]_0$ during two paradigms of the depolarizing pulses (100 msec pulses from -50 to $+10$ mV at 600 and 1,000 msec intervals) again shows the greater influx of Ca at intermediate compared with long interstimulus intervals. A maximum Ca influx at intermediate frequencies is thus evident from all three types of measurement.

Outward Currents

Potassium efflux can result from the activation of at least four distinct conductances in snail neurons. The fast outward potassium current has been described in detail (13,43). It inactivates quickly with depolarizations at the level of spike threshold but is probably active during the interspike interval in the slow burster. The contributions of the fast outward current to bursting itself may be negligible, however, since it is largely absent in the fast burster. The fast outward current of some molluscan neurons is affected more than others by extracellular 4-aminopyridine (57).

Delayed rectification consists of two components (21,41,57). One is primarily voltage dependent; the second needs Ca influx initiated by depolarization. It is possible to separate these two components pharmacologically. The voltage-dependent potassium current remains when the Ca current is blocked and displays many properties typical of other membrane currents (21). This K current ($I_{K(V)}$; $I_{Co\text{-}res}$) (21) is activated and inactivated by depolarizing voltage-clamp steps. Inactivation produced by depolarization is removed as a function of repolarization interval. Thus with repetitive depolarizations, the total outward current increases with increasing interstimulus intervals. This response is demonstrated by currents from the Co-treated cell in Fig. 5B. The I-V plot for Co-treated cells shows a monotonically increasing outward current with increasing membrane depolarizations. Measurements with K electrodes indicate that the frequency and voltage dependence of K efflux is similar to that of net outward currents (see Fig. 5). $I_{K(V)}$ is selectively blocked by intracellular (21) or extracellular (57) TEA ions and partially by intracellular cesium ions (35). It resembles the delayed rectifiers in such preparations as the squid giant axon. The responses of the outward current in Co-treated cells is quite similar to the behavior of the slow burster under voltage clamp (e.g., see Figs. 5 and 11 in ref. 21), suggesting that these are characteristics of a normally occurring membrane current and not some pharmacological action of Co Ringer.

The second major component of delayed rectification is activated by Ca influx and depressed by intracellular Ca. It therefore has voltage- and frequency-dependent properties unlike those described for other membrane currents. The I-V plots for many Helix neurons have an N-shape (20,21,41). This N is also seen in the relationship between K efflux and membrane potential (see Fig. 6A). Blocking Ca currents eliminates a component of the K current ($I_{K(Ca)}$) responsible for the hump of the N. That is, with in-

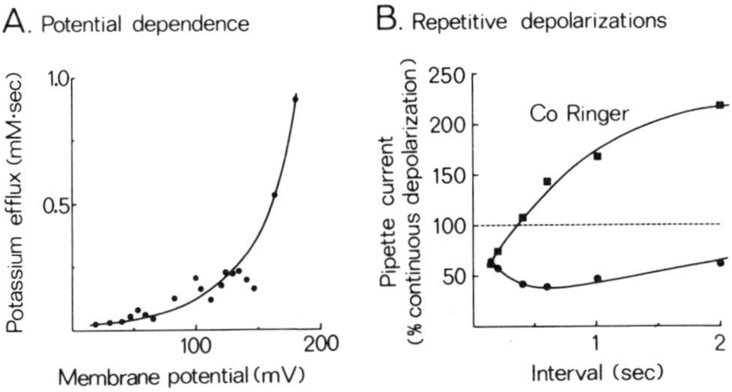

FIG. 5. Properties of the potential-dependent potassium ($I_{K(V)}$) revealed by treating a fast burster with Co Ringer. **A:** The N-shapes in I-V and K-V plots are abolished by Co Ringer, leaving a monotonic increase in K efflux with increasing membrane depolarization. **B:** Inactivation of $I_{K(V)}$ produced by depolarization is removed increasingly by longer repolarizations between pulses (*filled squares, upper curve*). This behavior is contrasted with the increasing depression of total current for increasing intervals up to about 500 msec in the neuron before Co treatment (*filled circles, lower curve*). Graph represents the sum of net outward currents with repeated pulses as a percentage of the total current during a sustained depolarization.

creasing membrane depolarizations, $I_{K(Ca)}$ (defined as that portion of K efflux abolished by Ca blockers, such as Co) increases to a maximum at potentials of about +50 to +70 mV and declines with further depolarization. An example of the potential dependence of $K_{(Ca)}$ is shown by the lower line in Fig. 6A. It should be noted that the difference curves for I-V (Fig. 8 in ref. 21) or K-V (Fig. 6A) plots before and during treatment with Co Ringer often do not go to zero at high potentials. Similar voltage-dependence curves

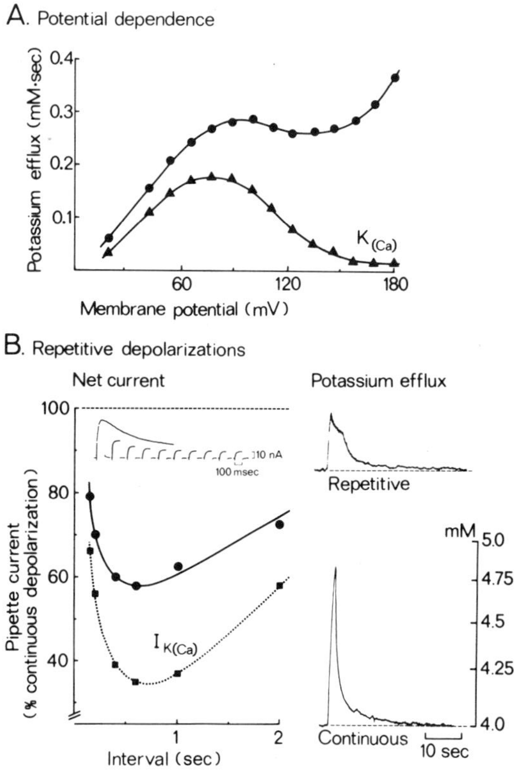

FIG. 6. Properties of the calcium-dependent potassium ($I_{K(Ca)}$). **A:** Ca-dependent K efflux (K_{Ca}) (*filled triangles*), i.e., that component of K efflux blocked by Co Ringer, increases and then decreases with increasing membrane potential; the difference curve does not go to zero. The potential dependence of the total K efflux (*filled circles, upper curve*) is included for comparison. **B:** Long-lasting depression of $I_{K(Ca)}$ with repetitive pulses is maximal at intervals of about 500 msec (*filled circles*) (tested for 10 100 msec pulses compared with a 1 sec pulse; −50 to +10 mV). This depression is obvious in the net current records, as shown in the inset to the graph. In this example, the short pulses were presented at 400 msec intervals and are compared with the current from a 1 sec depolarization. The depression represents a true decrease in K efflux as well as an increase in inward current; the K data on the right were measured for the same stimulus paradigms as the current inset on the left. There is less total K efflux during the continuous depolarization than there is for the sum of the repetitive depolarizations.

have been found for the Ca current measured as the deficit in net charge transfer (20) or with increases in Ca current implied from aequorin light emissions (35).

With repetitive depolarizing voltage-clamp pulses, a second characteristic of $I_{K(Ca)}$ becomes evident. When K efflux during sustained depolarizing pulses is compared with that during repeated short (100 msec) depolarizations of the same total duration, it is clear that for stimulus intervals of less than 2 to 5 sec the depression of $I_{K(Ca)}$ with repetitive pulses actually exceeds the inactivation with sustained depolarization. This depression is obvious in the depression of the outward current (Fig. 6B). The decrease in current does represent a true decline in K efflux (Fig. 6B, right column) as well as the increase in inward current. Indeed, when $I_{K(V)}$ is subtracted, the degree of depression is even greater (dotted lines in Fig. 6B representing $I_{K(Ca)}$.

The depression of $I_{K(Ca)}$ is caused by intracellular Ca. The frequency dependence of $I_{K(Ca)}$ depression is similar to, although not identical with, the frequencies of maximum Ca influx. Furthermore, the first and most reproducible effect of Ca injection (by pressure or iontophoresis, alone or with EGTA buffers) is to decrease the outward K current by selectively decreasing $I_{K(Ca)}$. As with Co treatment, increased $[Ca]_i$ from Ca entry during depolarization (R. Eckert and H. D. Lux, *unpublished observations*) or from Ca injection abolishes the hump in the I-V and K-V relationships (see Figs. 7A and 5B). Likewise, Ca injection can alter the frequency dependence of K efflux (Fig. 7B). Rapid removal of $[Ca]_i$ by injected EGTA buffers prevents the depression of K efflux at intermediate and long stimulus intervals without reducing the depression at short intervals and without abolishing the N in the K-V plot (Fig. 7C and D).

A fourth component of K efflux has been described for these neurons. Large Ca injections can cause an increased membrane conductance which becomes an increased K permeability (G. Hofmeier and H. D. Lux, *unpublished observations*). The response can be quite variable, often failing after more than one injection, and may depend on the metabolic state of the cell (36). It apparently consists of an increased instantaneous conductance and may correspond to the outward current remaining in Co Ringer after TEA injection (lower curve of Fig. 16 in ref. 21). This K conductance is differentiated from $I_{K(Ca)}$ by its lack of voltage dependence, its susceptibility to La^{3+} injection (36), and its probable behavior during a burst. $I_{K(Ca)}$ is depressed by repeated depolarizations; this decrease may contribute to spike elongation during the burst (see below). In marked contrast, the conductance increase caused by increased $[Ca]_i$ is proposed to reach its peak at the end of the burst, contributing to the postburst hyperpolarization (39). Thus the two currents show opposite time relationships within the burst and responses to increased $[Ca]_i$.

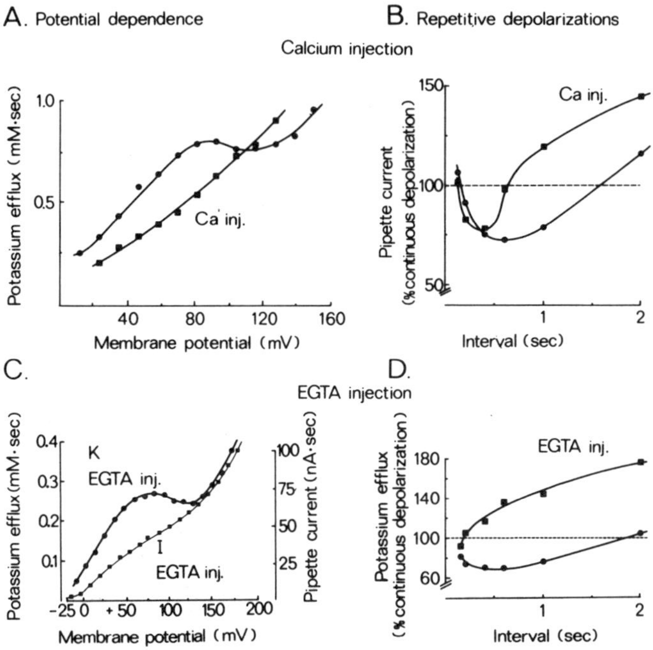

FIG. 7. Summary of the effects of increasing $[Ca]_i$ (by intracellular iontophoresis of $CaCl_2$ in **A** and **B**) or preventing the stimulus-induced accumulation of $[Ca]_i$ by pressure injection of EGTA (**C** and **D**). **A:** N-shape in the K-V curve in the untreated neuron (*filled circles*) is completely suppressed by Ca injection (*filled squares*). **B:** Frequency dependence of the summed current in 10 100 msec pulses compared with a continuous 1 sec pulse is altered by Ca injection (normal neuron, *filled circles;* after Ca injection, *filled squares*). **C:** A strong N-shape in the K-V plot remains after EGTA injection (*filled circles*); the change in the shape of the I-V plot with EGTA injection (*filled squares*) is then due to increased short-circuiting by the slow inward current. **D:** Depression of K efflux at intermediate and long intervals seen in normal neurons (*filled circles*) is removed by EGTA injection (*filled squares*). The total K efflux from 10 100 msec pulses (−50 to +10 mV) is compared with that from a continuous depolarization.

MEMBRANE CURRENTS AND BURSTING

There are many models for the relationships of membrane currents to burst patterns. These vary from ones based on simple two-current (Na, K) systems (7) to more complicated models using up to five different current components (57). The involvement of Na (56) and Cl pumps (55) has also been suggested. Since all of these can produce burst activity, we will not add to the growing list of models. However, several of the more recently described properties of membrane currents found in Helix neurons deserve comment as they might relate to repetitive firing and changes in action potential characteristics in neurons.

Slow Membrane Oscillations

Spike activity in bursting molluscan neurons appears to ride on slow oscillations of membrane voltage. As mentioned above, the slow inward Ca current does not inactivate even at membrane potentials corresponding to levels reached during the postburst hyperpolarization. Thus this current should provide a continuous depolarizing drive throughout the burst cycle. Coupled with gradual inactivation of the fast outward current, this could produce depolarization necessary for the onset of spike activity (16). Furthermore, the additional activation of $I_{in\ slow}$ by action potential generation undoubtedly adds to the depolarization during a burst (58).

Mechanisms for providing the periodic hyperpolarization are less obvious. In R_{15} of Aplysia, tetrodotoxin (TTX) abolishes the generation of action potentials; but the slow waves of membrane depolarization and repolarization remain (10,23). This suggests that models requiring massive Ca influx during action potentials to account for postburst hyperpolarization (39,40) need to be reevaluated. Ca currents during even very low levels of depolarization can activate $I_{K(Ca)}$ (Fig. 2). In addition, electrogenic pumps, including perhaps one involved in Ca extrusion (42), may contribute to the increase in membrane potential. Ouabain (10,23) and decreased temperature (50) can decrease or abolish the large periodic hyperpolarizations in bursters. Furthermore, extracellular recordings with K-sensitive microelectrodes indicate that the rate of decrease in $[K]_0$ following a burst may be more rapid than can be explained by simple diffusion and thereby also implicate active K uptake (34).

Changes in Action Potential Duration

One of the most striking characteristics of bursts in Aplysia R_{15} or the fast burster of Helix is the progressive increase in action potential duration within each cycle (e.g., Fig. 4B). Elongation of action potentials could be due to either a decrease in outward current or an increase in inward current during the repolarizing phase of the spike. Experimental verification of the first mechanism came from experiments with the squid giant axon. In this membrane, the slow inward current is probably negligible as far as action potential generation is concerned (4,5); likewise, $I_{K(Ca)}$ is probably almost entirely absent. Action potential repolarization is due to the inactivation of the fast inward Na and activation of the voltage-dependent outward K current similar in many ways to $I_{K(V)}$ of Helix. Internal perfusion of these axons with such blockers of $I_{K(V)}$ as TEA (2) or cesium (4) increases action potential duration. It is likely that the effect is primarily due to a decrease in the potential-dependent K current. Intracellular TEA also causes a moderate increase of action potential duration in Helix bursters primarily due to the blocking of $I_{K(V)}$ (21).

As discussed above, we have investigated the frequency dependence of currents in Helix neurons; several characteristics thus uncovered should be important during the repetitive spiking of bursts. The voltage-dependent K current is depressed with depolarization and recovers with hyperpolarization. Thus during a burst, the short depolarizations and relatively longer intervals between spikes should not produce strong inactivation of $I_{K(V)}$. Indeed, the fast afterhyperpolarization following each spike can remain relatively constant (58). The frequency-dependent depression of $I_{K(V)}$ in the slow burster is virtually identical to that in the fast burster. However, spike

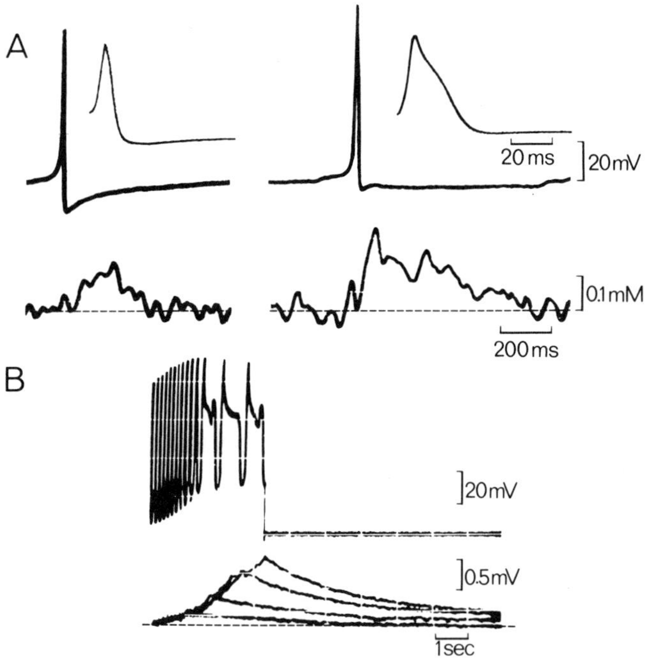

FIG. 8. Potassium efflux (*lower traces*) during spontaneous action potentials (*upper traces*). **A:** In an untreated neuron, the increase in K efflux between the first and last spikes is greater than the increase in action potential duration (differences in total depolarization during spikes can be seen in the expanded time scale in insets). **B:** Another neuron injected with EGTA and producing both apparently normal and subsequently prolonged plateau spikes when switched from voltage to current clamp. Four sweeps have been superimposed in the K data representing the K efflux during the first short spikes and then the K efflux of these action potentials plus one, two, or three plateau spikes. The K/D ratio (total K efflux/time integral of depolarization during the spikes) calculated for the plateau spikes is twice the K/D ratio for the first short action potentials. (H. D. Lux, *unpublished experiments*).

elongation is minimal or absent in the slow burster (Fig. 4B). The major difference between the two cells, then, is the much larger proportion of the coupled Ca-K system in the fast burster; therefore, changes in action potential duration during bursting should be due to properties of this coupled Ca-K system. Indeed, the prolongation of action potentials can appear as a distinct hump during repolarization, suggesting the involvement of relatively slow processes. Both the facilitation of the slow Ca influx and the depression of the associated K efflux show maximum frequency-dependent changes at frequencies (e.g., approximately 2 Hz) involved in bursting. Both effects will prolong and even increase action potentials and are undoubtedly involved in the normal behavior of the neuron. Our experiments indicate that the increase in inward current contributes significantly. Monitoring of K efflux during bursts shows that the amount of K leaving the cell per spike does not decrease but increases with repetitive activation. In the example of Fig. 8A, the time integral of depolarization during the last spike is about twice that of the first spike, but the comparable ratio of K effluxes is about three. Another treatment that causes very prolonged spikes in Helix neurons is the injection of EGTA, which only slightly reduces K current but greatly increases the frequency-dependent inward Ca current (36). This can result in the plateau spikes shown in Fig. 8B. Large K effluxes for single plateau spikes are obvious. Again, the ratio of K efflux/time integral of depolarization for the plateau spikes is twice that found for the time integral of depolarization for the sum of shorter spikes preceding the first plateau.

CONCLUSIONS

The study of molluscan neurons has added many concepts to our understanding of mechanisms which can underlie burst generation or changes in action potential characteristics. What remains is to see whether these concepts can be applied to other systems, for example, vertebrate central neurons. There are obvious limitations to any attempt to transfer information. For the most part, these studies have dealt with prescribed areas of hopefully uniform membrane. They have not been designed to investigate interactions within the morphologically and electrophysiologically complex forms of central neurons, nor have they dealt with the effects of transmitter substances. These areas await future investigation. Furthermore, it is obvious in many instances that molluscan cells differ pharmacologically from vertebrate neurons (26). Thus we may question whether the manipulations employed in the study of bursting or molluscan membrane currents are transferable to neurons in higher animals. There is evidence that some effects, in particular the Ca-K system described here, are present in other animals and may play important roles. The slow Ca current involved in transmitter release and the slow Ca system described here show many similarities (4,6,20). Indeed, the mechanisms of facilitation of synaptic release are not well understood and may actually require the presence of a facilitating Ca system (62). There is also evidence for slow Ca currents in postsynaptic vertebrate neurons (8,38,51). In some of these examples, the Ca currents are manifested as Ca spikes, probably on dendritic processes (29). It is clear that the induction of Ca action potentials in neuron processes should result in very different information processing by the affected cells. Ca-coupled K currents insensitive to TEA have been described for vertebrate neurons (8,11). Adequate data are not available to decide whether frequency and voltage dependencies of these Ca-K currents show properties similar to those of the Helix fast burster neuron. However, it is clear that the studies of this molluscan cell provide previously unrecognized mechanisms for plasticity which could result from the unusual properties of its Ca and K conductances.

REFERENCES

1. Alving, B. O. (1968): *J. Gen. Physiol.,* 51:29–45.
2. Armstrong, C. M., and Binstock, L. (1965): *J. Gen. Physiol.,* 48:859–872.
3. Arvanitaki, A., and Chalazonitis, N. (1955): *C. R. Acad. Sci.,* 240:462–464.
4. Baker, P. F., and Glitsch, H. G. (1975): *Philos. Trans. R. Soc. Lond. [Biol.],* 270:389–409.
5. Baker, P. F., Hodgkin, A. L., and Ridgway, E. B. (1971): *J. Physiol. (Lond.),* 218:709–755.
6. Baker, P. F., Hodgkin, A. L., and Shaw, T. I. (1962): *J. Physiol. (Lond.),* 164:355–374.
7. Barker, J. L., and Gainer, H. (1975): *Brain Res.,* 84:461–477.
8. Barrett, E. F., and Barrett, J. N. (1976): *J. Physiol. (Lond.),* 255:737–774.
9. Calvin, W. H., Sypert, G. W., and Ward, A. A., Jr. (1968): *Exp. Neurol.,* 21:535–549.
10. Carpenter, D. O. (1973): In: *Neurobiology of Invertebrates,* edited by J. Salánki, pp. 35–58. Akadémiai Kiadó, Budapest.
11. Christ, D. D., and Nishi, S. (1973): *Exp. Neurol.,* 40:806–815.
12. Chung, S. H., Raymond, S. A., and Lettvin, J. Y. (1970): *Brain Behav. Evol.,* 3:72–101.
13. Connor, J. A., and Stevens, C. F. (1971): *J. Physiol. (Lond.),* 213:21–30.
14. Cooley, J. W., and Dodge, F. A. (1966): *Biophys. J.,* 6:583–594.
15. Eckert, R., and Lux, H. D. (1975): *Brain Res.,* 83:486–489.
16. Eckert, R., and Lux, H. D. (1976): *J. Physiol. (Lond.),* 254:129–151.
17. Gola, M. (1976): *Experientia,* 32:585–587.
18. Guttman, R. (1969): *Biophys. J.,* 9:269–277.
19. Heinemann, U., Lux, H. D., and Gutnick, M. J. (1977): *Exp. Brain Res.,* 27:237–243.
20. Heyer, C. B., and Lux, H. D. (1976): *J. Physiol. (Lond.),* 262:319–348.
21. Heyer, C. B., and Lux, H. D. (1976): *J. Physiol. (Lond.),* 262:349–382.
22. Horn, G., and Wright, M. J. (1970): *J. Exp. Biol.,* 51:217–231.
23. Junge, D., and Stephens, C. L. (1973): *J. Physiol. (Lond.),* 235:155–181.
24. Kado, R. T. (1973): *Science,* 182:843–845.
25. Katz, B., and Miledi, R. (1971): *J. Physiol. (Lond.),* 216:503–512.
26. Kehoe, J., and Marder, E. (1976): *Ann. Rev. Pharmacol. Toxicol.,* 16:245–268.
27. Kerkut, G. A., and Meech, R. W. (1966): *Comp. Biochem. Physiol.,* 19:819–832.
28. Levitan, H., and Tauc, L. (1972): *J. Physiol. (Lond.),* 222:537–558.
29. Llinás, R., and Hess, R. (1976): *Proc. Natl. Acad. Sci. USA,* 73:2520–2523.
30. Llinás, R., and Nicholson, C. (1975): *Proc. Natl. Acad. Sci. USA,* 71:187–190.
31. Llinás, R., Steinberg, I. Z., and Walton, K. (1976): *Proc. Natl. Acad. Sci. USA,* 73:2918–2922.

32. Lux, H. D. (1974): *Nature,* 250:574–576.
33. Lux, H. D. (1974): *Neuropharmacology,* 13:509–517.
34. Lux, H. D., and Heyer, C. B. (1975): *Bioelectrochem. Bioenerg.,* 3:169–182.
35. Lux, H. D., and Heyer, C. B. (1977): *Neuroscience* 2:585–592.
36. Lux, H. D., and Heyer, C. B. (1978): *Neurosciences Study Program* (*In press.*)
37. Lux, H. D., and Neher, E. (1973): *Exp. Brain Res.,* 17:190–205.
38. Lux, H. D., and Schubert, P. (1975): *Adv. Neurol.,* 12:29–44.
39. Meech, R. W. (1974): *Comp. Biochem. Physiol.,* 48A:387–395.
40. Meech, R. W. (1974): *J. Physiol.* (*Lond.*), 237:259–277.
41. Meech, R. W., and Standen, N. B. (1975): *J. Physiol.* (*Lond.*), 249:211–239.
42. Moreno, J., Wald, F., and Mazzuchelli, A. (1969): *Experientia,* 25:826–828.
43. Neher, E. (1971): *J. Gen. Physiol.,* 58:36–53.
44. Neher, E., and Lux, H. D. (1969): *Pfluegers Arch. Ges. Physiol.,* 311:272–277.
45. Nicholson, C., Steinberg, R., Stöckle, H., and ten Bruggencate, G. (1976): *Neurosci. Lett,* 3:315–319.
46. Oehme, M., Kessler, M., and Simon, W. (1976): *Chimia,* 30:204–206.
47. Perkel, D. H., and Bullock, T. H. (1968): *Neurosci. Res. Prog. Bull.,* 6:221–348.
48. Rao, K. P., Babu, K. S., Ishiko, N., and Bullock, T. H. (1969): *J. Neurobiol.,* 2:233–245.
49. Sakkarov, D. A., and Salańki, J. (1969): *Acta Physiol. Hung.,* 35:19–30.
50. Salánki, J., Vadász, I., and Vero, M. (1973): *Acta Physiol. Acad. Sci. Hung.,* 43:115–124.
51. Schwindt, P., and Crill, W. E. (1977): *Brain Res.,* 120:173–178.
52. Segundo, J. P., Moore, G. P., Stenaas, L. S., and Bullock, T. H. (1963): *J. Exp. Biol.,* 40: 643–667.
53. Stefani, E., and Gerschenfeld, H. M. (1969): *J. Neurophysiol.,* 32:64–74.
54. Stein, R. B. (1967): *Proc. R. Soc. Lond.* [*Biol.*], 167:64–86.
55. Strumwasser, F. (1965): In: *Circadian Clocks,* edited by J. Aschoff, pp. 442–462. North-Holland, Amsterdam.
56. Strumwasser, F. (1968): In: *Physiological and Biochemical Aspects of Nervous Integration,* edited by F. D. Carlson, pp. 329–341. Prentice-Hall, Englewood Cliffs, New Jersey.
57. Thompson, S. H. (1977): *J. Physiol.* (*Lond.*), 265:465–488.
58. Thompson, S. H., and Smith, S. J. (1976): *J. Neurophysiol.,* 39:153–161.
59. Wald, F. (1972): *J. Physiol.* (*Lond.*), 220:267–281.
60. Walker, G. L., Jr. (1971): *Anal. Chem.,* 43:89–92A.
61. Williamson, T. L., and Crill, W. E. (1976): *Brain Res.,* 116:231–249.
62. Zucker, R. S. (1974): *J. Physiol.* (*Lond.*), 241:91–110.

Abnormal Neuronal Discharges, edited by N. Chalazonitis and M. Boisson. Raven Press, New York © 1978.

Changes in Extracellular Free Calcium and Potassium Activity in the Somatosensory Cortex of Cats

U. Heinemann, H. D. Lux, and *M. J. Gutnick

Department of Neurophysiology, Max Planck Institute for Psychiatry, Munich, Federal Republic of Germany

A causal relationship between potassium accumulation in the extracellular space of the central nervous tissue and the onset, spread, and termination of seizure activity has been suggested by Fertziger and Ranck (7) and others (5, 12,49). They have proposed that a threshold value of extracellular potassium activity (a_K) exists at which evolution of epileptiform discharge is inevitable. According to this hypothesis, a reduced potassium gradient across the neuronal membrane would produce a membrane depolarization (7) strong enough to cause repetitive discharges and to maintain the seizure as long as a_K was elevated. Ultimately, depolarization from a_K would inactivate membrane currents, an effect that would prevent further excitation and terminate seizure activity. Although observations in stimulus-induced seizures support the idea of a threshold value for seizure initiation (56), it is clear from recent work with K^+-sensitive microelectrodes that the hypothesis must be modified. Here, we report additional data showing that, as in the penicillin focus (42) and pentylenetetrazol (PTZ) seizures (34), K^+ accumulation plays at most a minor role in initiation of seizure discharge.

Failure to correlate a_K levels with initiation of seizure does not, however, exclude the possibility that K^+ released during the ensuing seizure contributes to spread of ictal discharge. Indeed, we find that increasing background a_K electrophoretically can support development of seizure discharge. Termination of seizure activity is probably not caused by depolarizing inactivation of neuronal excitability (19,42). Evidence is presented here in support of the idea that a K^+-activated electrogenic transport process contributes to termination of seizure discharge.

In a modified K^+ accumulation hypothesis, effects of other ions could be considered. Of special interest is Ca^{2+}, since a decreased extracellular Ca^{2+} activity (a_{Ca}) enhances neuronal excitability (9). In fact, reductions in a_{Ca}

* Present address: Unit of Comparative Medicine, Faculty of Medical Sciences, Ben Gurion University of the Negev, Beersheva, Israel.

were recently shown to accompany neuronal activity in the extracellular space of the cerebral (20) and the cerebellar cortex (44,55). A possible contribution of decreased a_{Ca} to seizure initiation is suggested by the finding that an accidental complete parathyroidectomy can lead to seizure activity in humans, probably by lowering a_{Ca} (10).

METHODS

Recordings of a_K, a_{Ca}, and field potential (FP) were taken from the somatosensory cortex and occasionally from the nucleus ventropostero-lateralis (VPL) thalami of cats. For measurements of a_K, a_{Ca}, and FP, we used exclusively double-barreled, ion-sensitive reference microelectrode pairs. Fabrication, characteristics, and testing of these electrodes were described elsewhere (20,35,40,58). Details of preparation, anesthesia, and supervision of the state of the animals can be found in references 18, 19, and 34. Changes in a_K and a_{Ca} were produced by repetitive stimulation of cortical surface (CS), underlying white matter, or the nucleus VPL. In some experiments, a_K was increased by electrophoresis of K^+ from a nearby phoresis electrode (distances between tips of phoresis and K^+-sensitive electrodes $\leq$ 100 μm). Seizures were induced by intense repetitive CS or VPL stimulation, by intravenous injection of 10 mg/kg bodyweight PTZ, or by topical application of penicillin (gelfoam pledget of 0.5 to 2 mm in diameter moistened with 25 mg/ml aqueous-buffered sodium penicillin G for 5 to 10 min).

Stimulus-Induced Increases in a_K and Onset of Self-Sustained Epileptiform Afterdischarges

Increases in a_K (Δa_K) from an average resting level of 2.9 mM/liter can be evoked by repetitive stimulation of the CS and appropriate afferent pathways (40,52). Transient changes in a_K usually consist of a rise to levels between 3 and 10 mM/liter, a subsequent fall to subnormal levels (undershoot), and a return to resting a_K. The amplitude and duration of these undershoots in a_K vary with the amplitude of preceding rise in a_K (17,18,26,28). Stimulus trains which increased a_K to levels higher than 4.5 mM/liter usually resulted in self-sustained epileptiform activity of clonic or tonic clonic type (Fig. 1 A–C). In Fig. 1B, it can be seen that a_K could further increase during self-sustained afterdischarges (SAD). The amplitude of this second rise in a_K was on the average largest when a_K was increased to levels between 5 and 5.5 mM/liter at the end of a stimulus train. Above this level, the amplitude of Δa_K during SAD is inversely related to the value of a_K at the end of stimulation and approaches zero when the level of a_K at the end of a stimulus train nears 10 mM/liter. This is apparently the upper limit of increases in a_K in neocortex, i.e., the "ceiling" level (18,19,34,42). Higher levels are reached

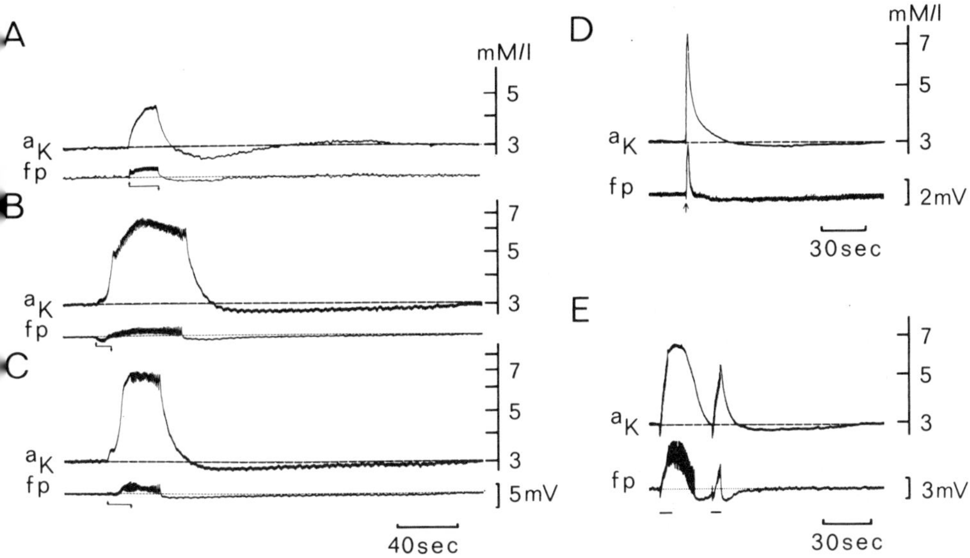

FIG. 1. Relationship between stimulus-induced rise in a_K and onset of SAD. **A:** Rise in a_K and change in FP evoked by 15/sec VPL stimulation with 0.5 mA. Recording depth 400 μm below CS. Beginning and end of stimulation in this and subsequent figures indicated by upward and downward arrows, respectively. **B:** Rise in a_K and change in FP evoked by 20/sec VPL stimulation with 0.8 mA. Recording depth 1,200 μm below cortical surface. Self-sustained epileptiform afterdischarges were evoked by these stimuli, during which a_K rose to a further elevated level. Note plateau in the rising slope of a_K. **C:** Rise in a_K and change in FP evoked by 20/sec VPL stimulation with 1.4 mA. Recording depth 800 μm below cortical surface. Note duration of SAD shorter than in **B,** although a_K rises during stimulation to a further enhanced level. Note also shoulder in the rising slope of a_K. **D:** 20/sec, 2 mA CS stimulation of 1 sec duration which did not evoke signs of paroxysmal discharge. **E:** Recording of a_K and FP in the nucleus VPL thalami. Changes in a_K and FP evoked by a 20/sec CS stimulus train (1 mA, 0.1 msec pulse width) which evoked SAD. It is followed by a second stimulus train with the same parameters which provoked a smaller rise in a_K and no seizure activity. Train duration indicated by horizontal bars. Similar observations can be made in the sensorimotor cortex, where rises in a_K to values beyond 5 mM/liter do not result in SAD when evoked during decay in a_K to subnormal levels. Note the large amplitude of FP recordings in VPL as compared with the DC recorded FP in the sensorimotor cortex.

only when active K^+ uptake mechanisms are impaired by anoxia (57), application of metabolic inhibitors (27), or spreading depression (43,45,57). The duration of SAD and the occurrence of a tonic phase in this type of ictal activity were not related to the amplitude of Δa_K during stimulation. Thus in Fig. 1B, a stimulus-induced rise in a_K to about 5 mM/liter resulted in an ictal episode of 50 sec duration, while a larger rise in a_K during stimulation, as in Fig. 1C, was followed by about 25 sec of paroxysmal afterdischarges.

The finding that stimuli which increased a_K to levels beyond 4.5 mM/liter can result in seizure activity is in agreement with a report of Sypert and Ward (56). They interpreted this as a confirmation of the K^+ accumulation hypothesis and argued that, as a rule, seizure is initiated when a_K is increased during stimulation to values beyond a "threshold level" of 4.5 to 5 mM/liter.

Since an increase in a_K is a result and within certain limits an index of nerve activity (34), the above rule may characterize only a stimulus as one intense enough to provoke seizure. If this is true, deviations from the rule that stimulus-induced increases in a_K to values beyond 4.5 mM/liter necessarily result in seizure activity are to be expected. Indeed, an increase in a_K to levels above 4.5 to 5 mM/liter could be separated from seizure induction by a variety of experimental paradigms. Thus short-lasting, very intense stimulus trains can increase a_K to levels of 6 to 9 mM/liter without evoking SAD. In the example of Fig. 1D, a_K rose to about 8 mM/liter during an intense cortical stimulus train of 1 sec duration without subsequent paroxysmal activity. During undershoots, stimuli intense enough to increase a_K above 4.5 mM/liter could also fail to induce SAD. In particular, SAD could not be evoked during the decay of a_K from increased to subnormal levels, nor were they triggerable during the first third of the undershoot (see also ref. 36). Levels of 7 mM/liter and more could be obtained without provoking paroxysmal afterdischarges. During recovery of a_K from subnormal levels, seizure susceptibility slowly recovered and was occasionally even increased above control during the last 20 sec before and also sometimes after the end of the undershoot. The same observations can also be made in the nucleus VPL thalami, as shown in Fig. 1E.

If the level of 4.5 mM/liter were a real threshold value for development of seizure discharge, one would expect antiepileptic drugs to prevent the increase in a_K to higher levels; this is not the case. After treatment with phenytoin (10 mg/kg bodyweight) (17), nembutal (5 mg/kg bodyweight), and diazepam (1 mg/kg bodyweight) (38), increases in a_K to levels between 4.5 and 8 mM/liter could be evoked by intense stimulus trains (1 to 3 mA, 20/sec). These stimuli frequently did not result in SAD.

No distinct threshold level of a_K can be found in the penicillin focus and in PTZ seizures. In the penicillin focus, no relationship was seen between Δa_K during interictal discharges and the probability of transition from interictal to ictal state (8,42,43). In PTZ seizures, ictal activity usually began at resting a_K (see Fig. 6) or, in visual cortex, with an initial fall to subnormal levels. Such results challenge the interpretation of a threshold value in a_K. However, K^+ accumulation may still contribute to propagation and spread of seizure discharge. To check for this possibility, a_K must be increased independently of neuronal activity. Such an increase in background a_K can be induced by local K^+ electrophoresis (40).

Diphasic Rises in a_K Indicate Epileptiform Activity

To test for a supporting effect of electrophoretically heightened a_K in development of seizure discharge, a borderline situation must be found where seizure is just initiated. Such a situation is probably met when diphasic rises are evoked by repetitive stimulation of VPL or CS (8 to 20 Hz, 0.3 to 1 mA) somewhat above the threshold for initiation of seizure activity. During such stimulation, a_K initially rose to a plateau level between 3.4 and 5 mM/liter, where it remained for up to 30 sec before rising to further increased levels, which were sometimes as large as 10 mM/liter. After these stimuli, SAD frequently developed (see Figs. 1B and 2A). Increasing the stimulus intensity or frequency shortened the intermediate plateau level; instead of a plateau, a shoulder was seen in the rising slope of a_K. The secondary rise in a_K during sustained stimulation is certainly due to increased neuronal excitation with an extra release of K^+. It might therefore indicate local initiation of seizure. If the diphasic rise were the required borderline situation and the secondary rise in a_K a sign of ensuing seizure, changes in field potentials and in the pattern of evoked neuronal spike activity would be expected. Indeed,

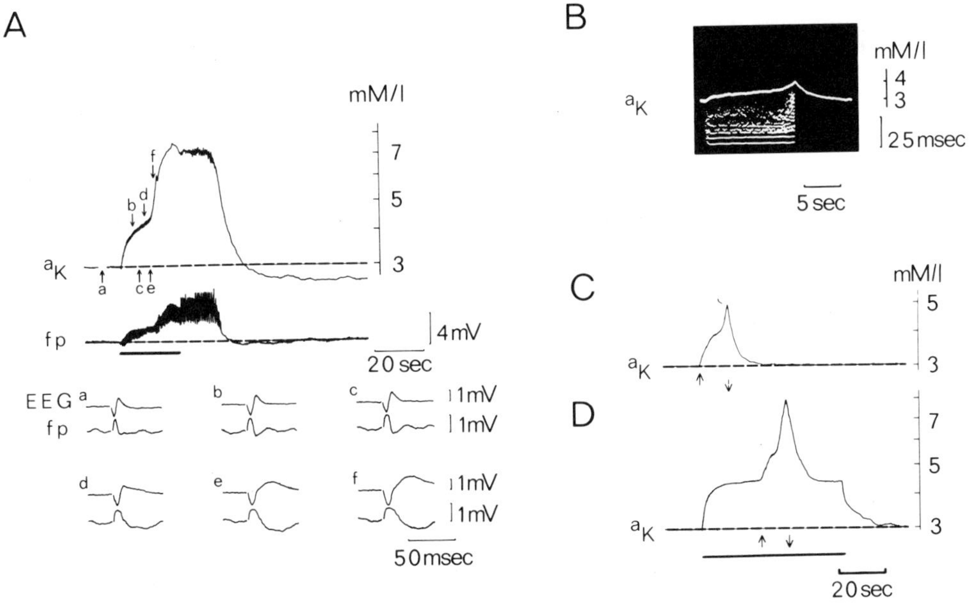

FIG. 2. Diphasic rises in a_K, neuronal excitability, and effects of electrophoretically increased a_K. **A:** Diphasic rise in a_K as evoked by 15/sec repetitive stimulation of VPL with 1 mA. Stimulus duration indicated by horizontal bars. Averages from five successive evoked field and EEG recordings are also shown. The various specimens were taken at times indicated by lettered arrows. Note the transition from a normal to an ictal response. **B:** Dot display of the response of a cortical neuron to 15/sec 0.5 mA cortical stimulation and simultaneously recorded changes in a_K. The response to each stimulus is plotted vertically. **C:** VPL stimulation, 15/sec, 0.8 mA. **D:** Electrophoretic elevation of K^+ with a current intensity of 150 nA. Duration indicated by horizontal bar. During electrophoresis, the same stimulus as in **C** was repeated. Note that latency to onset of secondary rise is shorter than in **C** and that the amplitude of the secondary rise is increased.

as shown in Fig. 2A and B, such changes do occur. In Fig. 2A, a_K was increased by repetitive stimulation of VPL. Specimen recordings were taken at times indicated by arrows. These potentials were averaged from five successively evoked potentials.

Figure 2A demonstrates that with the beginning of the secondary rise, the normal response changed into a pathologic one with an increased early positivity in the EEG recording and an increased and prolonged negativity. These changes in evoked EEG and FPs also occurred during monophasic rises in a_K caused by stimulus trains which provoke seizures (19). On the other hand, similar changes in EEG and FPs were not observed during stimulus trains which evoked rises in a_K to about 4 mM/liter only and which consequently did not elicit SAD. Associated with this change in EEG and focal potentials is an alteration of neuronal discharge pattern. With the onset of the secondary rise, high frequency bursts were evoked by the stimulus train. This is shown in the dot display of Fig. 2B. These observations suggest that the situation during diphasic rises is analogous to that during development and spread of seizure and that the secondary rise in a_K represents local onset of ictal activity.

K^{+} Electrophoresis Facilitates Epileptiform Activity

Increases in a_K to levels of up to 12 mM/liter were induced by local K^{+} electrophoresis. These elevations have a moderate excitatory effect (19); increases in neuronal discharge rates from 10 to 20 Hz at resting a_K to values between 40 and 50 Hz at 10 mM/liter a_K were reported for cortical neurons (see, e.g., Fig. 6 in ref. 19). This effect compares to that observed in the central nervous system of leech (3). Such data suggest that locally increasing a_K could facilitate seizure generation. The diphasic rise apparently represents a borderline case in which effects of electrophoretically increased a_K on development of ictal activity should be particularly obvious. A typical experiment is shown in Fig. 2C and D. The diphasic rise in Fig. 2C was evoked by 15/sec and 0.8 mA VPL stimulation. In Fig. 2D, the baseline a_K was electrophoretically increased to about 4.5 mM/liter. This rise did not effect the amplitude of the initial rise in a_K. However, the latency to the beginning of the secondary rise was shorter, sometimes by about 25%. Also significant is the finding that the amplitude of the secondary rise was larger by up to 45%. These effects were maximal when background a_K was increased to levels between 4 and 6 mM/liter. At higher levels, the amplitude of the secondary rise in a_K was less enhanced. Since the secondary rise in a_K represents local initiation of seizure discharge, these observations on an enhanced secondary rise in a_K suggest that increased a_K supports the development of seizure discharge. This is also suggested by the finding that the duration of ictal bursts increases with the rise in a_K (see Fig. 2B). Similar observations were made in experiments on spread of ictal discharge from a cortical focus to a thalamic nucleus (U. Heinemann and M. J. Gutnick, *in preparation*). A support of increased a_K in development of seizure discharge can also be concluded from experiments in which superfusion of neo- or allocortex was shown to increase seizure susceptibility (19) and to activate latent epileptic foci (60).

Termination of Seizure Activity

Evidence Against a Potassium-Dependent Depolarization Block of Neuronal Excitability

It is possible that a large accumulation of K^+ would eventually terminate seizure if, for example, the resultant membrane depolarization were so great that action potential-generating mechanisms were inactivated (7). This hypothesis suggests that seizures should be terminated at the peak of an increase in a_K, which is very rarely the case. Indeed, all reports on directly measured a_K in seizure agree with the original observation of Moody et al. (42) on penicillin seizures in which a_K fell slowly from its peak before the end of paroxysmal activity. Similar observations have been made in PTZ seizures (34). Also in stimulus-induced seizures, as shown in Figs. 1–4, a_K slowly fell from its peak before the end of seizures (19).

Moreover, in all these seizures, rise in a_K appears to be limited to a ceiling level of 10 mM/liter. Effects of K^+ electrophoresis suggest that neurons do not simply become inexcitable at these levels. Local K^+ electrophoresis did increase neuronal discharge rate at higher levels of a_K and, with a background of 12 mM/liter stimulation can increase the rate of neuronal discharge rate. Even levels of 10 mM/liter were rarely obtained during seizure; the mean peak a_K during SAD was 7.5 ± 1 mM/liter (mean ± SD). At the end of SAD, a_K had fallen to an average value of 6.5 mM/liter. At such a low level of a_K, K^+-dependent depolarizing block of neuronal excitability could not account for termination of seizure discharge. This hypothesis, therefore, is untenable.

Contribution of an Electrogenic Transport Mechanism

Alternatively, potassium accumulation could play a role in termination of ictal activity through the activation of an electrogenic K^+ uptake mechanism. This could lead to hyperpolarization of neurons and thus decrease excitability (1).

Several lines of evidence suggest that an active K^+ uptake mechanism is involved in limiting increases in a_K to 10 mM/liter and in generating the subsequent undershoots, whose amplitude and duration are related to the amplitude of preceding rise in a_K (17–19). Electrophoretically induced increases in a_K evoked by constant current steps are reduced in amplitude during stimu-

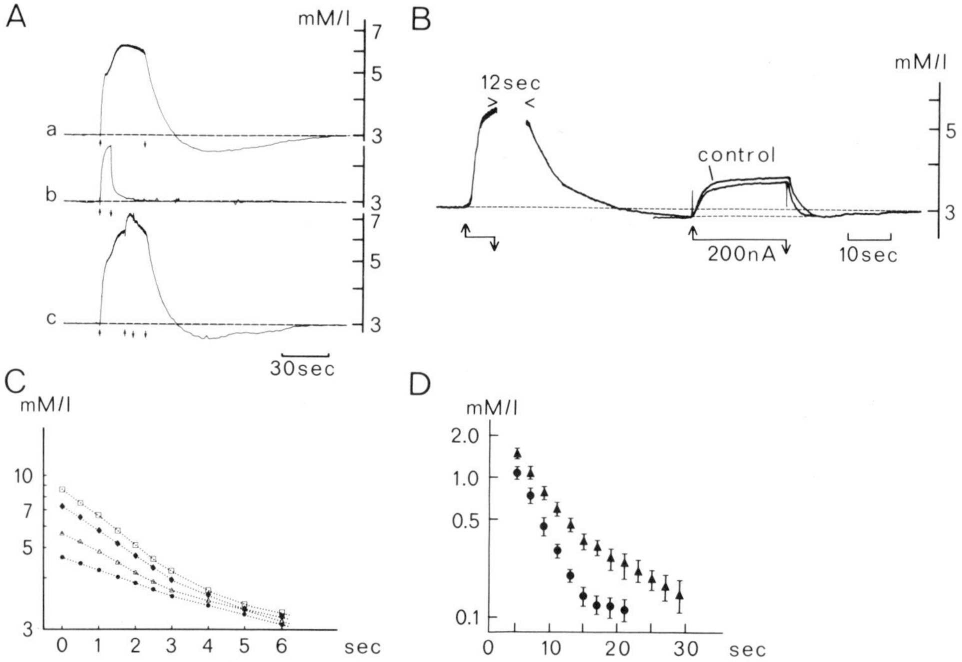

FIG. 3. Evidence for active K^+ uptake. **A:** *a,* Repetitive VPL stimulation, 20/sec, 2 mA; *b,* K^+ response induced by K^+ electrophoresis with 200 nA; *c,* repetitive VPL stimulation at 20/sec and 2 mA. At about peak level, K^+ electrophoresis with 200 nA was added to the stimulus. Note that the reduction of the phoretically induced K^+ response is about 40%. **B:** Comparison between two electrophoretically induced K^+ signals with a current intensity of 200 nA. *Upper,* designed with control, was evoked at resting a_K. *Lower,* induced during the undershoot which followed a stimulus-induced rise in a_K; 12 sec of seizure activity recording with an artifact is omitted. **C:** Plots of decay slopes from various levels of a_K obtained during VPL stimulus train which induced SAD. The curves were chosen because immediately with the onset of decay in a_K, neurons were silent. Thus deviances in the slopes from diffusion kinetics became apparent. **D:** Plots of decay phases of Δa_K evoked by electrophoretic K^+ ejection at baseline (*triangles*) level of a_K and during the undershoot (*circles*) of a_K. Mean values and SD of six trials in each situation. The electrophoresis current was 200 nA at a distance of about 50 μm between the tip of the K^+-sensitive and the electrophoresis electrode. The electrodes were located at a depth of 700 μm. A_K underceded resting activity by 0.2 mM/liter after a preceding rise of 2.5 mM/liter. The amplitude of Δa_K produced by electrophoresis of the same current strength decreased by 27 $\pm$ 2% (SD) during the undershoot. The halftime of these signals, if evoked during the undershoot of a_K, was 3.1 $\pm$ 0.3 (SD) sec ($\sim$ 28% reduction).

lus-induced increases (19) and during subsequent undershoots in a_K (18). In Fig. 3A, the amplitude of the K^+-test pulse was reduced by about 40% during stimulus-induced increase in a_K. We have observed decreases of more than 50%, whereas the reduction of electrophoretically induced K^+-test pulses during undershoots was usually less than 25% (see Fig. 3B and D). Changes in the decay of a_K following test pulses (Fig. 3D) and stimulus-induced increases in a_K (Fig. 3C) do not follow predictions based on

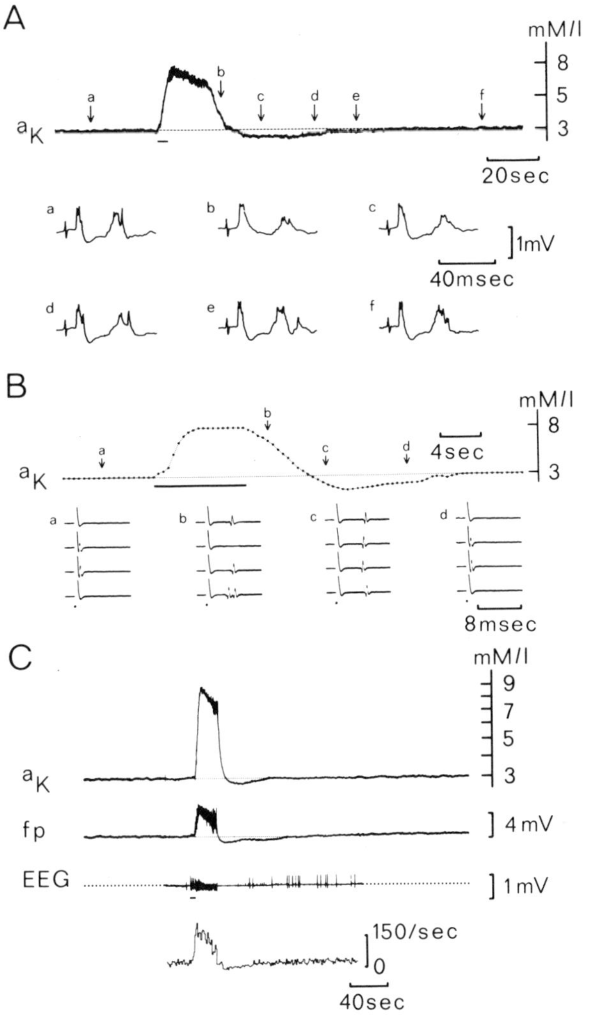

FIG. 4. Excitability of thalamocortical relay (TCR) cells. **A:** Change in a_K and excitability of a TCR neuron as tested with 1/sec cuneate nucleus stimulation. FP recordings (average from five successive stimuli) were taken at times indicated by lettered arrows; they were recorded with the reference electrode of the double-barreled, ion-sensitive reference electrode pair. Change in a_K was evoked by 20/sec CS stimulation. **B:** Change in a_K and antidromic invasion probability of action potential. Change in a_K evoked by 20/sec repetitive stimulation of CS. Before and immediately after the stimulus train 1/sec stimulation of the CS. Specimen recordings of action potentials as they invade to or propagate from the TCR neuron are taken at times indicated by letters. **C:** Change in a_K, FP, and neuronal discharge rate of a TCR cell during ictal activity in a cortical penicillin focus as indicated by the EEG recording. Seizures was triggered by five successive CS stimuli at 3/sec, as long as indicated by the bar.

diffusion equations (26,40) and are best explained by assuming active K^+ uptake (18,21).

An active uptake mechanism is also suggested by the finding that an increase in oxygen consumption (indicated by changes in NADH fluorescence) occurs during the rise in a_K and lasts until the end of undershoots (28,31). After poisoning the Na^+-K^+-ATPase with strophantidin (27), a_K rises to values beyond the ceiling level, and undershoots in a_K disappear. Thus the active K^+ uptake appears to contribute significantly to the equilibration time course of transient changes in a_K. An electrogenic action of this uptake mechanism is suggested by the finding that during the fall of a_K from a peak level, neuronal discharge rate decreases slowly and is significantly depressed at the end of seizure (see Fig. 4C). The duration of this inhibitory period varied considerably from cell to cell (see Fig. 7 in ref. 18). However, the duration of postictal depression of neuronal activity was relatively constant for a particular neuron and a given increase in a_K. It varied with the characteristics of transient changes in a_K, especially with the amplitude of rise in a_K and the amplitude and duration of subsequent undershoots. These latter parameters apparently indicate the degree of activation of the K^+ uptake mechanism during rise in a_K (18). Similar reductions in neuronal discharge rate were also observed during decay of a_K from increased to subnormal levels in the nucleus VPL thalami. An example is shown in Fig. 4C. The rise in a_K and the simultaneous change in FP was induced by stimulating a cortical penicillin focus with 2 Hz. This stimulus caused cortical seizure, as indicated by the EEG recording. The seizure spread to VPL where it resulted in large increases in a_K to levels as high as 11.4 mM/liter (16). During the rise in a_K, neuronal discharge rate increased. After termination of seizure, neuronal discharge rate fell to subnormal levels. Also, duration of this inhibitory period is related to duration and amplitude of a_K.

If an electrogenic pump were contributing to termination of seizure discharge (1), one would expect that inhibited neurons would still show postsynaptic potentials and would be able to generate action potentials. Recordings in VPL of FPs evoked by cuneate nucleus or contralateral forepaw stimulation show that all components of FPs are present during the decay and undershoot in a_K (Fig. 4A). Although excitability was decreased (U. Heinemann and M. J. Gutnick, *in preparation*), action potential generation was not completely depressed during this period (see Fig. 4B). Postictal hyperpolarizations of thalamocortical relay cells should reduce the probability of antidromic action potential invasion (13,14). As shown in Fig. 4B, the probability of antidromic invasion was significantly reduced during fall in a_K, although action potentials could be generated. A reduction in probability of antidromic invasion was also observed after ablation of the cortex. Thus excitability changes in the endings of thalamocortical neurons cannot explain this effect. The duration of periods with reduced probability of antidromic invasion varied with duration and amplitude of the undershoots in a_K.

This finding supports the idea that an activated K^+ uptake mechanism is linked to an electrogenic transport process which opposes the excitatory effects of increased a_K and which shifts the membrane potential in a hyperpolarizing direction (1). The antiepileptic effect of this mechanism can be demonstrated in experiments in which attempts were made to evoke SAD (17,36) during decay and undershoot of a_K. As already mentioned, seizure was not triggerable during the fall in a_K and the first third of the undershoot.

Ca^{2+} Changes During Epileptogenesis

Although K^+ accumulation in the extracellular space may support development and propagation of seizure discharge, it is obvious that changes in a_K can account for initiation of seizure activity only under very specific conditions. Such conditions are probably met when the allocortex is superfused with K^+-enriched solutions (60) or when K^+ regulation is disturbed, as after local application of ouabain (48). Disturbances in regulation of a_K were also assumed to be involved in seizure generation in a chronic focus (49; see also refs. 11 and 47).

Since extracellular calcium activity also controls excitability of neuronal membrane, changes in a_{Ca} might contribute to seizure generation. The recent development of a specific and selective Ca^{2+} ion exchanger for microelectrode application by Oehme et al. (46) made it possible to compare changes in a_{Ca} with changes in a_K during paroxysmal activity in the cortex of cats. With this electrode, the average resting a_{Ca} was found to be 1.25 mM/liter. Significant transient reductions in a_{Ca} to below this level were produced by natural stimulation of the contralateral forepaw. This reduction in a_{Ca} is probably caused by movement of Ca^{2+} into glial or neuronal elements. A predominant entry of Ca^{2+} into neuronal elements is suggested by the finding that decreases in a_{Ca} vary with the intensity and frequency of CS or VPL stimulation. During stimulus-induced seizures, a_{Ca} could fall to 0.7 mM/liter.

There appears to be no threshold value of lowered a_{Ca} for initiation of seizure discharge since decreases in a_{Ca} by 0.1 mM/liter, or even less, frequently accompanied stimulus trains which can produce SAD; other stimuli which did evoke larger decreases in a_{Ca} were not able to elicit paroxysmal activity. A supporting role of a_{Ca} in seizure generation, however, can be concluded from observations on seizure susceptibility during recovery of a_{Ca}. The recovery of a_{Ca} differs from that for a_K and can last as long as 5 min (20). During this period, a relative plateau or peak was obtained 10 to 30 sec after the end of stimulation or seizure; subsequently, a_{Ca} fell slightly and then slowly approached baseline. Frequently, it is still lowered long after a_K returned to baseline, Sometimes shortly before and after the end of undershoot in a_K, seizure susceptibility is enhanced, as mentioned above. This might be due to an excitatory action of lowered a_{Ca}.

Significant reductions of a_{Ca} occur also in a penicillin focus. Typical recordings are shown in Fig. 5. During interictal discharges, a_{Ca} could fall by 0.1 mM/liter, with the minimum a_{Ca} in 100 to 400 msec measured from onset of the interictal discharge. The recovery lasted between 0.5 and 2 sec. The changes in a_{Ca} varied with cortical depth. They were largest in about 200 μm (see Fig. 5A) while rise in a_K during interictal discharge was largest at 1000 μm (42). During interictal activity, a_{Ca} frequently slowly decayed and stabilized at levels of 1 to 1.2 mM/liter. Occasionally, spontaneous seizures oc-

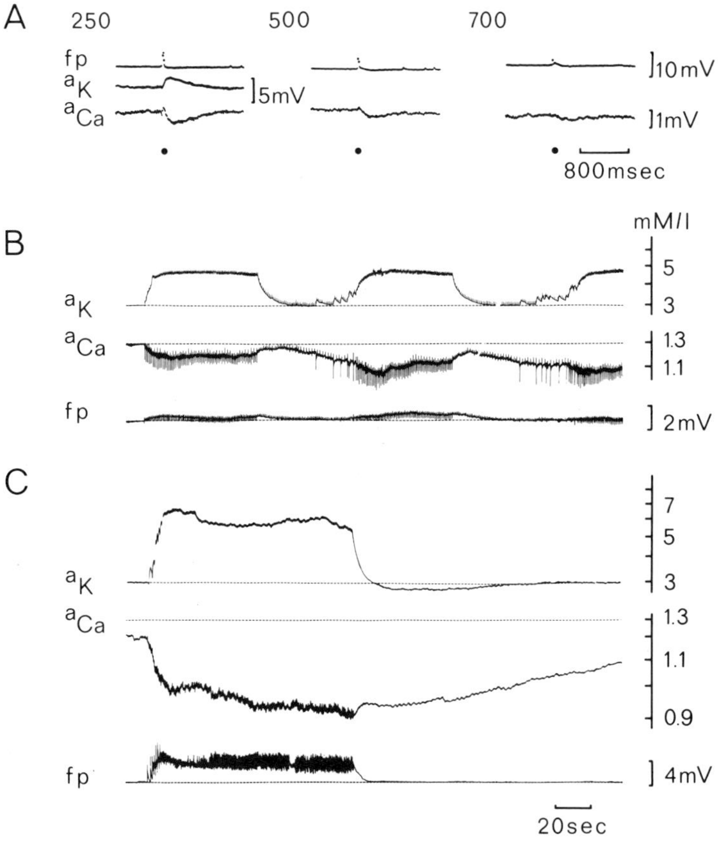

FIG. 5. Changes of a_K, a_{Ca}, and FP in a cortical penicillin focus. **A:** Per event averages of changes in a_{Ca} during interictal activity. Numbers refer to depth below CS. Interictal spikes recorded with the reference side of the Ca^{2+}-sensitive reference electrode pair. A_K recording at the border of the focus was not varied with depth. **B:** Cyclic spike driving in a penicillin focus evoked by 1/sec repetitive CS stimulation (pulse width 0.05 msec, intensity about 0.1 mA); FP recording from the reference side of the K^+-sensitive reference electrode pair. Recording depth of a_K and FP 200 μm below CS at the border of the penicillin focus. The a_{Ca} was recorded at 300 μm below CS in about the center of the focus. **C:** Change in a_K, a_{Ca}, and FP during full-blown seizure activity, evoked by successive CS stimuli. Seizure started from a lowered level in a_{Ca} due to previous ictal activity.

curred. These started from all levels between 1.3 and 0.8 mM/liter. During seizure, a_{Ca} could fall to about 0.7 mM/liter, while a_K could increase to about 10 mM/liter. Recovery slopes compared to that observed after stimulus-induced transient decreases in a_{Ca}. A typical recording is shown in Fig. 5C.

Prolonged, low frequency stimulation of the penicillin focus often induced a state of cyclically changing excitability in the epileptogenic neuron population, as previously described by Prince (50). Such cyclic spike driving is illustrated in Fig. 5B. In this experiment, 1/sec stimulation of the cortical surface evoked minute-long periods during which each stimulus was followed by generation of an interictal epileptiform event. These "active cycles" alternated with periods of approximately the same duration during which no epileptiform discharges were triggered and each stimulus evoked only a normal direct cortical response. Changes in a_K during the cycling activity closely followed variations in the excitability of the focus. Thus each interictal discharge led to an increase in a_K; when intervals between successive events were short, these increases summed to achieve levels as high as 8 mM/liter. Concomitant changes in a_{Ca} were also cyclic but quite different in their timing. During the first 15 to 20 sec of an active cycle, a_{Ca} fell steadily to levels as low as 1 mM/liter and then began to return slowly toward the baseline level. The gradual rise of a_{Ca} persisted for up to 20 sec after the end of the active cycle, whereupon a_{Ca} again began to fall. Thus the onset of a gradual decrease in a_{Ca} preceded the onset of the next active cycle by as much as 30 sec.

Similar reductions in a_{Ca} were also observed before seizures evoked by intravenous injection of 10 mg/kg bodyweight PTZ. During or shortly after injection of PTZ, a_{Ca} began to decrease, while a_K remained unaltered until the actual seizure. During ictal activity, a_{Ca} decayed further. The amplitude of the change in a_{Ca} during one ictal episode was maximally 0.5 mM/liter. During repeated seizures, reductions in a_{Ca} appeared to add (see Fig. 6B), and minimum activities of 0.7 mM/liter were obtained. The recovery slope corresponds to that observed after repetitive stimulation or penicillin seizures.

After repeated injection of PTZ, cyclically occurring ictal periods were seen. As during cyclic spike driving, decreases in a_{Ca} occurred before onset of seizure activity (see Fig. 6C).

These data demonstrate that synchronous activation of cortical neurons is accompanied by a focal decrease in a_{Ca}. Of particular significance is the finding that movement of Ca^{2+} from the extracellular space actually precedes the onset of seizure discharge and the enhancement of excitability of an epileptogenic focus. Such decreases could signify inward calcium currents in various neuronal compartments; they might be accompanied by increases in the excitability of a large cortical neuronal population. In presynaptic terminals, enhanced calcium flux would be expected to reflect increased transmitter release (24). The observed decreases in a_{Ca} before and during

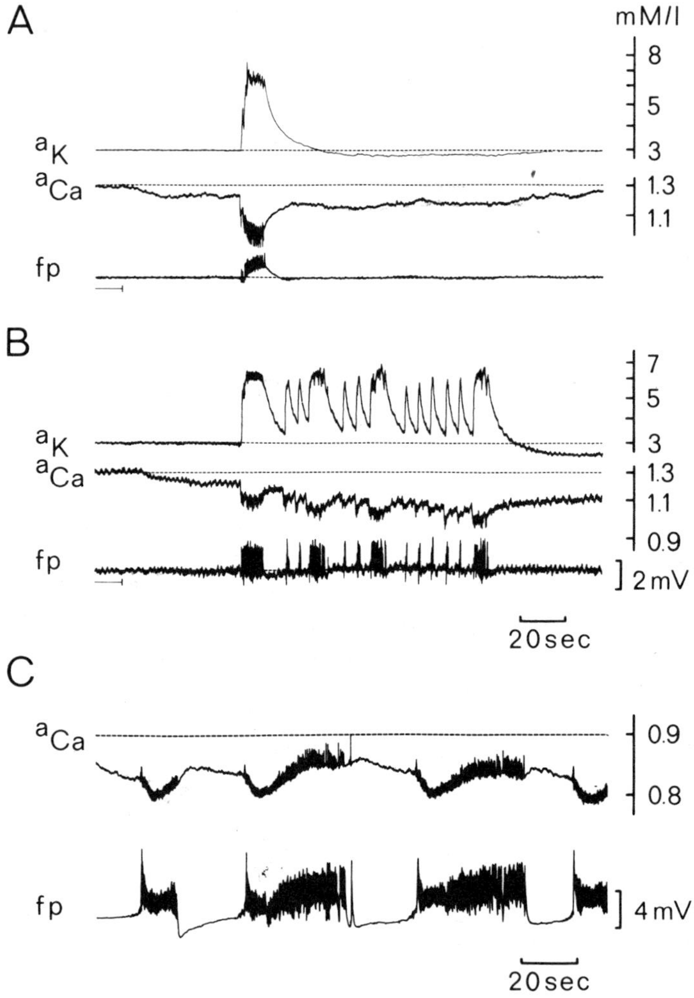

FIG. 6. Changes in a_{Ca}, a_K, and FP as evoked by intravenous injection of PTZ. End of injection in **A** and **B** indicated by bars; FP recordings with the reference side of the Ca^{2+}-sensitive electrode. Recording depth in **A**, 300 μm; in **B**, 1100 μm; and in **C**, 1000 μm. Recording depth of a_K in **A** and **B** is 500 μm. Recording in **C** was taken after repeated injections of PTZ which finally resulted in this type of recurrent seizure activity.

seizure could thus indicate a facilitating transmitter release due to a facilitating Ca^{2+} current, as described by Heyer and Lux (22,23,39; see also their contribution to this volume). The decrease in a_{Ca} might also reflect repetitive ectopic action potential generation in presynaptic endings with considerably enhanced transmitter release (13–15,51). Ectopic action potential generation would be involved in the generation of paroxysmal depolarization shifts if the latter are considered to be giant excitatory postsynaptic potentials and in any case result in an additional excitatory drive. Decreased a_{Ca} and increased a_K may be involved in ectopic action potential generation by depolarization of presynaptic endings (4,32,33,53) and by unspecific increase in their excitability. Therefore, the precise timing of change in a_{Ca}, a_K, and occurrence of ectopic action potential generation must be studied.

At the postsynaptic level, an inward calcium current may also be present (2), possibly related to modifications of membrane characteristics. For example, recent experiments of Lux and Eckert (6,37) strongly suggest that in snail neurons, anomalous inward-going rectification is caused by a slow inward calcium current. Other evidence suggests a role for calcium in inward going rectification of cat motoneurons (29,41). Greater inward-going rectification would increase postsynaptic potential amplitude and thus enhance efficacy of synaptic transmission. A K^+-dependent depolarization of neuronal membrane would be cooperative with this effect. It must also be recognized that Ca^{2+}-dependent dendritic action potential generation (30; see also Prince and Schwartzkroin, *this volume*) could participate in the generation of paroxysmal responses. It is tempting to speculate that such a change in neuronal excitability is caused by PTZ. Indeed, it has been shown that PTZ increases anomalous inward-going rectification in snail neurons (25,54,59). With PTZ, increased inward going rectification can also be seen in motoneurons (L. Liebl and H. D. Lux, *unpublished observation*). It is obvious that these mechanisms must be further elucidated.

CONCLUSIONS

Initiation of seizure activity is not a simple function of increase in a_K or decrease in a_{Ca}.

Changes in Ca^{2+}-dependent mechanisms may contribute to generation of seizure activity, as indicated by a fall in a_{Ca} before onset of seizure.

Increased a_K probably contributes to propagation and spread of seizure activity.

Termination of seizure cannot be explained by K^+-dependent depolarizing inactivation of neuronal excitability.

An electrogenic K^+ uptake mechanism, which by excess transport of cations out of the cell shifts the membrane potential in a hyperpolarizing direction, is possibly contributing to termination of seizure.

ACKNOWLEDGMENTS

This research was supported by grants Lu 158/7, Lu 158/8, and Lu 158/10 of the Deutsche Forschungsgemeinschaft. The skillful and experienced assistance of Ms. A. Bieber in these experiments, and her secretarial assistance in the preparation of the manuscript and figures, is gratefully acknowledged. We thank Dr. Carolyn Heyer for reading the manuscript and for helpful suggestions.

REFERENCES

1. Ayala, G. F., Matsumoto, H., and Gumnit, R. J. (1970): *J. Neurophysiol.,* 33:73–85.
2. Barrett, E. F., and Barrett, J. N. (1976): *J. Physiol.,* 255:737–774.
3. Baylor, D. A., and Nicholls, J. G. (1969): *J. Physiol. (Lond.),* 203:571–589.
4. ten Bruggencate, G., Lux, H. D., and Liebl, L. (1974): *Pfluegers Arch.,* 349:301–317.
5. Dichter, M. A., Hermann, J., and Selzer, M. (1972): *Brain Res.,* 48:173–183.
6. Eckert, R., and Lux, H. D. (1976): *J. Physiol.,* 254:129–151.
7. Fertziger, A. P., and Ranck, J. B., Jr. (1970): *Exp. Neurol.,* 26:571–585.
8. Fisher, R. S., Pedley, T. A., Moody, W. J., and Prince, D. A. (1976): *Arch. Neurol.,* 33: 76–83.
9. Frankenhaeuser, B., and Hodgkin, A. L. (1957): *J. Physiol.,* 137:218–244.
10. Gastaut, H., Rohmer, F., Cossette, A., and Kurtz, D. (1969): In: *The Physiopathogenesis of Epilepsies,* edited by H. Gastaut, H. Jasper, J. Baucaud, and A. Waltregny, pp. 5–25, Thomas, Springfield, Illinois.
11. Gloetzner, F. (1973): *Brain Res.,* 55:159–171.
12. Green, J. D. (1964): *Physiol. Rev.,* 44:501–608.
13. Gutnick, M. J., and Prince, D. A. (1972): *Science,* 176:424–426.
14. Gutnick, M. J., and Prince, D. A. (1974): *J. Neurophysiol.,* 37:1310–1327.
15. Gutnick, M. J., and Prince, D. A. (1975): *Exp. Neurol.,* 46:418–431.
16. Heinemann, U., Gutnick, M. J., and Lux, H. D. (1976): *Pfluegers Arch.,* 362:R36.
17. Heinemann, U., and Lux, H. D. (1973): *Electroencephalogr. Clin. Neurophysiol.,* 34:735.
18. Heinemann, U., and Lux, H. D. (1975): *Brain Res.,* 93:63–76.
19. Heinemann, U., and Lux, H. D. (1977): *Brain Res.,* 120:231–250.
20. Heinemann, U., Lux, H. D., and Gutnick, M. J. (1977): *Exp. Brain Res.,* 27:237–243.
21. Heinemann, U., Lux, H. D., and Zander, K. J. (1978): In: *Iontophoresis and Transmitter Mechanisms in the Mammalian Central Nervous System,* edited by J. Kelly and R. W. Ryall, 419–428.
22. Heyer, C. B., and Lux, H. D. (1976): *J. Physiol. (Lond.),* 262:319–348.
23. Heyer, C. B., and Lux, H. D. (1976): *J. Physiol. (Lond.),* 262:349–382.
24. Katz, B., and Miledi, R. (1970): *J. Physiol. (Lond.),* 207:789–801.
25. Klee, M. R., Faber, D. S., and Heiss, W. D. (1973): *Science,* 179:1133–1136.
26. Krnjevič, K., and Morris, M. E. (1975): *Can. J. Physiol. Pharmacol.,* 53:912–922.
27. Krnjevič, K., and Morris, M. E. (1975): *J. Physiol.,* 250:36P.
28. Lewis, D. V., and Schuette, W. H. (1975): *J. Neurophysiol.,* 38:405–417.
29. Liebl, L., and Lux, H. D. (1975): *Pfluegers Arch.,* 335:R80.
30. Llinás, R., and Hess, R. (1976): *Proc. Natl. Acad. Sci. USA,* 73:2520–2523.
31. Lothman, E. W., LaManna, J., Cordingly, G., Rosenthal, M., and Somjen, G. (1975): *Brain Res.,* 88:15–36.
32. Lothman, E. W., and Somjen, G. (1975): *J. Physiol.,* 252:115–136.
33. Lothman, E. W., and Somjen, G. (1976): *Electroencephalogr. Clin. Neurophysiol.,* 41: 253–267.
34. Lux, H. D. (1974): *Epilepsia,* 15:375–393.
35. Lux, H. D. (1974): *Neuropharmacology,* 13:509–517.
36. Lux, H. D. (1975): In: *Brain Work,* edited by D. H. Ingvar and N. A. Lassen, Munksgaard, Copenhagen, pp. 172–181.
37. Lux, H. D., and Eckert, R. (1974): *Nature,* 250:574–576.
38. Lux, H. D., and Heinemann, U. (1978): *EEG Clin. Neurophysiol.* (*In press.*)
39. Lux, H. D., and Heyer, C. B. (1977): *Neuroscience,* 2:585–592.
40. Lux, H. D., and Neher, E. (1973): *Exp. Brain Res.,* 17:190–205.
41. Lux, H. D., and Schubert, P. (1975): *Advances in Neurology, Vol. 12,* edited by G. W. Kreutzberg. Raven Press, New York.
42. Moody, W. J., Jr., Futamachi, K. J., and Prince, D. A. (1974): *Exp. Neurol.,* 42:248–263.
43. Mutani, R., Futamachi, K. J., and Prince, D. A. (1974): *Brain Res.,* 75:27–39.

44. Nicholson, C., ten Bruggencate, G., Steinberg, R., and Stöckle, H. (1977): *Proc. Natl. Acad. Sci. USA,* 74:1287–1290.
45. Nicholson, C., and Kraig, R. P. (1975): *Brain Res.,* 96:384–389.
46. Oehme, M., Kessler, M., and Simon, W. (1976): *Chimia,* 30:204–206.
47. Pedley, T. A., Fisher, R. S., Futamachi, K. J., and Prince, D. A. (1976): *Fed. Proc.,* 35:1254–1259.
48. Pedley, T. A., Zuckermann, E. C., and Glaser, G. H. (1969): *Exp. Neurol.,* 25:207–219.
49. Pollen, D. A., and Trachtenberg, M. C. (1970): *Science,* 167:1252–1253.
50. Prince, D. A. (1971): *Electroencephalogr. Clin. Neurophysiol.,* 31:469–484.
51. Prince, D. A. (1973): In: *Handbook of Electroencephalography and Clinical Neurophysiology,* edited by O. Creutzfeldt, pp. 2C56–2C70, Elsevier Scientific Publ. Co., Amsterdam.
52. Prince, D. A., Lux, H. D., and Neher, E. (1973): *Brain Res.,* 50:489–495.
53. Somjen, G. G., and Lothman, E. W. (1974): *Brain Res.,* 69:153–157.
54. Speckmann, E. J., and Caspers, M. (1973): *Epilepsia,* 14:397–408.
55. Stöckle, H., ten Bruggencate, G., Nicholson, C., and Steinberg, R. (1977): *Pfluegers Arch.,* 368:R37.
56. Sypert, G. W., and Ward, A. A., Jr. (1974): *Exp. Neurol.,* 45:19–41.
57. Vyskocil, F., Kriz, N., and Bures, J. (1972): *Brain Res.,* 39:255–259.
58. Walker, G. L., Jr. (1971): *Anal. Chem.,* 43:89–92A.
59. Williamson, T. L., Crill, W. E. (1976): *Brain Res.,* 116:231–249.
60. Zuckermann, E. C., and Glaser, G. H. (1968): *Exp. Neurol.,* 20:87–110.

Abnormal Neuronal Discharges, edited by
N. Chalazonitis and M. Boisson. Raven
Press, New York © 1978.

Mechanism of Action of Presynaptic Neurotoxins

M. Lazdunski, G. Romey, Y. Jacques, M. Fosset,
R. Chicheportiche, and M. Balerna.

Centre de Biochimie, Faculté des Sciences, Nice, France

Neurotoxins that provoke an abnormal neuronal response also are essential tools for analyzing the molecular aspects of the generation of an action potential by an excitable membrane. The purpose of this chapter is to review work recently carried out in our laboratory and to analyze, by electrophysiological and biochemical techniques, the mechanism of interaction of presynaptic neurotoxins with their membrane receptors.

TETRODOTOXIN

The best known and most widely used of all neurotoxins is tetrodotoxin (TTX). TTX has now been labeled with tritium (3,7) and bound to pure axonal membranes (2). Binding of the toxin to the membranes is very specific. The characteristics of [^{3}H]TTX binding to pure crab axonal membranes are the following: (a) there is a single family of binding sites; (b) the dissociation constant of the TTX-receptor complex is 2.9 nM at pH 7.4, 24°C, a value identical to those found by electrophysiological techniques; and (c) the content of TTX binding sites in the membrane is of 15 pmoles/mg membrane protein.

The complex formed between TTX and its receptor has now been solubilized in Lubrol WX. The molecular weight of the TTX receptor protein found by Agarose filtration is 230 to 260,000. An affinity column has been synthetized by oxidizing the TTX molecule in position 6 with periodic acid and grafting to glass beads with a phenylhydrazine arm. This affinity column retains specifically the TTX receptor (65 to 70% retention). The receptor can be eluted by an excess of free TTX or by a high concentration of NaCl. The molecular weight of the receptor subunit determined by SDS gel electrophoresis is approximately 52 to 63,000.

SEA ANEMONE TOXIN AND VERATRIDINE. INTERACTIONS AMONG SEA ANEMONE TOXIN, VERATRIDINE, AND THE TTX RECEPTOR

TTX specifically interacts with the selectivity filter of the Na^+ channel. The toxin does not alter the functioning of the gating system; i.e., it does not alter the rates of activation and inactivation of the channel (10). TTX is a competitive inhibitor of the Na^+ cation for the recognition of the mouth of the sodium channel. Other neurotoxins are required to analyze the properties of the molecular machinery which makes up the gating system. One particularly interesting toxin is the sea anemone toxin; a series has recently been isolated in the pure form from the sea anemone *Anemonia sulcata* (4). These toxins are small polypeptides; the sequence of one, ATX_{II}, has now been established. It is a miniprotein comprising only 47 amino acids crosslinked by three disulfide bridges (16). We have carefully analyzed the specificity of action of this neurotoxin and the physicochemical properties of its association with its receptor site (1,12).

Figure 1A shows that the thin axons of the crayfish periesophageal nerve are much more sensitive to ATX_{II} than giant axons. When the nerve is bathed in a solution containing 0.1 nM ATX_{II}, some of the thin axons begin to fire spontaneously. More axons are affected at a concentration of 1 nM ATX_{II}. The giant axon having the maximum diameter (about 100 μm), which has been used for microelectrode and voltage-clamp analysis, is sensitive to ATX_{II} at concentrations higher than 0.1 μM (Fig. 1C). Toxin action on this axon provokes a marked plateau phase of the action potential (Fig. 1B). The dose-effect curve (Fig. 1C) shows an all-or-none effect over a range of about 2 to 3 in the toxin concentration. At saturating concentrations of the neurotoxin (0.5 to 1 μM), for short incubation times of 1 or 2 min, one only observes a change of shape of the action potential with no change of the resting potential (Fig. 1B). For longer times (10 to 30 min), one also observes a depolarization of the axonal membrane accompanied by a decrease in the amplitude of the action potential. The toxin-induced depolarization can ultimately block nervous conduction completely. The effect of ATX_{II} on the resting potential for the giant axon is calcium dependent. At a concentration of 13.5 mM Ca^{2+}, the magnitude of the depolarization induced by 1 μM ATX_{II} is 10 mV, whereas at 1 mM Ca^{2+}, the depolarization can reach 40 mV. It is interesting that neither the effect of ATX_{II} on the action potential nor the effect on the resting potential can be reversed by a prolonged washing of 60 min with the van Harreveld solution free of toxin.

ATX_{II}, even at a concentration of 10 μM, has no effect on Sepia giant axons. It is positively charged at pH 7.5 (4). By the iontophoretic technique

→

FIG. 1. A: Spontaneous activity of the crayfish periesophageal nerve. *1,* Control; *2,* in the presence of 0.1 nM ATX_{II}; *3,* in the presence of 1 nM ATX_{II}. 18°C. Time scale (*horizontal bar*), 2 sec; voltage scale (*vertical bar*), 200 μV. **B:** ATX_{II} effect on the action potential of the crayfish giant axon. *1,* Control; *2,* 5 min after; *3,* 10 min after the application of 0.5 μM ATX_{II}. 18°C. Time scale for *1* and *2,* 2 sec; time scale for *3,* 5 msec; voltage scale, 50 mV. **C:** Dose-response curve of ATX_{II} action on the crayfish giant axon at 18°C. Do, spike duration in the control measured at half height of the action potential; D, spike duration measured after 40 min treatment with a given concentration of ATX_{II}. The different symbols represent different series of experiments.

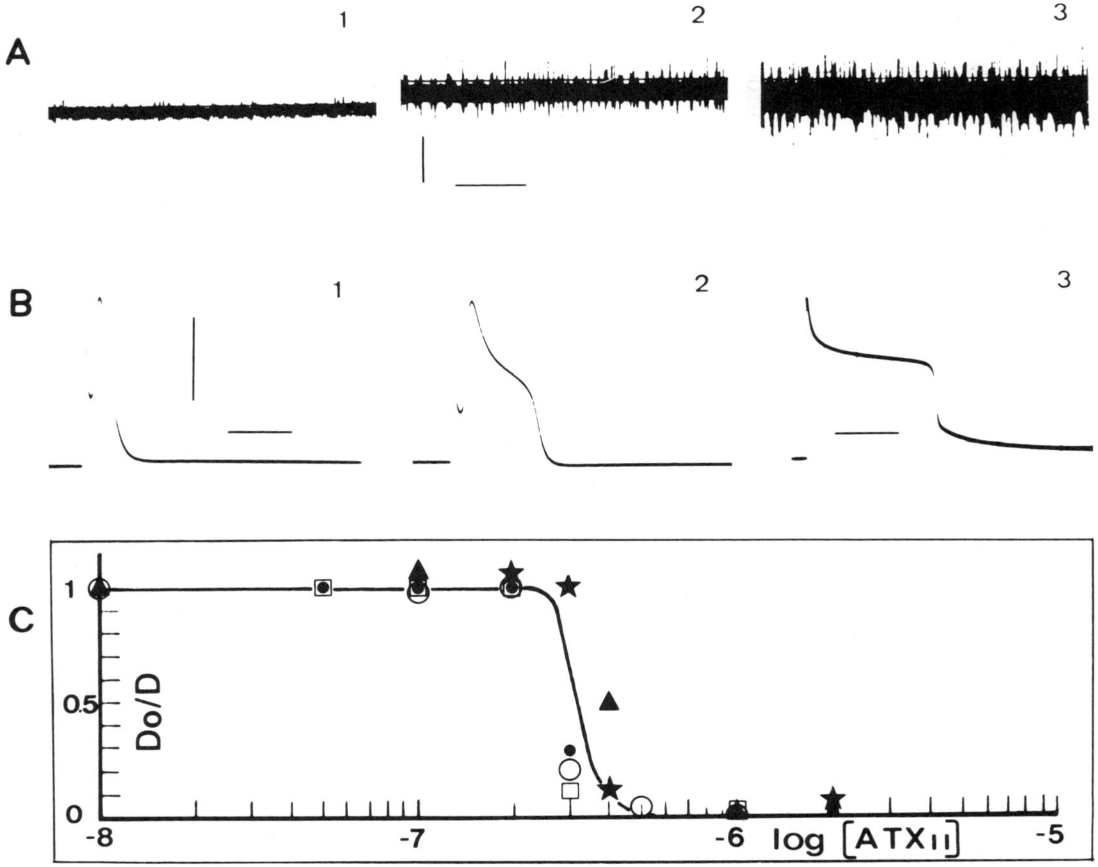
A
1
2
3
B
1
2
3
C
Do/D
1
0.5
0
-8
-7
-6
-5
log [ATXII]

with a glass microelectrode filled with 1 mM ATX_{II}, it is possible to apply the neurotoxin exclusively on the external or internal (cytoplasmic) face of the membrane. A typical plateau phase in the action potential developed in less than 1 min when the toxin was applied at the external surface; no effect developed when the toxin was applied from the cytoplasmic side. The ATX_{II} receptor is thus situated on the external face of the membrane.

TTX, which specifically blocks the sodium channel of untreated axons (10), also blocks sodium entry in axons treated with ATX_{II}. As shown in Fig. 2, the action potentials in both the axon used as a control and the axon treated with ATX_{II} are suppressed by TTX. TTX can also suppress the ATX_{II} effect on the resting potential. It easily repolarizes the membrane previously depolarized by 1 μM ATX_{II} in 1 mM Ca^{2+}. The dose-response curve gives an apparent dissociation constant of 10 nM.

The order of introduction of the two neurotoxins ATX_{II} and TTX is crucial. In the experiments presented in Fig. 2A, ATX_{II} was introduced first and TTX second. In this case, binding of ATX_{II} does not prevent TTX association with its receptor. Conversely, however, previous treatment of the axonal membrane with TTX prevents ATX binding (Fig. 2B). In this experiment, the crayfish axon is first treated with 50 nM TTX; this bathing solution is then replaced by a mixture of TTX (50 nM) and ATX_{II} (0.5 μM). The incubation period of 5 min would be sufficient for the manifestation of ATX_{II} action in the absence of TTX (Fig. 2A). The preparation is then washed with physiological solution. Since TTX binding is reversible, whereas ATX_{II} binding is irreversible, one would expect to observe an action potential with a plateau phase. In fact, the action potential reappears but without the plateau phase typical of ATX_{II} action. If the preparation is then incubated with an ATX_{II} solution of 0.5 μM, the plateau phase reappears after a few minutes. The experiments in Fig. 2B demonstrate that pretreatment of the membrane to block all Na^+ channels with TTX prevents binding of ATX_{II} to its specific sites. Actually, the same observation was made if instead of using TTX one replaces Na^+ by Tris or the cholinium ion. In other words, the toxin can only affect the membrane when the sodium channel is charged with sodium.

A typical series of voltage-clamp experiments presented in Fig. 3 shows that ATX_{II} does not interfere with the opening of the K^+ channel; it has little effect on the opening of the Na^+ channel but selectively and considerably affects the closing (inactivation) by slowing it down considerably. The Na^+ channel remains partially opened even after 6 msec (inward current, 2 mA/cm^2). Voltage-clamp analysis of the interaction of ATX_{II} with the node of Ranvier of myelinated fibers of *Rana esculenta* (12) has given results very similar to those just discussed for the crayfish axon. Until now, the most useful neurotoxin for the identification and analysis of the Na^+ channel has been TTX (10). ATX_{II} is complementary to TTX: (a) it specifically affects another mechanism of the Na^+ channel machinery, i.e.,

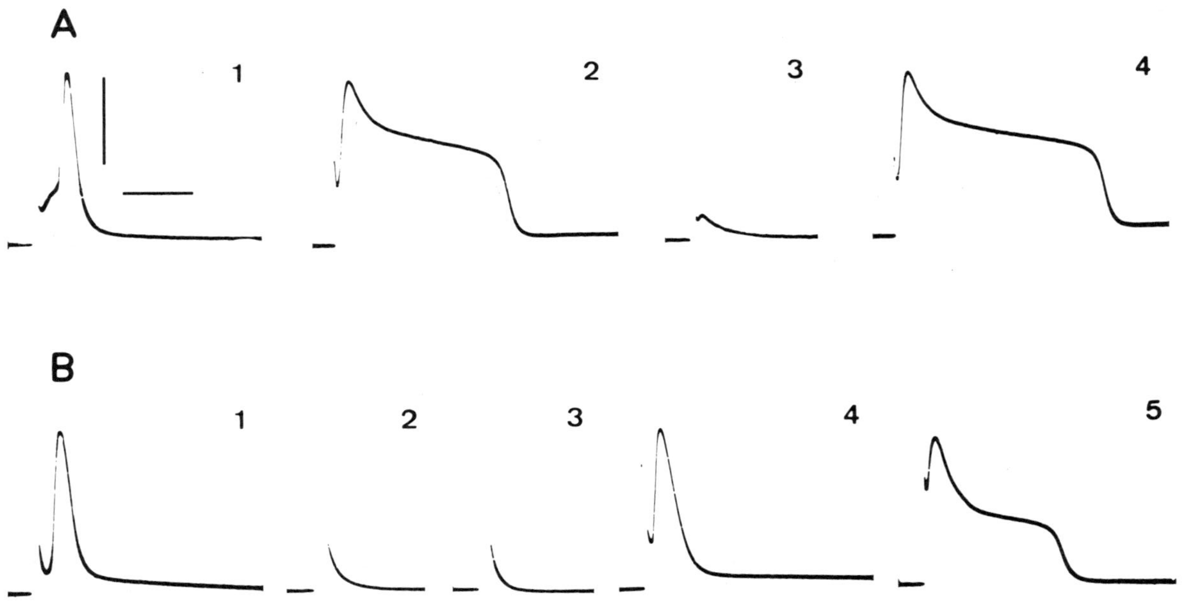

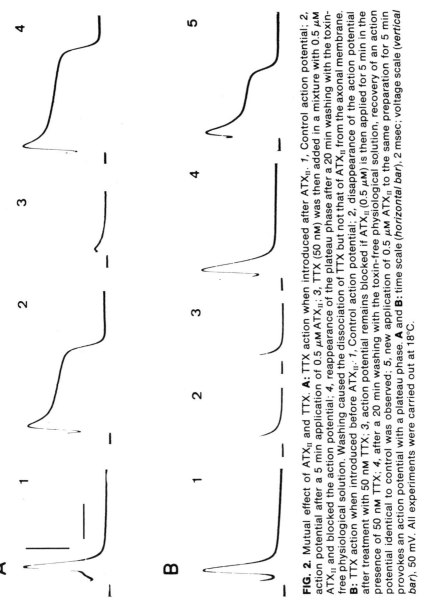

FIG. 2. Mutual effect of ATX_{II} and TTX. **A:** TTX action when introduced after ATX_{II}. *1,* Control action potential; *2,* action potential after a 5 min application of 0.5 μM ATX_{II}; *3,* TTX (50 nM) was then added in a mixture with 0.5 μM ATX_{II} and blocked the action potential; *4,* reappearance of the plateau phase after a 20 min washing with the toxin-free physiological solution. Washing caused the dissociation of TTX but not that of ATX_{II} from the axonal membrane. **B:** TTX action when introduced before ATX_{II}. *1,* Control action potential; *2,* disappearance of the action potential after treatment with 50 nM TTX; *3,* action potential remains blocked if ATX_{II} (0.5 μM) is then applied for 5 min in the presence of 50 nM TTX; *4,* after a 20 min washing with the toxin-free physiological solution, recovery of an action potential identical to control was observed; *5,* new application of 0.5 μM ATX_{II} to the same preparation for 5 min provokes an action potential with a plateau phase. **A** and **B:** time scale (*horizontal bar*), 2 msec; voltage scale (*vertical bar*), 50 mV. All experiments were carried out at 18°C.

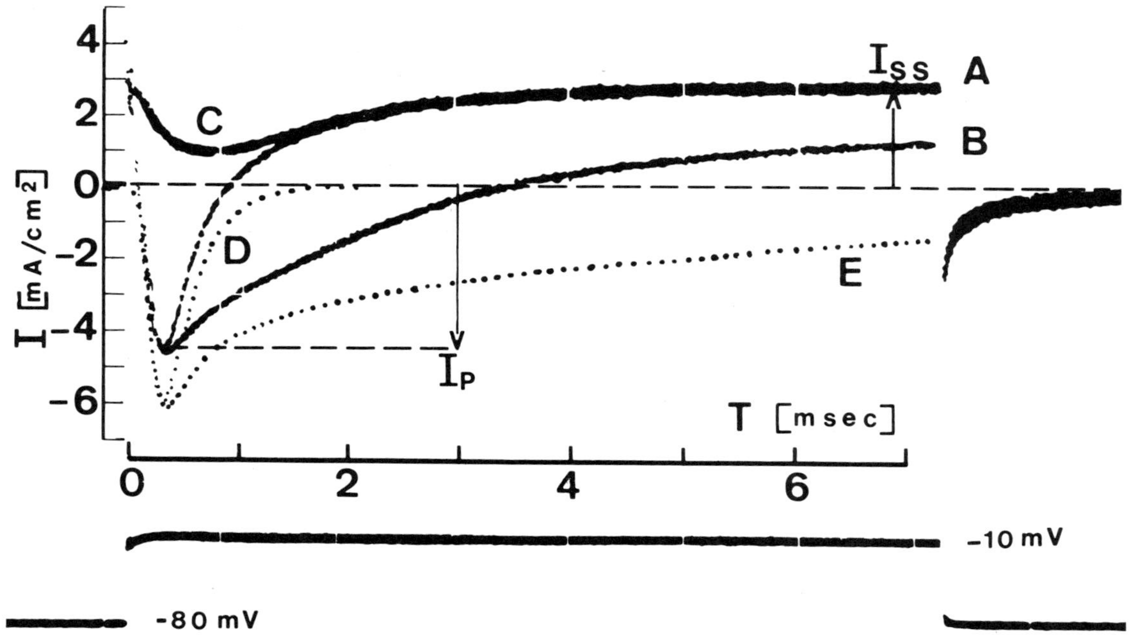

FIG. 3. Voltage-clamp analysis of ATX_{II} action on a crayfish giant axon. Membrane potential was clamped at −10 mV after a voltage jump from a holding potential of −80 mV. 15°C. Traces correspond to the membrane currents associated with the voltage jump. *A,* Without toxin; *B,* after 5 min application of 0.5 μM ATX_{II}; *C,* after treatment with 0.1 μM TTX as well as after treatment with 0.5 μM ATX_{II} followed by application of 0.1 μM TTX. The dotted curves *D* and *E* correspond to differences between *A* and *C* and *B* and *C*, respectively. They represent the time course of the Na^+ current in the absence and in the presence of ATX_{II}. Fast speed recordings (0.1 msec/division instead of 1 msec/division for the results shown in this figure) of Na^+ activation kinetics (i.e., the rate of opening of the Na^+ channel) are not shown here but were found to be superimposable in both the absence and presence of ATX_{II}.

the closing of the Na^+ channel; and (b) its receptor site is distinct from that of TTX.

A convenient approach to the analysis of interaction properties between neurotoxins and their receptors is that which uses nerve cells in cultures (6). Figure 4 describes the action of ATX_{II} on $^{22}Na^+$ influx into differentiated mouse neuroblastoma cells in culture (strain NIE 115). ATX_{II} stimulates ^{22}Na uptake by the cells by a factor of about 2. This observation fits with the fact that the toxin (a) maintains the Na^+ channel in an open form by preventing its closing, and (b) depolarizes the nervous membrane. The apparent dissociation constant of the interaction between toxin and receptor is $K_{0.5} = 0.2$ μM. This neuroblastoma system has been used to analyze in some detail the mutual interactions between different toxin receptors.

The first interaction to be analyzed was between veratridine and ATX_{II}. Veratridine is an interesting molecule that provokes membrane depolarization by forcing the opening of the Na^+ channel and preventing its closing (15). Some of the molecular aspects of its interaction with the axonal membrane have been reported (2). As shown in Fig. 4, veratridine at the low concentration of 13 μM does not alter ^{22}Na uptake in neuroblastoma

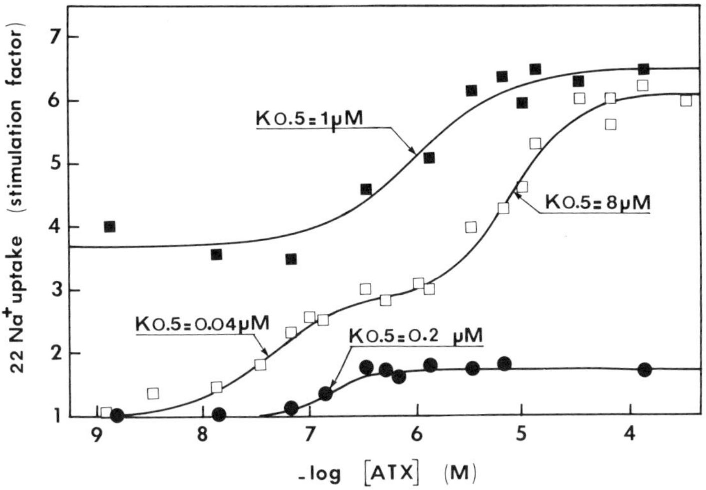

FIG. 4. Activation of $^{22}Na^+$ uptakes by neuroblastoma cells (NIE 115) using ATX_{II} in the absence or presence of veratridine. The initial rate of $^{22}Na^+$ uptake was measured after a period of uptake of 5 min at 37°C and pH 7.4 in the presence of 0.5 mM ouabain to prevent Na^+ efflux through the Na^+, K^+ ATPase system. The initial rate of uptake in the absence of any toxic compound was 15 nmoles of sodium × min^{-1} × mg^{-1} of total protein. The ATX_{II} concentration dependence of $^{22}Na^+$ uptake was measured in the absence (*filled circles*) or presence of veratridine at concentrations of 13 μM (*open squares*) and 133 μM (*filled squares*). Neuroblastoma cells used in this work were differentiated in dimethylsulfoxide as indicated by Kimhi et al. (8), with slight modifications. Their electrical properties were found to be the same as those published by Spector et al. (14).

cells. In the presence of veratridine at this concentration, however, there is clearly a great change in the dose-response curve of $^{22}Na^+$ uptake with respect to ATX_{II} concentration. There is a considerable synergistic effect between veratridine and ATX_{II}. Instead of a maximum stimulation of $^{22}Na^+$ influx by a factor of about 2 by ATX_{II} alone, we observe a maximum stimulation factor of about 6 in the presence of both veratridine (13 μM) and saturating concentrations of ATX_{II}. The other interesting observation is that veratridine reveals that there are two different families of sites for ATX_{II} on the neuroblastoma cell membrane. These two types of sites are mutually interacting; the property of this interaction is negative cooperativity. In the presence of 13 μM veratridine, the apparent affinity for the tight site ($K_{0.5} = 0.04$ μM) is about 200 times higher than that for the loose site ($K_{0.5} = 8$ μM). The affinity of ATX_{II} for the loose site is increased about eightfold if the veratridine concentration is increased from 13 to 133 μM (Fig. 4).

Figure 5 presents the veratridine concentration dependence of the $^{22}Na^+$ uptake by neuroblastoma cells in the absence and in the presence of ATX_{II}. Figure 5 also shows an important synergy between veratridine and ATX_{II} binding. Veratridine alone maximally stimulated $^{22}Na^+$ uptake by a factor of about 4, and the apparent dissociation constant for the interaction with the receptor is 44 μM. In the presence of 13 μM ATX_{II}, the affinity for veratridine is increased nearly 10-fold ($K_{0.5} \simeq 4.7$ μM), and the maximal ^{22}Na uptake stimulation is nearly 6. It is of a great interest to observe that these two toxic compounds, which both alter the gating system of the Na^+ channel, bind to different receptor structures and act in synergy.

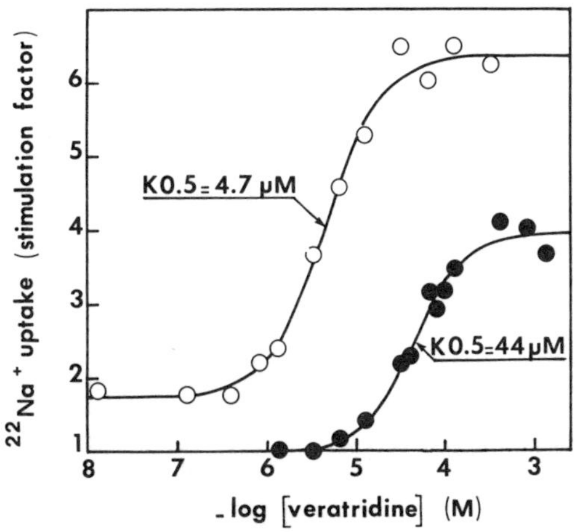

FIG. 5. Activation of $^{22}Na^+$ uptakes by neuroblastoma cells (NIE 115) using veratridine in the absence (*filled circles*) or presence (*open circles*) of ATX_{II} (13 μM). Standard experimental conditions for influx measurement were the same as in Fig. 4. $K_{0.5}$ values refer to the half-maximal effect of veratridine.

TTX at high enough concentrations inhibits the stimulation of $^{22}Na^+$ uptake provoked by ATX_{II} alone, veratridine alone, or ATX_{II} plus veratridine (Fig. 6). The apparent dissociation constants found for the TTX-receptor complex were $K_{0.5} = 5.6$ nM in 13 μM ATX_{II}, $K_{0.5} = 17.6$ nM in 133 μM veratridine, and $K_{0.5} = 38.9$ nM in 133 μM veratridine plus 13 μM ATX_{II}. This apparent dissociation constant is in fact considerably dependent on concentrations of ATX_{II} and veratridine (Fig. 6). $K_{0.5(TTX)}$ in the total absence of veratridine and TTX is found by extrapolation to be 2 nM (Fig. 7). When the Na^+ channel is kept open by a mixture of 13 μM ATX_{II} and of veratridine at concentrations higher than 50 μM, it displays a low affinity for TTX ($K_{0.5(TTX)} = 38$ to 39 nM). The presence of both ATX_{II} and veratridine at high enough concentrations decreases the affinity of TTX for its receptor by a factor of about 20. The open state of the Na^+ channel stabilized by a cocktail of both veratridine and ATX_{II} is a low affinity state for the TTX receptor.

The last way to analyze the binding properties of presynaptic neurotoxins to their receptor is to study transmitter accumulation and release by nerve terminals *in vitro* (1,11). Synaptosomes fulfill all of the criteria for stimulus-secretion coupling defined from the electrophysiological approach.

As shown in Figure 8, ATX_{II} stimulates GABA efflux from synaptosomes previously loaded with [^{3}H]GABA. The ATX_{II} concentration dependence of the release of GABA indicates a half-maximal effect at 0.02 μM. A similar type of effect has been observed with veratridine. The veratridine

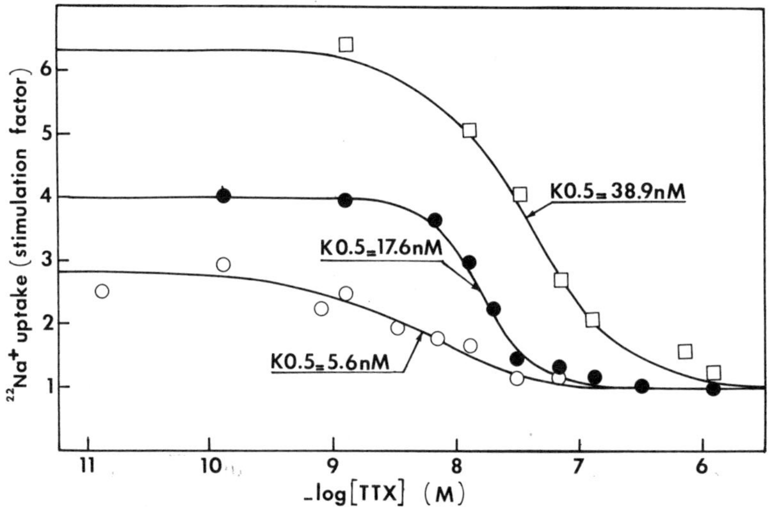

FIG. 6. TTX inhibition of ATX_{II} and veratridine-dependent $^{22}Na^+$ uptake in mouse neuroblastoma cells. *Open circles,* 13 μM ATX_{II}; *filled circles,* 133 μM veratridine; *open squares,* 133 μM veratridine and 13 μM ATX_{II}.

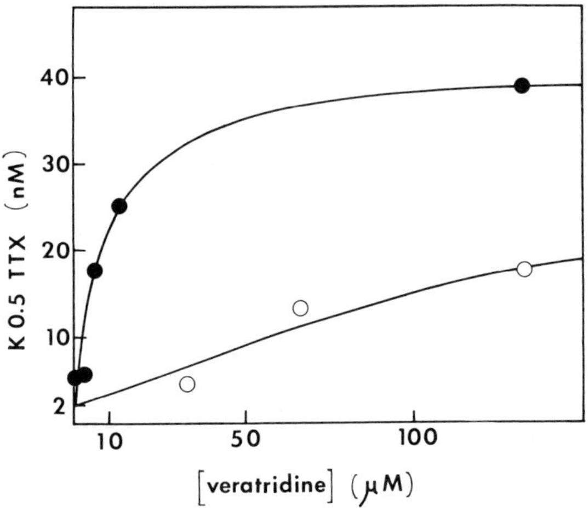

FIG. 7. The variation of the apparent dissociation constant ($K_{0.5(TTX)}$) of the complex formed between TTX and its receptor in the presence of veratridine and ATX_{II}. *Open circles,* no ATX_{II}; *filled circles,* 13 μM ATX_{II}. Values of $K_{0.5(TTX)}$ were determined as described in Fig. 6.

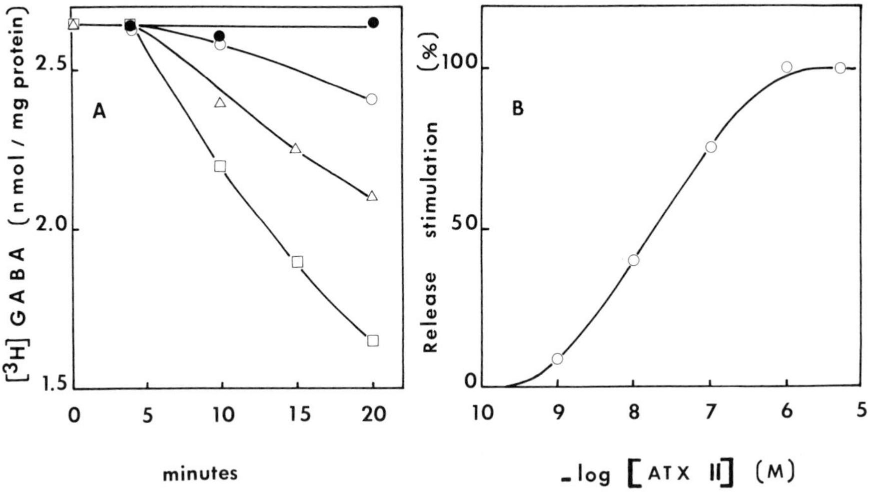

FIG. 8. A: Effect of ATX_{II} on the kinetics of [^{3}H]GABA release from preloaded synaptosomes. *Open circles,* control; efflux measured in the presence of 10 nM (*triangles*) and 3 μM (*squares*) ATX_{II}; *filled circles,* efflux measured in the presence of 3 μM ATX_{II} plus 1 μM TTX. **B:** ATX_{II} concentration dependence of the stimulation of [^{3}H]GABA release. Experimental points were obtained at pH 7.4, 22°C, after a release time of 20 min; 0% of release stimulation corresponds to the control measured in the absence of any toxin; 100% corresponds to the maximal stimulation which can be obtained at a high concentration of ATX_{II}.

concentration dependence of the release of [^{3}H]GABA indicated a half-maximum effect of the toxin at 10 μM at pH 7.4 and 22°C. Although veratridine and ATX_{II} both similarly provoke neurotransmitter release from rat brain synaptosomes, they bind to different receptor structures in the membranes. The effects of the two toxins have indeed been found to be additive.

In contrast to veratridine or ATX_{II}, TTX alone has no effect on GABA uptake. However, as shown in Fig. 8, TTX inhibits the effect of ATX_{II} on GABA release from loaded synaptosomes; it has the same action when ATX_{II} is replaced by veratridine. The $K_{0.5}$ values for TTX are 4 and 7.9 nM in the presence of 30 nM ATX_{II} and 12.5 μM veratridine, respectively.

SCORPION NEUROTOXIN

The chemical structure of several scorpion neurotoxins is now known. They are single-chain proteins containing 63 to 64 amino acids with eight half-cystine residues (9). The pharmacological mechanism of one of the pure toxins obtained from the venom of the scorpion *Androctonus australis hector* has been studied in much detail (5,11,13). The neurotoxin, similar to ATX_{II}, considerably slows down the closing of the Na^+ channel (there is also no sequence homology between the two types of toxins). However, in contrast to ATX_{II}, which was specific for the Na^+ channel, it also altered the opening of the K^+ channel (13). The effect of this toxin on axonal membranes is reversible. Both the electrophysiological analysis on crayfish muscular junction and the biochemical analysis using rat brain synaptosomes (11) indicated that the scorpion toxin stimulated neurotransmitter release, as previously described for ATX_{II} and veratridine. The apparent dissociation constant of the toxin-receptor complex is 0.1 to 0.2 μM at 22°C. The rate of dissociation of the toxin from its receptor was so slow that complex formation seemed quasiirreversible. TTX prevents scorpion toxin action if it is incubated with synaptosomes or with crayfish muscular junctions before scorpion toxin application. TTX does not reverse scorpion toxin action if it is added to the preparation after scorpion toxin.

CONCLUSION

In conclusion, neurotoxins, because they induce an abnormal neuronal behavior due to specific alterations of the selectivity filter or of the gating system of Na^+ and K^+ channels, are essential tools for solving the molecular mechanism responsible for the generation of an electrical signal by an excitable membrane.

ACKNOWLEDGMENTS

This work was supported by the Centre National de la Recherche Scientifique, the Commissariat à l'Energie Atomique, the Délégation

Générale à la Recherche Scientifique et Technique, and the Fondation pour la Recherche Médicale. The authors are grateful to Dr. M. C. Lenoir and N. Alenda for expert assistance and to Dr. H. Schweitz for kindly providing us with the sea anemone toxin.

REFERENCES

1. Abita, J. P., Chicheportiche, R., Schweitz, H., and Lazdunski, M. (1977): *Biochemistry,* 16:1838–1844.
2. Balerna, M., Fosset, M., Chicheportiche, R., Romey, G., and Lazdunski, M. (1975): *Biochemistry,* 14:5500–5511.
3. Benzer, T. I., and Raftery, N. A. (1972): *Proc. Natl. Acad. Sci. USA,* 69:3634–3637.
4. Beréss, L., Beréss, R., and Wunderer, G. (1975): *Toxicon,* 13:359–367.
5. Bernard, P., Couraud, F., and Lissitzky, S. (1977): *Biochem. Biophys. Res. Commun.,* 77:782–788.
6. Catterall, W. A. (1975): *Proc. Natl. Acad. Sci. USA,* 72:1782–1786.
7. Henderson, R., and Wang, J. H. (1972): *Biochemistry,* 11:4565–4569.
8. Kimhi, Y., Palfrey, C., Spector, I., Barak, Y., and Littauer, U. Z. (1976): *Proc. Natl. Acad. Sci. USA,* 73:2682–2686.
9. Kopeyan, G., Martinez, G., Lissitzky, S., Miranda, F., and Rochat, H. (1974): *Eur. J. Biochem.,* 47:483–489.
10. Narahashi, T. (1974): *Physiol. Rev.,* 54:813–889.
11. Romey, G., Abita, J. P., Chicheportiche, R., Rochat, H., and Lazdunski, M. (1976): *Biochim. Biophys. Acta,* 448:607–619.
12. Romey, G., Abita, J. P., Schweitz, H., Wunderer, G., and Lazdunski, M. (1976): *Proc. Natl. Acad. Sci. USA,* 73:4055–4059.
13. Romey, G., Chicheportiche, R., Lazdunski, M., Rochat, H., Miranda, F., and Lissitzky, S. (1975): *Biochem. Biophys. Res. Commun.,* 64:115–121.
14. Spector, I., Kimhi, Y., and Nelson, P. (1973): *Nature,* 246:124–126.
15. Ulbricht, W. (1969): *Ergeb. Physiol. Biol. Chem. Exp. Pharmakol.,* 61:18–71.
16. Wunderer, G., Fritz, H., Wachter, E., and Machleidt, W. (1976): *Eur. J. Biochem.,* 68: 193–198.

Abnormal Neuronal Discharges, edited by
N. Chalazonitis and M. Boisson. Raven
Press, New York © 1978.

Electrophysiological Studies of Molluscan Neurons Generating Bursting Pacemaker Potential Activity

J. L. Barker and T. G. Smith, Jr.

Laboratory of Neurophysiology, National Institute of Neurological and Communicative Disorders and Stroke, National Institutes of Health, Bethesda, Maryland 20014

A wide variety of studies on clinical and experimental epilepsy has demonstrated a number of electrophysiological events that correlate with and may underlie epileptic seizures. One particular type of event, found with intracellular recordings in epileptogenic foci in vertebrate CNS neurons, is a rapid and often prolonged depolarization called a depolarizing shift (DS) (1,3,51,52,56). These DSs, which often occur randomly or paroxysmally, are associated with a rapid burst of action potentials. Qualitatively similar paroxysmal burst activity (PBA), as well as periodic burst activity, can also be recorded with extracellular electrodes from other vertebrate CNS neurons during pathophysiological states. An example of such burst activity is that found in the magnocellular neurons of the hypothalamus during cardiovascular stress (22,23,27,35,41,50,62,63).

There is, of course, a keen interest on the part of scientists studying PBA in the vertebrate CNS in the neuronal mechanisms underlying such abnormal neuronal discharges. Indeed, this volume is a manifestation of that interest. Technical limitations, however, do not often allow the types of electrophysiological experiments necessary to investigate questions of mechanism.

On the other hand, recent research on experimentally more manageable invertebrate preparations has contributed to an understanding of the mechanisms underlying a form of neuronal activity associated with bursts of action potentials (2,10,12–15,18,24,28,29,33,36,38,39,55,57,58,60,66). Unlike the PBA in vertebrate CNS, the bursts on invertebrate neurons normally occur in a regular pattern (4–6,18,31,40,58). Moreover, since this activity has been shown to be endogenous to the cell generating regular bursts of spikes, it has been called "bursting pacemaker activity" (BPA) (2,18,40,58). In addition, the slow membrane potential oscillations that lead, alternately, to bursts of spikes and to silent periods has been called the "bursting pacemaker potential" (BPP) (2,18,40,58).

Clearly, the inclusion of this and similar chapters in this volume is, in part, an attempt to define the similarities and differences in phenomena and in mechanisms of vertebrate and invertebrate neurons that fire action potentials in a bursting manner. While we welcome such attempts, we would caution against hasty conclusions that assume that similar phenomena imply similar mechanisms.

For didactic purposes, the mechanism underlying bursting neuronal activity can be divided into two general categories: (a) those that involve the endogenous membrane properties of individual neurons, and (b) those that involve the interconnections between neuronal elements. While it is highly likely that both kinds of mechanisms are involved in bursting activity in complex neuronal systems, the cells we have studied (cell R_{15} in *Aplysia californica* and cell 11 in *Otala lactea*) allow an examination of the former category of BPA (30,31).

In this chapter, we focus on three areas of interest: (a) an identification of the ionic carriers of charge and hence the ionic pacemaker conductances underlying BPPs, (b) an analysis of the ionic and other compounds which regulate the pacemaker conductances, and (c) a presentation and analysis of events recorded in invertebrate neurons similar to events found in vertebrate CNS neurons that also fire in a bursting mode. We rely heavily on voltage-clamp data from our own and other laboratories.

RESULTS

Endogenous BPPs

The late Dr. Barbara Alving (2) first conclusively demonstrated that the BPPs (Fig. 1B, inset) generated in molluscan neurons can be endogenous rather than driven by synaptic inputs. She found that ligation of the axon and physical isolation of R_{15} in *Aplysia* did not abolish BPPs. Other important but less dramatic data supporting endogenous BPA are that BPPs could be abolished by applying sufficiently large exogenous depolarizing or hyperpolarizing current (6,18,20,40,58) and that, under voltage-clamp conditions, no oscillatory currents are observed at any constant membrane potential (28,29,33,55,66). There is little doubt, therefore, that BPP activity can be endogenous to a single cell and that the basis for BPP activity can result from the membrane properties of a single neuron.

Resting Potential

Cells that do not generate oscillatory membrane potential activity can be voltage clamped at a membrane potential that does not require the application of extrinsic clamp current. This membrane potential is the resting potential. There is no such potential in the range of the BPPs (−60 to −20 mV) in

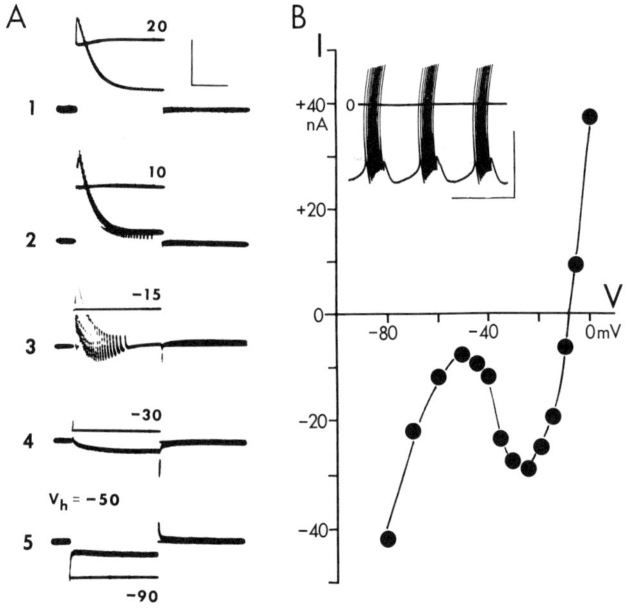

FIG. 1. Data from cell R_{15} of *Aplysia californica*. **A:** From voltage clamp. Voltage (*thin lines*) and current (*thick lines*) traces for 5 sec pulse commands from holding potential ($V_{holding}$, V_h) of −50 mV to potentials indicated by numbers adjacent to voltage trace. Traces begin with voltage and current traces superimposed. Positive-going potentials and (outward) currents are displayed upward. Calibration: 45 mV for voltage traces, 500 nA (*trace 1*); 200 nA (*trace 2*); 100 nA (*trace 3*); and 50 nA (*traces 4* and *5*) for current traces; 2.5 sec. **B:** Quasisteady state I-V curve from voltage-clamp data shown in **A.** *Inset,* spontaneous BPP oscillations in the unclamped cell. Calibrations: 50 mV, 26 sec. In this and some subsequent figures, zero membrane potential is indicated by a horizontal line marked 0. (Reproduced from ref. 55, with permission of *Nature*.)

molluscan neurons. When clamped to any steady potential more negative than, e.g., −20 mV, the membrane generates a steady inward current (Fig. 1B) (28,29,33,55,66). This steady inward current implies a constant depolarizing bias on the membrane potential. These observations, plus the obvious fact that the unrestrained (unclamped) membrane potential is always changing (Fig. 1B, inset), mean that cells with BPPs have no resting potential. In the unrestrained condition, the sum of the membrane currents are necessarily equal to zero, but the membrane potential continuously changes because the time-derivative of the currents, dI/dt, is never constantly zero.

N-Shaped Current-Voltage Curve

In every carefully examined molluscan preparation generating BPPs, voltage-clamp studies have shown a region of negative slope conductance, which roughly corresponds to the membrane potential range of the BPP

(14,21,24–26,28,33,36–38,55,56). The pertinent observation in this regard is that, unlike conventional neurons, depolarizing commands over a certain range of potential result in a steady increase in inward current (Fig. 1A4). Thus, when the currents associated with a wide range of membrane potential steps are examined, the steady state I-V curve is N-shaped (Fig. 1B) (14, 24,25,28,33,36,37,38,55,66). Systems with such curves are potentially unstable and, as is well known, any point on a negative slope is not a stable point in an unrestrained system. Stability only coincides with points on a positive slope (55).

Pacemaker Conductances

Having established the relationship between BPPs and the N-shaped I-V curve, the question arises as to which ions and, thereby, which ionic conductances underlie the measured currents. As is well known, the Na^+ and K^+ spike conductances can be separated and analyzed over a wide range of membrane potential because of their different time- and potential-dependent characteristics. No such differences exist over a wide range of membrane potential for the pacemaker conductances. Rather, one only has "windows" in time and in membrane potential to investigate these conductances.

Na^+ Pacemaker Conductance

Within the range of membrane potential between about −50 and −35 mV, the region of the negative slope, the steady inward current is the dominant one. We have found in *Otala* that the magnitude of this current, at a given membrane potential, varies directly with $[Na^+]_O$ (Fig. 2B and D). In addition, the magnitude of the negative slope also varies with $[Na^+]_O$ (Fig. 2C). Furthermore, the curve obtained by subtracting the I-V curve with no negative slope from that with maximum negative slope (Fig. 2E) shows that the Na^+-dependent currents are due to membrane conductances that are voltage independent (G_1) and voltage dependent (G_2) (55).

In *Aplysia,* replacement of $[Na^+]_O$ with $[Li^+]_O$ abolishes the region of negative slope (Fig. 3) but not the response of the membrane to ACh (Fig. 3, inset). Furthermore, the difference curve (Fig. 4) shows the Na^+-dependent currents. In the absence of strong evidence to the contrary, these data indicate that Na^+ ions are the main charge carriers for these steady inward currents and that the corresponding voltage-independent and voltage-dependent pacemaker conductance is a Na^+ conductance.

Other investigators (28,36,37) have suggested that, in *Helix,* a Ca^{2+} conductance underlies the persistent inward current. Figures 5 and 6 indicate that the main charge carrier is not Ca^{2+} in *Aplysia.* In Fig. 5, zero $[Ca^{2+}]_O$ increases the amplitude and duration of the BPP (cf. Av with BV) and does not abolish the inward current (cf. Fig. 5A and B). Moreover,

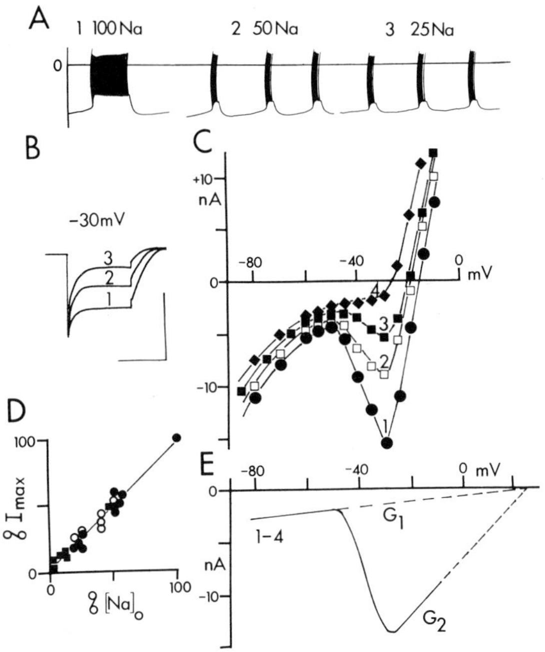

FIG. 2. Effects of altering $[Na^+]_O$ on membrane properties of BPP neurons. Data in **A–C** and **E** from cell 11 of *Otala lactea*. **D** contains data from both cell 11 and cell R_{15}. Cs^+ ions were iontophoresed intracellularly by means of a microelectrode in cell 11 before measurements. **A–C:** data labeled *1* (*filled circles*), *2* (*open squares*), and *3* (*filled squares*) were obtained when the $[Na^+]_O$ was, respectively, 100, 50, and 25 mM (Na^+ replaced with $Tris^+$). Other extracellular ions were kept constant (10 mM $SrCl_2$, 4 mM KCl). **A:** Effect of reducing $[Na^+]_O$ on the BPP. **B:** Line traces of voltage-clamp currents recorded in pulsing the membrane from holding of −50 to −30 mV for 5 sec in different $[Na^+]_O$s. **C:** Plots of voltage-clamp I-V curves. The curve linked by *filled diamonds* was taken in 25 mM NaCl, 80 mM $SrCl_2$, 4 mM KCl, and 5 mM Tris-Cl; V_h always −50 mV. **D:** Clamp current, at −30 mV, as a percentage of current required to clamp the membrane at −30 mV in maximum $[Na^+]_O$ against the percentage of the maximum $[Na^+]_O$. $[Na^+]_O$ replacements were Li^+ (*filled squares*), $Tris^+$ (*filled circles*), and sucrose (*open circles*). **E:** Curve *1* minus curve *4* of **C.** Solid parts of the curve are from data in **C;** dashed lines are straight-line extrapolations. Calibrations: 90 mV, 25 sec **(A);** 12.5 nA, 4 sec **(B).** (Reproduced from ref. 55, with permission of *Nature*.)

when the BPP, negative slope, and inward current are examined under stepwise changes in $[Ca^{2+}]_O$, they all vary inversely with $[Ca^{2+}]_O$ (Fig. 6) (see also ref. 59). Similar results have been found with *Otala* (*unpublished observations*).

A crucial consideration in determining the charge carrier and ionic conductance is the application of Ohm's law for an electrochemical system, namely, I = g(V-E). For an ion whose equilibrium potential (E) is less negative than membrane potential (V), the sign of the current (I) should be negative (or inward). Most important, however, is that the magnitude of the in-

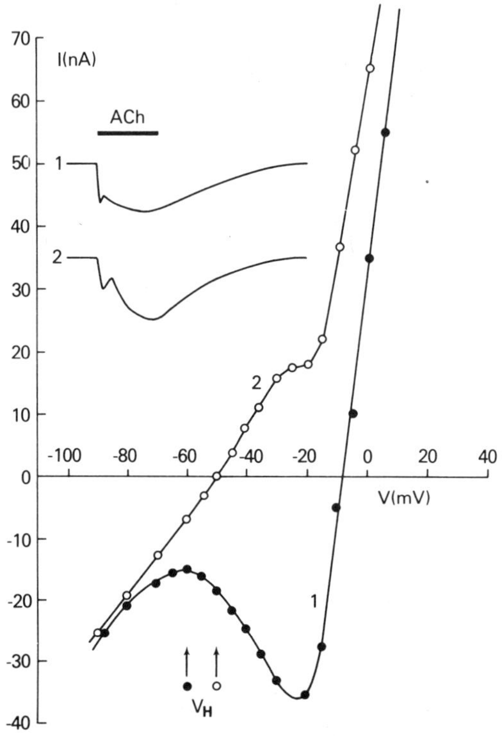

FIG. 3. Effect of replacement of $[Na^+]_0$ with $[Li^+]_0$ on membrane properties of cell R_{15} of Aplysia. Quasisteady state voltage-clamp I-V curves and bath-applied ACh current responses in the presence of : *1* (*filled circles*), $[NaCl]_0 = 500$ mM, $[LiCl]_0 = 0$; and *2* (*open circles*), $[NaCl]_0 = 0$, $[LiCl]_0 = 500$ mM. All other extracellular ions constant (10 mM $CaCl_2$, 60 mM $MgCl_2$, 10 mM KCl, 15 mM TrisCl, pH 7.8). V_h, holding potential; bar under ACh, 5 sec application of ACh. $[Li^+]_0$ replacement of $[Na^+]_0$ abolishes negative slope of I-V curve but not ACh response.

ward current should decrease as E approaches V; that is, I should decrease when the extracellular concentration of the current-carrying ion is decreased, assuming that the intracellular concentration of that ion is not significantly changed. This relationship, in *Aplysia* and *Otala,* corresponds to that found with changes in $[Na^+]_0$; the opposite is found with changes in $[Ca^{2+}]_0$. It is reasonable to conclude, therefore, that most of the inward current measured in *Aplysia* and *Otala* is carried by Na^+ ions and that one pacemaker conductance underlying these currents and, by extrapolation, the depolarizing phase of the BPP is a Na^+ conductance and not a Ca^{2+} conductance.

In addition to the voltage characteristics of this Na^+-pacemaker conductance, its characteristics in time are noteworthy. First, at constant potential, it shows little or no inactivation. Second, it changes rapidly with changes

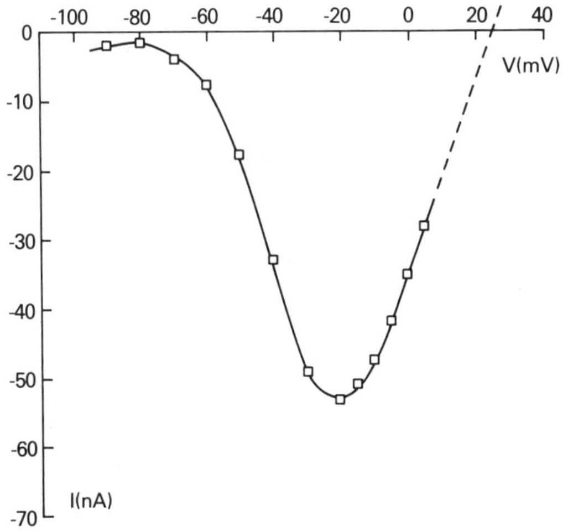

FIG. 4. $[Na^+]_O$-dependent I-V curve. Difference I-V curve of curve 1 minus curve 2 of Fig. 3. Curve shows Na^+-dependent currents from Na^+-pacemaker conductance. Circles and solid line from data. Dotted line is linear extrapolation from data curve. See text.

in membrane potential ($\leqslant$ 200 μsec), so rapidly, in fact, that we have been unable to measure its chord conductance.

K^+-Pacemaker Conductance

While there is a range of membrane potential over which the Na^+-pacemaker currents can be measured with little contamination with other currents, no such situation exists for the K^+-pacemaker conductance during a step command. The most reliable measures we have found of the K^+-pacemaker conductance are the slow tail currents observed at the holding potential following a long depolarizing step command (Fig. 7, left, I). These tail currents (I_t) are associated with an increase in membrane conductance (Fig. 7, right, I). Moreover, two-step command, voltage-clamp experiments (Fig. 8AV) show that the higher the $[K^+]_O$, the more depolarized the equilibrium potential (Fig. 8B, arrows, where $I = 0$). Furthermore, the peak magnitude of the late tail current, as a function of membrane potential, changes with $[K^+]_O$ in the direction expected if these currents are K^+. (Fig. 9). In addition, over a range of membrane potentials between −40 and −10 mV, there is a late outward-going current (Fig. 5AI; Fig. 6A, −36 mV). This end current (I_e) probably represents the turn-on of the K^+-pacemaker conductance, since it varies with $[K^+]_O$ and with the tail current (I_t) (Fig. 10). Quantitative measurements of this I_e are unreliable as a measure of the K^+-pacemaker conductance, however, since other currents are generated simultaneously from other ionic sources.

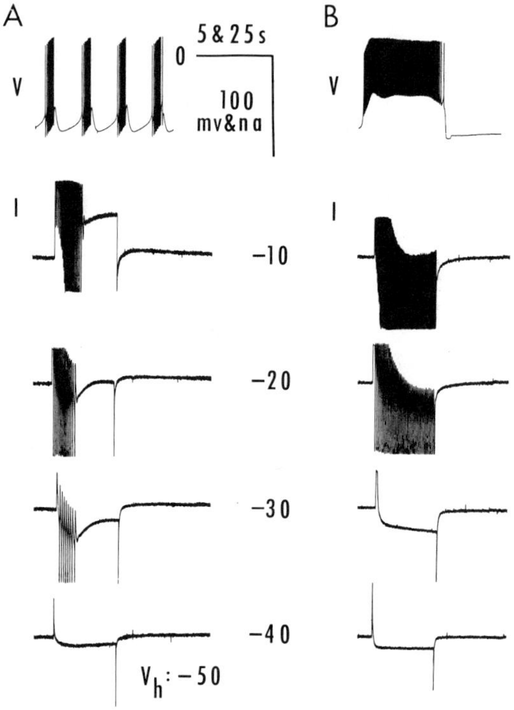

FIG. 5. Effect of $[Ca^{2+}]_0$-free seawater on BPPs and voltage-clamp currents in R_{15} of *Aplysia*. **A**: NaCl = 500 mM; $CaCl_2$ = 10 mM; $MgCl_2$ = 60 mM; KCl = 10 mM; TrisCl = 15 mM; pH 7.8. **B**: Same solution as in **A,** except $CaCl_2$ = 0 mM. Removal of $[CaCl_2]_0$ increases amplitude and duration of BPP and number of spikes per burst (compare AV with BV). In AI and BI, holding potential (V_h) = −50 mV. Traces show voltage-clamp currents evoked by 5 sec steps to potentials shown between **A** and **B.** $[Ca^{2+}]_0$-free solution does not abolish persistent inward current (cf. −40 and −30 in **A** and **B**) but does remove most of late outward-going current during the step and the slow outward tail currents (cf. −10, −20, and −30 in **A** and **B**). Calibrations: 5 sec and 100 nA for I-traces; 25 sec and 100 mV for V-traces.

Unlike the Na^+-pacemaker conductance, the K^+-pacemaker conductance changes slowly with time, following changes in membrane potential. The slowly developing outward current during depolarizing steps (I_e) and slowly relaxing tail currents following a return to the holding potential (I_t) have time constants of many seconds. As has been suggested, sufficient activation of this K^+ conductance provides the hyperpolarizing bias which terminates the burst of spikes and results in the silent, hyperpolarized phase of the BPP; its inactivation results in the early depolarizing phase of the BPP (11,15,18, 28,29,33,36,37,39,40,55,58,66). Clamping the membrane during various phases of the BPP shows the slowly decreasing outward current to be minimum at the onset of the bursts of spikes and maximum toward the end

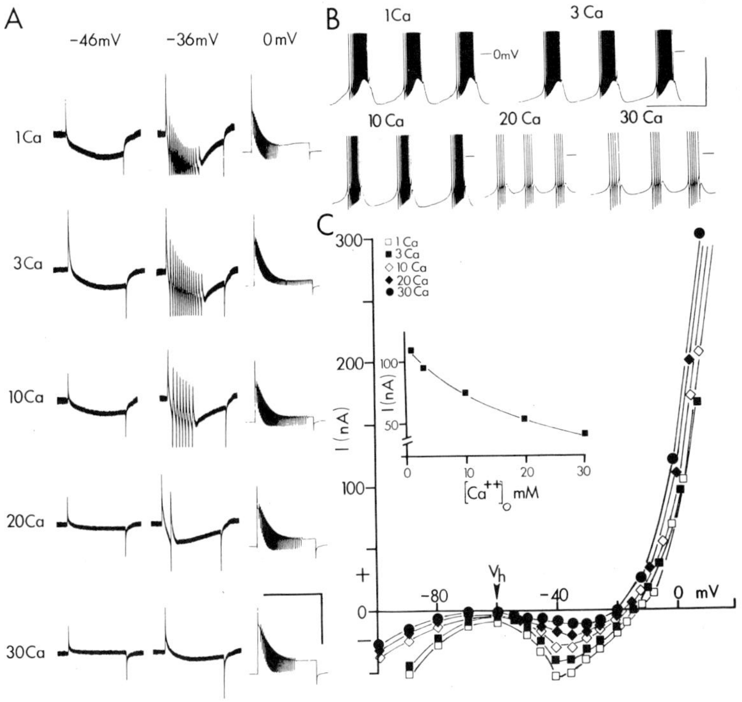

FIG. 6. Effects of stepwise changes in $[Ca^{2+}]_O$ on membrane properties of R_{15} in *Aplysia*. **A:** Voltage-clamp currents evoked by 5 sec steps from holding potential of −60 mV to potential above each row (−46, −36, and 0 mV) in various concentrations of $[Ca^{2+}]_O$ shown in column (1 to 30 Ca). Calibrations: 5 sec for all traces; 100 nA for −46 and −36 mV; 500 nA for 0 mV. **B:** Sample traces of BPPs in concentration of $[Ca^{2+}]_O$ shown above trace. Calibration: 50 mV and 25 sec. **C:** Steady state I-V curve in different concentrations of $[Ca^{2+}]_O$. Symbols at top of graph indicate $[Ca^{2+}]_O$. *Inset,* current at −40 mV minus leakage current as a function of $[Ca^{2+}]_O$. Decreasing $[Ca^{2+}]_O$ results in increase in inward current (**A,** inset in **C**), in increase in amplitude and duration of BPP and in spikes per burst **(B),** and increase in negative slope **(C).** With changes in $[Ca^{2+}]_O$, all other ions constant: 100 mM NaCl, 60 mM $MgCl_2$, 10 mM KCl, 15 mM TrisCl, pH 7.8.

of the burst (Fig. 11). Although it has yet to be shown, it would appear that the duration of the burst of spikes is determined by the turn-on kinetics and the duration of the interburst interval by the turn-off kinetics of the K^+-pacemaker conductance (see ref. 53).

Modulators of the Pacemaker Conductances

Ca^{2+} Ions

While we have shown in *Aplysia* and *Otala* that the pacemaker conductance is a Na^+ and not a Ca^{2+} conductance, it should not be construed that

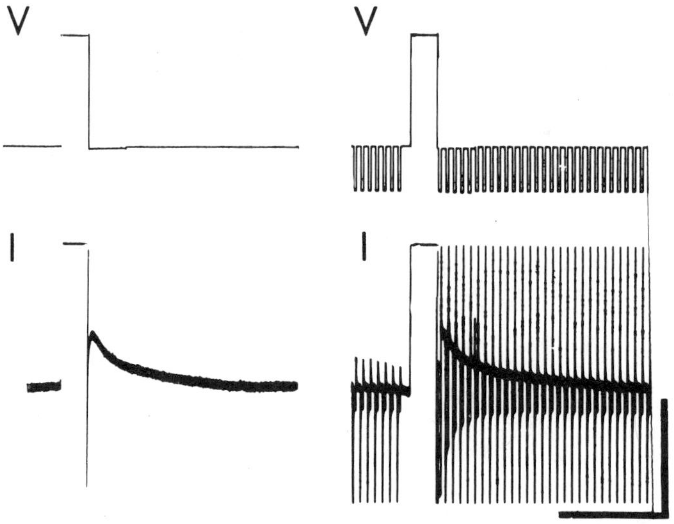

FIG. 7. Voltage-clamp tail currents from R_{15} of *Aplysia.* **Left:** I shows tail current when potential returned from 5 sec command to 0 mV (V). **Right:** I shows same as in **left,** plus the currents required for 0.5 sec, 20 mV hyperpolarizing pulses before and following 5 sec command to 0 mV (V). Increase in pulse currents during tail current indicates an increase in membrane conductance whose time-course tracks the tail current.

Ca^{2+} ions play no role in pacemaking. Several investigators have shown that $[Ca^{2+}]_i$ increases with the depolarizing phase of the BPP (Fig. 12C and D); the increases correspond in time to the occurrence of action potentials (Fig. 12A) (57,60). Furthermore, our conclusions should not indicate that there is no Ca^{2+} or divalent cation conductance in the membrane of pacemaker neurons. There clearly is such a conductance, as has been shown by others (24,26,28,34,36,37) and indicated by the data in Fig. 13. In Fig. 13A, D600, a blocker of divalent cation conductances, does not block the negative slope region of the I-V curve (cf. curves 1 and 2), but this negative slope is abolished when $[Ba^{2+}]_0$ ions replace both $[Na^+]_0$ and $[Ca^{2+}]_0$ (cf. curves 2 and 3). On the other hand, the magnitude of inward current is greatly increased when $[Ba^{2+}]_0$ ions replace $[Na^+]_0$ and $[Ca^{2+}]_0$ in the absence of D600 (Fig. 13, cf. curves 1 and 2) and the membrane can generate burst activity (similar to that shown in Fig. 5BV) (24,26,34). We interpret these results as indicating that burst activity can involve monovalent (Na^+) or divalent (Ba^{2+}) ions as the main charge carriers; but the inward currents under physiological conditions of BPP are caused mainly by Na^+ ions, since D600 does not block this current (Fig. 13A, curves 1 and 2). On the other hand, the oscillatory behavior of the membrane potential in the presence of $[Ba^{2+}]_0$ would appear to be an artifactual situation that bears no direct relationship to the physiological mechanisms of normal

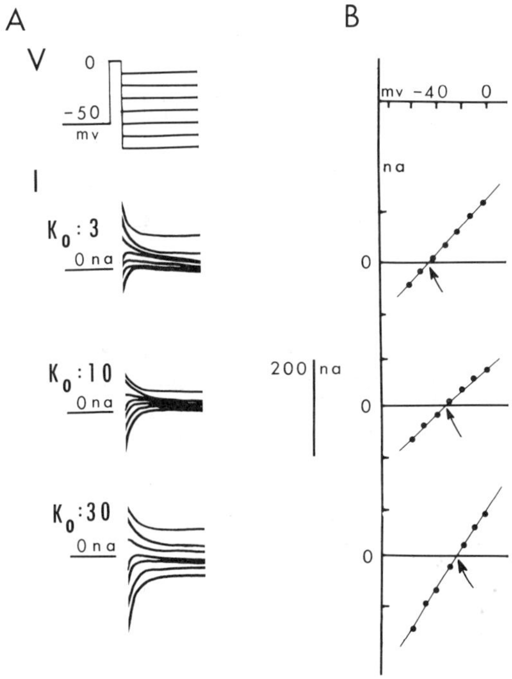

FIG. 8. Variations of tail currents and of driving potential as a function of $[K^+]_0$ in R_{15} of *Aplysia*. AV shows experimental procedure: 5 sec steps from holding potential of −50 to 0 mV, followed by steps to different membrane potentials. **A1** shows tail currents recorded at various potentials as a function of $[K^+]_0$ (3, 10, 30 mM KCl; all other ions constant: 500 mM NaCl, 10 mM $CaCl_2$, 60 mM $MgCl_2$, 15 mM Tris, pH 7.8). Zero current shown as baseline, marked "0 nA." **B:** Plot of initial tail current as a function of membrane potential (shown at top) and of $[K^+]_0$ (to right of corresponding tail currents). Calibration: 200 nA applies to AI and B. Equilibrium potential (arrow: membrane potential where I = 0), becomes less negative with increasing $[K^+]_0$, as expected if tail currents are K^+ currents. See text.

BPP generation. Apparently, a similar situation is found in *Aplysia* or *Helix* neurons that can be induced to generate paroxysmal bursts with PTZ or Ba^{2+}; namely, the PTZ-induced bursts involve Na^+ ions as the main charge carrier, whereas the Ba^{2+}-induced bursts utilize Ba^{2+} ions (24).

The question arises, then, what is the role of the pulsatile increase in $[Ca^{2+}]_i$ during the depolarizing phase of a BPP? In the following, we present evidence that this $[Ca^{2+}]_i$ activates the K^+-pacemaker conductance in a manner similar to that proposed by others and that found in other excitable cells (28,42–44,60). For example, Fig. 13A shows that the blockage of the Ca^{2+} entry by D600 results in a marked reduction of the outward K^+ current (cf. curves 1 and 2). In addition, the absence of $[Ca^{2+}]_0$ is associated with the absence of a late outward-going K^+ current during a depolarizing step and of long, slow tail currents (Fig. 5, cf. A and B at −30 and −40 mV).

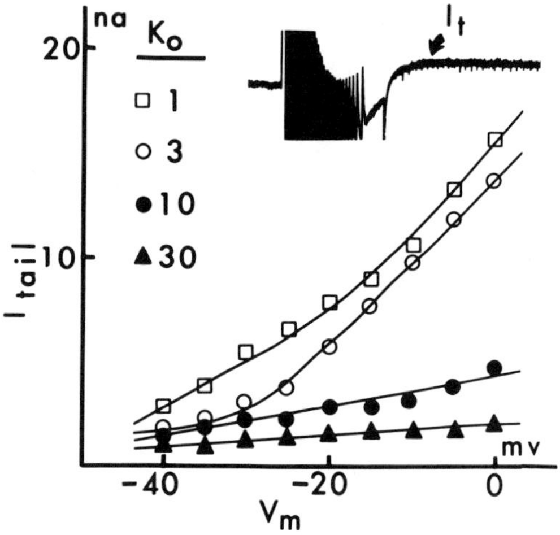

FIG. 9. Magnitude of outward tail currents as a function of membrane potential and $[K^+]_0$. Plot of magnitude of outward tail current (I_t in inset points to tail current) on ordinate as a function of potential to which membrane was clamped during a 5 sec depolarizing step. All currents recorded after the return of membrane to holding potential of −50 mV. Legend under "K_0" indicates symbols which correspond to each $[K^+]_0$. Graph shows tail currents increase with increase in driving force of K^+ battery, i.e., as equilibrium potential of K^+, E_K, becomes more negative and $-50\text{ mV} - E_K$ increases. Results consistent with interpretation that tail currents are mainly K^+ currents.

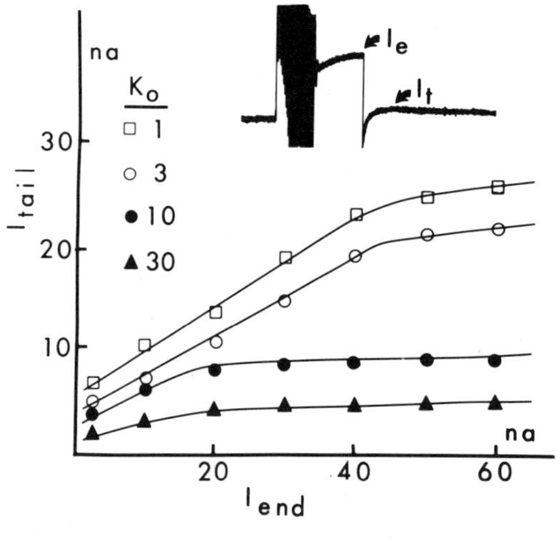

FIG. 10. Relationship between end-of-pulse currents and tail currents. Plot of magnitude of tail currents (I_t) versus the current recorded at the end (I_e) of a 5 sec depolarizing command. Legend under "K_0" indicates symbols which correspond to each $[K^+]_0$. Graph shows that I_t and I_e are directly related and hence consistent with slowly increasing I_e, mainly a K^+ current.

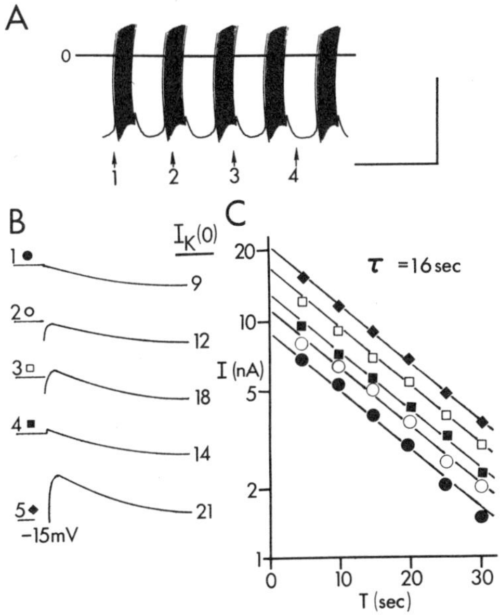

FIG. 11. Currents recorded by voltage clamping at different phases of the spontaneous BPP cycle. **A:** Arrows indicate times during the BPP cycle when the cell was clamped to −50 mV. **B:** *Traces 1–4,* line drawings of the currents recorded at the times numbered in **A.** Zero current indicated by the horizontal lines below the numbers in each trace. *Trace 5,* tail current recorded when the cell was returned to V_h (−50 mV) from a 5 sec step to −15 mV. **C:** Semilog plot of current time for currents shown in **B.** Slowly declining currents all fit an exponential with a time-constant of decay (τ) of about 16 sec. Zero time was the onset of the clamp for *1–4* and the beginning of the tail current in *5.* Zero time intercepts of each of these curves are listed under $I_K(0)$ in **B.** Calibration: 60 mV, 30 sec in **A;** 60 nA, 30 sec in **B.** (Reproduced from ref. 55, with permission of *Nature.*)

Furthermore, the development of significant late K^+ current during a step and of a slow K^+ tail current is dependent on the presence of spike currents during a depolarizing step (see Fig. 5A, −40 mV versus −30 and −20 mV; see Fig. 6A −46 mV versus −36 mV, for all values of $[Ca^{2+}]_0 \leqslant 20$ mm).

We have attempted a number of analyses of various membrane parameters to implicate a role for $[Ca^{2+}]_i$; however, the best relationship we have found is with the pulse end current (I_e) and the number of spikes in a burst (Fig. 14) (30,40). Since molluscan spikes have been shown to have a Ca^{2+} component (18,39), since $[Ca^{2+}]_i$ has been shown to increase in a pulsatile manner with spikes (Fig. 12A) (57,60), and since we have indicated a relationship between I_e and $[K^+]_0$ (Figs. 9 and 10), these data suggest that the increase in $[Ca^{2+}]_i$ during the spikes of a burst activates a component of the K^+-pacemaker conductance, which initiates the hyperpolarizing phase of the BPP and terminates the burst.

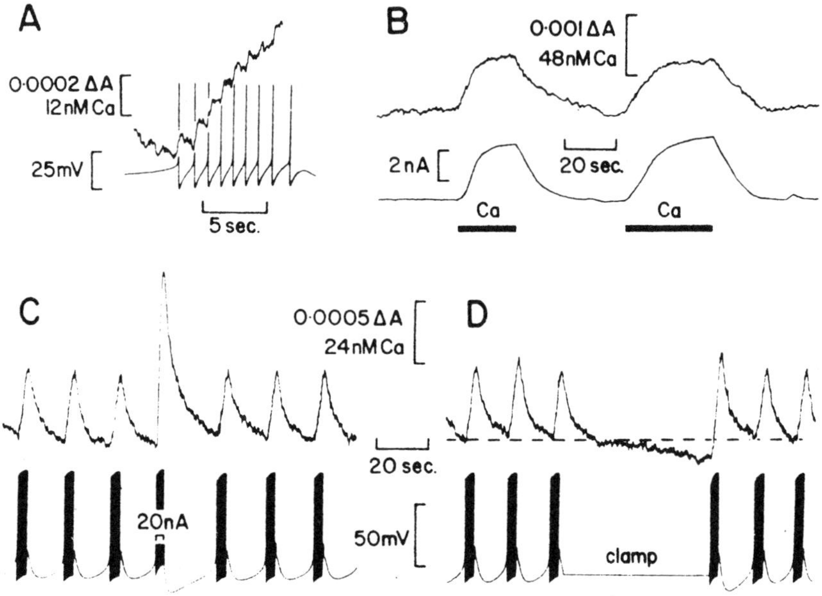

FIG. 12. Changes in free internal Ca^{2+} concentration under normal and experimentally manipulated conditions in R_{15} of *Aplysia.* **A:** Absorbance increase (*upper trace*) during the burst shown (*lower trace*) at high resolution. This result suggests that internal Ca^{2+} (*diagonal trace*) increases primarily in steps coincident with each action potential (*bottom trace*). **B:** Absorbance increase (*upper trace*) and outward membrane current seen when Ca^{2+} is injected under pressure into a voltage-clamped cell. **C:** Effect of a 20 nA depolarizing current, passed during a single burst, on the dye absorbance (*upper trace*) and membrane potential (*lower trace*). There is a larger membrane hyperpolarization, and the next burst is delayed until the absorbance has almost declined to its normal baseline value. Records **C** and **D** are from the same cell and share the same calibration. **D:** Effect of voltage-clamping R_{15} to the average value of the resting potential. The dye absorbance (*upper trace*) declines below its normal baseline value (*dashed line*) at the time when the next burst is expected. The next burst occurs immediately on release from the clamp, and the frequency of action potentials is somewhat higher than normal. (Reproduced from ref. 60, with permission of the authors and *Science.*)

Polypeptides

During the hibernating phase of its life cycle, the bursting pacemaker cell of *Otala,* cell 11, does not generate BPA in physiological Ringer solution, even with the application of extrinsic current. The cell is either electrically quiescent or generates random or semiregular spikes (Fig. 15, right, inset, control) (10,13). Furthermore, the voltage-clamp I-V curve has no region of negative slope (Fig. 15, right, control) (14). However, the addition of nanomolar to micromolar amounts of the antidiuretic hormone [a polypeptide also called lysine vasopressin (LVP)] initiates BPP activity within a matter of seconds (Fig. 15, right, inset, vasopressin) (10,13), and the I-V curve shows a region of negative slope (Fig. 15, right, vasopressin) (14). As

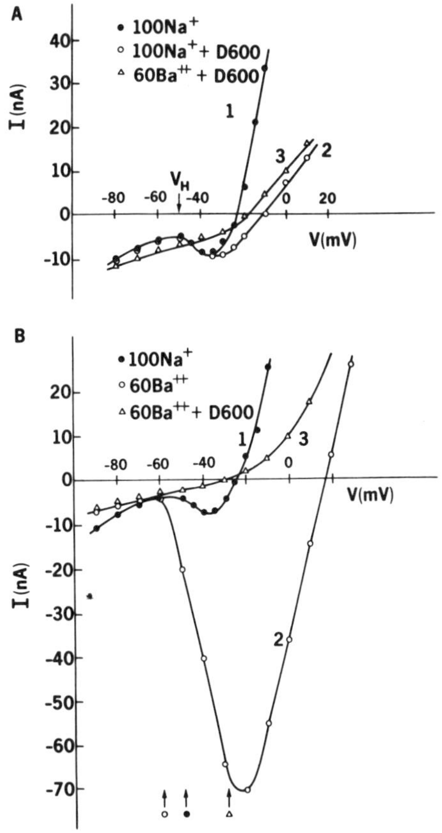

FIG. 13. Effects of D600 on voltage-clamp I-V curves on cell 11 of *Otala*. In **A** and **B**, symbols above graphs indicate experimental conditions. **A** and **B:** *1* (*filled circles*), I-V curve in normal Ringer solution (100 NaCl, 10 $CaCl_2$, 4 KCl, 5 Tris Cl, pH 7.8). **A:** *2* (*open circles*), D600, a blocker of Ca^{2+} conductance, does not abolish region of negative slope but does decrease outward current. When extracellular NaCl and $CaCl_2$ are replaced with 60 mM $BaCl_2$ in the presence of D600, the negative slope region is abolished. Holding potential (V_H) of −50 mV for all curves. **B:** *2* (*open circles*), when $BaCl_2$ replaces extracellular NaCl and $CaCl_2$, inward current is markedly increased and is abolished by D600. Arrows above symbols in lower part of **B** indicate holding potentials. See text.

in *Aplysia*, the inward current producing the negative slope is a Na^+ current; thus LVP appears to activate the Na^+-pacemaker conductance (55). At first glance, the shift of the outward current to the right along the voltage axis might suggest a decrease in the K^+ current of the K^+-pacemaker conductance. From an analysis of the slow tail currents, however, the K^+ currents are present and larger over a wider range of membrane potential with LVP than in control (Fig. 16B, cf. LVP versus CON). In addition, LVP shortens the time-constant of the K^+ inactivation at −40 mV (Fig. 16B,

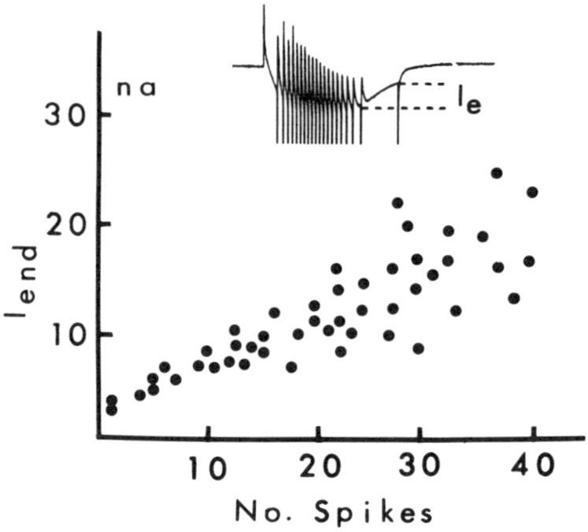

FIG. 14. Relationship between end-of-pulse current and number of spikes per burst. On the ordinate, the magnitude of the outward-going current at the end of a 5 sec depolarizing step (I_e, shown in inset) to various membrane potentials (not indicated) is plotted versus the number of action current spikes during command. See text.

bottom) and reduces the decay of poststimulus hyperpolarization to a fixed depolarizing stimulus (Fig. 16A).

In contrast, LVP does not affect the Na^+ and Ca^{2+} spike currents (not shown) or the magnitude and kinetics of the spike, delayed rectifier, K^+ currents (Fig. 17). Thus the progressive prolongation of the action potentials during a burst is probably more a consequence of a frequency- and time-dependent inactivation of the spike K^+ conductance than a direct action of the peptide on that conductance. Two other interesting characteristics of the action of LVP are: (a) the excitatory effect is specific to cell 11, and (b) the effect lasts for minutes to hours after removal of the peptide from the bathing medium (10,13).

Contrary to these excitatory effects of LVP on cell 11, another cell in *Otala* is inhibited by the peptide. The addition of LVP to the bathing solution results in a hyperpolarization of the membrane and a reduction of action potentials (Fig. 18A). Under voltage clamp, the I-V curve shows a marked increase in overall slope conductance (Fig. 18B). In addition, there is a direct reduction of the inward spike current (not shown). Also unlike the excitatory action, the inhibitory effects are rapidly reversed by removal of the peptide.

Although not central to the theme of this volume, the mode of action of LVP on molluscan membranes, which is quite different from that of conventional neurotransmitters, has led us to suggest that peptides might be neurohormones and that LVP in the molluscan system might be a model for the action of neuronally active peptides in the vertebrate CNS. We have developed these ideas more fully elsewhere (7–9,15), but Table 1 gives a

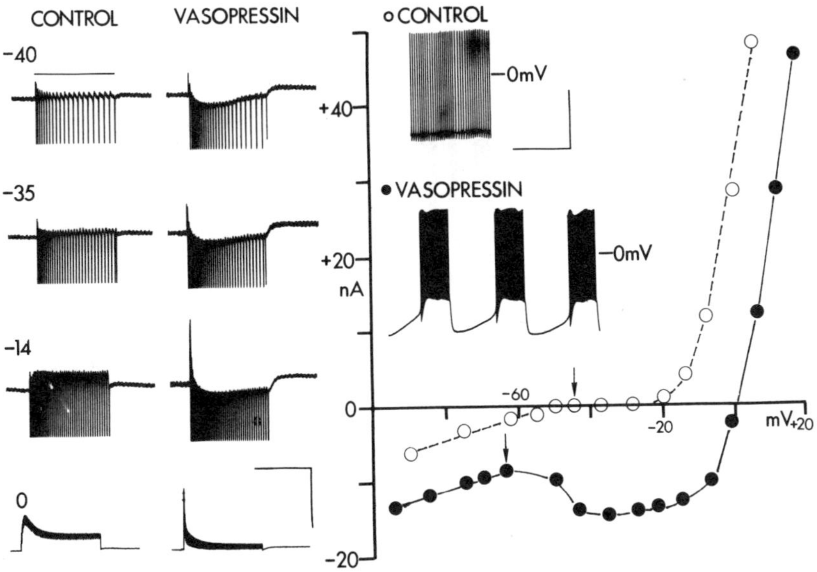

FIG. 15. Vasopressin alters steady state, voltage-clamp, current-voltage (I-V) curve. Recordings are from peptide-sensitive cell 11 from *Otala* before (CONTROL) and after (VASOPRESSIN) the bath application of 1 μM vasopressin. Membrane potential activity is illustrated in insets of I-V plot on right. Control trace shows beating pacemaker activity. Vasopressin induces BPA. Zero membrane potential, "0 mV." **Left:** Membrane of cell voltage-clamped and 5 sec voltage steps imposed (during time indicated by bar above current trace marked −40). Currents are shown at different depolarizing voltage steps (to membrane potentials indicated by numbers above traces) under control conditions and in the presence of vasopressin. Rapid downward current events represent action potential currents. Presence of slow inward current that decreases during the command is apparent in the vasopressin-treated membrane. **Right:** I-V curve derived from quasi-steady state currents using most negative or least positive current evoked after 1 sec during command. Current axis (nA), voltage axis (mV). Membrane of cell held at −45 mV in control and −65 mV in vasopressin (*downward arrows*). Calibrations: **Left:** 10 nA (*upper three traces*) and 40 nA (*lowermost traces*), 5 sec. **Right** (*inset*), 50 mV, 20 sec. (From ref. 14, with permission of *Brain Research*.)

comparison of synaptic transmission and neurohormonal communication. The potential relevance of such concepts to this volume is that appropriately placed, neuronally active peptides can evoke seizure-like electrical activity in the vertebrate CNS (61).

DISCUSSION

The basic conclusions to be drawn from our results are straightforward. (a) Molluscan neurons in *Aplysia* and *Otala,* which display endogenous BPA, have two pacemaker conductances, namely, a Na^+- and a K^+-pacemaker conductance, with voltage and temporal characteristics different from

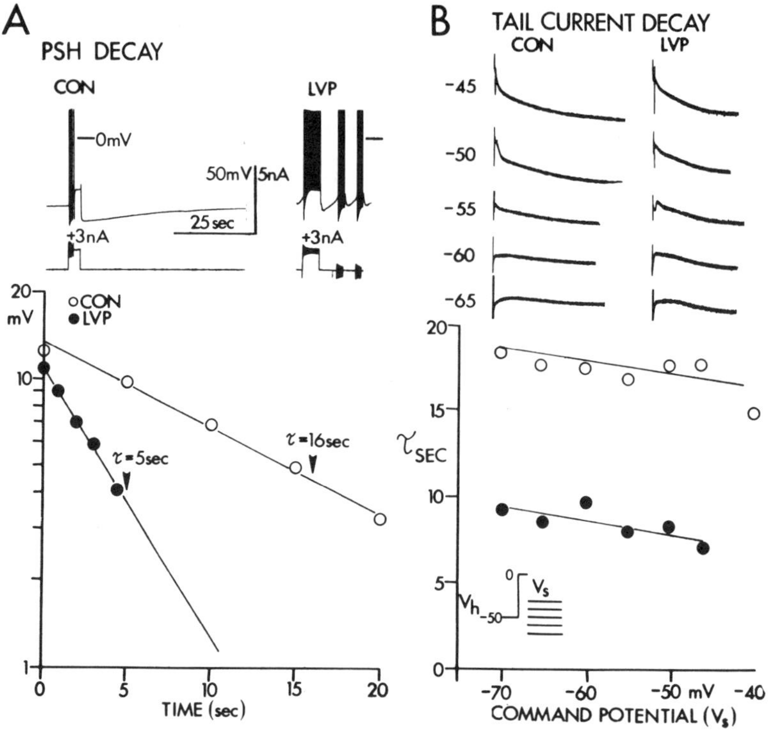

FIG. 16. LVP shortens K^+-pacemaker conductance decay. **A:** Recordings are from initially inactive cell in *Otala*. Under unclamped conditions, injection of 3 nA depolarizing current is followed by poststimulus hyperpolarization (PSH) that decays with a time constant of 16 sec (CON). Addition of LVP induces BPA. Now 3 nA depolarizing current injection is followed by PSH of approximately the same amplitude, but shorter time course (LVP) (time constant equals 5 sec). Resting potential, −50 mV. **B:** Decay of outward tail current evoked by double-step voltage commands from holding potential (V_h) of −50 mV first, to 0 mV for 5 sec, and then to various other potentials under control conditions and in LVP. Tail currents observed during second step potentials (indicated by numbers at left of traces) displayed above their time constants of decay. Decay is two phased and voltage dependent. The late, slow phase is 16 to 18 sec in control and 6 to 9 sec in LVP **(right),** approximating the membrane potential decay rates **(left)** of PSH in the two conditions.

those found in other excitable cells. (b) These pacemaker conductances can be modified and modulated by both divalent cations (12,28), normally Ca^{2+}, and polypeptides (10,13,14), which act like neurohormones rather than conventional neurotransmitters (7–9,15). Other researchers have reported that Ca^{2+} ions are a major charge carrier in *Helix;* thus there may be species differences in the ionic basis of the pacemaker conductances (28,36,37).

We have not yet put these ideas together in a precise and quantitative form (see ref. 53). However, the data shown in Fig. 19 are an attempt to convey a heuristic appreciation of (a) how the alternating Na^+ and K^+ currents, flowing through the Na^+- and K^+-pacemaker conductances, lead

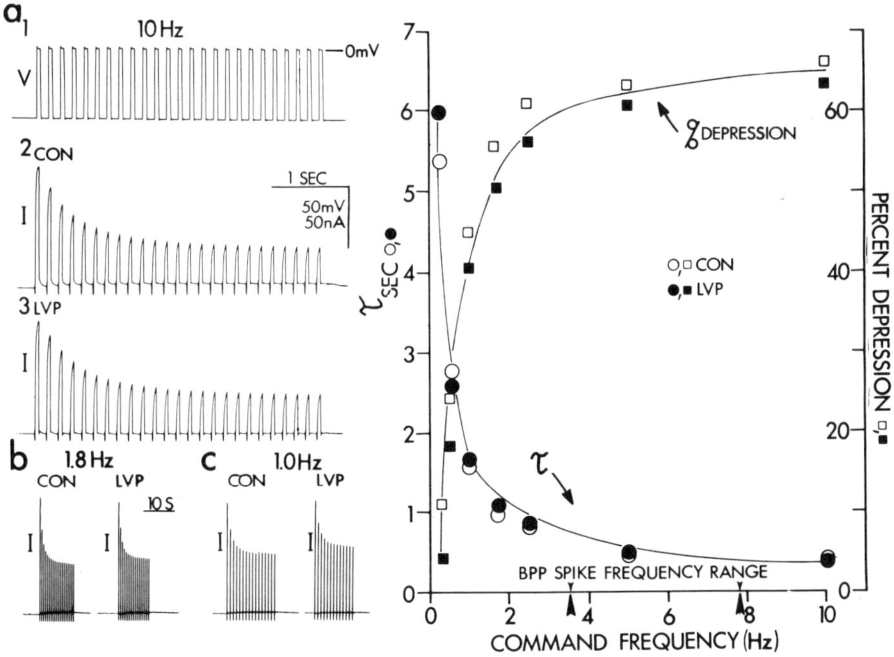

FIG. 17. Frequency-sensitive inactivation of action potential (delayed rectifier) K^+ conductance in *Otala.* Recordings are from a cell voltage-clamped under control conditions or bathed in LVP. Action potential train is simulated by holding membrane at −50 mV and repetitively stepping to 0 mV using 50 msec commands **(a1).** With a repetition rate of 10 Hz, outward currents evoked during commands decline rapidly (time constant is approximately 0.5 sec) to less than 40% initial value with same kinetics in control (**a2,** CON) and in LVP (**a3,** LVP). At slower repetition rates, rate of decline and final depression is less **(b,c).** Time constant of decline and percent depression of outward current amplitude are plotted as a function of command frequency in graph on right. A hyperbolic relationship between the two parameters and frequency is evident. The higher the frequency, the faster and greater the decline in outward current. The range of natural action potential frequencies present during a BPP is indicated by arrowheads on the abscissa.

to the depolarizing and hyperpolarizing phases of the BPP, and (b) the role of the Ca^{2+}-spike currents in BPP regulation. Figure 19A shows the spontaneous BPA in the unrestrained neuron; part B is a voltage-clamp recording in which a sinusoidal command is imposed on the membrane (Fig. 19BV) and shows the currents generated by such a command (Fig. 19BI; note zero current level: 0 nA).

During this simulated pacemaker potential activity, several time- and voltage-dependent changes in the magnitude and polarity of the total membrane current are evident. As the membrane is depolarized from its initially hyperpolarized level, at the beginning of the trace, net membrane current, which is inward (negative or depolarizing), gradually approaches and becomes zero about midway during the depolarizing phase of the simulated

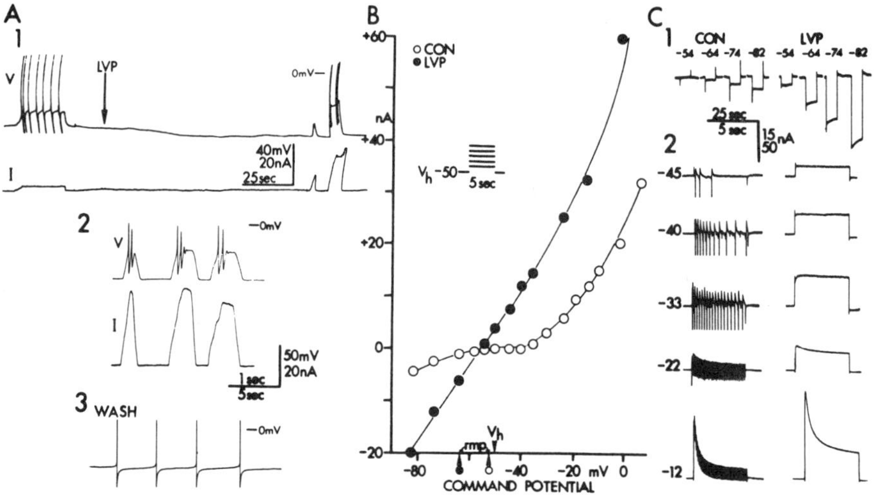

FIG. 18. Inhibitory effects of LVP. Recordings are from unidentified cell adjacent to cell (cell 11 in *Otala*) excited by LVP. **A1:** Potential trace (V) above current trace (I). Cell is electrically inactive but generates repetitive series of action potentials upon injection of depolarizing current, as seen in the early part of the trace. Bath application of 0.1 μM LVP (*arrow*) hyperpolarizes membrane and much more current is required to generate action potentials: late part of trace. Traces were recorded on curvilinear pen recorder. **A2:** Relatively fast, large depolarizing current pulses evoke small amplitude action potentials that rapidly accommodate. Rectilinear traces. **A3:** Washing in peptide-free solution leads to recovery of spikes within minutes. **B:** Steady state I-V curves are derived from currents **(C)** observed at different potentials under voltage clamp in control solution and in LVP. A holding potential (V_h) of −50 mV and 5 sec commands were used to generate data. Membrane conductance considerably increased and I-V curve changed from nonlinear to mainly linear in LVP; rmp, resting membrane potential. **C:** Currents evoked during hyperpolarizing (*1*) and depolarizing (*2*) commands. Note the absence of multiple action potential currents in LVP.

pacemaker waveform. The direction of the change in the current, that is, the time derivative of the current ($\dot{I}$), is positive. Because the membrane potential change is also positive, that is, its time derivative ($\dot{V}$) is positive, the slope conductance (dI/dV) during this phase is also positive. Further depolarization, however, leads to progressively more net inward membrane current until the peak depolarization of voltage waveform, at which potential net membrane current becomes progressively less negative. Thus during both the late phase of the depolarizing, positive-going voltage change ($\dot{V} > 0$) and the early phase of the hyperpolarizing, negative-going change ($\dot{V} < 0$), the corresponding current changes are of opposite sign: that is, when $\dot{V} > 0$, $\dot{I} < 0$, and when $\dot{V} < 0$, $\dot{I} > 0$. These results indicate that during these phases, the slope conductance is negative. Furthermore, these negative slope conductances probably correspond to the negative slope conductance region of the N-shaped I-V curve (Fig. 1).

The static, steady state I-V curve provides no conclusive information as

Table I. Synaptic Transmission and Neurohormonal Communication

	Synaptic Transmission	Neurohormonal Communication
Avenue	-between contiguous cells in synaptic contact -200 A synaptic cleft between adjacent pre- and post-synaptic neuronal membranes	-between remote cells without synaptic contact -extracellular space between one neuron and its target neurons
Distribution	-limited by axonal connections	-limited by appropriate receptors on target cells
Substance	-amino acid, catecholamine, acetylcholine, ?peptide	-peptide, ?acetylcholine, ?catecholamine
Release	-synaptic terminals -Ca^{2+}-dependent -facilitation -fatigue	-neurocirculatory terminals -Ca^{2+}-dependent
Effective concentration	-micromolar-millimolar	-nanomolar
Receptors	-sub-synaptic localization -couple to one or more voltage-independent conductances -cooperativity -desensitization	-extra-synaptic localization -couple to one or more voltage-dependent and voltage-independent conductances
Actions	-rapid kinetics (msec-sec) -change in voltage-independent membrane properties -additive factor in input-output relations	-rapid and slow kinetics (sec-min-hr) -change in voltage-independent and voltage-dependent membrane properties -additive and exponential factor in input-output relations
Function	-momentary regulation of single neuron excitability -simple and limited output	-sustained regulation of activity of multiple nerve cell aggregates -complex and concerted output

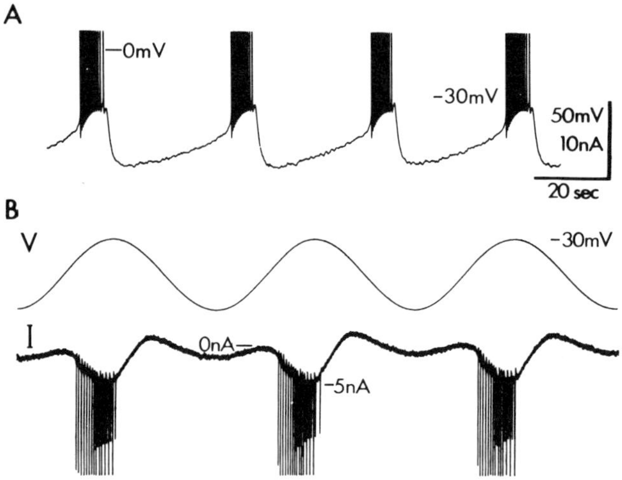

FIG. 19. Dynamics of net membrane currents during simulated BPA in *Otala*. **A:** Recordings were made with CsCl microelectrodes from cell bathed in LVP and 10 mM Sr^{2+}. Under unclamped conditions, pacemaker potential amplitude is about 40 mV, oscillating between −34 and −74 mV. **B:** After voltage clamping, a sinusoidal 50 mV command oscillating between −30 and −80 mV is imposed on the membrane (V), and the resulting net membrane current is recorded (I). At the beginning of the traces, voltage is −80 mV and net current is −2 nA. As membrane is depolarized, net current gradually becomes less negative until potential is about −40 mV, when net current becomes more negative, reaching −5 nA at peak of artificial pacemaker potential. While at peak of waveform, net current is rapidly reversed, momentarily becoming positive during the hyperpolarizing phase of oscillation. Current decays monophasically as potential is polarized to −80 mV. See text.

to whether its negative slope region is associated with the depolarizing or hyperpolarizing phases of the BPP. Figure 19 illustrates that the negative slope, as well as the negative I, is mainly related to the depolarizing phase of the BPP. Subsequently, following a series of spike currents, the polarity of the current is reversed, becoming outward (positive or hyperpolarizing). Eventually, the net outward current thus evoked decays monophasically ($\dot{I} < 0$) during the hyperpolarizing phase of the simulated pacemaker waveform, with a time constant that approximates the depolarization rate of the unclamped membrane during the interburst interval. Moreover, during this late hyperpolarizing phase, the slope conductance again becomes positive.

To summarize the events associated with polarization of the membrane through a potential range, as would occur during spontaneously generated pacemaker potentials, sufficient depolarization of the membrane from an initially hyperpolarized level activates a conductance causing inward current to dominate until further depolarization allows activation of another conductance that leads to outward current predominating. The former conductance is primarily Na^+ (55) and provides the sustained depolarizing phase of the pacemaker potential; the latter is a K^+ conductance that acts to hyperpolarize the membrane (18,28,33,39,40,55,57,58,66). Thus, under unclamped conditions, the membrane potential continually oscillates as the depolarizing Na^+-pacemaker conductance and then the hyperpolarizing K^+-pacemaker conductance are sequentially activated in an alternating manner. Since the changes in the Na^+ conductance are virtually instantaneous with changes in membrane potential, the periodicity of the pacemaker oscillations appears to be determined mainly by the kinetics of the K^+ conductance (53,55); the slower the activation, the longer the depolarizing phase; the slower the inactivation, the more prolonged the slow depolarization and the longer the period of oscillation.

The role of Ca^{2+} in this sequence of events is apparently to activate a portion of the K^+-pacemaker conductance (28,36,40,42,44,57,60). Thus the depolarizing phase of the BPP, which results from the activation of the Na^+-pacemaker conductance, brings the membrane potential to the level of spike initiation. The mainly pulsatile increase in $[Ca^{2+}]_i$ with the spikes (Fig. 12A) (57,60) leads to K^+-pacemaker activation, which then generates an outward-going current that hyperpolarizes the membrane (18,28,29,33, 39,40,55,57,58).

Several investigators (15,18,58,58) have shown that the duration of spikes progressively increases during a burst and that this increase correlates with a progressive, facilitated increase in $[Ca^{2+}]_i$ (57). One functional consequence of this prolonged spike-facilitated $[Ca^{2+}]_i$ process, when it apparently occurs at presynaptic terminals, is a facilitated increase in postsynaptic potentials and currents (Fig. 20B) (15). Indeed, the cumulative output of a system operating in the bursting mode can be much greater than one operating in a regularly spiking model (Fig. 20A).

Thus, while there is general agreement on the membrane mechanisms underlying normal BPP activity in molluscs, it is not entirely clear how these mechanisms relate to vertebrate neurons, such as epileptic cells and the vasopressin-magnocellular neurons that fire in a bursting or phasic mode. Phenomenologically, the known or presumptive large depolarizations leading to a burst of spikes are somewhat similar (1,3–5,18–20,22,23,27,31,35, 41,50–52,54,56,58,62,63). Other suggestive similarities are that known convulsive agents can induce BPPs and an N-shaped I-V curve in molluscan cells (21,25,38) and in some vertebrate spinal cord motoneurons (54). Endocrinologically, recent evidence (63) suggests that the repetitive bursts of the vasopressin-secreting cells may arise from endogenous mechanisms. And, of course, there is the enticing coincidence that vasopressin, produced by the vertebrate magnocellular neurons, induces BPP in the molluscan cells (10,13). A central question in this regard is whether the bursting activity in magnocellular elements is induced by a peptide or hormonal mechanism. It may be unlikely that vasopressin itself plays such a role; angiotensin is a more likely candidate (45,46).

From both following the literature and reading the chapters in this volume,

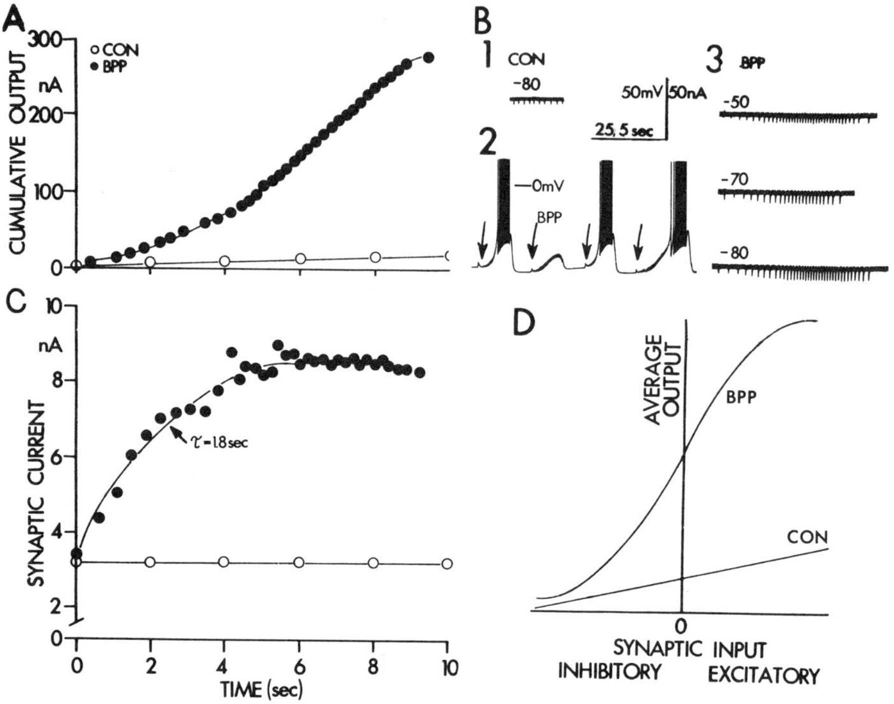

FIG. 20. Physiological consequences of BPP activity. Recordings are from R_{15} in *Aplysia,* which receives monosynaptic excitatory input from cell L_{10}. When L_{10} generates a beating pacemaker pattern of activity, unitary excitatory postsynaptic currents (EPSCs) are recorded in R_{15} voltage clamped at −80 mV (**B1**, CON). The constant-amplitude, constant-frequency EPSPs have been plotted in **C** and integrated over time in **A** as "cumulative output" of L_{10} (*open circles*). When L_{10} generates a bursting mode of pacemaker activity, a burst of EPSPs (**B2**, *arrows*) and EPSCs of increasing amplitude and frequency **(B3)** are recorded in R_{15}. Plotting the EPSC amplitude as a function of time (**C,** *filled circles*) reveals a single exponential increase in EPSC amplitude (time constant equals 1.8 sec), which plateaus at 1.5 times the amplitude of EPSCs recorded when L_{10} generates beating pacemaker activity. Integrating the EPSC over time (**A,** *filled circles*) shows 10-fold greater synaptic current accumulated during BPP activity. An approximately threefold increase in accumulated current would have been expected if each EPSC in the burst were of constant amplitude. Thus the output of L_{10} monitored by using EPSC amplitude in R_{15} is facilitated during the physiological BPP activity. **D:** Theoretical input-output relationships of a cell with almost linear membrane properties generating beating pacemaker activity, compared to one with nonlinear membrane properties generating BPP activity. Average output of neurotransmitter or neurosecretory product is considerably increased in cell generating BPPs because of its facilitated pulsatile character, and the effect of synaptic input is much greater in BPP cell due to the nonlinear membrane.

it appears that there are a number of similar electrophysiological events that occur both in vertebrate CNS and in molluscan neurons which fire action potentials in a bursting mode. From the molluscan research, it is possible to make certain statements or educated guesses about the mechanisms underlying these electrophysiological events. In the remainder of this chapter, we consider these matters. Our motivation is not to try to "explain" but rather to suggest to those who labor in the vertebrate CNS either membrane or other mechanisms they might consider or experiments they might perform that are consistent with the presence of those mechanisms.

Negative Slope and N-Shaped I-V Curve

The molluscan studies have shown that neurons that generate endogenous (i.e., not synaptically driven) BPA or PBA (K. Futamachi, *personal communication*) possess a region of negative slope and an N-shaped I-V curve (14,21,24–26,28,33,38,55,66). On the other hand, a region of negative slope is a necessary, but not a sufficient, condition for endogenous burst activity (55). Specifically, a cell will not show burst activity if the sum of the driving forces results in a membrane potential of that residing on either of the two positive slopes of the N-shaped I-V curve. For example, if the membrane potential is on the positive slope of the hyperpolarized limb of the I-V curve, and if this potential is below the threshold for action potentials, then the cell will be electrically silent. The application of a steady depolarizing current, however, which displaces the net driving forces into the region of the negative slope, can induce burst activity (see Fig. 3b in ref. 55). On the other hand, if the membrane potential rests on the positive slope of the depolarized limb of the N-shaped I-V curve, and if this potential is above threshold for action potentials, the cell will fire repetitive spikes. In this case, the application of a steady hyperpolarizing current can lead to a bursting mode of activity (Fig. 3a in ref. 55).

Burst Modification

Cells with a region of negative slope, and hence capable of endogenous bursting, can be altered by factors that can change membrane conductances that lie in parallel with the pacemaker conductances. Increases in these other conductances can sufficiently shunt the pacemaker conductances to prevent burst activity (55). There are several means of producing a shunt; for example, an increase in (a) $[K^+]_0$ which results in a depolarization, (b) a synaptically coupled K^+ conductance which results in a hyperpolarization, and (c) a synaptically coupled chloride conductance which can result in a depolarization or a hyperpolarization (depending on E_{Cl}). In these cases of burst activity abolition, such activity cannot be reinitiated by the application of extrinsic current (18–20,40,47,49,55,58). On the other hand, burst activity can be initiated or enhanced by factors that reduce shunting conductances that lie in parallel with the pacemaker conductances. Again, there are probably many ways to obtain such a result.

Triggered Versus Driven Bursts

For didactic purposes, it may be useful to make a distinction between "triggered" and "driven" bursts. A triggered burst might be defined as one whose depolarized phase lasts longer than the triggering event, for example, a short pulse of depolarizing current or an excitatory postsynaptic potential (EPSP) whose duration was brief with respect to the burst (16). Such triggered bursts can occur in cells capable of endogenous bursts; that is, they have a region of negative slope but rest on the positive slope of the hyperpolarized limb of an N-shaped I-V curve. The burst would, in this case, be all-or-none (K. Futamachi, *personal communication*). Thus the burst would result from regenerative displacement of the membrane potential across the region of negative slope in the depolarizing direction and last until either the slowly increasing K^+ conductance reaches sufficient magnitude to repolarize the membrane or another hyperpolarizing event regeneratively displaced the membrane back across the region of negative slope (65).

On the other hand, a driven burst would result from a depolarizing event (extrinsic current, EPSP) whose duration roughly coincides with the duration of the burst. Such driven bursts are not endogenous and the cells I-V curve need not have a region of negative slope conductance.

There is one variation on cells with N-shaped I-V curves and triggering worthy of note, namely, a group of such cells that are also connected by electrotonic junctions (17,32). Depending on the degree of coupling, bursts may or may not be triggered by the application of extrinsic current to one cell. In very tightly or very loosely coupled cells, bursts can probably be triggered electrically. With intermediate degrees of coupling, however, triggering may not be possible with extrinsic current delivered to one cell because the population of cells cannot be simultaneously depolarized across their regions of negative slope. Brief, synchronous EPSPs to all or most of the coupled cells, however, may give sufficient simultaneous depolarization to evoke a burst whose duration is longer than the EPSP (K. Futamachi, *personal communication*).

Continuous Repetitive Firing Prior to Bursting

In those cells that can be converted from a nonbursting to a bursting mode of activity, there can be a transition phase where the membrane generates only a rapid firing pattern of spikes (13,63). This may be due in part to the slow turn-on kinetics of the K^+-pacemaker conductance to a level sufficient to lead to a hyperpolarized, quiescent period.

Na^+- and K^+-Pacemaker Conductances and Anomalous Rectification

Prior to the discovery of the negative slope conductance and the Na^+-pacemaker conductance, the temporal changes in conductance during each cycle of the BPP were assessed by applying constant current pulses and measuring the evoked IR drops. These experiments led to the conclusion that there is a steady decline in K^+ conductance during the silent depolarizing phase of the BPP (12,18,39,40,58,64). The voltage-clamp data substantiate that conclusion; however, this K^+ conductance is not the same as the anomalous rectification found in striated muscle but a different conductance, which should be called a pacemaker conductance (39). In addition, in the constant current experiments, the IR drops are maximum during the early phase of the burst; however, they do not necessarily represent a decrease in K^+ conductance but may reflect a regenerative displacement of the membrane across the negative slope, due to potential-dependent changes in the Na^+-pacemaker conductance. Evidence consistent with this interpretation is seen in Fig. 11B2 (55). Clamping the membrane early in the burst requires a large, initially negative current, indicating the flow of ions from a source whose equilibrium potential is positive to the membrane potential, e.g., Na^+. If only a K^+ conductance were involved, the initial current should be outward.

K^+ Pacemaker Conductance

Provisional evidence for the presence of a K^+-pacemaker conductance can be found with intracellular recordings in situations where voltage clamping is impracticable. A long, slowly declining hyperpolarization, following the removal of a depolarizing current sufficient to evoke action potentials, is presumptive evidence of the K^+-pacemaker conductance (Fig. 17A). In our experience, cells without this phenomenon cannot be made to generate BPP activity.

SUMMARY

In this chapter, we present evidence for the existence of two pacemaker conductances, one to Na^+ and the other to K^+. Sequential alterations on the activation and inactivation of these potential-dependent conductances underlie the BPP; the BPP kinetics are determined mainly by the kinetics of the K^+-pacemaker conductance. These conductances, and thereby BPP activity, can be modulated by the divalent cations and by peptides. Ca^{2+} ions appear to affect primarily the K^+-pacemaker conductance, while peptides alter both pacemaker conductances.

A number of electrophysiological events are discussed in terms of their possible membrane mechanisms.

REFERENCES

1. Ajmone-Marsan, C. (1969): In: *Basic Mechanisms of the Epilepsies,* edited by H. H. Jasper, A. Pope, and A. A. Ward, pp. 299–319. Little, Brown, Boston.

2. Alving, B. O. (1968): *J. Gen. Physiol.,* 51:29–45.
3. Andersen, P., Gjerstad, L., and Langmoen, I. A. (1978): *This volume.*
4. Arvanitaki, A., and Cardot, H. (1941): *C. R. Seances Soc. Biol. Fil.,* 135:1207–1211.
5. Arvanitaki, A., and Chalazonitis, N. (1964): *C. R. Seances Soc. Biol. Fil.,* 158:1119–1122.
6. Arvanitaki, A., and Chalazonitis, N. (1973): In: *Neurobiology of Invertebrates,* edited by J. Salanki, pp. 169–199. Hungarian Academy of Sciences, Budapest.
7. Barker, J. L. (1976): *Physiol. Rev.,* 56:435–452.
8. Barker, J. L. (1977): In: *Central Actions of Angiotensin,* edited by J. P. Buckley and C. Ferrario, pp. 24–51. Pergamon Press, New York.
9. Barker, J. L. (1977): In: *Peptides in Neurobiology,* edited by H. Gainer, pp. 295–343. Plenum Press, New York.
10. Barker, J. L., and Gainer, H. (1974): *Science,* 184:1371–1373.
11. Barker, J. L., and Gainer, H. (1975): *Brain Res.,* 84:461–477.
12. Barker, J. L., and Gainer, H. (1975): *Brain Res.,* 84:479–500.
13. Barker, J. L., Ifshin, M., and Gainer, H. (1975): *Brain Res.,* 84:501–513.
14. Barker, J. L., and Smith, T. G. (1976): *Brain Res.,* 103:167–170.
15. Barker, J. L., and Smith, T. G., Jr. (1977): In: *Approaches to the Cell Biology of Neurons, Society for Neuroscience Symposia, Vol. II,* edited by W. M. Cowan and J. A. Ferrendelli, pp. 340–373. Society of Neuroscience, Bethesda.
16. Benjamin, P. R. (1978): *This volume.*
17. Bennett, M. V. L. (1966): *Ann. NY Acad. Sci.,* 188:242–269.
18. Carpenter, D. O. (1973): In: *Neurobiology of Invertebrates, Mechanisms of Rhythm Regulation,* edited by J. Salanki, pp. 35–58. Akademiai Kiado, Budapest.
19. Chalazonitis, N. (1978): *This volume.*
20. Chalazonitis, N. (1978): *This volume.*
21. David, R. J., Wilson, W. A., and Escueta, A. V. (1974): *Brain Res.,* 67:549–554.
22. Dreifuss, J. J., Gähwiler, B. H., and Sandoz, P. (1978): *This volume.*
23. Dreifuss, J. J., and Kelly, J. S. (1972). *J. Physiol., Lond.,* 220:105–118.
24. Ducreux, C. (1978): *This volume.*
25. Ducreux, C., and Gola, M. (1975): *Pfluegers Arch. Ges. Physiol.,* 361:43–53.
26. Ducreux, C., and Gola, M. (1977): *Brain Res.,* 123:384–389.
27. Dyball, R. E. J. (1971): *J. Physiol. (Lond.),* 214:245–256.
28. Eckert, R., and Lux, H. D. (1976): *J. Physiol. (Lond.),* 254:129–151.
29. Faber, D. S., and Klee, M. R. (1972): *Nature [New Biol.],* 240:29–31.
30. Frazier, W. T., Kandel, E. R., Kupfermann, I., Waziri, R., and Coggeshall, R. E. (1967): *J. Neurophysiol.,* 30:1288–1351.
31. Gainer, H. (1972): *Brain Res.,* 39:403–418.
32. Getting, P., and Willows, A. O. D. (1974): *J. Neurophysiol.,* 37:858–868.
33. Gola, M. (1974): *Pfluegers Arch. Eur. J. Physiol.,* 352:17–36.
34. Gola, M. (1978): *This volume.*
35. Hayward, J. N., and Jennings, D. P. (1973): *J. Physiol. (Lond.),* 232:545–572.
36. Heyer, C. B., and Lux, H. D. (1976): *J. Physiol. (Lond.),* 262:349–382.
37. Heyer, C. B., and Lux, D. (1978): *This volume.*
38. Hoyer, J., Park, M. R., and Klee, M. R. (1978): *This volume.*
39. Junge, D., and Stephens, C. L. (1973): *J. Physiol. (Lond.),* 235:155–173.
40. Kandel, E. R. (1976): *Cellular Basis of Behavior.* Freeman, San Francisco.
41. Koizumi, K., and Yamashita, H. (1972): *J. Physiol. (Lond.),* 221:638–706.
42. Meech, R. W. (1972): *Comp. Biochem. Physiol.,* 42A: 493–499.
43. Meech, R. W., and Brown, H. M. (1976): In: *Perspectives in Experimental Biology, Vol. 1,* edited by P. Spencer Davis, pp. 331–351. Pergamon Press, New York.
44. Meech, R. W., and Standen, N. B. (1975): *J. Physiol (Lond.),* 249:211–239.
45. Nicoll, R. A., and Barker, J. L. (1971): *Brain Res.,* 35:501–511.
46. Nicoll, R. A., and Barker, J. L. (1971): *Nature [New Biol.],* 233:172–174.
47. Parnas, I., Armstrong, D., and Strumwasser, F. (1974): *J. Neurophysiol.,* 37:594–608.
48. Parnas, I., and Strumwasser, F. (1974): *J. Neurophysiol.,* 37:609–620.
49. Pinsker, H., and Kandel, E. R. (1967): *Physiologist,* 10:279.
50. Poulain, D. A., Wakerley, J. B., and Dyball, R. E. J. (1977): *Proc. R. Soc. Lond. [Biol.],* 196:367–384.

51. Prince, D. A. (1969): In: *Basic Mechanisms of the Epilepsies,* edited by H. H. Jasper, A. Pope, and A. A. Ward, pp. 320–328. Little Brown, Boston.
52. Prince, D. A., and Schwartzkroin, P. A. (1978): *This volume.*
53. Roberge, F. A., Gulrajani, R. M., Jasper, H. H., and Mathieu, P. A. (1978): *This volume.*
54. Schwindt, P. C., and Crill, W. E. (1976): *Sixth Annu. Meeting Soc. Neurosci., Toronto,* p. 266.
55. Smith, T. G., Barker, J. L., and Gainer, H. (1975): *Nature,* 253:450–452.
56. Somjen, G., Lothman, E., Dunn, P., Dunaway, T., and Cordingley, G. (1978): *This volume.*
57. Stinnakre, J., and Tauc, L. (1973): *Nature,* 242:113–115.
58. Strumwasser, F. (1973): *Physiologist,* 16:9–42.
59. Takeuchi, H. (1978): *This volume.*
60. Thomas, M. V., and Gorman, A. L. F. (1977): *Science,* 196:531–533.
61. Urca, G., Frenk, H., Liebeskind, J. C., and Taylor, A. N. (1977): *Science,* 197:83–86.
62. Vincent, J. D., Arnauld, E., and Bioulac, B. (1972): *Brain Res.,* 44:371–384.
63. Vincent, J. D., Poulain, D., and Arnauld, E. (1978): *This volume.*
64. Waziri, R., Frazier, W. T., and Kandel, E. R. (1965): *Physiologist,* 8:300.
65. Williamson, T. L., and Crill, W. E. (1976): *Brain Res.,* 116:217–229.
66. Wilson, W. A., and Wachtel, H. (1974): *Science,* 186:932–934.

Abnormal Neuronal Discharges, edited by
N. Chalazonitis and M. Boisson. Raven
Press, New York © 1978.

Ionic Mechanisms for Rhythmic Activity and Bursting in Nerve Cells

F. A. Roberge, R. M. Gulrajani,
H. H. Jasper, and P. A. Mathieu

Biomedical Engineering Group, Department of Physiology, University of Montreal, Montreal, Quebec, Canada

One of the characteristic features of certain excitable cells is their ability to spontaneously generate rhythmic and bursting discharges which, in turn, are often used to control or drive other cells. The problem of the ionic mechanisms underlying these discharges is of fundamental importance to our understanding of both normal and abnormal function of the nervous system. Most nerve cells in the central nervous system appear to be in a state of more or less continuous activity in the waking alert animal, and many continue firing, although with a changed pattern, even during deep sleep. Patterns of firing are continuously changing in the waking animal, depending on its background excitatory state or the role being played by a given cell in the transfer or processing of information and in the control of postural tone or behavior. Spontaneous rhythmic activity adds an additional parameter to the transfer of information by decreased firing during inhibition, and postures or movements may result from this inhibition of a continuous repetitive discharge. Obviously, disturbances in the ionic mechanisms which regulate rhythmic or bursting discharges may have a profound effect on the normal integrative functions of the nervous system, even when they do not produce such gross disturbances as an epileptic seizure.

Our study of these mechanisms is an offshoot of previous work with electronic analogs of the squid axon membrane (30) and of the *Aplysia* bursting cell R_{15} (32,33). These analogs are useful for qualitative studies of membrane behavior, since many of their parameters are easily adjustable, with the resultant changes in simulated membrane activity simultaneously observable on an oscilloscope. Accordingly, following a brief description of their principle of operation, the mechanisms determining rhythmic activity and bursting in these analogs are discussed and compared with experimental observations.

FIELD-EFFECT TRANSISTOR ANALOG OF EXCITABLE MEMBRANE

The simulation of the voltage- and time-dependent membrane conductances via the drain-source conductance of junction field-effect transistors (FETs) was first described by Roy (52). The same technique was later used to closely simulate the Hodgkin-Huxley (H-H) equivalent electrical circuit for the squid axon membrane (30). This electrical circuit (39) consists of four parallel branches connecting the inside and outside membrane surfaces. These branches represent, respectively, the membrane capacitance (C_M), the membrane conductance (g_{Na}) to sodium ions, the membrane conductance (g_K) to potassium ions, and a lumped membrane conductance (g_L) to all other ions besides Na^+ and K^+. The ionic conductances are each placed in series with an electromotive source, whose magnitude equals the Nernst equilibrium potential for the respective ion species: E_{Na}, E_K, E_L.

Using the voltage-clamp technique, Hodgkin and Huxley (36–38) found that depolarization caused an activation or increase in both g_{Na} and g_K, with the g_{Na} increase being only temporary (Fig. 1A). The peak g_{Na} and g_K values reached varied with the amount of depolarization (Fig. 1B). The g_{Na} and g_K transients of Fig. 1A may be characterized by their rise and decay times, the second being the time for the conductance to reach its resting value when the applied depolarization is removed (the dotted transient in Fig. 1A). In addition, the g_{Na} transient is characterized by an inactivation time constant which governs the decay of g_{Na} with maintained depolarization and a recovery-from-inactivation time constant that governs the rate of removal of the effects of this decay once the depolarization is removed. Also, both the sodium and potassium rise times and the sodium inactivation time vary with the level of applied depolarization. The increase in g_{Na} and g_K with membrane potential, as well as the relative magnitudes of the different parameters characterizing their transients, are responsible for the classic action potential waveform.

The modeling of the squid axon membrane involved using a junction FET and series resistor to simulate the variable sodium and potassium conductances (Fig. 1C). Since the drain-source conductance of a FET is controlled by the voltage applied to the gate terminal, the dependence of g_{Na} and g_K on membrane depolarization (Fig. 1B) was realized by feeding back the simulated membrane potential (V_M) to this terminal. An accurate match required a careful selection of FET, series resistor, feedback gain, and the size of the membrane patch to be simulated (30). The correct transients in g_{Na} and g_K were realized by interposing a waveshaping network in the feedback loop. Correct adjustment of this network was done empirically such that for a step change in V_M, the waveform at the gate resulted in the desired conductance transients. The results obtained showed that the four-branch FET analog was capable of faithfully reproducing the more important characteristics observed in the real squid axon membrane.

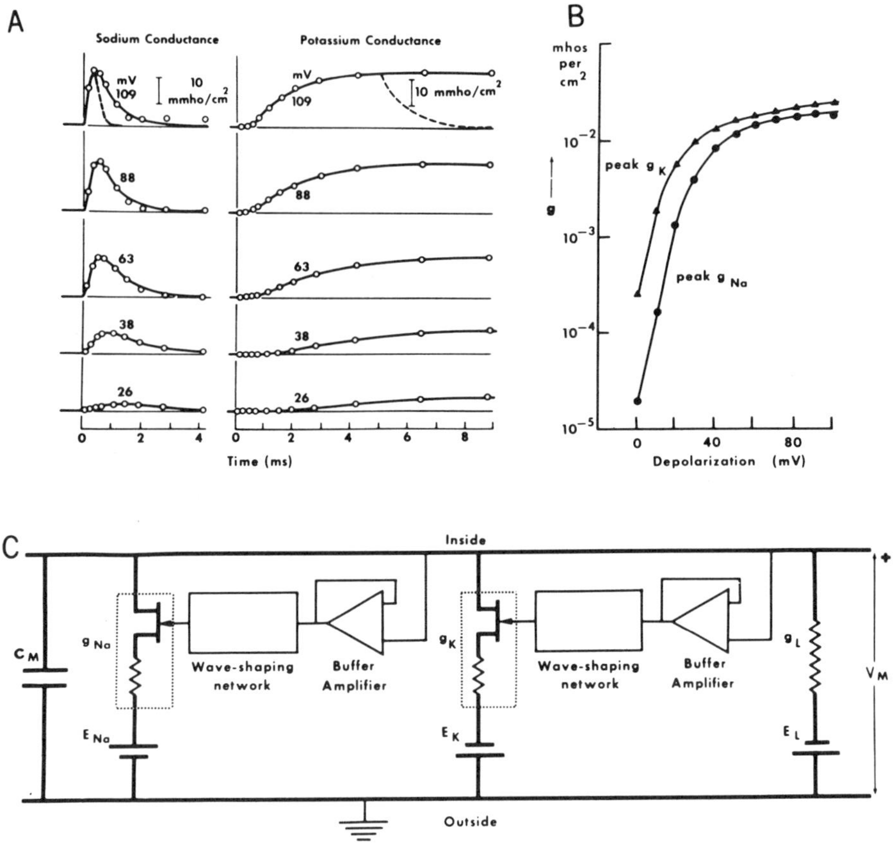

FIG. 1. Conductance characteristics of the squid axon membrane and their simulation with FETs. **A:** Sodium and potassium conductance transients for various clamp voltages. **B:** Variation in peak sodium and peak potassium conductance with depolarization. (Modified from refs. 36–38.) **C:** FET analog of the squid axon membrane. (From ref. 30.) (Figure reproduced, with permission, from ref. 32.)

This analog was later modified (32) to simulate the endogenous bursting cell R_{15} in the mollusc *Aplysia californica* (27). This cell generates bursts of action potentials due to an underlying slow membrane potential oscillation (Fig. 2) (3,45). Current thought implicates two conductances in the generation of the slow oscillations (29,31,56,64). The depolarizing phase is due to an increase in a sodium conductance (g'_{Na}) and the hyperpolarizing phase to a subsequent increase in a potassium conductance (g'_K). Both conductances are presumed to increase with membrane depolarization. Accordingly, in addition to changing the various parameters of the four-branch analog to correspond to values measured in *Aplysia,* it was also essential to add two more branches in parallel, to simulate the variations in g'_{Na} and g'_K and hence

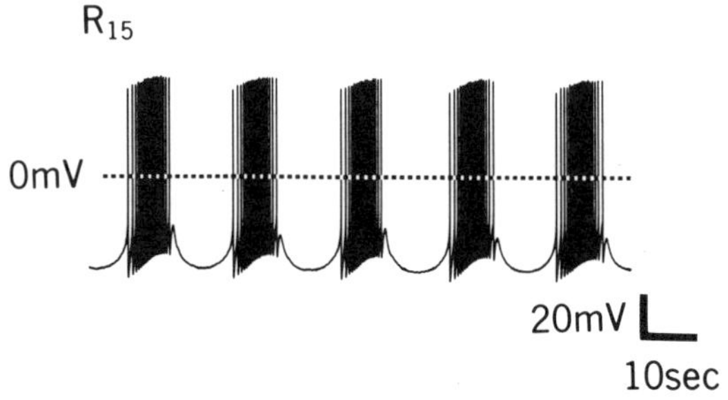

FIG. 2. Endogenous bursting activity, recorded intracellularly from cell R_{15} in the abdominal ganglion of *Aplysia californica*.

generate the slow oscillations. Each of these additional branches also consisted of a FET in series with a resistor and the appropriate Nernst potential, namely, E_{Na} and E_K. Furthermore, an auxiliary feedback loop simulated the decrease in the potassium Nernst potential (E_K) that has been shown to occur in R_{15} following long-duration voltage-clamp depolarization (21). This decrease occurs because of accumulation of the outward-flowing K^+ ions just outside the cell membrane and, in the analog, it was made proportional to the quantity $\{\ln (\int I_K dt)\}$, where I_K represents the outward potassium current through the g_K branch. This expression assumes that the extracellular potassium concentration change is proportional to the time integral of I_K. A resistor placed across the integrating capacitor allowed for gradual recovery to the resting E_K level once I_K returned to its resting value.

MECHANISMS UNDERLYING RHYTHMIC ACTIVITY

Studies with the squid axon membrane analog have revealed two modes of rhythmic firing. In mode 1, the potassium conductance decay time must exceed the sodium conductance recovery-from-inactivation time. The mechanism responsible for rhythmic firing is easily understood. Following an action potential, the resultant increase in potassium conductance gradually decays, leading to the slow depolarization commonly seen in pacemaker preparations. This slow or pacemaker depolarization can effectively act as a ramp trigger for the succeeding action potential, provided the recovery from sodium inactivation is sufficiently complete; hence the condition stated above. It must be stressed, however, that even when this condition is satisfied, rhythmic firing can be inhibited in the analog by lowering the resting sodium conductance in order to result in an inward current that is less than the sum of the outward currents. As is well known, this would prevent depolarization in a space-clamped membrane, unless compensated for by an external depolarizing current. Thus the above condition is necessary, but not sufficient, for rhythmic firing in mode 1. At the other extreme, too large a potassium conductance decay time, by allowing the membrane to accommodate during the pacemaker depolarization, would also inhibit rhythmic firing.

In mode 2, rhythmic firing is achieved by introducing an additional noninactivating or steady state sodium conductance that increases with depolarization. This new conductance now underlies the pacemaker depolarization. However, constant current pulses injected during the pacemaker depolarization in various preparations (9,54,62) demonstrate a larger membrane voltage response as the depolarization progresses (Fig. 3). This would seem to indicate a progressive decrease in membrane conductance during the pacemaker depolarization and thus make mode 1 a more likely candidate for rhythmic firing in biological preparations. Recently, however, several authors (31,54) have cautioned that the membrane response to constant current pulses is an indication of the variation in membrane slope conductance which, for a membrane with voltage-dependent conductances, may not correctly reflect the ongoing changes of the individual membrane chord conductances.

The equations characterizing the H-H equivalent circuit could provide a basis for further study of these two modes of rhythmic firing. Unfortunately, few systematic investigations of the causes of rhythmic firing as predicted by the equations have been published. This is undoubtedly because the equations are complex to begin with, several parameters are indirectly interdependent through V_M, and changes in one parameter can therefore have multiple effects on the overall behavior. It is known that the equations do predict repetitive firing in response to a constant current stimulation (12,57) and that this behavior may be justified on mathematical grounds (24,50).

It is worthwhile to note that the analog simulates the squid axon membrane and not the H-H equations. The main difference lies in the simulation of g_{Na}. In the equations, this is achieved by multiplying the activation and inactivation variables m and h, whose steady state values $m_\infty(V)$ and $h_\infty(V)$,

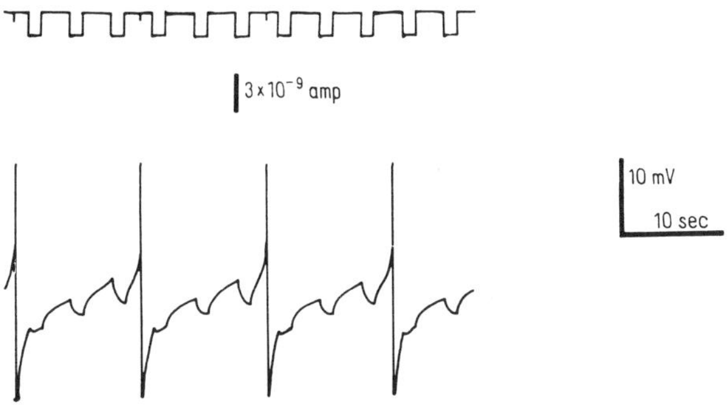

FIG. 3. Variation in membrane conductance during pacemaker discharge in cell R_{14} in Aplysia. *Upper trace*, injected intracellular current pulses; *lower trace*, membrane potential changes in response to these pulses. Action potential peaks have been clipped during recording. (Modified from ref. 9.)

respectively, increase and decrease with membrane depolarization. Faster m kinetics ensure activation before inactivation. In the analog, there is in effect only one variable—the equivalent of m. Depolarization causes a rapid activation of m, followed by a slow shift of the $m_\infty(V)$ characteristic in the depolarizing direction, thus resulting in a reduction of m and hence inactivation. Recovery from inactivation takes place as $m_\infty(V)$ shifts back in the hyperpolarizing direction. Accordingly, which of the above two modes of rhythmic firing observed with the analog best characterizes rhythmic activity as predicted by the H-H equations? Inspection of the H-H curves for $m_\infty(V)$ and $h_\infty(V)$ indicates an appreciable overlap between −60 and −30 mV, where both variables are greater than zero (48). This is tantamount to a steady state sodium conductance over this range of membrane potentials, which encompasses the pacemaker depolarization region. Furthermore, single-spike experiments with the analog revealed a postspike afterdepolarization ("rebound") that was absent with no steady state sodium conductance component. As is well known, single spikes predicted by the H-H equations also exhibit this afterdepolarization (57). Thus it would appear that, in the H-H equations, a mode 2 mechanism predominates in generating repetitive rhythmic activity.

Both modes observed with the analog embody the general principle of a slow depolarizing mechanism following the action potential, which underlies all present-day explanations of pacemaker activity. The ubiquitous nature of the principle may be gauged by a brief literature survey. It was originally proposed for the cardiac Purkinje fiber (20,62). Here, the pacemaker depolarization was due to a decrease in a potassium conductance, although these early reports attributed pacemaker activity more to the presence of a high resting sodium conductance. McAllister et al. (47) identified this potassium conductance as distinct from the conductance responsible for the termination of the plateau in the Purkinje fiber action potential and reasserted its importance in pacemaker activity. On the other hand, in atrial and possibly in sinus node cells, the pacemaker depolarization is governed by the decay of the same potassium conductance responsible for the termination of the action potential (6) and is thus very similar to mode 1. As in the analog, a sufficiently large resting inward current is also needed (44).

In the nodes of frog motor nerves rendered capable of rhythmic activity by exposure to reduced calcium and tetraethylammonium chloride, Bergmann et al. (5) claim that the recovery from sodium inactivation following an action potential is responsible for the pacemaker depolarization. This would occur through an increase in h and hence in the product m^2h in the Frankenhaeuser-Huxley (25) expression for sodium permeability. This is similar to mode 2 found in the analog in that it relies on a significant overlap of the $m_\infty(V)$ and $h_\infty(V)$ curves over the range of pacemaker depolarization, which as pointed out earlier is tantamount to the presence of a steady state sodium conductance. A depolarizing shift of the $h_\infty(V)$ curve, and hence an increased overlap of the $m_\infty(V)$ and $h_\infty(V)$ curves, has also been observed to correlate well with enhanced rhythmic firing in single myelinated nerve fibers of *Xenopus laevis* (63).

In pacemaker neurons of the marine gastropod *Anisodoris*, Connor and Stevens (14) have identified an early current I_A, presumably carried by potassium ions, that is implicated in rhythmic activity. The conductance g_A mediating this current I_A becomes completely inactivated at potentials above −50 mV. Removal of this inactivation requires a conditioning hyperpolarization. During normal rhythmic activity, the spike undershoot can provide this hyperpolarization if it crosses −50 mV. At high firing rates, this does not happen; repetitive activity is determined by the decay of the conventional potassium conductance g_K, exactly as in mode 1 for the analog. Indeed, from plots of the time constants of potassium activation and sodium inactivation (15), it would appear that the condition that the potassium decay time exceed the sodium recovery-from-inactivation time is well satisfied. At low firing rates, since the inactivation of g_A is removed by the spike undershoot, the outward current I_A slows down the rate of pacemaker depolarization. Eventually, however, g_A again inactivates, the pacemaker depolarization speeds up, and the following action potential results. Thus in this case, two potassium conductances, namely, the decay of g_K and the activation-inactivation of g_A, act simultaneously to control the pacemaker depolarization. Connor (13) has also experimentally determined the presence of the current I_A in repetitively firing crustacean axons. More recently (16), it was also implicated in the low frequency discharge of these crustacean axons.

In cat spinal cord motoneurons, it was suggested (7) that repetitive activity is not due to intrinsic membrane properties but to a spatial interaction between the spike-originating region (e.g., the initial segment of the axon) and the spike-replication region (e.g., the soma). This was based on observations that the initial segment spike (or A spike) lacked an afterhyperpolarization, which was not the case for the soma spike (or B spike) (17). Thus, according to this hypothesis, the occurrence of rhythmic firing requires soma invasion by the A spike, since only then will the decaying afterhyperpolarization of the B spike induce, via electrotonic conduction, a pacemaker depolarization at the initial segment or spike trigger zone. The afterhyperpolarization has been shown to be due to an increase in a potassium conductance (17), so that the net effect is similar to mode 1. One word of caution is in order. The A spike was shown to lack an afterhyperpolarization in single spike experiments and not under conditions of rhythmic firing. Although there is little doubt that electrotonic spread of the B spike afterhyperpolarization can play a very important role in controlling the firing rate, much more hard evidence in rhythmically firing motoneurons is needed before it can be asserted that the initial segment membrane lacks an intrinsic pacemaker depolarization mechanism and is therefore incapable of rhythmic activity without the invasion of the B spike afterhyperpolarization (7).

An alternative hypothesis for rhythmic firing in motoneurons, not based on the foregoing spatial interaction mechanism, was suggested by Kernell (43). This postulates an increase in a potassium conductance that takes place during or immediately following the action potential and which subsequently decays exponentially to generate the pacemaker depolarization. Baldissera and Gustafsson (4) attempted to measure this conductance decay during the afterhyperpolarization following a single evoked spike by the familiar technique of noting the membrane voltage response to constant current pulses. They found an exponential decay of the membrane conductance over most of the afterhyperpolarization except at the beginning, immediately after the spike, and accordingly suggested a modification of Kernell's simple exponential expression for the conductance decay.

Schwindt and Calvin (54), in a similar study, found that the membrane conductance measured during the afterhyperpolarization following a single spike differed significantly from that measured during the interspike interval when the cell was beating rhythmically. They attributed the difference to the activation of a sodium conductance toward the end of the interspike interval, just prior to the following spike. Thus, apparently, both the decay of a potassium conductance and the activation of a sodium conductance act in concert to produce rhythmic activity. This corresponds to a combination of modes 1 and 2 of rhythmic firing. As pointed out earlier, however, conductance measurements with constant current pulses are not expected to reveal the voltage-dependent conductances involved during the afterhyperpolarization. For this reason, Schwindt and Calvin's assertion of the involvement of two conductances in rhythmic activity may seem somewhat uncertain. On the other hand, support for the same hypothesis comes from the appearance of a negative slope conductance region in the quasisteady state I-V curves obtained upon subjecting cat motoneurons to a slowly increasing ramp voltage-clamp command (53). Furthermore, neurons that exhibited this negative slope conductance region were also, in general, repetitively active. This negative slope conductance region was attributed to the activation of a slow inward current, probably carried by both sodium and calcium ions. The conductance mediating this slow inward current is probably the second conductance underlying rhythmic activity suggested by Schwindt and Calvin. These recent voltage-clamp results tend to diminish the importance of spatial interaction mechanisms for rhythmic activity.

BURSTING MECHANISMS

This section is essentially limited in scope. The bursting cell analog described earlier mimics an endogenously active cell. Studies with the analog therefore focus on endogenous bursting; accordingly, synaptic activity or a system of interconnected neurons as a cause for bursting is not discussed.

In particular, the analog utilizes the two-conductance mechanism to generate the slow membrane potential oscillations and hence bursts. Thus the presence of the g'_{Na} and g'_K branches is essential; without these branches, the residual four-branch analog is only capable of, at best, regular rhythmic activity. It must be pointed out that in bursting cells in related biological species, e.g., *Helix pomatia,* a calcium conductance g'_{Ca} may replace g'_{Na} (22). Also in these cells, the increase in g'_K, instead of being voltage dependent, may be triggered by the influx of calcium ions (22).

It has been observed that bursting cells under voltage clamp often exhibit a negative slope conductance region in the steady state (approximately 5 sec after clamp onset) I-V curve (Fig. 4A) (8,29,56,64). The inward current peak is presumably due to the g'_{Na} (or g'_{Ca}) increase and is followed by an outward current due to the g'_K increase at larger depolarizations. The inward current is partially masked at later times by the slower time development of the outward current. Since a negative slope conductance region in an I-V characteristic can lead to instability, these observations were used to support the two-conductance theory. That this negative slope conductance region does occur as a result of the g'_{Na} and g'_K branches was verified in the analog using slow depolarizing and hyperpolarizing ramp voltages as clamp commands (Fig. 4B).

The strong correlation between the presence of bursting activity and the existence of a negative slope conductance region in the steady state I-V curve has led some groups to assert that the presence of the latter is necessary for the former (56,64). Partridge (49), however, demonstrated one instance of pentylenetetrazol-induced bursting in a *Tritonia* cell in the absence of a negative slope conductance region in the steady state I-V curve. As he correctly pointed out, the latter characteristic does not reflect the temporal relationships between the two conductances, and a transitory negative slope region, which is subsequently masked because of a developing g'_K, could underlie the slow oscillations. Thus, while the presence of a negative slope conductance region over some time interval during the oscillations is essential, it is by no means certain that the steady state I-V curve will necessarily exhibit a negative slope region when bursting is present. However, it must be pointed out that efforts to generate bursts with the analog in the complete absence of a negative slope conductance in the steady state I-V curve were not successful.

Further experiments suggested that adding the g'_{Na} and g'_K branches to a residual four-branch analog that was capable of regular rhythmic activity on its own resulted in richer bursts that were more representative of those seen in R_{15}. By analogy, it is likely that the slow oscillation in R_{15} does not "drive" the cell to burst but rather simply modulates the activity of what would otherwise be a regular pacemaker. This viewpoint concurs with the observation that upon depolarization, R_{15} often generates regular rhythmic activity (45). It follows that the parabolic frequency modulation of action potentials seen within an R_{15} burst (Fig. 2) is a consequence of the mecha-

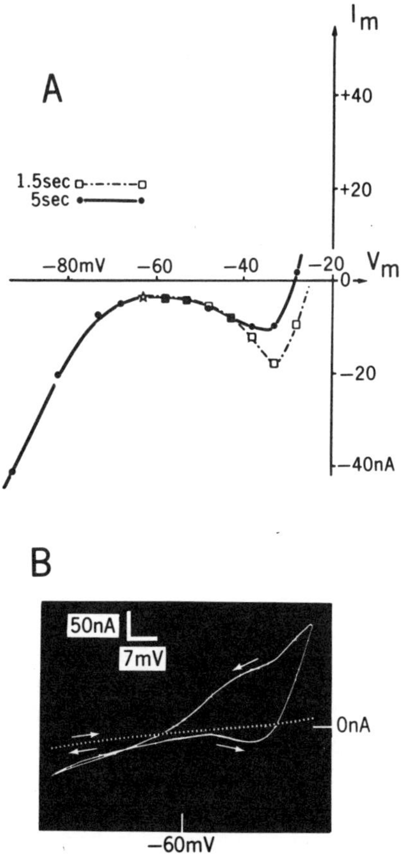

FIG. 4. A: I-V curves obtained for cell R_{15}, via voltage-clamp steps from a holding potential of −63 mV (*star*). Currents were measured at 1.5 and 5 sec following onset of step commands. (Modified from ref. 8.) **B:** I-V curves obtained for the bursting cell analog (*solid traces*), in a voltage-clamp experiment using depolarizing and hyperpolarizing ramp voltages (of slope 0.6 mV/sec) as command inputs. The negative slope conductance region is present during the depolarizing sweep but not during the return phase because of the large outward current caused by g'_K activation. The dependence of the negative slope conductance region on the g'_{Na} and g'_K branches is demonstrated by repeating the experiment with these branches removed (*dotted trace*). Note also that now the depolarizing and return traces overlap. (From ref. 31.)

nism of burst generation. The bursts generated by the analog (Fig. 5A, B) demonstrate this frequency modulation property.

An interesting consequence of the E_K shifts due to K^+ accumulation mentioned earlier was observed with the analog. It was found that the introduction of these E_K shifts invariably led to a larger slow oscillation amplitude and, consequently, to a greater number of action potentials per burst (compare Fig. 5A, B). This accentuation of the oscillations occurs because the

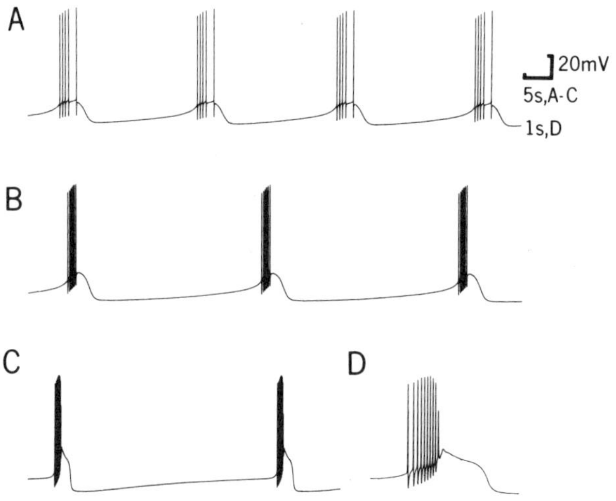

FIG. 5. Bursts generated by the R_{15} analog with no E_K shifts **(A)** and progressively increasing amounts of E_K shifts due to K^+ accumulation **(B, C)**. **D:** More detailed view of the burst in **C.** (Figure reproduced, with permission, from ref. 31.)

ongoing decrease in E_K during the depolarizing phase, caused by action potential activity, results in a reduced repolarizing potassium current (I'_K) through the g'_K branch. Indeed, excessive K^+ accumulation leads to simulated bursts (Fig. 5C, D) that resemble those caused by the paroxysmal depolarization shifts (PDS) observed in neurons within an epileptic focus (46). Thus K^+ accumulation per se, although not directly responsible for bursting activity, would seem to play an important part in its manifestation by the accentuation of initially minor membrane potential oscillations. K^+ accumulation has already been implicated previously in the appearance of epileptic seizures (18,23,65,66), and it is possible that it also underlies the abnormal PDS mentioned above.

Finally, passing reference should be made to an early hypothesis put forward to explain the slow oscillations in R_{15}. This involved a high resting sodium conductance that led to the depolarizing phase of the oscillations, following which an electrogenic sodium pump, coupled to chloride ions, was activated by the increased internal Na^+ concentration and led to the hyperpolarizing phase (58,59). Subsequent findings have contradicted this hypothesis. A more encompassing review of endogenous bursting in invertebrate cells than space permits here is to be found elsewhere (31).

DISCUSSION

This chapter distinguishes two classes of spontaneous activity, namely, regular rhythmic firing and burst firing. Separate mechanisms have been suggested to underlie each of these: a slow pacemaker depolarization following the action potential mediating rhythmic activity, and a slow membrane potential oscillation mediating bursts. Occasionally, however, by triggering a lone action potential during each depolarizing phase, the slow oscillations can evoke a rhythmic firing pattern. In such a case, rhythmic firing should be considered as a special form of bursting, with a single action potential in each burst. Schwindt and Calvin's (54) assertion of the involvement of both a potassium and a sodium conductance in rhythmic activity in motoneurons described earlier, particularly since backed by the observation of a negative slope conductance region in the steady state I-V curve (53), might fall into this category.

The principle of a slow depolarizing mechanism following the action potential also ties in with the observation that a faster accommodating preparation tends to be less repetitive (40,63). Presumably, the pacemaker depolarization is less effective as a ramp trigger in such preparations. Indeed, Adrian (1) postulated negligible or slow accommodation as an underlying cause for repetitive activity. Furthermore, according to him, the frequency of firing of the cell was mainly controlled by the time course of refractoriness, the cell discharging again after it had recovered sufficiently from the refractoriness caused by the previous spike. This second assertion, however, was questioned by Hodgkin (35) who pointed out that in the crab nerve, the frequency of firing was much slower than that dictated by the refractory period and that the time taken by the membrane potential to reach threshold determined the firing frequency. It is useful to briefly discuss here the involvement of accommodation in repetitive activity and the opposing points of view on the factors determining the firing rate.

As mentioned above, faster accommodating preparations tend to be less repetitive. Katz (40) also showed this theoretically on the basis of Hill's (34) two-factor theory of excitation. If slower accommodation is responsible for rhythmic firing, it might be useful to look at the factors underlying accommodation. Hodgkin and Huxley (39) suggested that accommodation to a slow depolarization occurs because the latter, by slowing down the normally rapid activation of the sodium conductance, permits the accompanying sodium conductance inactivation and potassium conductance activation processes to inhibit action potential generation. Thus changes in not only the kinetics but also the voltage dependence of these last two processes would be expected to alter the rate of accommodation. Vallbo (63) attempted to verify this by measuring both the accommodation and the ionic currents under voltage clamp in single myelinated nerve fibers of *Xenopus*. He found that faster accommodation was accompanied by a shift in the steady state curve for sodium inactivation, $h_\infty(V)$, in the hyperpolarizing direction, i.e., by a greater degree of resting sodium inactivation. This shift was due to a smaller rate constant (α_h) governing the kinetics of the inactivation variable h.

Vallbo (63) did not investigate the correlation between the turning-on of the potassium conductance and accommodation in the experimental preparation. Instead, Frankenhaeuser and Vallbo (26) used the Frankenhaeuser-Huxley (25) equations for myelinated nerve to study the factors responsible for accommodation. They verified that faster accommodation led to a loss of rhythmic activity and that, while changes in all constants affected the rate of accommodation, the latter was most sensitive to changes in α_h. This was again probably due to the effect of α_h on the steady state sodium inactivation curve. More generally, it was found that changes that led to an increase in inward current and a decrease in outward current resulted in slower accommodation and a greater tendency to rhythmic activity. Their results seem to suggest that slower accommodation is an accompanying correlate, but not necessarily the cause, of rhythmic activity.

Additional insight into the factors determining the firing frequency (f) may be gained by considering the variation in this frequency with steady injected current (I). The shape of this f-I curve (more precisely, the steady state f-I curve, after early and late adaptation, if present, have taken place) is sometimes linear ($f \alpha I$) and sometimes logarithmic ($f \alpha \ln I$). The latter is characteristic of the H-H equations (2,57), the former of the sensory receptor (28) and the motoneuron (41). Moreover, in the sensory receptor and motoneuron, repetitive activity commences at a low rate and is capable of considerable modulation; in the H-H axon, however, there is an abrupt onset of repetitive activity at a high firing rate and with little frequency modulation possible because of the logarithmic dependence on current.

Attempts to realize the sensory receptor f-I characteristic with relatively slight modifications of the H-H equations have met with limited success (19,57). A more ambitious, as well as successful, effort to realize a linear f-I relationship, a lower threshold frequency, and a wider modulation range was made by Shapiro and Lenherr (55). This was achieved by increasing the time constant of g_K in the H-H equations by a factor (γ_o) at the resting membrane potential. With depolarization, however, γ_o decreases linearly to become unity at E_{Na}. This artifice recognized the paramount importance of the g_K kinetics around the resting membrane potential in determining the rate of pacemaker depolarization and hence the frequency of firing. These authors concluded that there are two categories of rhythmic firing. The first corresponds to that of the classic squid axon, which fires at a high rate with a nonlinear f-I curve and little frequency modulation, and which is useful perhaps for mediating a rapid escape response. The second category has a linear f-I curve, lower frequencies, and extensive modulation capabilities, and characterizes the sensory receptors, where sensitive resolution is necessary. It is interesting to note that for the crayfish stretch receptor, the f-I curve, while linear for low firing rates, becomes logarithmic at higher frequencies (60). A similar effect may be noted for the class I crustacean axon (11), which is akin to the stretch receptor in having a wide range of firing frequencies (5 to 150 Hz). This would seem to indicate that, irrespective of functional requirements, a logarithmic f-I relationship is characteristic of rapid firing, whereas a linear f-I curve is typical at a low firing rate.

Why does this happen? Qualitatively, it appears that at the higher frequencies it is principally the sodium recovery-from-inactivation time that determines the firing rate. This sets a definite upper bound on frequency and possibly leads to the logarithmic dependence on current. Thus, over this range, Adrian's (1) suggestion of the refractory period determining frequency is valid. On the other hand, at low frequencies the f-I curve is linear, apparently because in this range it is the decay of a potassium conductance or the increase in a sodium conductance that eventually leads to an action potential. This agrees with Hodgkin's assertion of the time to threshold determining the frequency of firing, and models based on this principle generally predict a linear f-I curve at the lower frequencies, e.g., that of Shapiro and Lenherr (55), Connor and Stevens (15), and Kernell (43).

The f-I curve for the H-H equations (and for that matter the squid axon) does not exhibit the earlier linear regions because lower frequencies are not realizable via these equations. Stein (57) has suggested that the threshold frequency is most likely determined by the reciprocal of the time-to-peak afterdepolarization, following single-spike generation by a brief pulse. He has also shown that the excitability following a conducted action potential peaks at approximately the same time as the afterdepolarization. Thus, for a maintained current step, the frequency jumps abruptly to this threshold frequency, which, because of a relatively small time-to-peak afterdepolarization, is well within the logarithmic portion of the f-I curve. A similar explanation was given by Hodgkin (35) to account for the higher firing frequencies of class II crustacean axons vis-à-vis class I axons. He also demonstrated the presence of the supernormal excitability phase in class II axons and its absence in class I types. On the other hand, a much larger time-to-peak afterdepolarization for motoneurons results in a much lower threshold frequency and, consequently, a linear f-I curve. The strong correlation between the threshold frequency and the reciprocal of the time-to-peak afterdepolarization in motoneurons has also been demonstrated experimentally by Kernell (42).

In bursting cells, it would appear that the frequency of the underlying oscillations would also determine the burst frequency. Moreover, since the duration of the hyperpolarized phase of these oscillations is usually larger than that of the depolarized phase, the time constant governing the decay of the potassium conductance (g'_K) would play an important part in determining burst frequency. This has been verified qualitatively in the analog. However, the effect of the action potentials is not to be discounted. For instance, in the analog, K^+ accumulation caused by the action potentials, by increasing the slow oscillation amplitude, also increases its period. An increase in oscillation amplitude, by current injection or by any other means, will always result in an increase in period. This occurs because a larger amplitude results in an increase in the amount of g'_K activation and, consequently, in its subsequent decay time. A similar mechanism can also operate in the real cell where an exponential increase in oscillation period with amplitude has been demonstrated (45).

Thomas and Gorman (61) recently suggested that the interburst period depends on the rate at which Ca^{2+} ions that enter the cell during the action potentials are transported out. This was based on observations that depolarization during a burst, by increasing the number of action potentials, increased Ca^{2+} ion entry and prolonged the interburst period. With the action potentials blocked by tetrodotoxin, an increase in slow oscillation amplitude still leads to an increase in period (45), an observation that suggests that our hypothesis of the decay in g'_K controlling the period is the more important one. During normal bursting, however, Ca^{2+} entry could augment the activation of g'_K and thereby further enlarge the period.

Finally, current research centers on unraveling the biochemical processes underlying the ionic mechanisms for rhythmic and bursting discharges described here. A recent review (51) suggests an important role for catecholamines in cardiac pacemaking. Bursting has also been induced in normally beating *Aplysia* neurons by the administration of selected enzyme activators (10). In this case, it was suggested that the activated enzymes constitute a substrate cycle in which ATP is hydrolyzed and H^+ ions are produced. The changes in ATP and H^+ concentration near the membrane then affect the membrane conductances, with the H^+ ions activating the inward current and ATP the outward current. These biochemical studies, by exploring the mechanisms behind the conductance changes themselves, are likely to yield immensely useful information on the molecular basis of membrane behavior.

SUMMARY

A few possible ionic mechanisms underlying rhythmic and bursting discharges in nerve cells are described, based on studies with electronic analogs of excitable membranes. One such analog models the four-branch Hodgkin-Huxley equivalent circuit for the squid axon membrane. The variable sodium and potassium conductances are each simulated by a FET in series with a resistor. Studies with this analog revealed two possible modes of rhythmic activity. In the first, the potassium conductance decay time must exceed the sodium conductance recovery-from-inactivation time. Following an action potential, the increase in potassium conductance gradually decays, leading to the slow depolarization seen in pacemaker preparations. This slow depolarization acts as a ramp trigger for the succeeding action potential, provided the recovery from sodium inactivation is sufficiently complete. In the second mode, rhythmic discharge is achieved by introducing an additional noninactivating or steady state sodium conductance that increases with depolarization and underlies the pacemaker depolarization. The experimental evidence supporting each of the two modes is reviewed.

The above analog was modified to simulate the endogenous bursts seen in cell R_{15} in *Aplysia*. These bursts result from slow oscillations of the transmembrane potential, which are caused by variations in two membrane conductances: an increase in a sodium conductance underlying the depolarizing phase, with a subsequent increase in a potassium conductance underlying the hyperpolarizing phase. The observation of a negative slope conductance region in the membrane I-V characteristic supports this theory. Modification of the analog simply involved adding these two conductance branches in parallel. Several facets of the two-conductance theory are verified. The analog also simulated the phenomenon of K^+ ion accumulation outside the cell membrane during a burst. Excessive K^+ accumulation led to simulated bursts that resembled those caused by the PDS observed in neurons within an epileptic focus.

Finally, the role played by accommodation in rhythmic activity and the factors determining both the firing frequency and the shape of the frequency-current relationship in repetitive cells are discussed.

ACKNOWLEDGMENT

This work was supported by the Medical Research Council of Canada.

REFERENCES

1. Adrian, E. D. (1928): *The Basis of Sensation.* Christophers, London.
2. Agin, D. (1964): *Nature,* 201:625–626.
3. Arvanitaki, A., and Chalazonitis, N. (1965): *C.R. Acad. Soc.,* 261:548–552.
4. Baldiserra, F., and Gustafsson, B. (1974): *Acta Physiol. Scand.,* 91:512–527.
5. Bergmann, C., Nonner, W., and Stämpfli, R. (1968): *Pfluegers Arch. Ges. Physiol.,* 302:24–37.
6. Brown, H. F., Clark, A., and Noble, S. J. (1976): *J. Physiol.,* 258:521–545.
7. Calvin, W. H., and Schwindt, P. C. (1972): *J. Neurophysiol.,* 35:297–310.
8. Carnevale, N. T. (1974): *Voltage Clamp Analysis of the Slow Oscillations in Bursting Neurons Reveals Two Underlying Current Components.* Ph.D. Thesis. Dept. of Physiology, Duke University, Durham, N.C.
9. Carpenter, D. O. (1973): In: *Neurobiology of Invertebrates,* edited by J. Salanki, pp. 35–58. Publishing House of the Hungarian Academy of Sciences, Budapest.
10. Chaplain, R. A. (1976): *Brain Res.,* 106:307–319.
11. Chapman, R. A. (1966): *J. Exp. Biol.,* 45:475–488.
12. Cole, K. S., Antosiewicz, H. A., and Rabinowitz, P. (1955): *J. Soc. Ind. Appl. Math.,* 3:153–172.
13. Connor, J. A. (1975): *J. Neurophysiol.,* 38:922–932.
14. Connor, J. A., and Stevens, C. F. (1971): *J. Physiol.,* 213:21–30.
15. Connor, J. A., and Stevens, C. F. (1971): *J. Physiol.,* 213:31–53.
16. Connor, J. A., Walter, D., and McKown, R. (1977): *Biophys. J.,* 18:81–102.
17. Coombs, J. S., Eccles, J. C., and Fatt, P. (1955): *J. Physiol.,* 130:291–325.
18. Dichter, M., Herman, C. J., and Selzer, M. (1972): *Brain Res.,* 48:172–183.

19. Dodge, F. A., Jr. (1972): *Int. J. Neurosci.,* 3:5–14.
20. Dudel, J., and Trautwein, W. (1958): *Pfluegers Arch. Ges. Physiol.,* 267:553–565.
21. Eaton, D. C. (1972): *J. Physiol.,* 224:421–440.
22. Eckert, R., and Lux, H. D. (1976): *J. Physiol.,* 254:129–151.
23. Fertziger, A. P., and Ranck, J. B., Jr. (1970): *Exp. Neurol.,* 26:571–585.
24. Fitzhugh, R. (1961): *Biophys. J.,* 1:445–466.
25. Frankenhaeuser, B., and Huxley, A. F. (1964): *J. Physiol.,* 171:302–315.
26. Frankenhaeuser, B., and Vallbo, A. B. (1965): *Acta Physiol. Scand.,* 63:1–20.
27. Frazier, W. T., Kandel, E. R., Kupfermann, I., Waziri, R., and Coggeshall, R. A. (1967): *J. Neurophysiol.,* 30:1288–1351.
28. Fuortes, M. G. F., and Mantegazzini, F. (1962): *J. Gen. Physiol.,* 45:1163–1179.
29. Gola, M. (1974): *Pfluegers Arch. Ges. Physiol.,* 352:17–36.
30. Gulrajani, R. M., and Roberge, F. A. (1976): *Med. Biol. Eng.,* 14:31–41.
31. Gulrajani, R. M., and Roberge, F. A. (1978): *Fed. Proc., in press.*
32. Gulrajani, R. M., Roberge, F. A., and Mathieu, P. A. (1977): *Biol. Cybern.,* 25:227–240.
33. Gulrajani, R. M., Roberge, F. A., and Mathieu, P. A. (1977): *Proc. IEEE,* 65:807–809.
34. Hill, A. V. (1936): *Proc. R. Soc. Lond* [*Biol.*], 119B:305–355.
35. Hodgkin, A. L. (1948): *J. Physiol.,* 107:165–181.
36. Hodgkin, A. L., and Huxley, A. F. (1952): *J. Physiol.,* 116:449–472.
37. Hodgkin, A. L., and Huxley, A. F. (1952): *J. Physiol.,* 116:473–496.
38. Hodgkin, A. L., and Huxley, A. F. (1952): *J. Physiol.,* 116:497–506.
39. Hodgkin, A. L., and Huxley, A. F. (1952): *J. Physiol.,* 117:500–544.
40. Katz, B. (1937): *J. Physiol.,* 88:239–255.
41. Kernell, D. (1965): *Acta Physiol. Scand.,* 65:74–86.
42. Kernell, D. (1965): *Acta Physiol. Scand.,* 65:87–100.
43. Kernell, D. (1968): *Brain Res.,* 11:685–687.
44. Lenfant, J., Mironneau, J., and Aka, J.-K. (1972): *J. Physiol.* (*Paris*), 64:5–18.
45. Mathieu, P. A., and Roberge, F. A. (1971): *Can. J. Physiol. Pharmacol.,* 49:787–795.
46. Matsumoto, H., and Ajmone-Marsan, C. (1964): *Exp. Neurol.,* 9:286–304.
47. McAllister, R. E., Noble, D., and Tsien, R. W. (1975): *J. Physiol.,* 251:1–59.
48. Palti, Y. (1971): In: *Biophysics and Physiology of Excitable Membranes,* edited by W. J. Adelman, Jr., pp. 168–182. Van Nostrand Reinhold, New York.
49. Partridge, L. D. (1975): *Brain Res.,* 94:161–166.
50. Plant, R. E. (1976): *Comput. Programs Biomed.,* 6:85–91.
51. Pollack, G. H. (1977): *Science,* 196:731–738.
52. Roy, G. (1972): *IEEE Trans. Biomed. Eng.,* 19:60–63.
53. Schwindt, P., and Crill, W. E. (1977): *Brain Res.,* 120:173–178.
54. Schwindt, P. C., and Calvin, W. H. (1973): *J. Neurophysiol.,* 36:955–973.
55. Shapiro, B. I., and Lenherr, F. K. (1972): *Biophys. J.,* 12:1145–1158.
56. Smith, T. G., Barker, J. L., and Gainer, H. (1975): *Nature,* 253:450–453.
57. Stein, R. B. (1967): *Proc. R. Soc. Lond* [*Biol.*], 167B:64–86.
58. Strumwasser, F. (1968): In: *Physiological and Biochemical Aspects of Nervous Integration,* edited by F. D. Carlson, pp. 329–341. Prentice-Hall, Englewood Cliffs, New Jersey.
59. Strumwasser, F. (1971): *J. Psychiatr. Res.,* 8:237–257.
60. Terzuolo, C. A., and Washizu, Y. (1962): *J. Neurophysiol.,* 25:56–66.
61. Thomas, M. V., and Gorman, A. L. F. (1977): *Science,* 196:531–533.
62. Trautwein, W., and Kassebaum, D. G. (1961): *J. Gen. Physiol.,* 45:317–330.
63. Vallbo, A. B. (1964): *Acta Physiol. Scand.,* 61:429–444.
64. Wilson, W. A., and Wachtel, H. (1974): *Science,* 186:932–934.
65. Zuckermann, E. C., and Glaser, G. H. (1968): *Exp. Neurol.,* 20:87–110.
66. Zuckermann, E. C., and Glaser, G. H. (1970): *Arch. Neurol.,* 23:358–364.

Abnormal Neuronal Discharges, edited by N. Chalazonitis and M. Boisson. Raven Press, New York © 1978.

Ionic Distribution Changes During Bursting Activity Induced by Pentylenetetrazol in a Single Isolated Snail Neuron: Tentative Application of Electron Probe X-Ray Microanalyzer to a Single Isolated Neuron

Eiichi Sugaya, Minoru Onozuka, *Masayoshi Usami, and **Aiko Sugaya

*Department of Physiology, Kanagawa Dental College, Yokosuka, Japan; *Research Laboratory, Chugai Pharmaceutical Co., Ltd. Takatanobaba, Tokyo; and **Faculty of Pharmaceutical Sciences, Josai University, Sakado, Saitama, Japan*

The role of divalent cations, particularly of calcium and magnesium, is of increasing interest as a charge carrier and in processes such as regulation of enzymatic activities, secretion of humoral transmitters, and initiation of muscle contraction. The role of calcium has become very important in manifesting the bursting activity by various electrophysiological techniques, especially by voltage clamping (1–6,8).

The distribution of ions compared with the ultrastructure of the nerve cells, however, has not been widely investigated. Analysis by the electron probe X-ray microanalyzer (EPXMA) makes this type of research easier. To determine the relationship between ionic localization within the neuron and cellular function, we tried to examine the calcium and magnesium distribution in a single, isolated, identified neuron of the snail by EPXMA.

MATERIALS

The D-neurons of the subesophageal ganglion of the Japanese land snail *Euhadra peliomphala* were used. The neurons of Euhadra can be classified into three types according to their response to ACh, as in the case of other molluscan neurons, the D-, H-, and I-neurons (10).

In this experiment, we used D-neurons exclusively because they are the most sensitive to pentylenetetrazol (PTZ) (10).

APPLICATION OF EPXMA TO SINGLE ISOLATED NEURON

EPXMA is a method of elemental ultramicroanalysis based on the spectrometry of the characteristic X-rays emitted when a sample is excited by an electron beam. Application of EPXMA to the single isolated neuron is a promising method with many advantages, but also some limitations in investigating the intracellular ionic mechanism of bursting activity. The most powerful advantage is the possibility of analysis concerned with the cellular ultrastructure, and the most troublesome restriction is the limitation of sampling methods and damage of samples during analysis.

It is said that in the case of preparation of specimens for EPXMA, the conventional epon embedding method frequently leads to the wrong conclusion because of redistribution of ions during fixation and repeated dehydration (9). At present, the most reliable method for preparing EPXMA specimens is freeze drying (9). We therefore used freeze-dried or frozen ultrathin-sectioned and freeze-dried specimens. The ganglion was frozen rapidly in Freon cooled by liquid nitrogen after various experimental procedures. The frozen ganglion was placed in a specially made ganglion holder and dried by vacuum within a deep-freezing box, without passing through a liquid phase. The completely dried ganglion was dissected with fine forceps under a binocular microscope, and the D-neurons were selected according to the identifiable localization and their characteristic orange color. The isolated, dried, single neuron was placed on a hand-made carbon mesh, fit into the EPXMA holder, and coated with carbon in a vacuum chamber. The JEOL 50A disperse type X-ray microanalyzer was used.

MEASURING DEPTH OF EPXMA

To apply the EPXMA to the biological specimen, it is necessary to determine the measuring depth of the EPXMA. We performed the following experiment to determine the analyzing depth.

Ultrathin-sectioned freeze-dried specimens embedded in epoxy resin of various thicknesses were laid on a single copper grid. The analysis was performed to determine to what extent the thin-sectioned specimen prevents underlying copper detection. In Fig. 1Aa and b, the secondary electron image and the X-ray image, respectively, of the copper grid itself are shown. In Fig. 1Ac–f, the X-ray images of copper through the ultrathin-sectioned specimens of various thicknesses are shown. This demonstrates that with specimens of over 7.5 μm, the underlying copper is scarcely detected. The table in Fig. 1B shows a spot analysis of the same specimen. This also demonstrates that the analyzing depth is about 5.0 to 7.5 μm.

CALCIUM AND MAGNESIUM DISTRIBUTION PATTERN OF SNAIL NEURON IN NORMAL STATE

The calcium and magnesium distribution of an isolated single neuron is uneven. As shown in Fig. 2D and E, the X-ray image shows a denser distribution of calcium in the axon hillock region and a denser distribution

A

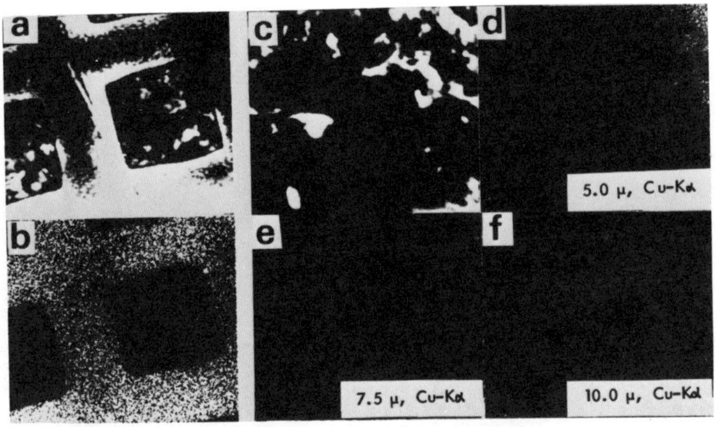

B

thickness \ counts	Cu	sample
5.0 μ	12962	1196
7.5 μ	13261	140
10.0 μ	14621	105

back ground: 112

FIG. 1. A: Examination of measuring depth by X-ray image. *a:* Secondary electron image of copper grid. *b:* Ca K_α X-ray image of copper grid. *c:* Secondary electron image of thin-sectioned specimen. *d–f:* Ca K_α X-ray image of copper grid through 5.0 (*d*), 7.5 (*e*), and 10 (*f*) μm thick specimen. Between 5.0 and 7.5 μm, the Cu K_α is scarcely detected. **B:** Examination of measuring depth by spot analysis. Cu K_α count through thin-sectioned specimen with accelerating voltage of 25 KV, absorbed current 1×10^{-8} A for 10 sec. Background count is 112.

of magnesium in the cell body region. Figure 2F is the simultaneous line analysis of calcium K_α and magnesium K_α through the cell body to the axon hillock, which also shows that the calcium is denser in the axon hillock and the magnesium is denser in the cell body.

To obtain more precise information, we carried out a spot analysis, which uses a more concentrated electron beam. To obtain the same measuring conditions, we prepared several isolated, freeze-dried cells on a single sample disk made of carbon and measured successively under exactly the same conditions as shown in Fig. 2A. Figure 3 shows the calcium and magnesium distribution by spot analysis with an accelerating voltage of 25 KV and absorbed current of 10^{-8} A for 10 sec. The measured points were point *a* on the axon hillock, point *b* near the axon hillock, point *c* far from the axon hillock, and point *d* on the cell body. The data are expressed by the counts of the secondary X-ray in the upper part. The lower part is the ratio of counts when the axon hillock is 1.0.

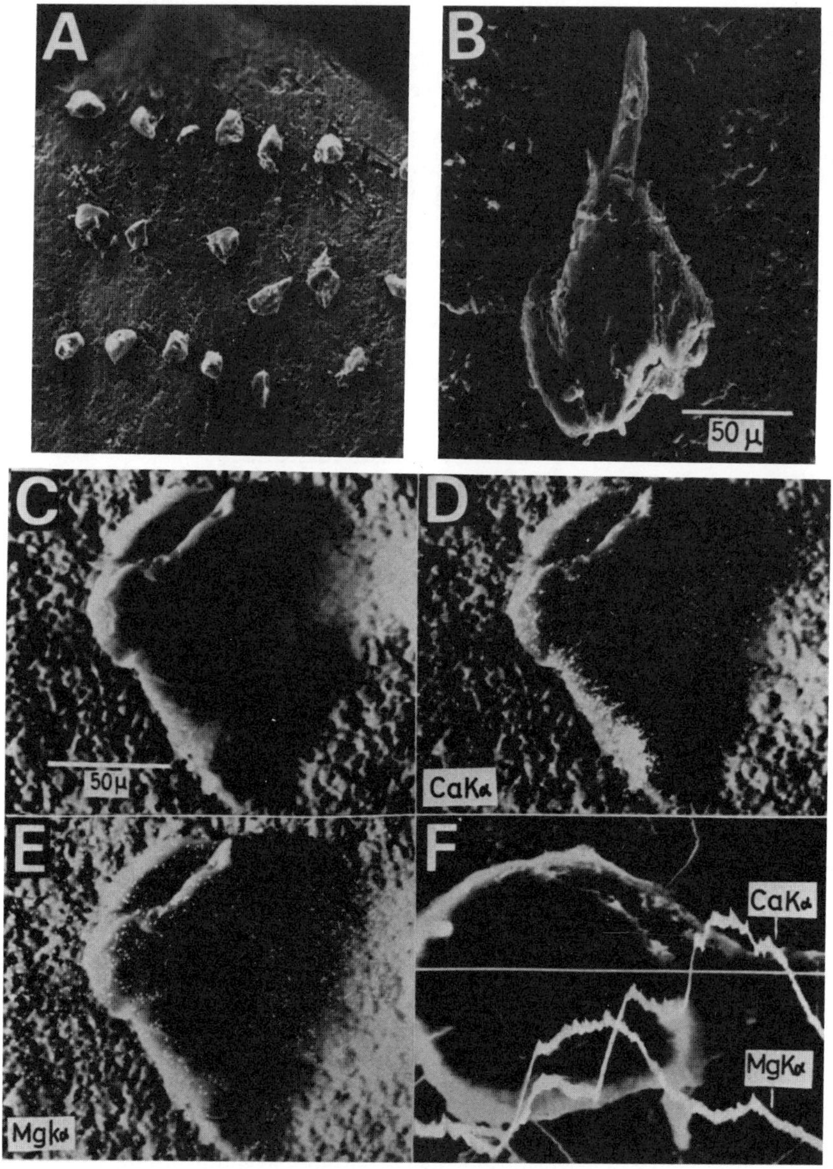

FIG. 2. A: Secondary electron image of single isolated freeze-dried cells on a carbon disk. **B:** Single isolated freeze-dried neuron of Euhadra. **C:** Single isolated freeze-dried D-neuron of Euhadra. **D:** Ca K_α X-ray image of **C. E:** Mg K_α X-ray image of **C. F:** Simultaneous line analysis of Ca K_α and Mg K_α through the axon hillock to the cell body.

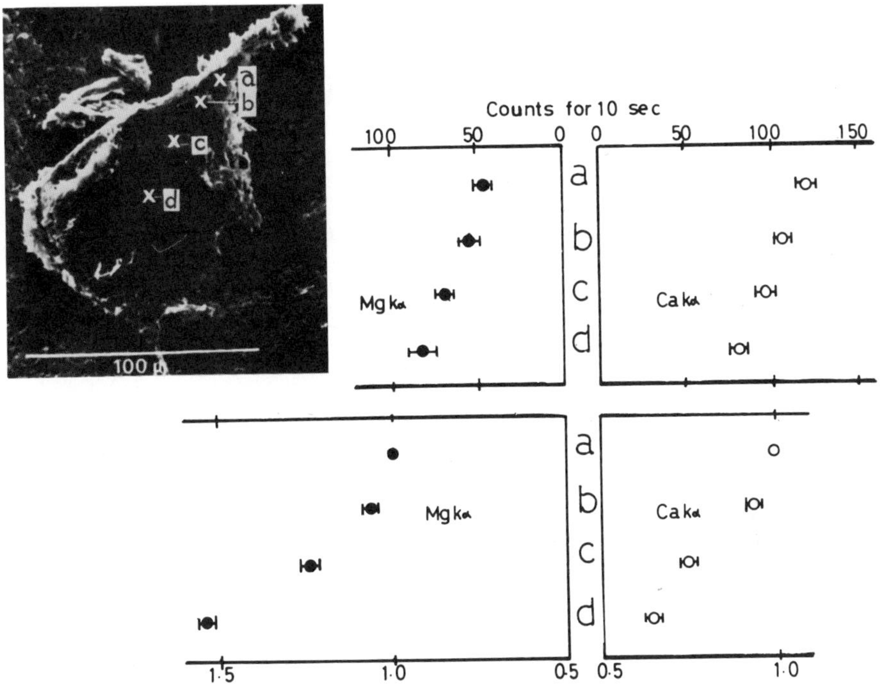

FIG. 3. Calcium and magnesium distribution in single isolated neuron. **Upper:** Counts for 10 sec of Ca K_α (*right*) and Mg K_α (*left*). **Lower:** Plot when the axon hillock region is 1.0. *a–d,* correspond to the analyzing spot demonstrated in the inset. Accelerating voltage 15 KV and absorbed current 1×10^{-8} A.

INTRACELLULAR CALCIUM DISTRIBUTION CHANGE DURING BURSTING ACTIVITY INDUCED BY CONVULSANT

The neurons from RC-1 to 5 show characteristic bursting activity by applying PTZ, as in Fig. 4A and B, and show the negative resistance characteristic in the current-voltage relationship curve by voltage clamping, as shown in Fig. 4C and D. We examined the calcium distribution changes after treatment with PTZ by means of EPXMA.

Convulsant drugs, such as strychnine, picrotoxin, and PTZ, change the distribution pattern of calcium and magnesium in isolated neurons. After 15 min incubation in 1×10^{-2} moles strychnine and 5×10^{-3} moles picrotoxin, calcium increased to some extent; but the remarkable change occurred in the case of PTZ. The calcium content after 15 min incubation with 5×10^{-2} moles PTZ increased to about double that of the normal state; but the distribution pattern did not change. Moreover, by the application of PTZ, the magnesium distribution pattern was reversed; i.e., it became rich in the axon hillock region and poor in the cell body. In the

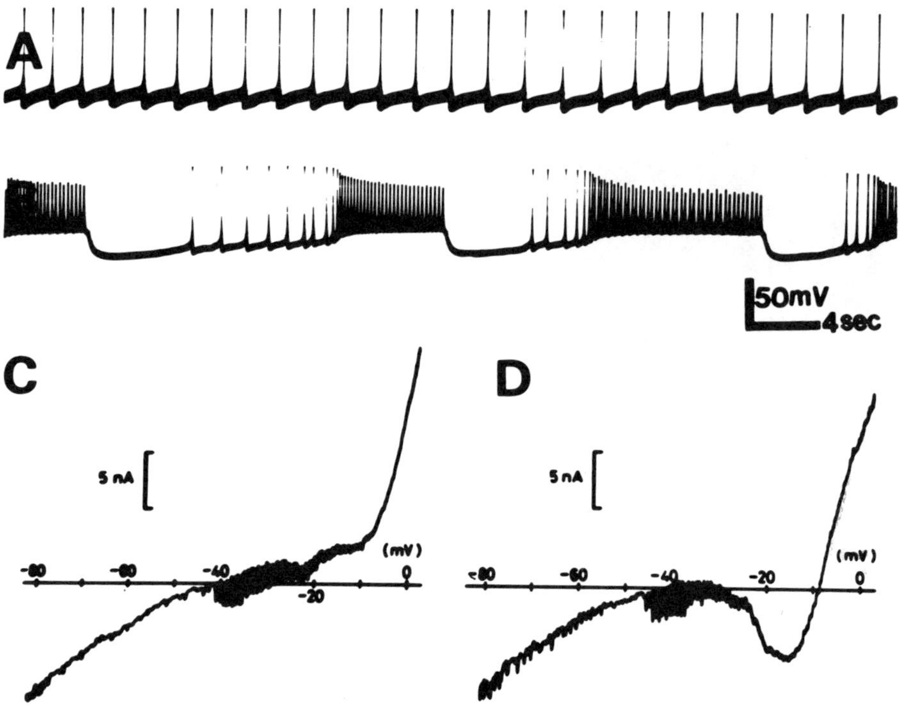

FIG. 4. Effect of PTZ on D-neuron of Euhadra. **A:** Spontaneous regular firing in normal state. **B:** Bursting activity induced by PTZ. **C:** I-V relationship by voltage clamping using a triangular command of D-neuron. **D:** Appearance of negative resistance characteristic in I-V relationship curve after application of PTZ.

case of strychnine and picrotoxin, the magnesium distribution pattern did not change remarkably. From the viewpoint of intracellular potential change, PTZ application induced remarkable bursting activity, but strychnine and picrotoxin induced few changes.

The EPXMA spot analysis is a better method than line analysis and X-ray image analysis. There is, however, a limitation in the analysis; i.e., the overall analysis of the entire cell is impossible because of the sample damage and difficulty of regular time-scheduled stepwise movement of the sample. To solve these problems, we used the computer-operated method for EPXMA. Samples were mounted on stages which are moved by micrometers in three dimensions stepwise by computer control. The electron beam was stopped during the time of movement from one spot to another to minimize the sample damage. The count of calcium K_α by secondary X-ray was stored and the numbers from 0 to X which determine the range of counts were automatically typed out. Figure 5 shows the calcium and magnesium distribution map of 5 micrometer steps of X-axis advance and 20 micrometer

LEVEL	-	0	1	2	3	4	5	6	7	8	9	X
Mg	15	34	52	71	89	100	126	145	163	182	200	
Ca	20	33	46	59	72	85	98	111	124	137	150	

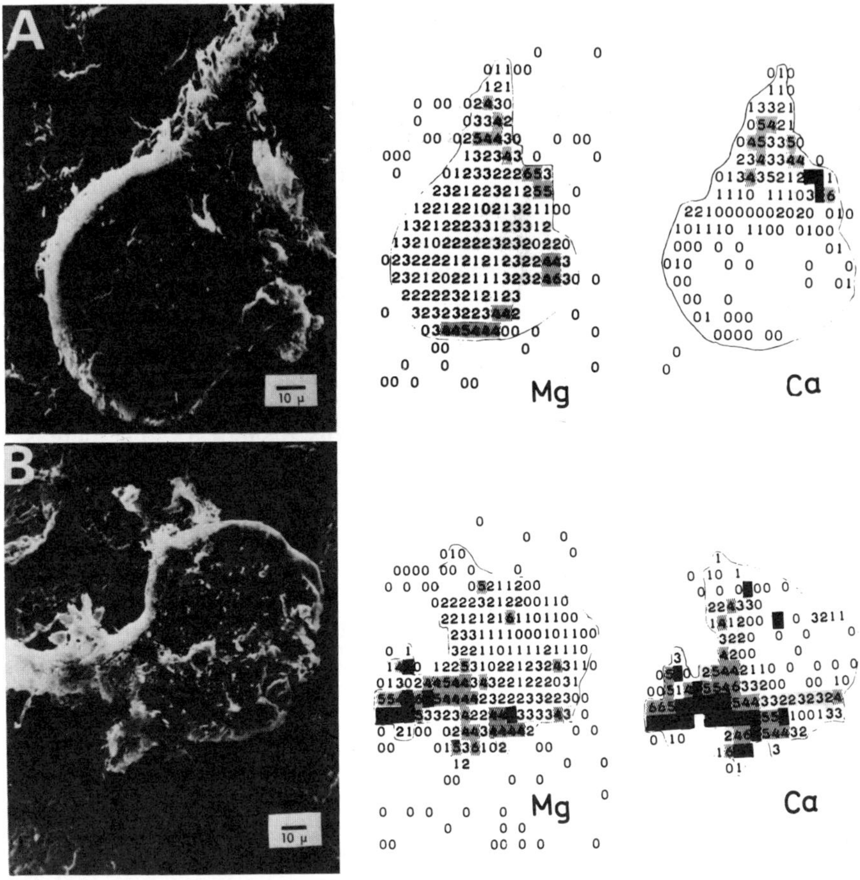

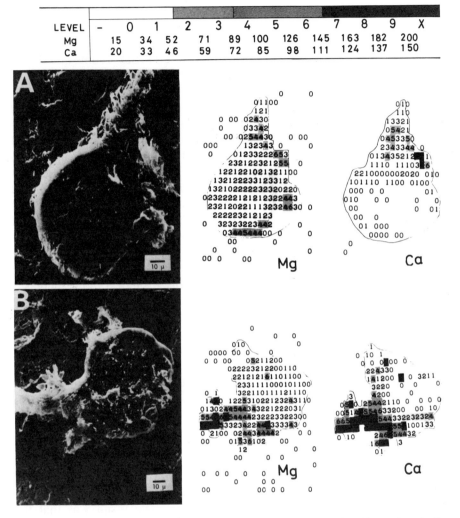

FIG. 5. Calcium and magnesium distribution map of neuron before (*upper*) and after (*lower*) PTZ. *Left column,* secondary electron image of measured samples; *middle column,* magnesium distribution map; *right column,* calcium distribution map. Secondary X-ray count levels are divided from 0 to X according to the counts shown at the top. Calcium is increased by PTZ, but the distribution pattern did not change and the magnesium distribution pattern is reversed.

steps of Y-axis movement. The left part is the secondary electron image of the measured sample and the next column is the magnesium distribution pattern; the right part is the calcium distribution pattern. 0 is the lowest count and X is the highest; between 0 and X, the count range is determined from 1 to 9, as described in the uppermost lines of the map. The results are the same as for the spot analysis without computer control.

To examine more precisely the calcium increase in and near the cell membrane, the electron beam was enlarged so that the whole cell body is enveloped, and the whole cell counts of the calcium K_α were measured for 10 sec. The result is that the control was 404 ± 24 (SD) counts and that of after PTZ was 718 ± 56 (SD) counts; the PTZ increases the calcium counts about double the control by this method.

To know if the intracellularly stored calcium moves to the membrane by application of PTZ, the following experiment was performed. The D-neurons incubated in normal snail Ringer and the PTZ-containing snail Ringer were dipped into liquid Freon cooled by liquid nitrogen and ultra-thin-sectioned to about 10.0 μm thick in a freezing chamber. Both ultrathin-sectioned specimens were freeze-dried without passing through the liquid phase in a deep freezer. These specimens were mounted on a carbon disk and analyzed by computer-controlled EPXMA. Figure 6 is an example of such an analyzing map; Fig. 6A is the control and Fig. 6B is PTZ treated. The left photographs are the secondary electron image, and the right are the calcium-analyzing map. These maps clearly show that the densely scattered calcium in the cytoplasm in controls disappears and increases in or near the cell membrane in the PTZ-treated specimen.

Then, to determine the origin of the increased calcium more precisely, i.e., whether the increased calcium is from the extracellular solution or of intracellular origin, the same procedures were performed with a cobalt chloride-containing solution and lanthanum chloride-containing solution. Figure 7A is the calcium-analyzing map with PTZ in normal snail Ringer; Fig. 7B is with PTZ in 30 mmoles cobalt chloride-containing Ringer; Fig. 7C is PTZ with 5 mmoles lanthanum chloride Ringer solution. These maps show almost the same calcium distribution pattern; i.e., increased calcium by PTZ in or near the cell membrane is mainly of intracellular origin.

To obtain more quantitative data, we performed the following experiment. Using an electron beam concentrated to 1 to 2 μm, analysis was performed for 15 to 20 spots just on the border of the cell membrane of an ultrathin-sectioned and freeze-dried neuron. The result was that the incubation in 4 × calcium with PTZ, calcium-free with EGTA and PTZ, cobalt chloride with PTZ, and lanthanum chloride with PTZ were not changed remarkably when compared with incubation in PTZ only.

If the increased calcium by PTZ is a result of calcium binding to the membrane structure, the calcium content of an isolated single neuron should be remarkably decreased when the surface of the freeze-dried cell is etched to about 100 Å. Therefore, we tried to perform the etching of the freeze-dried cell surface by the ion shower milling machine, which is used to etch the surface of transistors and ICs at the Ångstrom level. The electron microscopic observation of etched cell surface clearly demonstrated the loss of cell membrane compared with the intact cell. The intact cell and the supposed membrane-etched and deprived cell were weighed

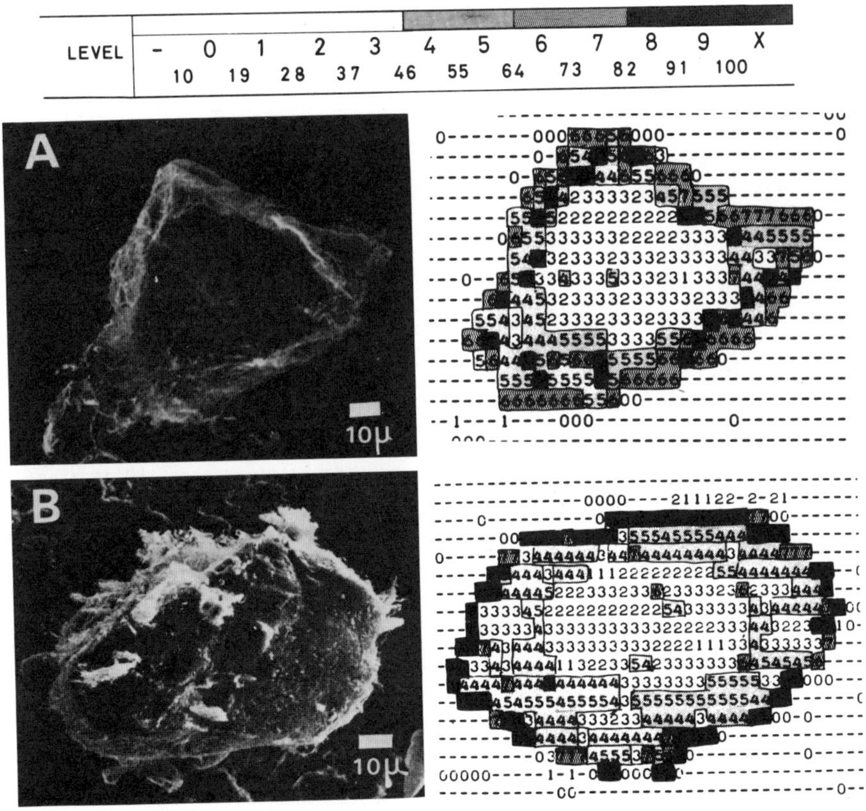

FIG. 6. Intracellular calcium distribution change during bursting activity induced by PTZ. **A:** Before PTZ. **B:** After 15 min incubation in PTZ. *Left column,* secondary electron image of measured sample; *right column,* calcium distribution map of ultrathin-sectioned freeze-dried D-neuron. Secondary X-ray count levels are divided from 0 to X according to the counts shown at the top. Note the increased calcium distribution in the cell membrane region after PTZ.

by the fish pole balance according to Lowry's (7) method and then by the aid of a flameless atomic analyzer; the calcium content was measured by placing the sample on a tungsten boat for heating and atomizing. The content of the calcium of the etched cell is clearly decreased in both normal and PTZ-treated cells. Especially in the PTZ-treated cell, the calcium content of the etched cell was about half that of the intact cell (compare the left-most white column to the middle hatched column of Fig. 8).

To find some clue as to where the membrane-increased calcium comes from intracellularly, we investigated some subcellular ultrastructural changes. In Euhadra neurons, especially in D-cells, there are electron-dense, lysosome-like granules of about 5,000 Å in diameter. By PTZ treatment, these electron-dense granules change into lamella-like granules. In

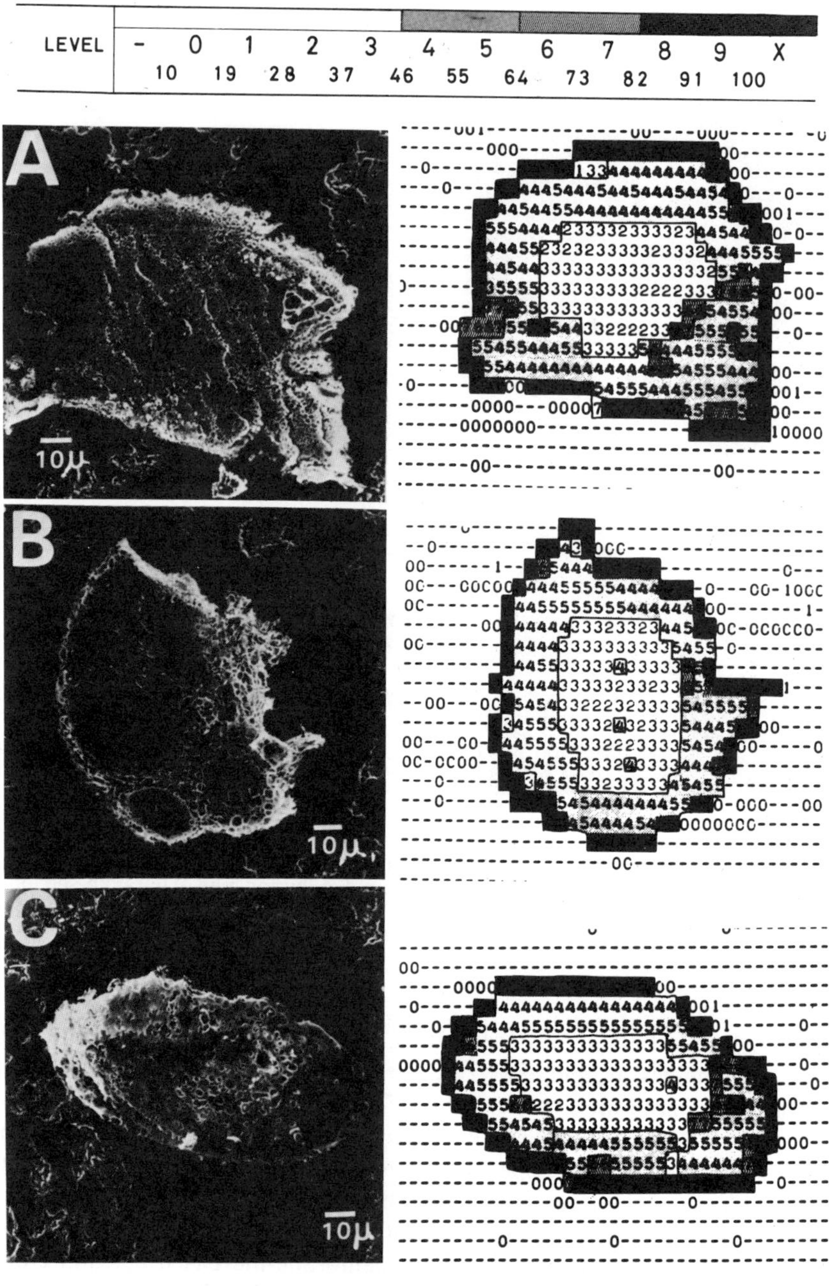

FIG. 7. Intracellular calcium distribution change induced by PTZ **(A)**, PTZ with cobalt chloride **(B)**, and PTZ with lanthanum chloride **(C)**. Ultrathin-sectioned and freeze-dried specimen. *Left column,* secondary electron image of measured samples; *right column,* calcium distribution map. Numbers and marks of count level are the same as in Fig. 6. Note almost the same calcium distribution pattern among the three cases.

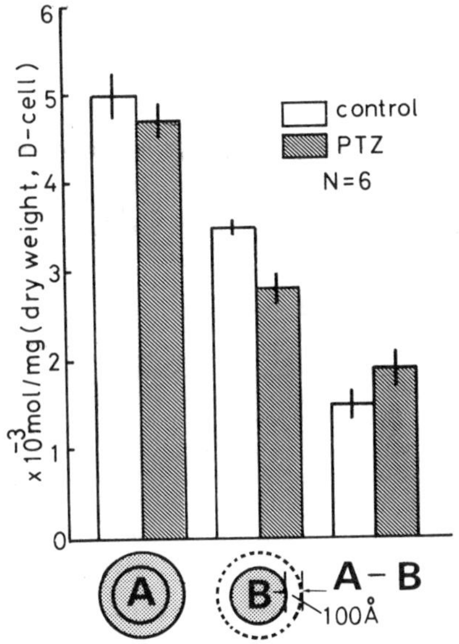

FIG. 8. Calcium content change of single isolated neuron after etching of cell membrane by ion shower milling machine. *A*, calcium content of cell before, and *B*, after etching the cell membrane by about 100 Å. White columns are controls and hatched columns are after PTZ.

the normal state, there are some lamella-like granules; but in PTZ-treated neurons, the ratio increased about 3.5 times.

We tried to compare the calcium density of these granules in both states. Using a nondisperse type EPXMA, we measured the relative calcium content in both types of granules. The results showed a reduction of calcium content in lamella-like granules.

CONCLUSION

In the D-cell of the Euhadra ganglion, remarkable bursting activity is produced by PTZ treatment, and the negative resistance characteristic I-V relationship appears. At the same time, from the viewpoint of intracellular calcium movement, the intracellular calcium probably bound to some granule or to endoplasmic reticulum and mitochondria will be released and moves toward the cell membrane. Then, binding to the inner or outer surface of the cell membrane occurs; this in turn triggers changes of ion channels for potassium and sometimes for calcium. This is the intracellular mechanism of the bursting activity. The EPXMA provides a promising technique to study intracellular ion movement compared with ultrastructural changes.

ACKNOWLEDGMENTS

This study was supported in part by a grant from the Japanese Ministry of Education, Science and Culture.

We thank Mr. S. Okudera and Mr. Y. Ono for help in computer programming, and Mr. M. Hotta for his help in using the ion shower milling machine.

REFERENCES

1. Baker, J. L., and Gainer, H. (1975): *Brain Res.,* 84:461–477.
2. Baker, J. L., and Gainer, H. (1975): *Brain Res.,* 84:479–500.
3. Eckert, R., and Lux, H. D. (1976): *J. Physiol.,* 254:129–151.
4. Gola, M. (1976): *Experientia,* 32:585–587.
5. Heinemann, U., Lux, H. K., and Gutnick, M. J. (1977): *Exp. Brain Res.,* 27:237–243.
6. Heyer, C. B., and Lux, H. D. (1976): *J. Physiol.,* 262:319–348.
7. Lowry, O. H., and Passonneau, J. V. (1972): The quartz fiber fishpole balance. In A Flexible System of Enzymatic Analysis, pp. 237–249, Academic Press, New York.
8. Smith, T. G., Jr., Barker, J. L., and Gainer, H. (1975): *Nature,* 253:450–452.
9. Sugaya, E., and Onozuka, M. (1974): *IRCS,* 2:1571.
10. Sugaya, A., Sugaya, E., and Tsujitani, M. (1973): *Jpn. J. Physiol.,* 23:261–274.

Subject Index